Enantiomers, Racemates, and Resolutions

Enantiomers, Racemates, and Resolutions

Jean Jacques **André Collet**

Centre National de la Recherche Scientifique
Collège de France
Paris

Samuel H. Wilen

The City University of New York
The City College
New York, NY

KRIEGER PUBLISHING COMPANY
MALABAR, FLORIDA
1991

Original Edition 1981
Reprint Edition 1991

Printed and Published by
KRIEGER PUBLISHING COMPANY
KRIEGER DRIVE
MALABAR, FLORIDA 32950

Library of Congress Cataloging-In-Publication Data
Jacques, Jean.
Enantiomers, racemates, and resolutions / Jean Jacques, Andre' Collet, Samuel H. Wilen.
p. cm.
Reprint. Originally published: New York : Wiley, c1981.
Includes bibliographical references and index.
ISBN 0-89464-618-4 (acid-free paper)
1. Stereochemistry. 2. Racemization. 3. Chirality. 4. Enantiomers. I. Collet, Andre'. II. Wilen, Samuel H. III. Title.
QD481.J26 1991
547.1'224--dc20 91-15940
CIP

10 9 8 7 6 5 4 3 2

Preface

Experimental stereochemistry and, in particular, the synthesis of chiral stereoisomers enriched in one enantiomer by a variety of methods is the subject of a multitude of contemporary studies. The reasons for the enormous effort in producing optically active molecules are well known to chemists but they bear repeating. Such molecules, whether small or large, have inherent interest, that is, their physical and chemical properties merit study; they are useful adjuvants in the elucidations of chemical reactions and of reaction mechanisms; and, above all, their biological properties have driven chemists to seek them out and to create them for utilitarian purposes.

It would seem that all of this activity would have given rise to several book-length summaries and overviews of the properties of racemates and of their constituent enantiomers and of the principles that underlie their separation from one another. While these optics have indeed been dealt with in review articles, briefly in textbooks as well as in comprehensive multivolume works, to the best of our knowledge no full-length monograph dealing with the properties of enantiomers exists. If this absence was earlier justified by an insufficient understanding and knowledge of racemate properties, this is no longer the case. We believe that a monograph on these subjects is timely, especially in view of the present-day interest and activity in asymmetric synthesis, in enzymatic synthesis, and in chromatographic resolutions.

We have also been concerned to some extent by errors we continue to find in the literature in connection with separation of enantiomers and their purification. These are often conceptual errors that derive from a lack of awareness of cause-and-effect relationships between planned experiments and properties of systems. These errors are in part forgivable; no clear and easily accessible statement of these properties and their consequences exists. Some of these properties have not been known heretofore; reports of others have been scattered in the literature and have not been easy to find. The time is thus ripe for a book in which the properties of enantiomers and racemates are brought together and examined critically.

Characteristics unique to enantiomers manifest themselves principally in the solid, that is, crystalline, state. This has led us to place special emphasis on their properties in this state and to pay special attention to the separation of enantiomers and, by extension, of diastereomers by physical methods, namely, by crystallization with and without a solvent, and by sublimation. While chromatographic resolutions are indeed treated here, we have placed less emphasis on such methods since the theory of chromatographic techniques is adequately discussed elsewhere.

We have been monitoring and commenting upon developments in nonchromatographic resolutions for well over a decade. In spite of statements and views to the effect that traditional resolution is passé, we have yet to witness the predicted demise of even the most classic variant of this useful unit operation in organic chemistry. This is attested to by the very large number of resolutions reported in the contemporary journal literature, and the substantial number of patents dealing with resolutions attests to their considerable economic importance. Yet we have in no way conceived of this book, or any part thereof, as a defense of classical resolution as against the modern chromatographic variants or against other methods of optical activation such as asymmetric synthesis. Indeed, what is wrong with the traditional way of carrying out resolutions is a lack of real understanding of it. It is this very void we attempt to fill. Such an understanding is equally necessary for classical resolutions and for the study of systems enriched in one enantiomer produced by other methods.

The book is divided in two parts. The first treats racemic systems and enantiomers: the nature of the crystals, enantiomer mixtures and their properties and energetics, and solution properties of enantiomer mixtures. The second part treats the resolution of enantiomers: (1) by direct crystallization, (2) by formation and separation of diastereomer mixtures, and (3) experimental aspects of resolutions. While the examples in this book are virtually all organic, the principles are equally applicable to inorganic compounds; the chapters on resolution are therefore also suggestive of approaches to the resolution of inorganic compounds.

The readership for which we have geared this book is research workers in the area of stereochemistry in general, synthetic organic chemists who are concerned with asymmetric synthesis, medicinal and pharmaceutical chemists in universities and in industry, as well as others who must prepare chiral substances in nonracemic form, and graduate students in the corresponding fields. The volume is designed as a sourcebook and general reference and not as a textbook. It contains much that is of practical use to working chemists in the identified fields. The necessary theory is given in a treatment that is rigorous though largely nonmathematical.

This book is the outcome of an unusual collaborative effort between researchers working on both sides of the Atlantic Ocean. Many of the results and interpretations reported herein are original and derive from work carried out at the Collège de France, where studies of optical activity and of chiral substances follow the long tradition which began when J. B. Biot discovered optical activity in organic compounds there in 1815. It will be remembered that it was Biot who required Louis Pasteur to repeat the very first resolution – that of tartaric acid – in his laboratory at the Collège de France. We trust that we may be forgiven an element of pride in our ability to present this work in continuity of the classic studies in stereochemistry.

Finally, we would be remiss if we did not acknowledge the contributions of our associates to the successful completion of this work. Without their efforts this volume could not have seen light of day. We have in particular greatly benefited from the experimental results (often unpublished) and from the many suggestions and ideas of Drs. Martine Leclercq and Marie-Josèphe Brienne. Mrs. Cécile Bertrand and Mrs. Rosamond Lewis Wilen both labored in more than one capacity over our bilingual manuscripts. It is with considerable pleasure and in the spirit of friendship that we express our thanks to them.

JEAN JACQUES
ANDRÉ COLLET
SAMUEL H. WILEN

Paris
New York

Contents

PART 1

Racemates and Their Enantiomer Constituents

The separation of racemates into their chiral constituents rests upon a number of physicochemical principles and data which are the essential subject of the first part of this volume.

Chemists who are occasionally called upon to reproduce a resolution described in the literature or who have (or will have) succeeded in carrying out a resolution "without problem" may feel that we attach greater importance than is warranted to a particular body of knowledge and to facts that do not appear to be immediately useful.

In fact, the goal of Part 1 is a double one. We wanted first of all to collect data and interpretations which, when understood and applied, would lead those who do encounter difficult resolutions to be less hesitant in facing them. We also wished to describe, in as complete a way as possible, the actual state of an area of physical chemistry which, after having been much worked over at the end of the last century, has for the most part lain fallow over several decades.

Yet the study of the properties of enantiomers and of their mixtures must not be thought of as an abstruse subject falling outside the mainstream of physical chemistry; it is an essential component of contemporary problems of general interest. The problems of molecular packing in solids, of cocrystallization, of polymorphism, and of the thermodynamics of solutions are but specific facets of larger problems which arise from the symmetry properties of the systems which are taken up in this book. The reader ought not to be too surprised to find herein, more often than at first supposed, ideas and notions familiar in other contexts.

CHAPTER 1

Types of Crystalline Racemates

We begin our examination of chiral substances with the definition of terms, follow this with an analysis of those physical properties which distinguish the principal types of racemates, and conclude with a discussion of concepts applicable to racemates and their constituents at the interface of crystallography, symmetry, and thermodynamics.

1.1 DEFINITIONS

Chirality is a concept well known to organic chemists and, indeed, to all chemists concerned in any way with structure. It has numerous implications ranging from those affecting physical properties of matter to those related to biological mechanisms. These implications extend far beyond the borders of "pure" chemistry. Since it is likely that readers concerned with those aspects of chirality examined in this volume may belong to a wide variety of disciplines, we believe that it is worthwhile, if not essential, to define the terminology employed in this book in as precise and unequivocal a manner as possible.

The geometric property that is responsible for the nonidentity of an object with its mirror image is called *chirality*. A *chiral* object may exist in two *enantiomorphic* forms which are mirror images of one another. Such forms lack *inverse symmetry elements*, that is, a center, a plane, and an improper axis of symmetry. Objects that possess one or more of these inverse symmetry elements are superposable on their mirror images; they are *achiral*. All objects necessarily belong to one of these categories; a hand, a spiral staircase, and a snail shell are all chiral, while a cube and a sphere are achiral.

According to Kelvin,[1] "two equal and similar right hands are *homochirally* similar. Equal and similar right and left hands are *heterochirally* similar . . .". All of the foregoing definitions remain valid at the molecular level; there are achiral as well as chiral molecules. The latter exist in two *enantiomeric* forms (the adjective enantiomorphic is more generally applied to macroscopic objects). The term enantiomer is used to designate either a single molecule, a *homochiral collection* of molecules, or even a *heterochiral collection* that contains an excess of one enantiomer and whose composition is defined by its enantiomeric purity p, or the *enantiomeric excess* e.e. which is equivalent to p.

The oldest known manifestation of molecular chirality is the *optical activity*, or *rotatory power*, the property that is exhibited by the rotation of the plane of polarization of light. The two enantiomers of a given compound have rotatory powers of equal absolute value but of opposite *sign*, or *sense*. One is *positive*, or *dextrorotatory*, while the other is *negative*, or *levorotatory*. The absolute designations of sign are arbitrary inasmuch as they are wavelength, temperature, and solvent dependent, but the relative designations are always valid. That is, a given enantiomer may be (+) at one wavelength and (−) at another. The other enantiomer will always have the opposite sign at the corresponding wavelength.

While we shall use as often as possible the (+) and (−) symbols to designate a pair of enantiomers, we shall occasionally employ the letters *d* and *l* or D and L for convenience.

The expression *optically active substance* may signify a pure enantiomer or a mixture containing an excess of one of the two. The composition of a mixture of two enantiomers may be characterized by its *optical purity*, which may in turn be determined from the ratio of the optical rotation of the mixture to that of the pure enantiomer. The optical purity (experimental value) is generally equal to the enantiomeric purity, which reflects the real composition. A pure enantiomer is often called *optically pure*.

The *absolute configuration* of a chiral substance is known when an enantiomeric structure can be assigned to an optically active sample of a given sign. We have little need to concern ourselves with absolute configurations as such in this book. Recall that absolute configurations are designated by means of an alphabetic symbolism (*R*, *S* for *rectus* and *sinister*) whose application is determined by the rules of Cahn, Ingold, and Prelog.[2] However, the D and L descriptors of Rosanoff[3] are still used for carbohydrates. Care should be exercised so as not to confuse these with the sign of the optical activity.

An equimolar mixture of two enantiomers whose physical state is unspecified or unknown is called a *racemate*;[4] the corresponding adjective is racemic, as in a racemic substance, for example. This word is derived from racemic acid, a name used to designate one of the isomeric tartaric acids during the last century and which Pasteur demonstrated to be a mixture of dextro- and levorotatory forms. Racemates, which we generally designate by the symbol (±), are evidently optically inactive by external compensation.

The separation of the two enantiomers that constitute a racemate is called a *resolution*, or an *optical resolution*. When the separation is not complete, a mixture is obtained which is often called either a partially resolved racemate or a partially resolved enantiomer.

We shall see that crystalline racemates may belong to one of three different classes. In the first, the crystalline racemate is a *conglomerate*, that is, a mechanical mixture of crystals of the two pure enantiomers. A conglomerate is formed as a result of a *spontaneous resolution*. The expression "racemic mixture" has generally been used in the literature[4] to designate this type of racemate. However, we shall not use this expression because we believe it to be imprecise and subject to confusion.

The most common type of crystalline racemate is that in which the two enantiomers are present in equal quantities in a well-defined arrangement within the crystal lattice. The resultant homogeneous solid phase corresponds to a true crystalline addition compound which we call *racemic compound*.[4] We prefer this name to "true racemate", which is sometimes used. The former is less ambiguous in that it is less easily confused with the general term racemate.

The third possibility corresponds to the formation of a solid solution between the two enantiomers coexisting in an unordered manner in the crystal. We use the term *pseudoracemate* to designate this case, which is rather rare.

REFERENCES 1.1

1 Lord Kelvin, in *Baltimore Lectures*, C. J. Clay and Sons, London, 1904, p. 619; also see K. Mislow and P. Bickart, *Israel J. Chem.*, 1976/77, **15**, 1.

2 R. S. Cahn, C. K. Ingold, and V. Prelog, *Angew. Chem.* 1966, **78**, 413; *Angew. Chem. Int. Ed.*, 1966, **5**, 385.

3 M. A. Rosanoff, *J. Am. Chem. Soc.*, 1906, **28**, 114.

4 IUPAC Tentative Rules for the Nomenclature of Organic Chemistry, Section E, Fundamental Stereochemistry, *J. Org. Chem.*, 1970, **35**, 2849, 2857.

1.2 THE CRYSTALLIZATION OF RACEMATES AND ENANTIOMERS

The subject of this book has a special place in the history of chemistry. It begins with a *crystallographic* enigma, that of the special isomerism of the tartaric and racemic acids, whose solution was provided by Pasteur.[1] While it has not been especially evident to chemists for the better part of a century, the crystallographic distinctions between stereoisomers observed by Pasteur have, or ought to have, much relevance to stereochemical problems of the present.

As a matter of fact, as we shall see, the choice of resolution methods as well as their probability of success depend to a large extent upon the type to which the racemates belong in the *solid state*. These two considerations confer an unusual importance on the characterization of crystalline enantiomers and racemates.

It was by observing the crystal morphology that Pasteur was able to recognize right, left, and racemic tartaric acids and their salts. This diagnosis is possible when features associated with enantiomorphism such as *hemihedral faces* (Fig. 1) are apparent in the crystal. This is not an easy determination to carry out, however, and requires a size and quality of crystals that are not too often available. Moreover, the *hemihedrism* of enantiomorphous crystals which would permit visual differentiation of dextro and levo crystals is not always apparent in the crystal aspect.

Fortunately, crystallographic differences between racemic and enantiomorphous crystals are generally easier to recognize.

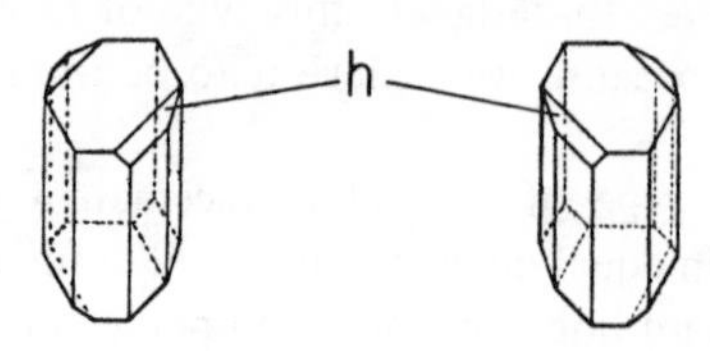

Figure 1 Dextro and levo sodium ammonium tartrates as described by Pasteur. In the absence of hemihedral facets such as *h*, the enantiomorphism of the crystals is not detectable from their aspect. L. Pasteur, *Ann. Chim. Phys.*, 3rd series, 1850, 28, 56. Reproduced by permission of Masson, S.A., Paris.

For a long time, it was primarily by comparison of the morphologies of crystals of enantiomers and of the racemate, their symmetry, and crystal classes that forms we now call conglomerate, racemic compound, and pseudoracemate (see Section 1.1) were distinguished or intuitively differentiated with some precision. By way of example, we illustrate in Fig. 2 the type of highly detailed drawing of crystalline forms of newly characterized substances that was so common in the chemical literature toward the end of the nineteenth centruy. It was only after the work of Roozeboom,[2a] in 1899, that the different types of racemates were more precisely characterized through the phase diagrams of binary mixtures of the constituent enantiomers.

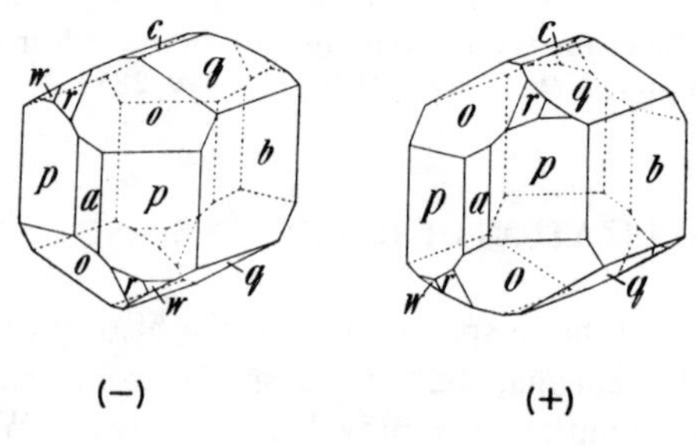

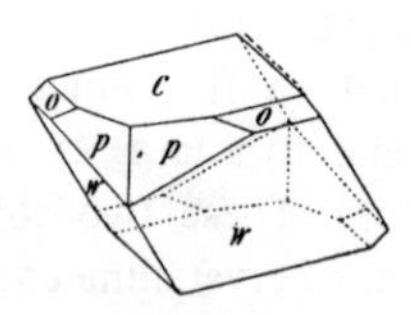

Figure 2 Crystals of (—)-, (+)-, and (±)-tetrahydroparatoluquinaldine as drawn by W. J. Pope and E. M. Rich, *J. Chem. Soc.*, 1899, **75**, 1093. Reproduced by permission of the Royal Society of Chemistry.

Today, visual observation of crystalline morphology by means of a goniometer is no longer carried out. Chemists now possess more rapid means of identifying and of describing compounds. As a matter of fact, these new techniques and, in particular, those which permit the determination of crystal structures through X-ray diffraction reveal, this time on a molecular scale, the very differences between enantiomers and racemates which were earlier so well described by our predecessors by the esthetically more satisfying drawings of their crystals. Nevertheless, problems and questions related to the crystallization of racemates and enantiomers continue to appear often in unpredictable, new, diverse forms.

1.2.1 Chirality and Packing Modes in Crystals

The packing of chiral molecules differs for pure enantiomers on the one hand and for racemic crystals on the other simply as a consequence of geometry. In the former case, the crystal lattice is necessarily enantiomorphous. In the latter case, the crystal generally possesses inverse elements of symmetry which transform an object into its mirror image.

This idea was expressed first by Pasteur in 1860 even before the three-dimensional structure of carbon compounds had been clearly recognized and firmly associated with optical activity. Pasteur wrote,[1b] "Lorsque les atomes élémentaires des produits organiques sont groupés dissymétriquement, la forme cristalline du corps manifeste cette dissymétrie moléculaire par l'hémiédrie non superposable."*

A half-century later, in 1919, Pope could be much more specific:[2b] "Parmi les 32 classes cristallines, il y en a 11 qui sont énantiomorphes ou non superposables à leur image dans un miroir . . . Toute substance douée de pouvoir rotatoire cristallise dans une de ces 11 classes énantiomorphes."†

Translating all of this into modern terms, there are 230 ways of arranging objects repetitively in a three-dimensional network. These 230 *space groups* may be divided among the 32 crystal classes according to their symmetry. The 11 enantiomorphous crystal classes encompass 66 space groups which are devoid of inverse symmetry elements. Thus, an enantiomer may only crystallize in one of these 66 groups. As we shall see, only a few of these groups are highly preferred.

While an enantiomer necessarily crystallizes in an enantiomorphous system, the inverse of this statement is not true. A racemate may in principle crystallize in *any* group, even a chiral one; the optical activity of the crystal does not necessarily imply any optical activity of the molecules in the liquid state (a circular staircase may be built out of achiral bricks). In fact, in almost all cases racemates crystallize in those space groups that possess elements of inverse symmetry. There are 164 possibilities of which only a small number are common. (±)-*o*-Tyrosine is one of the rare examples of racemic crystals that belong to an enantiomorphous class.[3] The same holds true for (±)-*erythro*-phenylglyceric acid,[4,5] (±)-camphoroxime,[6] and (±)-α-methylsuccinic acid[21] (see Section 1.2.3).

* "When the elementary atoms of organic products are arranged dissymmetrically, the crystal form of the body manifests this molecular dissymmetry through a nonsuperposable hemihedry."

† "Among the 32 crystal classes, there are 11 which are enantiomorphic or nonsuperposable upon their mirror images. . . All substances endowed with rotatory power crystallize in one of these 11 enantiomorphic classes."

1.2.2 Preferred Space Groups for Crystalline Enantiomers and Racemic Compounds

Many of the theoretically possible three-dimensional arrangements have been observed by crystallographers; however, the frequency with which the space groups are found varies greatly. Some statistical data concerning the types of crystals formed from chiral molecules – mostly organic ones – are given in Table 1; these have been extracted from the inventories of Hägg,[7] Zorkii,[8, 9] and Cesario.[4c] It appears that 70 to 90% of enantiomers crystallize in space groups $P2_12_12_1$ and $P2_1$, while 60 to 80% of racemic crystals belong to groups $P2_1/c$, $C2/c$, and $P\bar{1}$.

Table 1 Most frequently encountered space groups for crystalline enantiomers and racemic crystals

Type of Structure and Space Groups	Hägg[7]	Zorkii[8, 9]	Cesario[4c]
Enantiomorphous			
$P2_12_12_1$	46%	67%	50%
$P2_1$	23%	27%	33%
$C2$		1%	10%
Number of cases studied	79[a]	430	109[c]
Racemic			
$P2_1/c$	38%	56%	50%
$C2/c$	6%	15%	10%
$P\bar{1}$	8%	13%	8%
Number of cases studied	79[b]	792	40[c]

[a] Substances optically active in solution.
[b] Racemic crystal structures including cases of achiral compounds crystallizing in a chiral conformation (see Section 1.2.3).
[c] Amino acids and their derivatives.

These common space groups are described in Figs. 3 through 7 in representations familiar to crystallographers. The unit cell, that is, the smallest geometric unit from which the periodic arrangement that constitutes the crystal lattice is constructed or built up, is drawn according to normal conventions.[10] This finite figure allows for a variety of symmetry operations called *point groups* which generate the *space groups* through additional symmetry operations that are combined with translations. Each space group is symbolized by a letter which defines the type of lattice (P = primitive, C = centered, etc.), followed by a group of numbers and/or of small letters which designate the symmetry and translation operators, for example, 2_1 = twofold screw or binary screw axis, c = glide plane parallel to c-axis, and so forth.

For cases that are pertinent here, the asymmetric unit, from which the unit cell may be reconstituted by means of the symmetry operations of the space group may consist either of a single chiral molecule, of an array (conceivably chiral) o

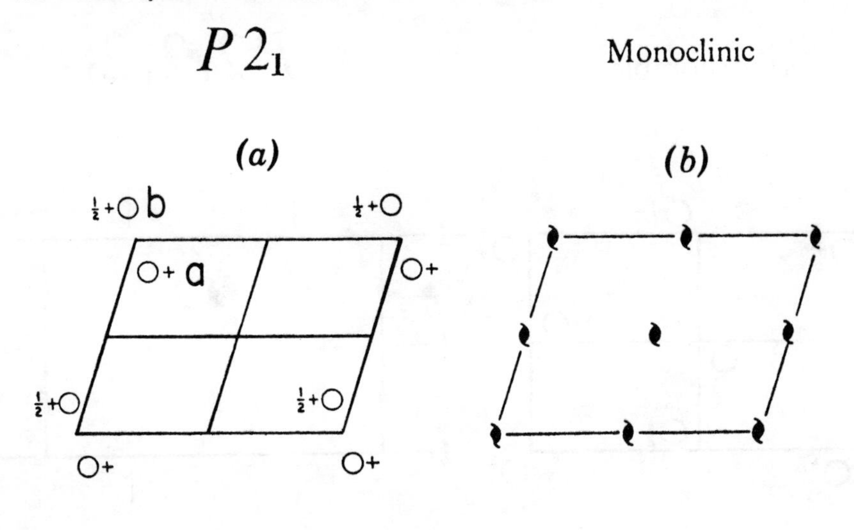

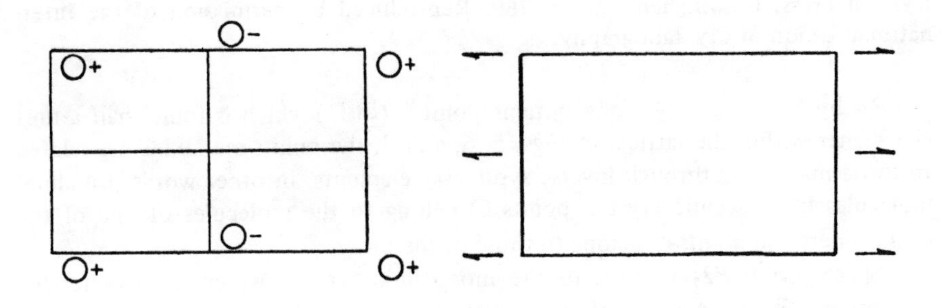

Figure 3 Representation of the enantiomorphic space group $P2_1$. There are two asymmetric units per unit cell which corresponds most often to two homochiral molecules ($Z = 2$). *International Tables for X-ray Crystallography*, Kynoch Press, Birmingham, Ala., 1969. Reproduced by permission of the International Union of Crystallography.

several molecules, or occasionally even of part of a molecule (as when the latter possesses a special symmetry, e.g., half of a molecule of binary symmetry). The number of molecules in the unit cell is designated by the letter Z.

The two space groups $P2_1$ and $P2_12_12_1$ (Figs. 3 and 4) represent the majority of homochiral crystals. In the figures, one focuses on a given point in the unit cell (designated ○) and allows the elements of symmetry to operate which leads to all the points that correspond to one another. The plus and minus signs define the positions of these points relative to the plane of the paper. For example, when point *a* (○+) (Fig. 3*a*) located above the plane is translated along axis 2_1 (shown in

Orthorhombic $P2_12_12_1$

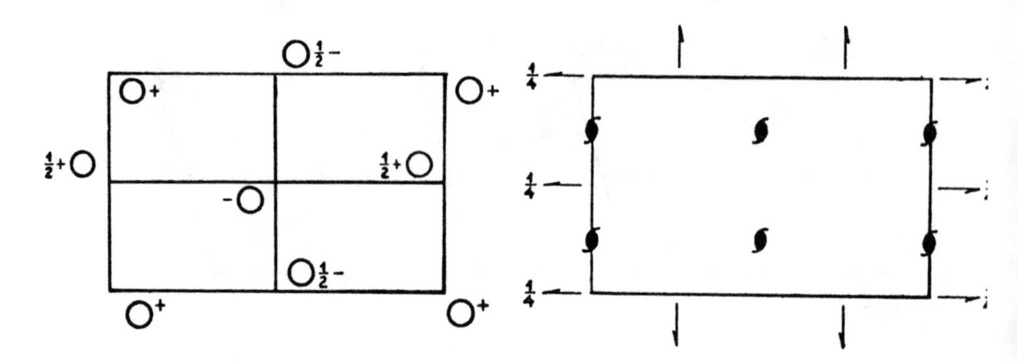

Figure 4 Representation of the enantiomorphic space group $P2_12_12_1$. There are four asymmetric units per unit cell which corresponds most often to four homochiral molecules per cell ($Z = 4$). *International Tables for X-ray Crystallography*, Kynoch Press, Birmingham, Ala., 1969. Reproduced by permission of the International Union of Crystallography.

Fig. 3*b* by the symbol), one obtains point *b* ($\frac{1}{2}$+O) which is found half a unit cell higher within the lattice. In Figs. 5, 6, and 7, the points marked ⊙ are related to those marked O through inverse symmetry elements. In other words, for chiral molecules in a racemic crystal, points O belong to the molecules of one of the enantiomers and points ⊙ belong to those of the other.

Space group $P2_1$ belongs to the monoclinic system, which is characterized macroscopically by a plane of symmetry and a twofold or binary axis. The cell generally contains two molecules related by a binary screw axis that is either normal to (first setting) or parallel (second setting) to the plane of the paper.

Space group $P2_12_12_1$ is orthorhombic, which on a macroscopic level translates to a system consisting of three orthogonal binary axes. Most often, the unit cell consists of four homochiral molecules that are related to one another by three binary screw axes (one perpendicular and two parallel to the plane of the paper). In the exceptional cases in which a racemate crystallizes in these enantiomorphic systems, the unit cell consists of two (in the case of $P2_1$) or four (in the case of $P2_12_12_1$) groups comprised of one or several *pairs* of oppositely handed molecules.

Among the 164 space groups possessing at least one element of inverse symmetry, three groups ($P2_1/c$, $C2/c$, and $P\bar{1}$) contain the great majority of racemic crystals (Figs. 5, 6, and 7). Group $P2_1/c$ belongs to the monoclinic system, as does $P2_1$ (macroscopic plane of symmetry and binary axis). The cell contains two of each of the oppositely handed molecules which are related to one another by a center of symmetry and a binary screw axis.

Several authors[7, 11–13] have commented on the fact that the most frequently

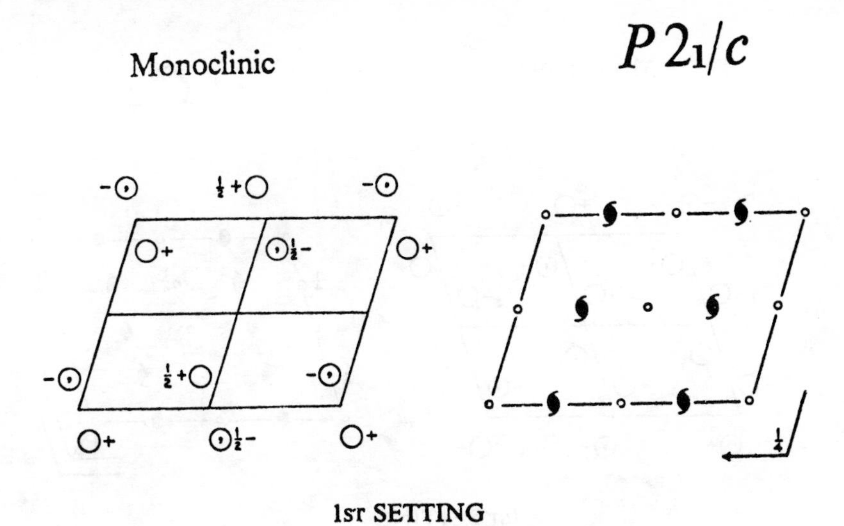

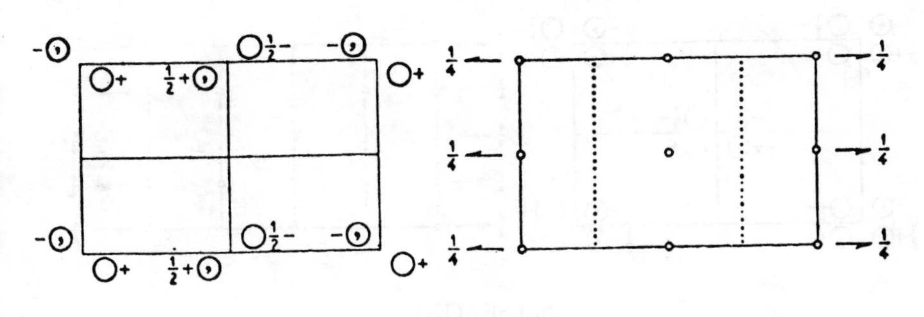

Figure 5 Representation of the centrosymmetric space group $P2_1/c$. There are four asymmetric units per unit cell which corresponds most often to two dextro and two levo molecules ($Z = 4$). *International Tables for X-ray Crystallography*, Kynoch Press, Birmingham, Ala., 1969. Reproduced by permission of the International Union of Crystallography.

encountered space groups possess one or more binary screw axes 2_1. Among these, Wittig[13] has pointed out the advantages associated with a binary screw axis, a glide plane, or a center of symmetry all on the basis of simple electrostatic considerations. Also, Wittig has identified the electrostatic incompatibilities created by a mirror during the packing of molecules. These are illustrated in Fig. 8.

This matter has been treated in a thorough way by Kitaigorodskii and his associates.[14,15] He has emphasized the fact that molecular crystals are assemblages in which invariably compactness tends toward the maximum compatible with the

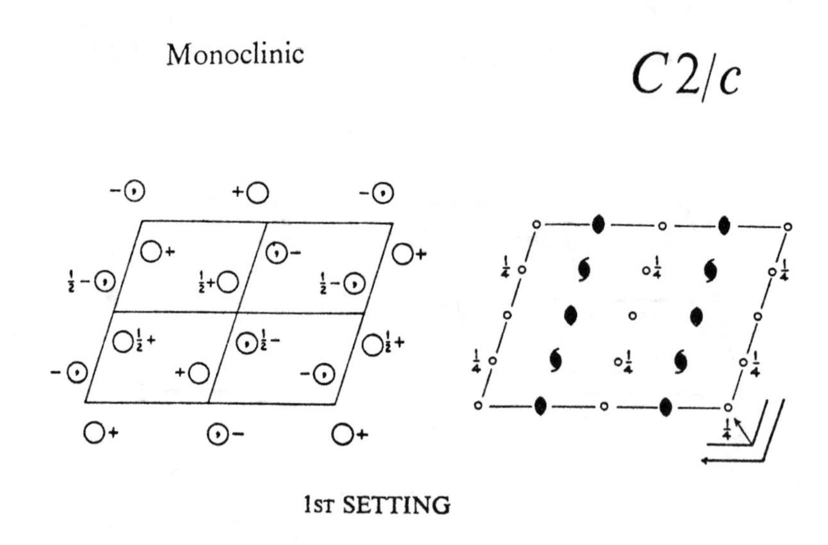

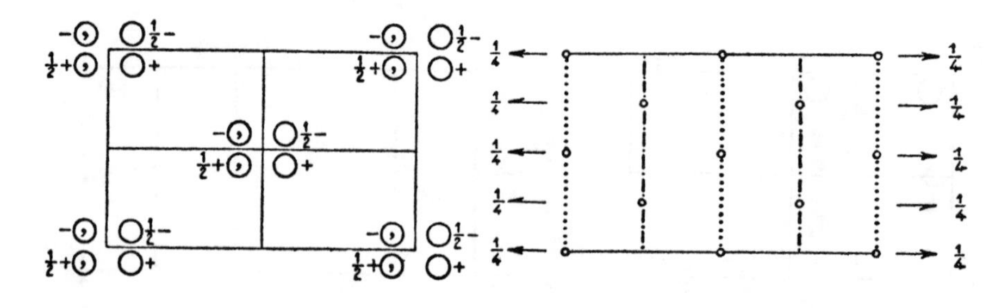

2nd SETTING

Figure 6 Representation of the centrosymmetric space group $C2/c$. There are eight asymmetric units per unit cell which corresponds most often to four dextro and four levo molecules ($Z = 8$). *International Tables for X-ray Crystallography*, Kynoch Press, Birmingham, Ala., 1969. Reproduced by permission of the International Union of Crystallography.

molecular geometry. In this way, the space-filling or packing coefficient in crystals always lies between 0.65 and 0.77. This is of the same order as that which corresponds to the regular packing of spheres or ellipsoids (0.74). (According to Kitaigorodskii,[15] molecules that cannot attain a packing coefficient at least equal to 0.6 as a consequence of their geometry do not crystallize; upon cooling they form glasses. The packing coefficient p is defined by $p = vZ/V$, where v is the volume of the molecule, V is the volume of the cell, and Z is the number of molecules per cell.)

In order to fill space in the most compact way possible with objects of any geometry, one must first allow a plane to be filled in a compact manner. There are

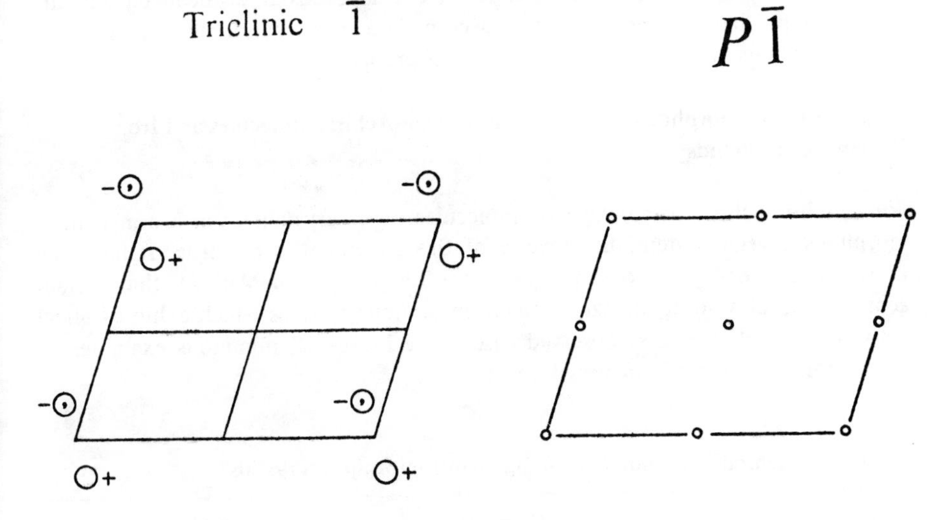

Figure 7 Representation of the centrosymmetric space group $P\bar{1}$. There are two asymmetric units per unit cell corresponding most often to one dextro and one levo molecule ($Z = 2$). *International Tables for X-ray Crystallography*, Kynoch Press, Birmingham, Ala., 1969. Reproduced by permission of the International Union of Crystallography.

only a limited number of *two-dimensional space groups*[10] in which each object may be in contact with six neighbors, a necessary condition for the optimal close-packing of the molecular assembly. Examination of the several possible ways of stacking these molecular layers in a compact manner reveals the favorable influence of the binary screw axis simply as a consequence of *geometry*. Finally, consideration of various possible combinations of stacking leads to a limited number of space groups which fulfill the requirements of three-dimensional close-packing.

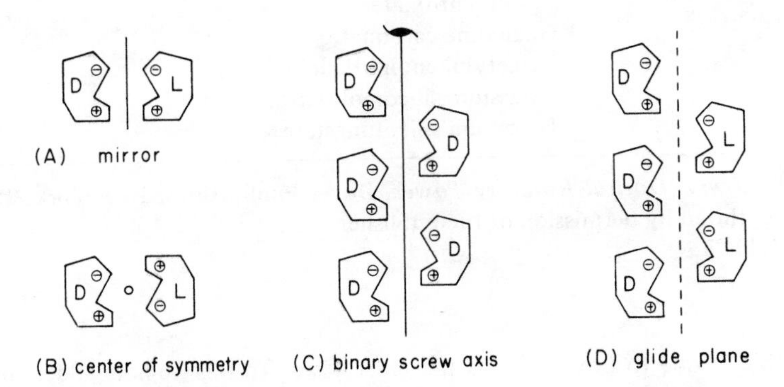

Figure 8 Electrostatic interactions of chiral molecules in a crystal depending on their mode of packing: ⊕ and ⊖ represent electric charges; (A) unfavorable electrostatic repulsion; (B), (C) and (D) electrostatically favorable dispositions.

For the special case of chiral molecules, Kitaigorodskii has deduced that the space groups that are most likely to accomodate pure enantiomers are $P2_1$ and $P2_12_12_1$, whereas racemic crystals would most probably belong to the $P2_1/c$ system.

1.2.3 Enantiomorphous Crystals Derived from Achiral Molecules and from Racemic Compounds

While a homochiral assemblage of molecules never crystallizes in a nonenantiomorphous crystal system, the inverse of this statement is not true. It has been known since 1854 as a consequence of the observation of Marbach that certain achiral molecules do crystallize in enantiomorphous systems which exhibit optical activity. Though the first observed cases were inorganic, numerous examples of such organic compounds are now known.

Table 2 Achiral compounds forming enantiomorphous crystals[a]

Isotropic	Uniaxial	Biaxial
Sodium chlorate	Quartz	Hydrazine sulfate
Sodium bromate	Cinnabar	Strontium formate
Sodium sulfantimoniate	Sodium periodate	Barium formate
Sodium uranyl acetate	Potassium dithionate	Lead formate
	Rubidium dithionate	Iodic acid
	Cesium dithionate	Ammonium oxalate
	Calcium dithionate	Ammonium potassium oxalate
	Strontium dithionate	Sodium arsenate
	Lead dithionate	Lithium sulfate
	Ethylenediamine sulfate	Magnesium sulfate
	Benzil[b]	Nickel sulfate
	Potassium lithium sulfate	Magnesium chromate
	Potassium lithium sulfochromate	Sodium dihydrogen phosphate
	Guanidine carbonate	
	Diacetylphenolphthalein	
	Potassium silicomolybdate	
	Potassium silicotungstate	

[a] T. M. Lowry, *Optical Rotatory Power*, Dover Publications, New York, 1964, p. 338. Reprinted by permission of the publisher.
[b] See Table 3.

Table 2, due to Lowry,[56] lists the older cases. To this we have added, in Table 3, more recent examples representative of the variety of structures and space groups exhibiting the phenomenon. The occurrence of such optically active crystals may have several origins:

Table 3 Nonresolvable substances and racemic compounds that form enantiomorphous crystals

Name	Space Group	Refs.
Hydrogen peroxide	$P4_12_12$	24
Dibenzoyl peroxide	$P2_12_12_1$	18
Di-(*o*-chlorobenzoyl) peroxide	$P2_12_12$	25
p-Nitroperoxybenzoic acid	$P2_12_12_1$	18
Hexaphenylcarbodiphosphorane	$C2$	27
2,7-Dimethylthianthrene	$P2_12_12_1$	26
Phenol	$P222_1$	28
o-Cresol	$P3_1$	29
2.4-Dinitrophenol	$P2_12_12_1$	38
ω-Bromoacetophenone	$P2_12_12_1$	30
Succinic anhydride	$P2_12_12_1$	31
Benzil	$P3_12$	32
Phenylsulfinylacetic acid	$P2_12_12$	33
2-Butynoic acid	$P2_1$	34
1,2,5,6-Dibenzanthracene	$P2_1$	17
3,4-Benzophenanthrene	$P2_12_12_1$	19
Difluorenylidene	$P2_12_12_1$	35
o-Quinone	$P2_12_12$	36
p,p'-Dimethylchalcone	$P2_12_12$	37
Triphenylene	$P2_12_12_1$	40
2-Amino-1,1,3-tricyanopropene	$P2_1$	39
1*e*, 2*e*, 3*e*, 4*e*, 5*e*, 6*e*-Hexachloro-1*a*-phenylcyclohexane	$P2_1$	43
Glycine	$P2_1$, $P3_1$	44
Methanesulfinic acid	$P2_12_12_1$	45
Tri-*o*-thymotide	$P3_12_1$	41
Urea	$C6_12$	42
(±)-*erythro*-Phenylglyceric acid	$P2_1$	4, 5
(±)-α-Methylsuccinic acid	$P2_1$	21
(±)-*o*-Tyrosine	$P2_1$	3
(±)-Camphoroxime	$P2_1$	6

1 The achiral molecules are rigid and not subject to deformations. They are arranged in the crystal lattice in such a way that no inverse symmetry element appears. This is the case for urea (see Section 5.1.8) and even for centrosymmetrical molecules[16] such as 1,2,5,6-dibenzanthracene (**1**).[17]

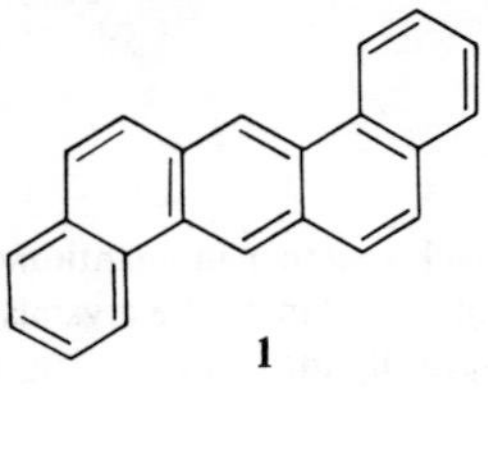

1

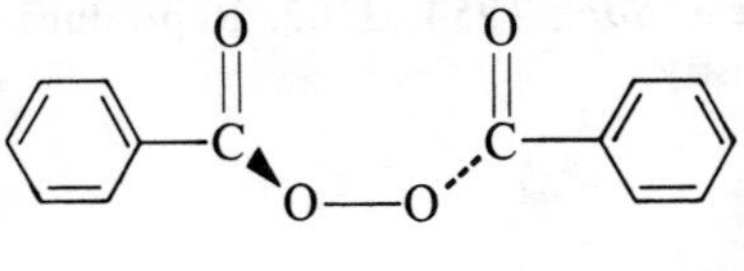

2

2 The molecules, while revealing no chirality in their planar formulas, adopt *chiral conformations* in the crystal. Selection of identical conformations is then equivalent to a spontaneous resolution.* The case of dibenzoyl peroxide (2) constitutes a good example of this possibility; the conformation adopted in the crystal contains a binary axis which makes the molecule chiral. The boundary between conformation and configuration may on occasion be difficult to define since it depends only upon the height of the energy barrier which itself is temperature dependent. Thus, 3,4-benzophenanthrene ([4]-helicene) crystallizes in the enantiomorphic point group $P2_12_12_1$ in a chiral conformation (see Fig. 9). However, the occurrence of a spontaneous resolution is not demonstrable through the measurement of optical activity since the conformers interconvert very rapidly in solution.[19] On the other hand, in the case of tri-*o*-thymotide (see Section 5.1.8), which is a flexible molecule that crystallizes in an enantiomorphous crystal system,[20] a finite optical activity is observed which rapidly vanishes ($t_{1/2}$ = ca. 2.4 minutes at 21°) when a crystal is dissolved.

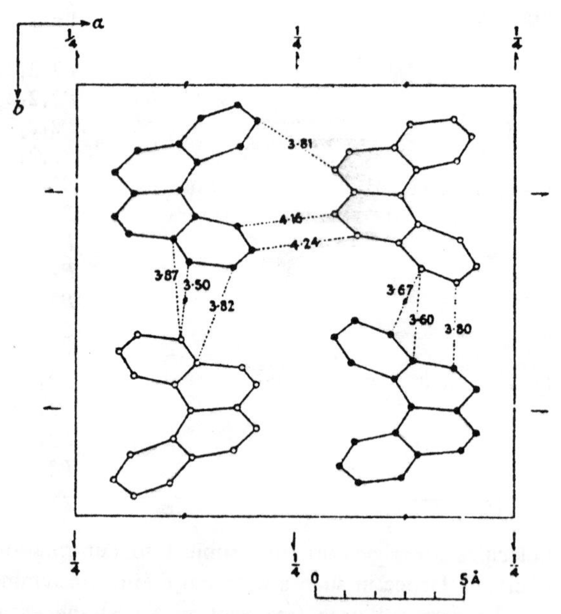

Figure 9 Crystal structure of 3,4-benzophenanthrene. The *C*2 chiral conformation of the molecules is readily apparent. F. H. Herbstein and G. M. J. Schmidt, *J. Chem. Soc.*, 1954, 3302. Reproduced by permission of the Royal Society of Chemistry.

* However, in the majority of cases, chiral conformations pack together in enantiomeric pairs mimicking the formation of racemic compounds. Thus, in most instances the crystals of molecules whose preferred conformations are chiral are, in fact, optically inactive.

3 A mixture of equal numbers of dextro and levo molecules may arrange itself in a crystal in such a way as to form a chiral stack. This may occur in two ways which correspond to one or the other of the preceding cases. In the first of these cases, the enantiomeric molecules arrange themselves two by two about a center of symmetry so as to form an "achiral dimer." This is equivalent to case (1). The centers of symmetry of these "dimers" are not part of the symmetry elements of the crystal lattice. This relatively rare type of structure is illustrated by (±)-*o*-tyrosine[3] (Fig. 10) and by (±)-α-methylsuccinic acid.[21] In the second case, a dextro and a levo molecule together form a chiral "dimer" equivalent to case (2). This is observed with (±)-*erythro*-phenylglyceric acid.[4]

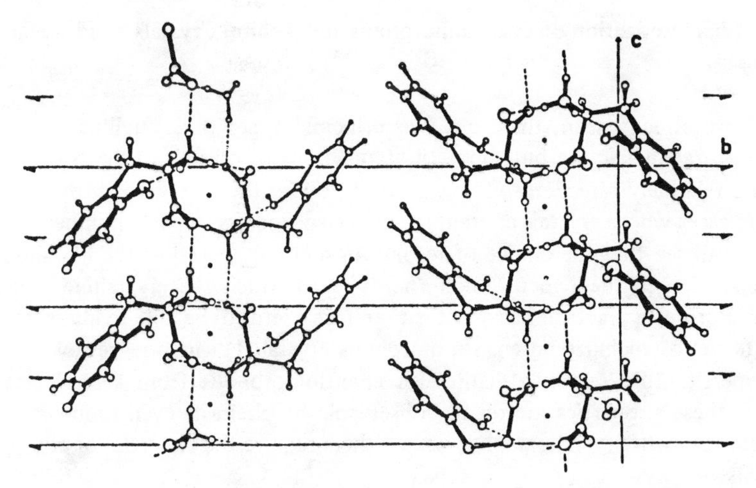

Figure 10 The crystal structure of *o*-tyrosine as viewed down the *a*-axis. The noncrystallographic centers of symmetry are drawn as small circles. A. Mostad, C. Rømming, and L. Tressum, *Acta Chem. Scand.*, 1975, **B29**, 171. Reproduced by permission of the editor.

These examples of chiral arrangements in the crystal would not warrant our attention were it not for the fact that they actually provide access to optically active molecules, that is, they allow for optical activation. For example, G. M. J. Schmidt and his associates have described a remarkable study of reactions which may take place on the surface of or within enantiomorphous crystals and which, in some favorable cases, allow asymmetric syntheses to take place. Thus, reaction of gaseous bromine at the surface of enantiomorphous crystals of dimethylchalcone (**3**) leads to a partially resolved *erythro* adduct (**4**) with an optical yield as high as 25%.[22,23] Other examples of such optical activations have been described by B. S. Green, M. Lahav, and their co-workers.[46]

We also call attention to the formation of enantiomorphous crystal lattices which contain cavities sufficiently large to stereoselectively accommodate certain chiral molecules. We later examine, in particular, the possibility of carrying out

CH_3 CH_3 $+ Br_2 \longrightarrow$ CH_3 H Br Br H O CH_3

O

Monocrystal $P2_12_12_1$ xD, $(1-x)$L $x \neq 0.5$

3 **4**

resolutions employing inclusion compounds of urea and tri-*o*-thymotide (Section 5.1.8).

1.2.4 Characterization of Enantiomorphous and Achiral Crystals by Physical Methods

As we have already seen, there are two principal types of crystalline racemates. *Conglomerates*, which are but simple juxtapositions of crystals of the two enantiomers, are relatively rare. *Racemic compounds*, by far the most frequently observed type, are cases whose crystals contain the two enantiomers in equal number.

We shall see that the choice of resolution method as well as the likelihood of success are largely dependent upon the type to which the crystalline racemate belongs. For quite practical reasons it is therefore useful to be able to characterize a racemate or to recognize an enantiomorphous crystal. Let us now review some of the properties that serve to distinguish enantiomorphous from achiral crystals. Some of these properties are readily observable by chemists even though they be little familiar with crystallography, while others require the expertise of competent physicists.

For the contemporary chemist, the modern equivalent of the crystallographic observations which were routine during the last century no doubt is the single crystal X-ray crystallographic determination of the space group. However, this method generally requires the intervention of crystallographers. On the other hand, solid state infrared spectroscopy can be used and the results interpreted without difficulty by any organic chemist. While this method is less elegant and visually less informative in that infrared spectra do not allow direct recognition of a chiral system, its virtue is simplicity which is independent of the size and of the quality of the sample crystals.

Simple comparison of an infrared spectrum of the racemate in the solid state with that of one of the enantiomers (in suspension in Nujol or other mulling agent or in a KBr pellet but, of course, not in solution) constitutes a very convenient diagnosis of the nature of the racemate. If the racemate is a conglomerate, its infrared spectrum is superposable on that of the enantiomers, while it is different in the case of a racemic compound. The difference is sometimes small; more often the spectra are so different that, in the absence of other information, one would be tempted to believe that two different substances are being compared (Fig. 11). This is due to the fact that the crystals of the enantiomers and of the racemic compound possess

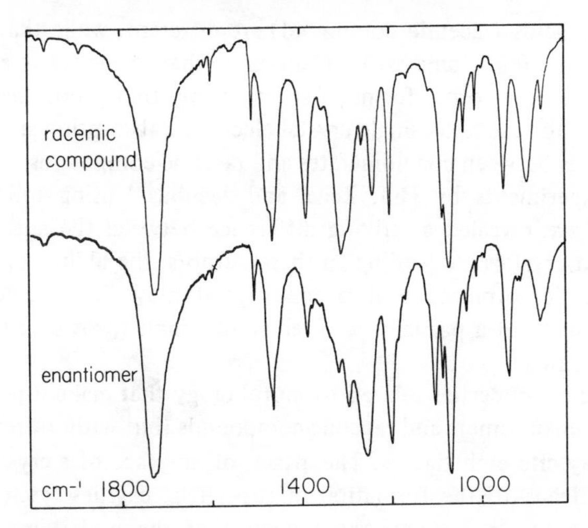

Figure 11 Solid-state infrared spectra of racemic and active mandelic acid in Nujol mull.

different organizations (lattice symmetry and relative arrangement of the molecules), which gives rise to differences in vibration modes, that is, in their frequencies and their intensities. Moreover, the stronger the intermolecular interactions in the solid state, the more the infrared spectra of the crystals differ from those observed in solution[47] which, to reiterate, are identical for the enantiomers and the racemate.

When comparing the spectra of the two forms in the solid state, a doubling of certain absorption bands either in the racemate[48] or in the enantiomer[49] has occasionally been observed. For example, for long-chain carboxylic acids methylated in various positions,[49] the methylenic bands in the vicinity of 720 and 1470 cm^{-1} are simple in the racemate spectra and resolved into doublets in the spectra of the enantiomers. This feature has been related to the disposition of the polymethylene chains in the crystals: they are parallel to one another in the racemates and perpendicular in the resolved substances. These differences are by no means limited to the fingerprint region of infrared spectra. Thus, in comparing the carbonyl and hydroxyl stretching frequencies of the hydrogen phthalate esters of (+)- and (±)-*p*-ethylphenylmethylcarbinol, Eliel and Kofron have shown that these acids exist as dimers in the crystals of the racemic compound but not so in the crystals of the enantiomers. The enantiomers do dimerize in benzene or chloroform solution, however, as is evident from cryoscopic data.[50] Similarly, the infrared spectra of serine show differences which are consistent with the presence of stronger hydrogen bonding between the hydroxyl and both the carboxylate and ammonium groups in the racemic compound than in the enantiomers.[47] In most cases, however, the interpretation of the spectral differences is far from obvious.

What has just been stated for infrared spectra remains true also for Raman spectroscopy. As has been shown by Mathieu,[51,52] the solid-state spectra of the (+)-

and (±)-tartaric acids (racemic compound) are different, while that of (±)-sodium ammonium tartrate (conglomerate) is identical to that of the active compound.

It is supposed that other forms of spectroscopy that probe the internal structure of crystals, for example, single crystal electronic absorption spectra would also reveal differences between conglomerates and racemic compounds.

Recent experiments by Hill, Zens, and Jacobus[53] using solid-state ^{13}C-nmr spectroscopy have revealed a striking difference between the spectra of (+)- and (±)-tartaric acid crystals. According to these authors, the ability of this method to distinguish enantiomorphous from racemic crystals may allow the determination of enantiomeric purity of a partially resolved solid sample (providing that the latter is not a conglomerate).

Among other properties related to morphology that make it possible to distinguish between enantiomers and racemic compounds (but without revealing which is which) one may cite etch figures. The attack of one face of a crystal by an appropriate solvent leads to the formation of superficial cavities (etch figures) which possess a geometry and symmetry indicative of the underlying internal structure.*[54] Along the same lines, the superficial attack of crystals of a conglomerate by a resolved reagent would allow one to distinguish dextro from levo species to separate them by triage. Curtin et al. have demonstrated this with the vapor of an optically active amine attacking crystals of the two enantiomers of a carboxylic acid at different rates.[55]

The refraction of light by a crystal also furnishes information on the system to which it belongs without, however, giving any indication of the possible chirality of its constituent molecules. Cubic crystals are optically *isotropic*, that is, the refraction is simple and the index of refraction is a constant which is independent of the orientation of the crystal relative to the light beam (that is, there is only a single emerging beam no matter what the orientation of the crystal). In the case of *uniaxial* (trigonal, tetragonal, and hexagonal systems) and *biaxial* crystals (orthorhombic, monoclinic, and triclinic systems), one generally observes the phenomenon of double refraction, that is, the incident beam is split and emerges as *two* beams except when the light travels along privileged directions (optic axes). In the case of biaxial crystals, the two optic axes do not coincide with the symmetry axes of the crystal and may even vary with the wavelength of the light utilized.

When the incident light beam is linearly polarized, the change in polarization of the light which traverses the crystal depends both upon its structure and upon the possible optical activity of the constituent molecules. In cubic crystals, which are isotropic, one only observes the rotation of the plane of polarization, and this rotation is relatively easy to measure. In uniaxial and biaxial crystals, linear polarization is usually transformed into elliptical polarization except when the orientation of the beam coincides with an optic axis, in which case a rotation can be measured. With biaxial crystals, the measurement is a delicate one due to the

* The attack of a face of an achiral crystal by a resolved reagent may also yield different etch figures according to the chirality of the reagent. For example, one observes this when (+)- and (—)-tartaric acids, respectively, attack calcite crystals in aqueous solution.

difficulties in orienting the light beam relative to the two optic axes and the rotations are generally different for the two axes.

In the case of optically active molecules, all of whose crystals are themselves optically active, the rotatory power of the crystals is usually greater than that observed in solution. For example, according to Longchambon,[57] the specific rotatory power of crystalline sucrose (biaxial) is $[\rho]_1 = +336°$ for one of the axes and $[\rho]_2 = -100.6°$ for the other (both measured at 579 nm); the specific rotation in solution is $[\alpha] = +69°$ (also at 579 nm). Note that $[\rho]$ and $[\alpha]$ need not even have the same sign, nor do the $[\rho]$ values for the two axes.*

Finally, certain crystals which lack centers of symmetry exhibit the phenomena of piezo- and pyroelectricity, that is, the appearance of surface electric charges under the influence, respectively, of mechanical constraints or heat. These effects are observed on enantiomorphous crystals as well as on achiral crystals which possess planes of symmetry. The pyroelectricity of dextro and levo tartaric acid crystals was observed by Pasteur;[1a] the heat of the hand suffices to demonstrate it. The observation of these two phenomena do not, however, allow one to determine whether the absence of a center of symmetry has a molecular origin or whether it is due only to the arrangement of chiral or achiral molecules within the crystal.

The modern literature reveals little application of the last cited phenomena (rotatory power of crystals or piezo- and pyroelectricity) to the study of properties of enantiomers and of racemates. An exception to this is the work of Schlenk, Jr.,[58] who showed that the measurement of piezoelectricity permits the characterization of certain substances whose rotatory power is too weak to reveal their chirality. This is particularly so for natural triglycerides such as 1-lauro-2,3-dipalmitin and 1-palmito-2-oleo-3-stearin.

REFERENCES 1.2

1 (a) L. Pasteur, *Ann. Chim. Phys.*, 3ème Sér., 1850, **28**, 56. (b) L. Pasteur, *Leçons de Chimie professées en 1860*, Société Chimique de Paris, 1861, p. 25.

2 (a) H. W. Roozeboom, *Z. Phys. Chem.*, 1899, **28**, 494. (b) W. J. Pope, *Bull. Soc. Chim. Fr.*, 1919, **25**, 427. (c) W. J. Pope and E. M. Rich, *J. Chem. Soc.*, 1899, **75**, 1093.

3 A. Mostad, C. Rømming, and L. Tressum, *Acta Chem. Scand.*, 1975, **B29**, 171.

4 (a) C. N. Riiber and E. Berner, *Ber.*, 1917, **50**, 893. (b) S. Furberg and O. Hassel, *Acta Chem. Scand.*, 1950, **4**, 1020. (c) M. Cesario, Thesis No. 289, Orsay, 1976.

5 M. Cesario, J. Guilhem, C. Pascard, A. Collet, and J. Jacques, *Nouv. J. Chim.*, 1978, **2**, 343.

6 F. Baert, Thesis No. 349, Lille, 1976.

7 G. Hägg, in *The Svedberg 1884 30/8 1944*, A. Tiselius and K. O. Pedersen, Eds., Almqvist and Wiksell, Uppsala, 1944, p. 140.

8 V. K. Bel'skii and P. M. Zorkii, *Krystallographyia*, 1970, **15**, 704. *Soviet Phys. Crystallogr. (Engl. Transl.)*, 1971, **15**, 607.

* The optical rotatory power of the crystal $[\rho]$ is expressed in the same units as the specific rotation in solution $[\alpha]$. It is defined as $[\rho] = 100\rho/d$, where ρ is the rotation of the crystal per mm and d is its density.[57]

9 P. M. Zorkii, *Krystallographyia*, 1968, **13**, 26. *Soviet Phys. Crystallogr. (Engl. Transl.)* 1968, **13**, 19.

10 *International Tables for X-Ray Crystallography*, The International Union of Crystallography, Ed., Kynoch Press, Birmingham, Ala., 1979.

11 A. Reis, *Ber.*, 1926, **59**, 1547.

12 K. Weissenberg, *Ber.*, 1926, **59**, 1526.

13 G. Wittig, *Stereochemie*, Akademische Verlagsgesellschaft, Leipzig, 1930, 309.

14 A. I. Kitaigorodskii, *Organic Chemical Crystallography*, Consultants Bureau, New York, 1961.

15 A. I. Kitaigorodskii, *Molecular Crystals and Molecules*, Academic Press, New York and London, 1973.

16 F. H. Herbstein and F. R. L. Schoening, *Acta Crystallogr.*, 1957, **10**, 657.

17. J. M. Robertson and J. G. White, *J. Chem. Soc.*, 1956, 925.

18 G. A. Jeffrey, R. K. McMullan, and M. Sax, *J. Am. Chem. Soc.*, 1964, **86**, 949.

19 F. H. Herbstein and G. M. J. Schmidt, *J. Chem. Soc.*, 1954, 3302.

20 A. C. D. Newman and H. M. Powell, *J. Chem. Soc.*, 1952, 3747.

21 Y. Schouwstra, *Acta Crystallogr.*, 1973, **B29**, 1626.

22 (a) E. Hadjoudis, E. Kariv, and G. M. J. Schmidt, *J. Chem. Soc., Perkin II*, 1972, 1056. (b) K. Penzien and G. M. J. Schmidt, *Angew. Chem.*, 1969, **81**, 628. *Angew. Chem. Int. Ed.*, 1969, **8**, 608.

23 B. S. Green and M. Lahav, *J. Mol. Evol.*, 1975, **6**, 99.

24 S. C. Abrahams, R. L. Collins, and W. N. Lipscombs, *Acta Crystallogr.*, 1951, **4**, 15.

25 J. Z. Gougoutas and J. C. Clardy, *Acta Crystallogr.*, 1970, **B26**, 1999.

26 Chin Hsuan Wei, *Acta Crystallogr.*, 1971, **B27**, 1523.

27 A. T. Vincent and P. J. Wheatley, *J. Chem. Soc., Dalton*, 1972, 617.

28 H. Gillier-Pandraud, *Bull. Soc. Chim. Fr.*, 1967, 1933.

29 C. Bois, *Acta Crystallogr.*, 1972, **B28**, 25.

30 M. P. Gupta and S. M. Prasud, *Acta Crystallogr.*, 1971, **B27**, 1649.

31 M. Shahat, *Proc. Pharm. Soc. Egypt*, 1953, **35**, 57.

32 C. J. Brown and R. Sadanaga, *Acta Crystallogr.*, 1965, **18**, 158.

33 L. Leiserowitz and M. Weinstein, *Acta Crystallogr.* , 1975, **B31**, 1463.

34 V. Benghiar and L. Leiserowitz, *J. Chem. Soc., Perkin II*, 1972, 1763.

35 E. Harnik, F. H. Herbstein, G. M. J. Schmidt, and F. L. Hirschfeld, *J. Chem. Soc.*, 1954, 3288.

36 A. L. MacDonald and J. Trotter, *J. Chem. Soc., Perkin II*, 1973, 476.

37 D. Rabinovich and Z. Shakked, *Acta Crystallogr.*, 1974, **B30**, 2829.

38 F. Iwasaki and Y. Kawano, *Acta Crystallogr.*, 1977, **B33**, 2455.

39 B. Klewe, *Acta Chem. Scand.*, 1972, **26**, 317.

40 A. Klug, *Acta Crystallogr.*, 1950, **3**, 165.

41 S. Brunie, A. Navaza, G. Tsoucaris, J. P. Declercq, and G. Germain, *Acta Crystallogr.*, 1977, **B33**, 2645.

42 A. E. Smith, *Acta Crystallogr.*, 1952, **5**, 224.

43 C. A. De Mey, A. J. De Kok, J. Lugtenburg, and C. Romers, *Rec. Trav. Chim.*, 1972, **91**, 383.

44 Y. Iitaka, *Nature*, 1959, **183**, 390.

45 K. Seff, E. G. Heidner, M. Meyers, and K. N. Trueblood, *Acta Crystallogr.*, 1969, **25**, 350.

46 B. S. Green, M. Lahav, and D. Rabinovich, *Accts. Chem. Res.*, 1979, **12**, 191.

47 (a) C. N. R. Rao, *Chemical Application of Infrared Spectroscopy*, Academic Press, New York and London 1963, pp. 62, 585. (b) M. Avram and G. D. Matecscu, *Infrared Spectroscopy*, Wiley, New York 1972, pp. 482–3.

48 J. Vaissermann, *C.R. Acad. Sci.*, 1970(B), **270**, 948.

49 I. Fischmeister, *Ark. Kemi*, 1963, **20**, 353.

50 E. L. Eliel and J. T. Kofron, *J. Am. Chem. Soc.*, 1953, **75**, 4585.

51 J. P. Mathieu, *C.R. Acad. Sci.*, 1972(B), **274**, 880.

52 J. P. Mathieu, *J. Raman Spectrosc.*, 1973, **1**, 47.

53 H. D. W. Hill, A. P. Zens, and J. Jacobus, *J. Am. Chem. Soc.*, 1979, **101**, 7090.

54 F. C. Phillips, *An Introduction to Crystallography*, 3rd ed., Longmans, London and Wiley, New York 1963, p. 152.

55 C. -T. Lin, D. Y. Curtin and I. C. Paul. *J. Am. Chem. Soc.*, 1974, **96**, 6199.

56 T. M. Lowry, *Optical Rotatory Power*, Dover, New York and London, 1964, p. 338.

57 Quoted by T. M. Lowry, *ibid.* p. 340.

58 W. Schlenk, Jr., *Angew. Chem.*, 1964, **76**, 161. *Angew. Chem. Int. Ed.*, 1965, **4**, 139.

1.3 RELATIONSHIPS BETWEEN ARRAYS OF ENANTIOMERS AND THE CORRESPONDING RACEMIC COMPOUND IN CRYSTALS

1.3.1 Comparison of Crystal Structures of Enantiomers and of the Racemic Compound

Although they are different, the crystal structures of the enantiomers and of the racemic compound may nonetheless exhibit similarities in the packing of the constituent molecules. These structural similarities were observed for the first time by Simpson and March[2] in the case of alanine.

Table 1 lists examples of compounds for which the crystal structures of the

Table 1 Crystallographic data for enantiomers and the corresponding racemic compounds

Compound	Racemic Compound		Enantiomer		Ref.
Alanine	$Pna2_1$*	$z = 4$†	$P2_12_12_1$	$z = 4$	1, 2
trans-1,2-Cyclohexanedicarboxylic acid	$C2/c$	$z = 4$	$P2_1$ (or $C2_1$	$z = 2$ $z = 4$)	3, 4, 21
trans-1,2-Cyclopentanedicarboxylic acid	$C2/c$	$z = 4$	$P2_1$ (or $C2_1$	$z = 2$ $z = 4$)	5, 21
p-Fluorophenylhydracrylic acid	$Pna2_1$	$z = 4$	$P2_1$	$z = 2$	6, 7
erythro-Phenylglyceric acid	$P2_1$	$z = 2 \times 2$	$P2_12_12_1$	$z = 2 \times 4$	8, 9, 29
Isoleucine	$P\overline{1}$	$z = 2$	$P2_1$	$z = 4$	10, 11
Valine	$P2_1/c$	$z = 4$	$P2_1$	$z = 4$	12, 13
Mandelic acid	$Pbca$	$z = 8$	$P2_1$	$z = 2 \times 2$	14, 15
trans-1,2-Dibromoacenaphthene	$Pbc2_1$	$z = 4$	$P2_12_12$	$z = 2$	16, 17
trans-2,3-Tetralindiol	Pc	$z = 16$	$P2_1$	$z = 4$	18
Tyrosine	$Pna2_1$	$z = 4$	$P2_12_12_1$	$z = 4$	19, 20

* Space group.
† z = number of molecules per unit cell.

enantiomers and the racemic compound have been determined. Pedone and Benedetti[21] have indicated that the crystal structure of an optically active compound frequently may be deduced from that of the racemic compound, and vice versa. In general, the structures of the two forms have in common homochiral layers or columns of molecules arranged in a compact way. The repetition of the layers or columns allows the building of either the structure of the enantiomer or that of the racemic compound, due account being taken of the different symmetry elements possible. This process may be envisaged[16] as follows. Beginning with columns of homochiral molecules one can successively construct first a layer and then a three-dimensional structure, either through the application of *direct* symmetry elements (for instance, a binary axis [labeled 2]) or through the application of *inverse* symmetry elements (like a center of symmetry [labeled *i*]).

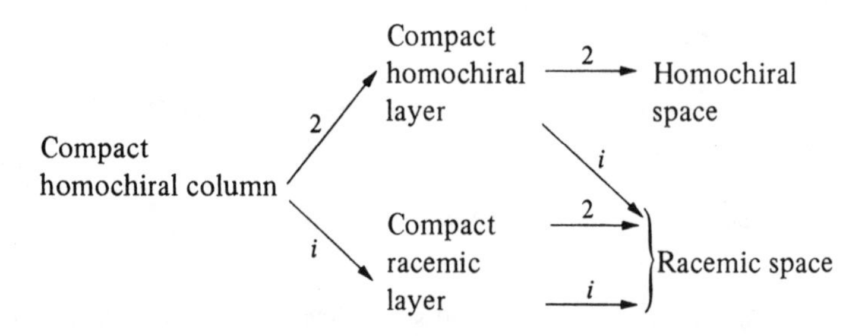

There are three alternative ways of building up the crystal to form the three-dimensional "racemic space," whereas for construction of the "homochiral space" the route is unique. According to whether the racemate is obtained by one or the other of the three routes, the two species (enantiomer and racemic compound) possess in common either a linear packing (columns), or a molecular plane (layers). The following examples are illustrative.

In the crystal structures of *trans*-1,2-dibromoacenaphthene,[16,17] virtually identical columns are formed in both the enantiomer and the racemic species. These columns are composed of homochiral molecules stacked parallel to a crystallographic axis with a periodicity of 8.47 Å in the optically active crystal and of 8.55 Å in the racemic crystal. In the case of the latter, a reflection followed by a translation of 4.55 Å is required to pass from a homochiral column to its mirror image counterpart (Fig. 1). The packing of the racemic compound is less compact ($d = 1.945$) than that of the enantiomer ($d = 1.973$).

The crystals of optically active *p*-fluorophenylhydracrylic acid[6,7] contain alternating layers of phenyl groups and oxygen atoms bound by hydrogen bonds. The layers are separated from one another by a distance of 10.03 Å, which corresponds to the largest cell parameter. The sense of twist of the phenyl groups alternates about a perpendicular binary screw axis, while the carbon backbone of the side chain points alternately toward one or the other neighboring layer (Fig. 2*a*).

The layers themselves are bound to one another through a network of hydrogen bonds which are responsible for the cohesion of the whole. The same layered

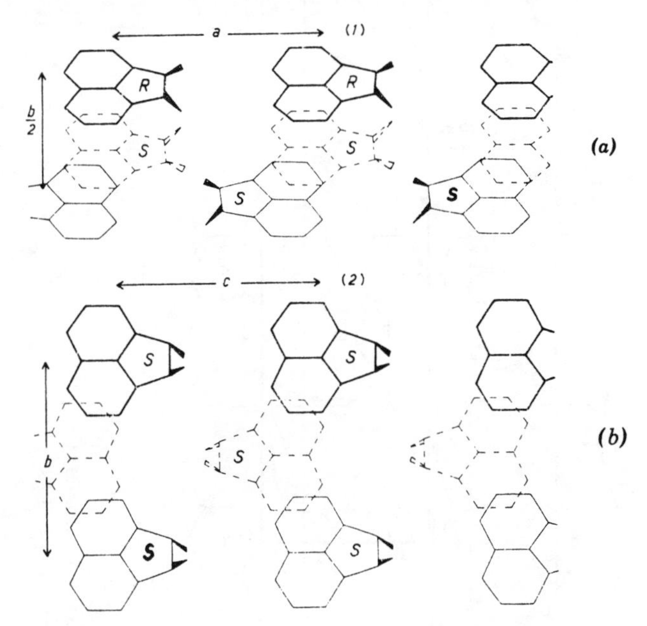

Figure 1 Comparison of projections of racemic (a) and of enantiomeric (b) *trans*-1,2-dibromoacenaphthene which show the common homochiral "columns." M. C. Perucaud, J. Canceill and J. Jacques, *Bull. Soc. Chim. Fr.*, 1974, 1011. Reproduced by permission of the Société Chimique de France.

structure is found in crystals of the racemic compound, but here the layers alternate. Layers of R and S molecules, 9.53 Å apart, are bound to one another by a network of hydrogen bonds (Fig. 2*b*). Note that the racemic network ($d = 1.39$) is slightly less compact than is that of the enantiomer ($d = 1.42$).

Both the active and the racemic forms of *trans*-1,2-cyclohexanedicarboxylic acid have a homochiral molecular plane in common. In this case, the structure of the enantiomer was deduced from that of the racemate.[3,4,21] Here, it is the racemic network ($d = 1.43$) that is more compact than the enantiomer network ($d = 1.38$).

It should also be mentioned that cases are known where no obvious relationship or analogy exists between the crystal structures of optically active and racemic forms. Mandelic acid[14,15] (Fig. 3) and *erythro*-phenylglyceric acid[8,9,22] are examples of such cases. Moreover, the latter is exceptional in that the racemic compound structure, which crystallizes in an enantiomorphous system, is extremely complex.

Tris(pentane-2,4-dionato)chromium(III) is another example of not too closely related racemic and enantiomeric crystal structures; it is discussed in detail by Kuroda and Mason:[23] The packing mode of the racemic *D*3 complex shows homochiral columns of molecules which pack parallel to one another, with adjacent columns being of opposite configurations. In this case, a parallel packing of such homochiral columns with the same configuration is likely to be sterically disad-

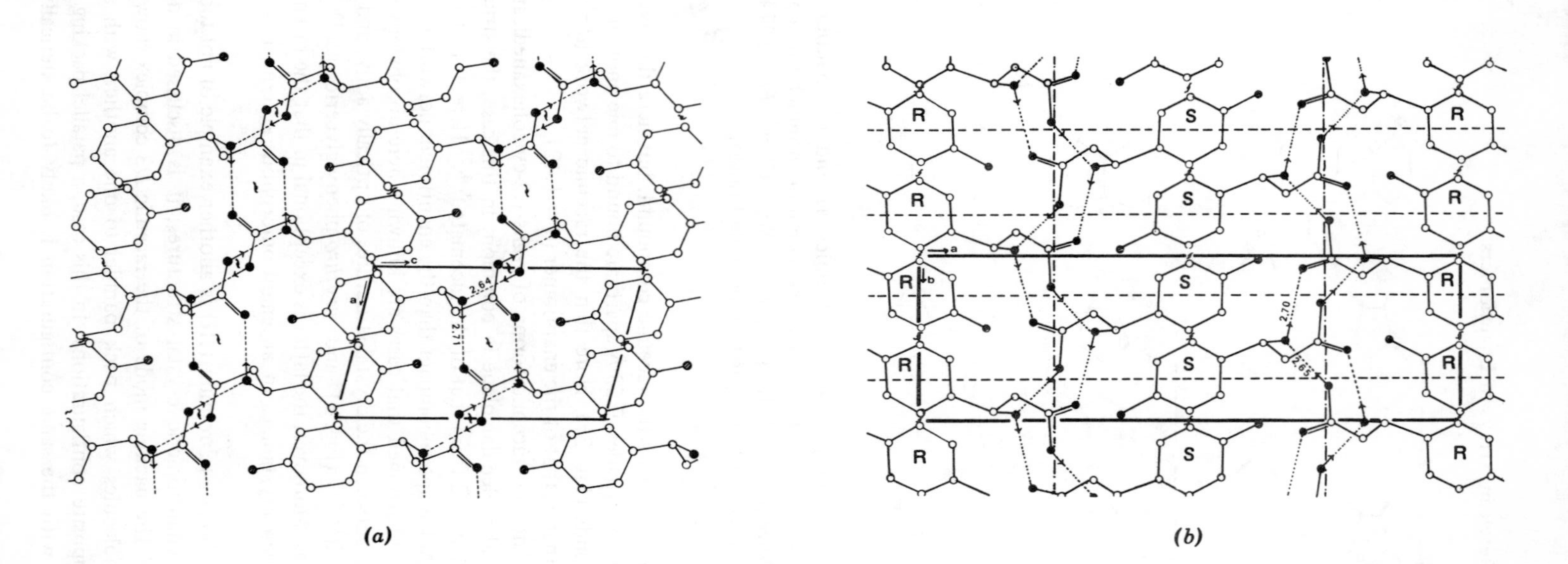

Figure 2 Crystal structures of the *p*-fluorophenylhydracrylic acids: (a) enantiomer; (b) racemic compound. M. Cesario and J. Guilhem, *Cryst. Struct. Comm.*, 1974, 127 and 131. Reproduced by permission of the editors and authors.

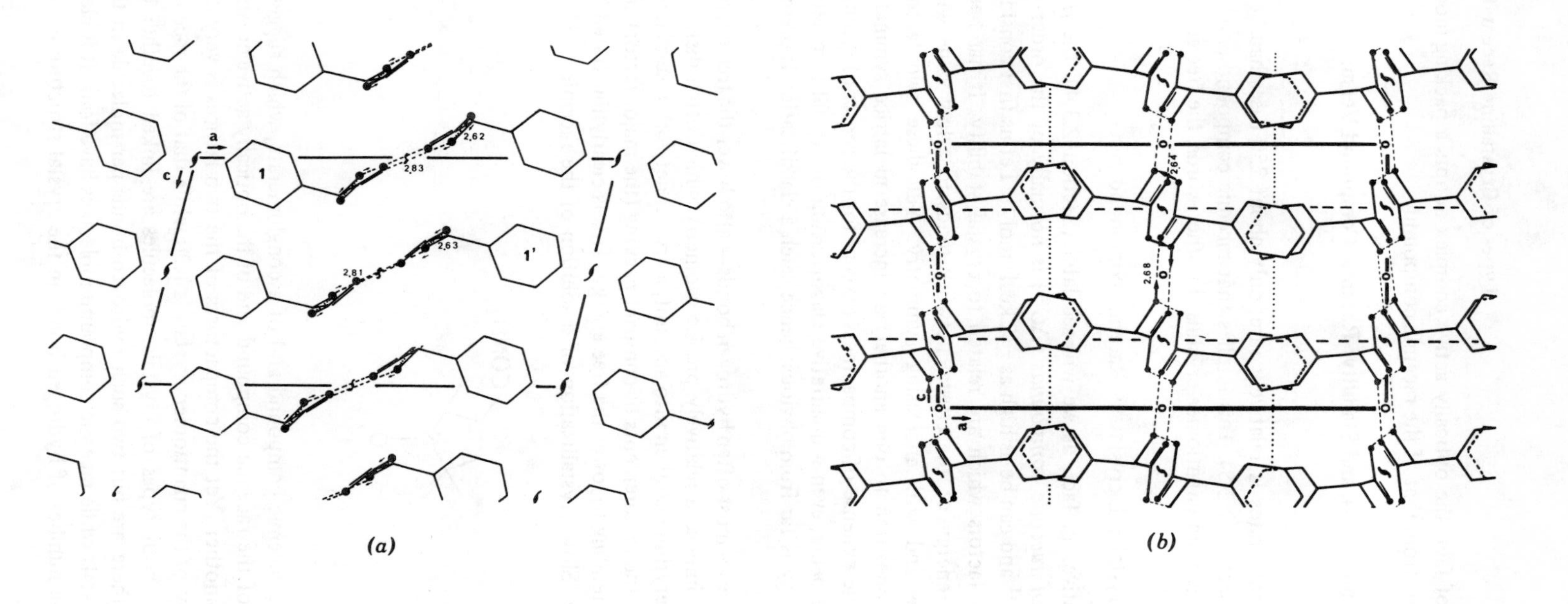

Figure 3 Crystal structures of the mandelic acids; (a) enantiomers; (b) racemic compounds. T. S. Cameron and M. Duffin, *Cryst. Struct. Comm.*, 1974, 539. Reproduced by permission of the editors and authors.

vantaged; as a matter of fact, the optically active complex adopts a packing mode which is largely different from that of the racemic compound.

1.3.2 Symmetry, Compactness, and Stability: Racemic Compound Versus Conglomerate

The incidence of conglomerates is relatively rare; only about 250 cases had been inventoried by 1979 (Section 2.2.5). This implies that racemic compound crystals are generally more stable than enantiomer crystals. In other words, the free energy change for the "reaction"

$$\text{D crystal} + \text{L crystal} \rightarrow \text{Racemic compound}$$

is almost always negative. In fact, as we will see later (Section 2.3.4), the *free energy of formation of racemic compounds* (ΔG^{ϕ}) is normally of the order of -0.2 to -1 kcal·mol^{-1} and can be as high as -2 kcal·mol^{-1}. Let us first consider some of the principal factors which are related to crystal stability. It has been stated[24] that the free energy of the crystal network tends to a low value with increasing compactness and with increasing symmetry. Yet these two factors generally operate in opposition to one another: an increase in lattice symmetry most often occurs at the expense of compactness (except for the center of symmetry case). For the time being, even a qualitative classification of crystal symmetry types according to energy is far from obvious; hence, such a classification is premature.

Intermolecular forces – most often hydrogen bonds – also lower the free energy of a system, but they impose a relatively precise geometry which often disfavors compactness. When everything is taken into account, a crystal network results from a compromise between the requirements of compact packing (the major factor) and those of other factors tending to lower the free energy. This is clearly illustrated by the following example. Slow crystallization of a solution of the racemic amide **1**

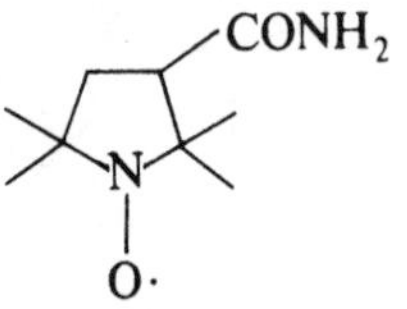

1

leads to a mixture of a racemic compound and of a conglomerate, which suggests that the free energies of the racemic compound and of the optically active crystals are very close to one another. Yet the compactness of the two lattices is very different since the density of the enantiomer crystals is 1.20 while that of the racemic compound is 1.14. In both types of crystals, molecules are linked together by hydrogen bonds; yet there are but two such hydrogen bonds per molecule in the case of enantiomer crystals while racemic compound molecules have four. It is clear that the increase in the number of hydrogen bonds in the crystal structure of the

Table 2 Comparison of densities of optically active and racemic substances[a]

Products	Densities		% Diff.	Refs.
	Enant.	Rac.	$e \to r$	
Tartaric acid	1.759	1.788	+1.7	26
Malic acid	1.595	1.601	+0.4	26
Camphoric acid	1.188	1.228	+3.4	26
Isocamphoric acid	1.243	1.249	+0.5	26
Limonene tetrabromide	2.134	2.225	+4.3	26
Fenchonoxime	1.117	1.142	+2.2	26
β-Isofenchonoxime	1.134	1.180	+4.1	26
Sobrerol	1.128	1.131	+0.3	26
Chlorosuccinic acid	1.687	1.679	−0.5	26
Bromosuccinic acid	2.093	2.073	−1.0	26
Glutamic acid	1.538	1.511	−1.8	26
Mandelic acid	1.341	1.300	−3.1	26
trans-2,3-Tetralindiol	1.287	1.252	−2.8	18
trans-1,2-Dibromoacenaphthene	1.973	1.945	−1.4	18
Alanine	1.374	1.393	+1.4	21
trans-1,2-Cyclohexanedicarboxylic acid	1.38	1.43	+3.6	3, 4
trans-1,2-Cyclopentanedicarboxylic acid	1.340	1.341	0	5
trans-2*a*-Bromo-3*a*-hydroxydecalin	1.490	1.457	−2.2	28
trans-2*a*-Chloro-3*a*-hydroxydecalin	1.299	1.198	−8.1	28
trans-2*a*-Dimethylamino-3*a*-hydroxydecalin	1.095	1.058	−3.4	28
p-Fluorophenylhydracrylic acid	1.417	1.390	−1.9	6, 7
Tyrosine	1.414	1.436	+1.5	11
Valine	1.261	1.326	+5.0	11
Isoleucine	1.195	1.155	−3.4	11
erythro-Phenylglyceric acid	1.40	1.365	−2.5	8, 9
Mandelic acid	1.35	1.32	−2.2	14, 15

[a] The first part of the data in this table is taken from the direct volume and mass measurements of Liebisch and Walden.[26] The data in the second part were obtained by X-ray diffraction measurements. Comparison between the two types of measurements is possible only for the case of mandelic acid. The agreement is remarkably good.

racemic compound may be associated with a reduction in compactness of its lattice without an appreciable free-energy change.[25]

Given the above, what are the factors which may account, in general, for the greater stability of racemic compounds? We shall successively examine three of these: compactness, symmetry, and thermodynamics.

Table 2 summarizes data on the densities of racemates and enantiomers. Those in the upper part of the table cited by Landolt were obtained by Liebisch and Walden[26] in 1895. These measurements, which were carried out at the instigation of

Wallach,[27] led the latter to formulate the following rule (sometimes erroneously ascribed to Liebisch): *In most cases, combination of two enantiomers into a racemate takes place with contraction*. This rule, which is based upon analysis of but a very small number of examples, probably contributed to the establishment of the widespread notion that the greater stability of racemic compounds is due to their greater compactness. In fact, examination of all the results summarized in Table 2, taking into account the more recent data derived from X-ray diffraction measurements, shows that only in half the cases does formation of the racemic compound lead to an *increase* in compactness; it follows that their greater stability cannot be ascribed solely to closer packing of the racemic crystals.

The second factor of importance to be considered is symmetry. From statistical data extracted from the inventory carried out above (Section 1.2.2) we can deduce that in most cases (+) and (−) molecules are paired in the racemic compound lattice about a *center of symmetry*. Inasmuch as this symmetry element is also always compatible with close-packing, it is tempting to ascribe part of the increase of stability of racemic compounds to this connection.

There is also a thermodynamic argument to explain the preferred formation of racemic compounds. The crystallization of a conglomerate is a spontaneous resolution. This separation of enantiomers alone is responsible for a reduction in entropy of $R \ln 2$ (Section 2.2.4) which corresponds to an increase in free energy of $RT \ln 2$ equaling ca. 0.4 $\text{kcal} \cdot \text{mol}^{-1}$ at room temperature. This means that, all other things being equal, conglomerates have an initial handicap of 0.4 $\text{kcal} \cdot \text{mol}^{-1}$ relative to racemic compounds. It would seem that the origin of the stability of racemic compounds whose free energy of formation is less than 0.4 $\text{kcal} \cdot \text{mol}^{-1}$ is essentially entropic. Among 36 representative racemic compounds whose free energies of formation have been measured (Section 2.3.3, Table 2), one-third do in fact exhibit $|\Delta G^{\phi}| \leqslant 0.4 \text{ kcal} \cdot \text{mol}^{-1}$. And, in three-fourths of the examples studied, this entropic contribution accounts for more than half of the value of ΔG^{ϕ}.

Recent progress in the calculation of lattice energies[24] may provide a different insight into these problems. Given a chiral molecule, it should be possible to predict through calculation of the packing mode which of the two species – conglomerate or racemic compound – corresponds to the minimum of free energy.

According to Zorkii,[29] the general approach to the resolution of this problem is as follows: (a) It is possible to show, from the symmetry of the molecule, that only a small number of space groups are permitted, among them variants corresponding to racemic or enantiomeric structures; this can be demonstrated with the aid of the potential symmetry function.[30] (b) For each variant chosen, the free energy is calculated as a function of structural parameters (i.e., lattice parameters and parameters that characterize the location and orientation of the molecule in the lattice). However, so far as we are aware no calculation of this type has yet been carried out.

REFERENCES 1.3

1 J. Donohue, *J. Am. Chem. Soc.*, 1950, **72**, 949.
2 H. J. Simpson, Jr., and R. E. Marsh, *Acta Crystallogr.*, 1966, **20**, 550.
3 E. Benedetti, P. Corradini, and C. Pedone, *J. Am. Chem. Soc.*, 1969, **91**, 4075.
4 E. Benedetti, P. Corradini, C. Pedone, and B. Post, *J. Am. Chem. Soc.*, 1969, **91**, 4072.
5 E. Benedetti, P. Corradini, and C. Pedone, *J. Phys. Chem.*, 1972, **76**, 790.
6 M. Cesario and J. Guilhem, *Cryst. Struct. Comm.*, 1974, **3**, 127.
7 M. Cesario and J. Guilhem, *Cryst. Struct. Comm.*, 1974, **3**, 131.
8 M. Cesario and J. Guilhem, *Cryst. Struct. Comm.*, 1975, **4**, 197.
9 M. Cesario, Thesis No. 289, Orsay, 1976.
10 K. Torii and Y. Iitaka, *Acta Crystallogr.*, 1971, **B27**, 2237.
11 E. Benedetti, C. Pedone, and A. Sirigu, *Acta Crystallogr,*, 1973, **B29**, 730.
12 K. Torii and Y. Iitaka, *Acta Crystallogr.*, 1970, **B26**, 1317.
13 M. Mallikarjunan and R. Thygaraja, *Acta Crystallogr.*, 1969, **B25**, 296.
14 T. S. Cameron and M. Duffin, *Cryst. Struct. Comm.*, 1974, **3**, 359.
15 T. C. Van Soest, unpublished results. M. Cesario, Thesis, Orsay, 1976.
16 M. C. Perucaud, J. Canceill, and J. Jacques, *Bull. Soc. Chim. Fr.*, 1974, 1011.
17 M. T. LeBihan and M. C. Perucaud, *Acta Crystallogr.*, 1972, **B28**, 629.
18 M. C. Perucaud, Thesis, Université de Paris VI, 1973.
19 A. Mostad and C. Rømming, *Acta Chem. Scand.*, 1973, **27**, 401.
20 A. Mostad, H. M. Nissen and C. Rømming, *Acta Chem. Scand.*, 1972, **26**, 3819.
21 C. Pedone and E. Benedetti, *Acta Crystallogr.*, 1972, **B28**, 1970.
22 M. Cesario, J. Guilhem, C. Pascard, A. Collet, and J. Jacques, *Nouv. J. Chim.*, 1978, **2**, 343.
23 R. Kuroda and S. F. Mason, *J. Chem. Soc., Dalton*, 1979, 273.
24 A. I. Kitaigorodskii, *Molecular Crystals and Molecules*, Academic Press, New York and London, 1973.
25 B. Chion and J. Lajzerowicz, *Acta Crystallogr.*, 1975, **B31**, 1430.
26 See H. Landolt, *Das optische Drehungsvermögen*, 2nd ed., F. Vieweg und Sohn, Braunschweig, 1898, p. 70.
27 O. Wallach, *Liebigs Ann. Chem.*, 1895, **286**, 140.
28 C. Cabestaing, Thesis, Montpellier, 1969.
29 P. M. Zorkii, private communication to J. Jacques, 1971.
30 P. M. Zorkii, *Kristallographiya*, 1968, **13**, 26. English translation, *Soviet Phys. Crystallogr.*, 1968, **13**, 19.

CHAPTER 2

Binary Mixtures of Enantiomers

If there is one central theme to this book, it is that the properties of mixtures of enantiomers, particularly those affecting the solid state, are special and even different from those of achiral molecules. It is these properties, their measurement and their interpretation, that we examine in detail in this chapter.

2.1 CHARACTERIZATION OF RACEMATE TYPES BY MEANS OF BINARY (MELTING POINT) PHASE DIAGRAMS

The distinction among the several types of racemates remained imprecise and even controversial as long as it was limited to the visual or microscopic comparison of crystal form or of densities of racemates with those of the enantiomers. The development of more decisive criteria became possible only when it was recognized that enantiomer mixtures were no more than a specific type of binary system (or ternary system when in the presence of solvent) whose properties may be described by the phase rule. It is on this basis that, in 1899, Roozeboom[1] characterized the three fundamental types of enantiomers mixtures by their melting point (fusion) diagrams, which are illustrated in Fig. 1: (a) conglomerate; (b) racemic compound; (c) solid solution: 1, ideal; 2, with a maximum; or 3, with a minimum.

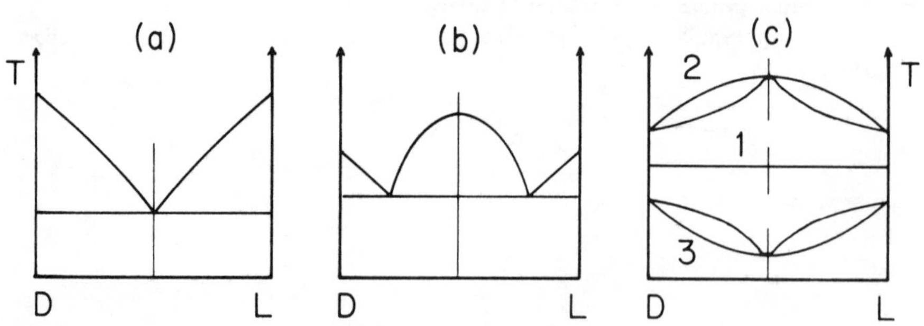

Figure 1 Binary phase diagrams illustrating the three fundamental types of crystalline racemates.

These different categories of enantiomer mixtures may be defined equally well by the *binary* fusion diagram of mixtures of enantiomers D and L and by the solubility diagram that corresponds to the *ternary* system, D, L, and solvent, which we describe in Chapter 3.

2.1.1 Representation of Melting Point Phase Diagrams

Before examining the several types of binary systems that characterize enantiomer mixtures, let us briefly review the significance of phase diagrams. In a general way, phase diagrams describe the behavior of systems that may consist of one or more components distributed in one or more phases (solid, liquid, etc.) as a function of variables such as temperature, pressure, concentration, and so on.) In the present case, in order to describe the behavior of two components, enantiomers D and L, on melting (or its converse, solidification) the representation described by Fig. 2 is used in which the temperatures of beginning and of termination of melting (T_1, T_2) are given as a function of composition M.

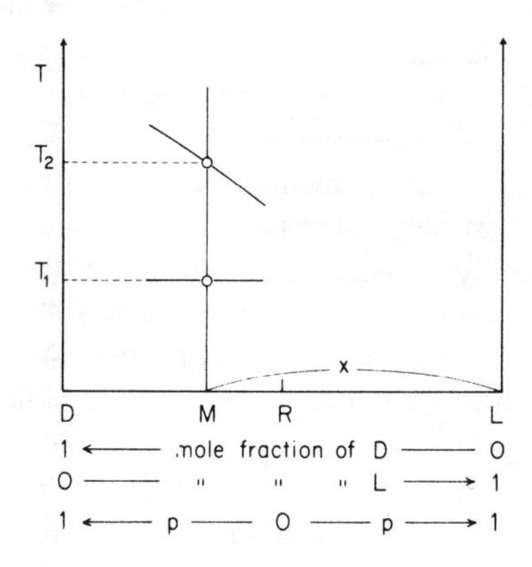

Figure 2.

Diagrams which describe solid–liquid equilibria are designated as "melting point phase diagrams." Similarly, those which describe solid–vapor and liquid–vapor equilibria are, respectively, designated as "sublimation phase diagrams" and "boiling point phase diagrams." The composition of a mixture is most often defined in terms of the mole fraction of one of its components. In mixture M, if the mole fraction of D is x, that of L is $1-x$. These concentrations are sometimes expressed as weight percent. For enantiomer mixtures in which the component molecular weights are obviously identical, mole fractions are equivalent to weight percent (except for the factor 100).

In the case of mixtures of D and L, it is quite common to define compositions through the *enantiomeric purity* p, which is equivalent to the *enantiomeric excess*, *e.e*. The enantiomeric purity of mixture M is equal to MR/DR (Fig. 2). Given this, it is not difficult to derive the following useful relationships between x and p:

$$p = 2x - 1 \tag{1}$$

and

$$x = \frac{1+p}{2} \tag{2}$$

where x represents the mole fraction of the predominant enantiomer. We see that p may vary from 0 to 1 as x varies from 0.5 to 1.

Let us also recall that the enantiomeric purity of a sample is generally equal to its optical purity, *o.p.* The latter is an experimental quantity defined as follows:

$$o.p. = \frac{[\alpha_M]}{[\alpha]} \tag{3}$$

where $[\alpha]$ is the specific rotation of the pure enantiomer and $[\alpha_M]$ is that of the mixture M, both being measured under identical conditions. Polarimetry is thus a good means of determining the composition of enantiomer mixtures provided only that $[\alpha]$ is known. Otherwise one must have recourse to direct experimental methods for determining x. These are summarized in Chapter 7.

Finally, note that in common usage enantiomeric and optical purities and sometimes mole fractions are expressed as percentages, which requires that eqs. (1) to (3) be multiplied by 100. In this book, capital letters are used when dealing with percentages; for example, $x = 0.75$ corresponding to $p = 0.5$ will be equivalent to $X = 75\%$ and $P = 50\%$.

Let us now turn to the experimental methods employed in the construction of phase diagrams. The first phase diagrams of mixtures determined toward the end of the last century generally depended upon data furnished by *cooling curves*. Such curves were obtained through use of a conventional freezing point apparatus.[2] In this device, a pure substance or a mixture is heated until a homogenous liquid is obtained; the latter is allowed to cool in an insulated bath and the temperature is recorded at regular time intervals.

If the substance studied is pure, one first observes a steady drop in temperature until the first crystals appear. The sample is stirred to reduce the possibility of supercooling. From that point on, the temperature remains constant (melting point T^f of the pure compound) while the entire sample solidifies, after which the temperature decreases once again (Fig. 3, curve *a*).

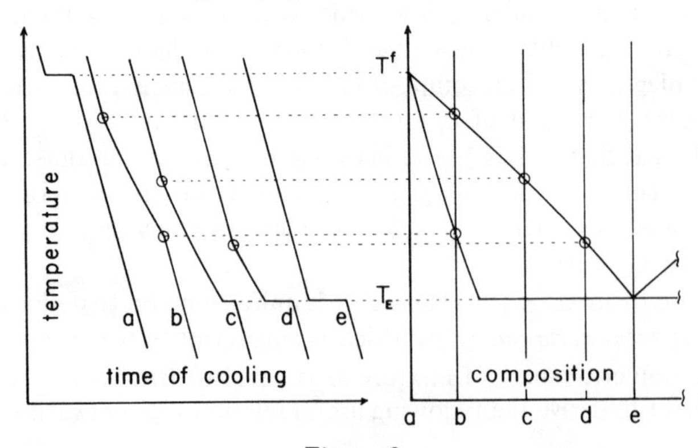

Figure 3.

In the case of mixtures, cooling curves can be of two types. In the first type, the appearance of crystals is attended initially by a decrease in the rate of cooling of the sample (change in slope in the cooling curve). This is followed by a halt in temperature change until the end of crystallization (Fig. 3, curves *c* and *d*). Analysis of the crystals formed throughout the halt would show that their composition does not change; we shall see later that this corresponds to the *eutectic*. The length of the horizontal halt is proportional to the quantity of eutectic present in the mixture: for equal weights of binary mixtures, one observes curves *c* and *d* (Fig. 3) for different proportions of the components. The horizontal part of the cooling curves is called the *eutectic halt*. When a mixture has the composition corresponding to the eutectic itself, its cooling curve is identical to that of a pure substance (Fig. 3, curve *e*).

In the second type of cooling curve, this temperature halt is not observed. From the moment when crystallization begins, one observes only a change in slope in the cooling curve corresponding to a slowing down of the rate of cooling of the sample until the whole sample has crystallized (Fig. 3, curve *b*). This behavior is indicative of the formation of a *solid solution* between the mixture constituents.

From the inflection points observed, the phase diagram may be constructed to relate the observed temperatures to the compositions of the respective binary mixtures (see Fig. 3).

We shall have occasion to reexamine the significance of these temperature halts and changes in slope of cooling curves which reflect the proportions of crystals and of liquid as well as the number of phases present during the course of the solidification process.

It should be evident that the phenomena just described, which occur during the cooling and *solidification* of a sample, can just as well be observed during melting or *fusion*. To observe and study the latter process, however, a much more sophisticated device and technique would be required than those just described.

While this classic way of obtaining binary phase diagrams still remains valuable and instructive, it does require samples of mixtures of the order of several grams. For enantiomers, whose separation is too often time consuming, such large samples are hard to come by, however. We will now examine other methods for the establishment of phase diagrams which overcome this limitation.

2.1.2 Preparation of Mixtures for the Construction of Binary Phase Diagrams

In order to construct a binary phase diagram for a system of two enantiomers, several mixtures of known composition (containing various quantities of the two enantiomers, or one of the enantiomers and the racemate) are required. However, the possibility of *calculating* phase diagrams from the enthalpies of fusion and the melting points of mixture constituents may, as we shall see later in detail, reduce the number of determinations substantially. All that needs to be done is to spot check for coincidence of experimental melting points with the theoretical curve.

The essential requirement is to measure the beginning and termination of fusion of each of the mixtures. Before examining techniques that allow the observation of these temperatures, let us look at methods that are specifically applicable to the preparation of binary mixtures of organic compounds.

When relatively large amounts of compounds are available, each component can be weighed directly, for example, in the same small test tube. The mixture is then melted, allowed to crystallize (if necessary while mixing), and crushed so as to obtain a sample which is as homogeneous as possible. With this technique, each point on the diagram may require as much as 20 to 30 mg of each compound. When optical micromethods are used that involve capillary melting point tubes or the hot stage microscope, or when microcalorimetry is used, quantities of material of the order of several milligrams suffice, and these can be weighed in small test tubes or on small watch glasses. The procedure of Rheinboldt[3] for the preparation of mixtures directly in melting point capillaries may also be used.

In our experience, mixtures are obtained most conveniently from *solutions* prepared in volatile solvents (most commonly acetone) which are then evaporated as thoroughly as possible. Such solutions can be prepared from weighed samples of components to which solvent is added *ad libitum*. Or, one may mix known volumes of solutions containing known concentrations of a racemate and one enantiomer. The solutions are evaporated and, if necessary, crystallization is induced by scratching. The crystalline residue is then briefly placed under vacuum to free it from the last traces of solvent.

Whatever the method employed in the preparation of the mixtures, it is advisable to "anneal" the samples by warming them for several hours at an intermediate temperature (in a sand bath, for instance) below the melting point. The purpose of this treatment is to promote equilibration of different crystal forms so as to reduce problems associated with polymorphism.

Finally, we emphasize that the need to attain reasonable precision (ca. 1%) in the composition of a sample weighing 10 mg or less requires the use of a high-precision balance, namely, a microbalance or a torsion balance.

2.1.3 Visual Measurement of Transition Temperatures

The temperature measurements required to establish phase diagrams do not especially differ in kind from the routine determinations carried out in the laboratory. What does differ is the need to determine with good precision the temperature at which fusion begins and that at which it ends. This is in contrast to common practice in which the fusion *interval*, which is taken as a criterion of purity, is more often estimated than actually measured.

The classical measurement of melting points in capillary tubes is by far the simplest and most economical method possible for the establishment of a phase diagram. Numerous devices are available for this purpose; most are well known to organic chemists. Some incorporate heating baths fitted with stirrers, with temperature regulators, and with magnifying and lighting systems. All of these facilitate observation of the sample and permit precise, reproducible determination of the fusion events. These devices are not further described here. None of the accessories or refinements entirely eliminates the difficulties inherent in the visual detection of the beginning of melting.

In general, it is observed that a colorless compound consisting of shiny crystals becomes dull, that is, it loses its shine, at the moment when melting begins. According to Rheinboldt,[3] this contrast is reminiscent of the difference between glazed and unglazed porcelain. If the substance is slightly colored, a deepening of the color is observed at the onset of fusion.

Use of a hot stage microscope is also familiar to many organic chemists. One of the best known of these is the Kofler hot stage manufactured by Reichert.[4a] In this device, temperatures are measured usually by means of a thermometer whose bulb is located as close as possible to the sample, which is viewed under magnification through a hole in the stage. The thermometer is calibrated with pure reference compounds. This has the advantage of directly furnishing temperatures that are both corrected for emergent thermometer stem and calibrated for thermometer accuracy.

It is well known that the measurement of melting points depends to a considerable extent upon the rate of heating. Kofler and Kofler recommend a heating rate of 4°/minute.[5a] However, there is general agreement that a slower heating rate (1 to 2°/minute) furnishes more reproducible data.

Matell has observed[6] that the metallic plate which serves as sample support and which protects the thermometer in the classic Reichert microscope can induce errors of as much as 2° in the melting point. This is due to overheating of the thermometer stem and is particularly serious when the melting process takes a long time (as in making measurements to construct a complete fusion phase diagram).

Figure 4 illustrates that it is possible to construct very serviceable and accurate phase diagrams using visual methods. Hundreds of such diagrams have been recorded and published, notably the very large number by A. Fredga and his school. The figure suggests that the beginning of fusion need not actually be observed to establish either the form of the phase diagram or the composition of the eutectic. The procedure described by Kofler and Kofler also requires observation only of the end of melting (which they significantly call "point of primary crystallization").[5c]

The methods just described depend upon visual observation and have the advantage of simplicity. However, they are tiring to the eye and somewhat subjective. In spite of the best intention, the initiation of fusion is often difficult to detect; and while the end of fusion is generally easier to see, this too can be "invented." It is clear why one would prefer to automate such determinations.

2.1.4 Automated Procedures

The reviews of Jucker and Suter[7] and Skau and Arthur[2] provide well-documented summaries of modern methods for the automated determination of melting points. We limit ourselves here to the essential aspects of these methods, particularly as they apply to the determination of phase diagrams.

The best-known commercial device is one that employs a visual method where the eye of the observer has been replaced by a photoelectric cell. The Mettler melting point apparatus measures the change in the transparency of a sample as it is heated. In this device, a light beam is directed through the sample and onto a

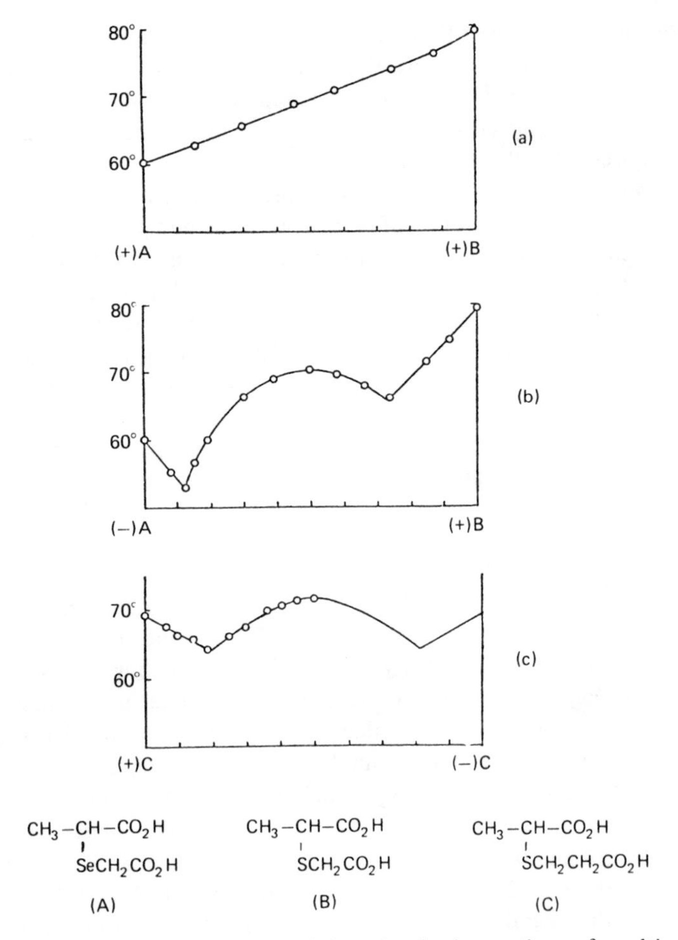

Figure 4 Phase diagrams constructed by visual observation of melting points. A. Fredga, *Ark. Kemi, Mineral. Geol.*, 1943, **17A**, No. 17. Reproduced by permission of the Royal Swedish Academy of Sciences and the author.

calibrated photocell. As soon as the sample begins to melt, during a programmed heating run, the intensity of the light detected by the cell rises abruptly. In the most common model of the Mettler device the detector cell is adjusted in such a way that it responds only if the light intensity crosses a given threshold, whereupon it releases a signal which stops the heating program. The digital temperature readout indicates the temperature at which heating ceased, which corresponds to the melting point. This is particularly useful for the analysis of pure substances.

An alternative mode of use, more pertinent to the construction of phase diagrams, requires the use of a recorder to continuously record the temperature as a function of transmittance. This is equivalent to a heating curve which allows the "objective" determination of the beginning and termination of fusion. A representative recorder trace obtained with the Mettler device is shown in Fig. 5.

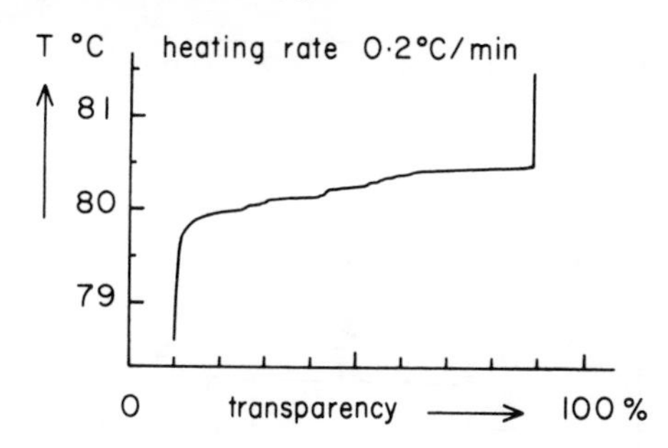

Figure 5 Melting point curve obtained with the Mettler melting point apparatus.

A second category of automated devices is based on thermal analysis which, in all of its variants, measures the absorption (or the evolution) of heat by the sample as a function of temperature as the sample is heated (fusion) or cooled (crystallization). The techniques best adapted to the construction of phase diagrams are *differential scanning calorimetry* (DSC) and *differential thermal analysis* (DTA). Differential microcalorimeters require only very small amounts of substance – of the order of 0.1 to 10 mg. The most widely used commercial instruments are those made by Perkin–Elmer, du Pont, and Mettler.

These instruments contain a sample cell and a reference cell which are heated (or cooled) simultaneously at a constant rate. Two different operating modes are possible. In the first, each cell is supplied with the same amount of energy per unit time and one records the *temperature difference* ΔT between the two (DTA). In the second, more sophisticated approach, one measures the energy difference required to maintain the two cells at the same temperature while the latter rises or decreases linearly (Perkin–Elmer DSC-2). In this case, the output is a curve of *energy* absorbed or evolved by the sample as a function of temperature. DSC melting curves are actually easier to analyze in a quantitative way than those furnished by DTA, which require further conversion of temperature differences into energy.

A differential microcalorimeter detects the exchange of energy per unit time; consequently, the instrument response must be proportional to the (scanning) rate of heating or cooling. In the temperature range where no phase change occurs, the heat energy supplied by the calorimeter is entirely used to satisfy the heat capacity of the sample and of the reference cell. A nearly horizontal curve is obtained whose deflection relative to the isothermal baseline (i.e., without heating) corresponds to the difference in heat capacity between the two cells (see Fig. 6). In a transition region such as melting, for example, the instrument must additionally furnish the energy corresponding to the change. This is recorded as a signal whose area gives access to the transition enthalpy and which, as a consequence of its position and shape, gives information about the temperature range in which the transition takes place and about the type of phase diagram involved.

When a pure substance or a pure addition compound is melted, a very narrow peak is obtained (Fig. 7); the melting point is given by the intersection of the baseline and a line drawn along the leading edge (ascending) of the fusion peak. The melting point curves of binary mixtures exhibiting a eutectic (Fig. 7*a*) or forming

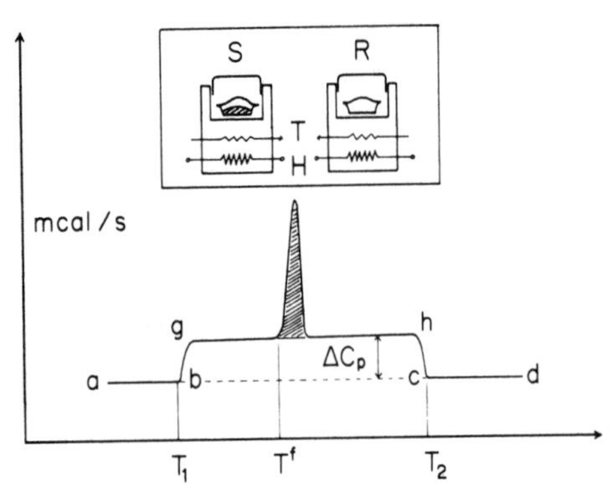

Figure 6 Principle of operation of a differential microcalorimeter (Perkin-Elmer DSC system). The cells containing the sample (S) and the reference (R) are heated simultaneously and their temperatures are monitored (thermocouples). Interpretation of curves: ab, cd isothermal baseline; gh baseline during heating (interval between the two baselines gives the difference in specific heats between S and R); shaded area gives the enthalpy of the transition, for example, melting which takes place at temperature T^f.

addition compounds (Fig. 7*b*) have characteristic shapes which vary with the composition as is shown in the figure (the first peak represents the melting of the eutectic).

The temperatures of the beginning and the end of fusion are read off the melting curves as shown in the figure. Comparison should be made with runs carried out on well-known systems. Where an addition compound exists, the same type of curve may equally well correspond to mixtures found on either side of the eutectic (Fig. 7*b*, α and β).

Solid solutions give rise to more or less broad peaks according to the width of the gap between the solidus and the liquidus curves (Fig. 7*c*). It is in such solutions that the determination of the beginning and the termination of fusion is the most problematic. We return to this problem in Section 2.4.

Finally, the measurements carried out by the various methods described allow one to construct diagrams in which the temperatures of the various events observed during heating are plotted as a function of the composition of the mixtures. The curve corresponding to the beginning of melting is the *solidus*, while that corresponding to the termination of fusion is the *liquidus*.

2.1.5 The Contact Method of Kofler

When a complete phase diagram is not needed but the type of system to which an enantiomer pair belongs must be determined qualitatively, the necessary information may be simply obtained through use of the contact method of Kofler.[4b,4c,5b,8] Moreover, no mixtures are needed and the amount of each substance required is only of the order of milligrams. While the method is applicable to any two-component mixture, we limit our description to the behavior of a pair of enantiomers.

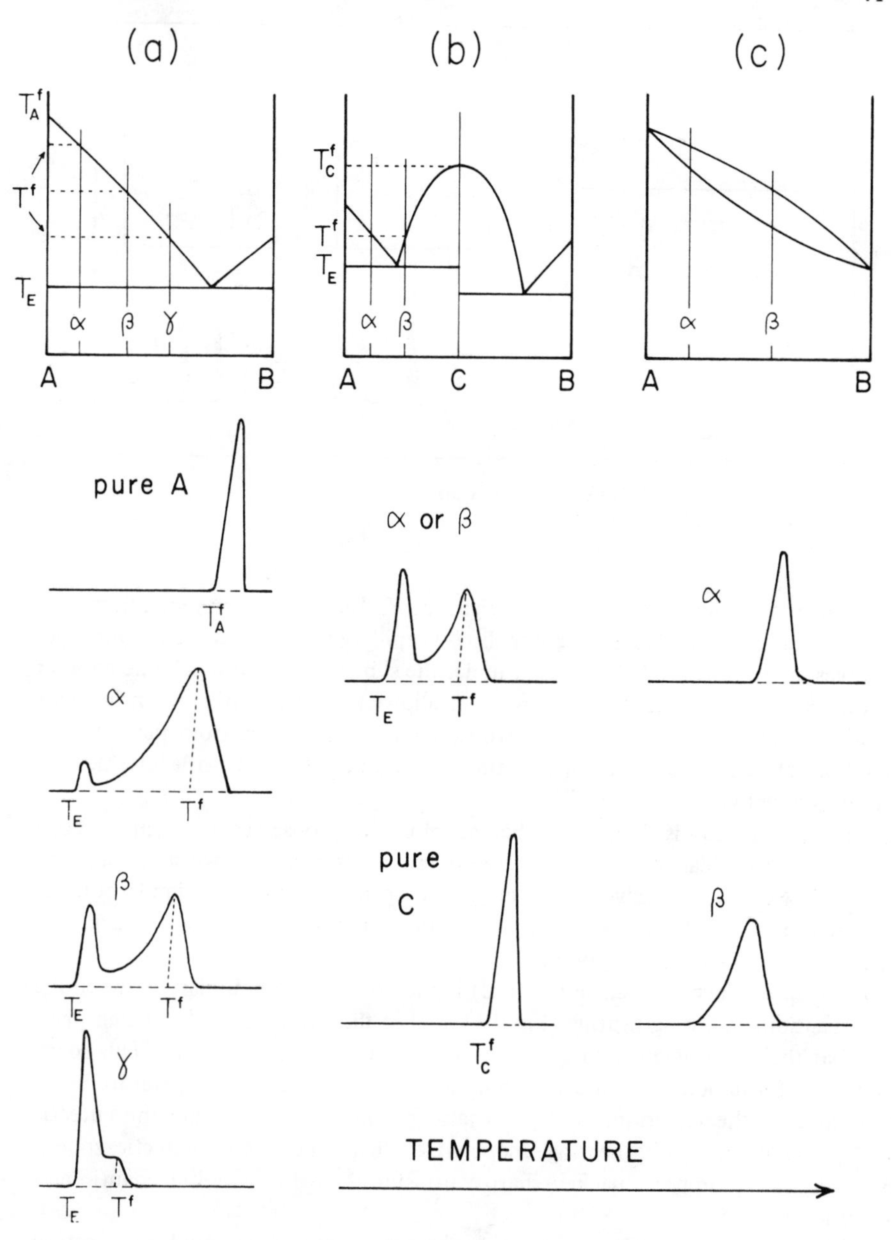

Figure 7.

The observations are carried out on the heating stage of a microscope. Small samples of the two enantiomers are examined while being heated in contact with one another under the cover slip of a microscope slide. One enantiomer is placed on the microscope slide at the left of the cover slip (*A* in Fig. 8*a*). The other

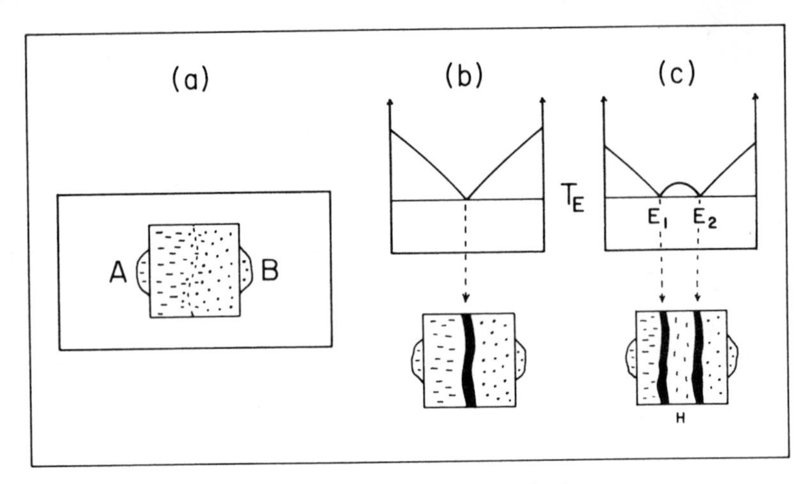

Figure 8 The contact method.

enantiomer is similarly placed to the right, at B. The two samples are allowed to melt and gradually run under the cover slip by capillary action until each enantiomer occupies approximately half the area under the slip and the two touch one another in the middle. The enantiomers are then allowed to cool until the entire mass crystallizes. We now have a sample whose composition varies from pure A (one enantiomer) at left to pure B (the other enantiomer) at right, with intermediate mixtures in between.

The preparation is then slowly heated while being observed in the microscope preferably with polarized light and between crossed Nicols. The two simplest cases, to which we limit ourselves here, are those corresponding to a simple eutectic exhibited by a conglomerate and to an addition compound resulting from enantiomers forming a racemic compound.

The appearance of a single (vertical) liquid zone at the boundary between the two enantiomers during heating (Fig. 8*b*) implies that no reaction has taken place and that the two isomers exhibit no mutual solubility in the solid state. This corresponds to formation of a eutectic which necessarily melts at a temperature lower than that of the enantiomers. It may also be possible to estimate the eutectic melting point during the heating. In polarized light, the molten eutectic zone is especially clear; it appears black, in sharp contrast to the light, unmolten enantiomers on either side.

When the two enantiomers react to form a racemic compound, two vertical liquid zones appear at the interface during heating (E_1 and E_2 in Fig. 8*c*). They form simultaneously inasmuch as the two eutectics have equal melting points in the case of enantiomers. The area (H in Fig. 8*c*) intermediate between the eutectics (E_1 and E_2) tends to be turbid and opaque. It may also be possible to estimate the melting point of the racemic compound or at least to determine whether this melting point is higher or lower than that of the enantiomers.

REFERENCES 2.1

1 H. W. B. Roozeboom, *Z. Phys. Chem.*, 1899, **28**, 494.

2 E. L. Skau and J. C. Arthur, Jr., "Determination of Melting and Freezing Temperatures," in *Physical Methods of Chemistry*, Part V of *Techniques of Chemistry*, Vol. 1, A. Weissberger and B. W. Rossiter, Eds., Wiley–Interscience, New York, 1971, Chapter 3, p. 137ff.

3 H. Rheinboldt and M. Kircheisen, *J. Prakt. Chem.*, 1926, **113**, 348.

4 (a) W. C. MacCrone, Jr., *Fusion Methods in Chemical Microscopy*, Interscience, New York, 1957, Chapter 2, pp. 15–36; (b) pp. 94–101; (c) pp. 148–157.

5 (a) L. Kofler and A. Kofler, *Thermomikro Methoden*, Verlag Chemie, Weinheim, 1954, p. 7; (b) 151; (c) p. 191.

6 M. Matell, Stereochemical Studies on Plant Growth Substances, Inaugural Dissertation, Uppsala, 1953, p. 15.

7 H. Jucker and H. Suter, *Fortschr. Chem. Forsch.*, 1969, **11**, 430.

8 H. Rheinboldt, in *Methoden der Organischen Chemie* (*Houben-Weyl*), 4th ed., Vol. 2, Part 2, E. Müller, Ed., Georg Thieme Verlag, Stuttgart, 1953, p. 862.

2.2 CONGLOMERATES

The first of the three types of racemates that may be defined on the basis of a melting point phase diagram is called a *conglomerate*.

A conglomerate is an equimolecular mixture of two crystalline enantiomers that are, in principle, mechanically separable. This mixture melts as if it were a pure substance and thus fits the definition of an eutectic. The phase diagram for a conglomerate is illustrated in Fig. 1. Before interpreting this diagram, let us review some of the qualitative and quantitative aspects of the "phase rule."

2.2.1 The Phase Rule

A system consisting of one or more substances in the process of melting or while crystallizing (with or without solvent) constitutes a heterogeneous system. After some time, and under defined conditions of temperature and pressure, the system ceases to change, whereupon it may be said to have attained equilibrium. The different portions of the system, each of which is homogeneous, physically distinct, and mechanically separable, are called phases. These phases need not be chemically homogeneous, that is, composed of single pure constituents; for example, a solution constitutes a phase. Moreover, a given constituent may be present in more than one phase in different concentrations.

The relationship between the number of phases and the number of components present at equilibrium is given by the phase rule.* Note that a distinction is made between constituents and components; the latter refers to constituents whose concentrations are independently variable. Providing that in a particular system of

* More extensive analyses of heterogeneous equilibria and on the phase rule may be found, for example, in the monographs of Prigogine and Defay,[217] Ricci,[218] and Findlay.[219]

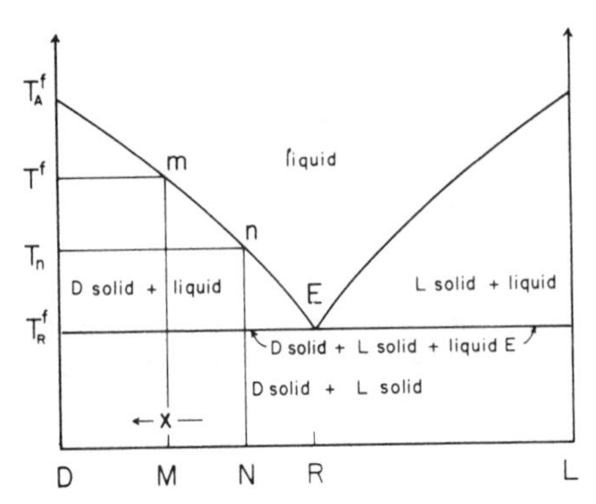

Figure 1 Binary melting point diagram of a conglomerate. The mole fraction of the more abundant enantiomer (D) in mixture M is x.

interest the equilibrium depends only upon temperature, pressure, and the concentrations of the components in the different phases, the phase rule gives the number of the factors that may be varied simultaneously without altering the number of phases of the system. This is the variance or the degree of freedom of the system at equilibrium.

In a system consisting of C components and comprising Φ phases and which may be submitted to changes of temperature and pressure, the variance v is given by the familiar relationship $v = C - \Phi + 2$ which defines the phase rule. In condensed systems – those composed only of solid and liquid phases – which are of particular interest here, the effect of pressure on equilibria may be neglected and we may then write the phase rule in the form

$$v = C - \Phi + 1$$

2.2.2 Analysis of the Binary Phase Diagram

Let us now return to the diagram of Fig. 1. Let us examine a system M composed of 1 mole of a mixture of two enantiomers in which the mole fraction of the more abundant component (D here) is x. Thus, the mixture M contains x moles of D and $1 - x$ moles of L. Since this system consists of two components, its variance is $v = 2 - \Phi + 1 = 3 - \Phi$.

When the temperature lies below T_R^f, the system consists of two solid phases (D and L crystals); that is, $\Phi = 2$. Its variance is then equal to 1 (univariant system). The concentration of each component in these two solid phases is fixed, since they are necessarily pure components. The choice of factors that may be varied without altering the state of the system is therefore limited to temperature, but only up to T_R^f at which a liquid phase appears (beginning of melting). The variance then becomes zero, and the three phases (i.e., D and L crystals and the liquid) may coexist only as long as the temperature and the composition of the liquid phase are fixed. This equilibrium is defined by $T = T_R^f$, and, by symmetry, the liquid phase must necessarily be racemic (point E).

If the temperature of the mixture is now raised to a temperature greater than T_R^f, only the crystals of the less abundant L enantiomer disappear; if on the other hand the temperature is lowered, the liquid phase crystallizes. And if, at T_R^f, one tries to change the composition of the molten phase, for example, through addition of enantiomer L, the only consequence is an increase in the quantity of crystalline L present.

Let us now examine in greater detail what actually happens when a solid mixture of composition M is heated; the analysis follows the vertical line Mm. The sample begins to melt at temperature T_R^f. Since the liquid phase that appears is racemic, it is only the $2(1-x)$ moles of racemate – alternatively put, of eutectic – present in the original sample which melts. As long as the melting of the racemate takes place, the temperature of the mixture remains constant and equal to T_R^f; this is the eutectic halt described in Section 2.1.1.

When all of the racemate has melted at T_R^f, two phases remain – one solid [$2x-1$ moles of D] and the other liquid [$2(1-x)$ mole of racemate]. At this moment, the system becomes once again univariant and one enters another part of the diagram. When we continue to heat the sample, the temperature rises and the solid phase (pure D) progressively dissolves in the liquid phase whose composition moves along the liquidus curve from E toward m. Since the system is univariant, only one of the factors affecting the equilibrium may be altered at a time. If one stipulates a temperature, for example, T_n, then the liquid phase is fixed at point n corresponding to composition N ($T_n \rightarrow N$). Conversely, if one fixes the composition of the liquid, then the temperature of the system is fixed ($N \rightarrow T_n$). We shall see in the following pages that the relation between the temperature of the system and the composition of its liquid phase may be mathematically expressed (equation of Schröder–Van Laar).

It is worth recalling that, in addition to their compositions, the relative quantities of the phases present may be deduced from the phase diagram with the aid of the lever rule. For a mixture of overall composition M brought to temperature T_n, the quantity s of pure solid L at equilibrium is given by MN/DN moles, while the quantity l of the liquid (of composition N) is given by DM/DN.

When one continues to heat the sample, the temperature T^f is attained at which the last crystal of L disappears (m on the liquidus curve). The liquid phase m of course has the same composition as that of the original mixture (M).

Above the liquidus curve, only one phase (liquid), exists and consequently its variance is equal to 2. Hence, both the temperature *and* composition of the liquid may be independently varied without changing the state.

From Fig. 1 it follows that the eutectic is present whenever one of the enantiomers contains as little as a trace of the other. This is in fact an idealized picture; a range of composition can exist in which there is some miscibility of the enantiomers in the crystals, that is, a small amount of one enantiomer can incorporate into the crystal lattice of the other. However, the miscibility range is as a rule very narrow, almost negligible. Instances nevertheless exist in which binary systems of enantiomers exhibit large areas in which such solid solutions are formed. This point is examined in Section 2.4.8.

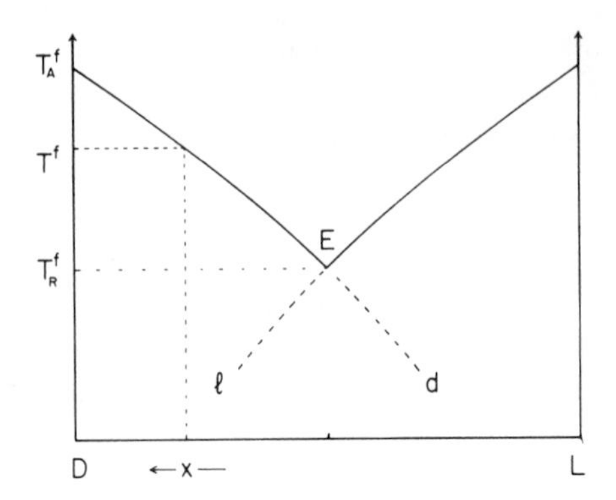

Figure 2 Melting point phase diagram for a conglomerate. A. Collet, M. J. Brienne, and J. Jacques, *Chem. Rev.*, 1980, **80**, 215. Reproduced by permission of the editor. Copyright 1980, American Chemical Society.

2.2.3 The Equation of Schröder–Van Laar

A binary system consisting of two enantiomers forming a conglomerate is amenable to a general thermodynamic treatment to which one may submit any pair of compounds which is fully miscible in the liquid state and fully immiscible in the crystalline state.

Schröder,[220] Van Laar,[221] and Le Chatelier[222] almost simultaneously proposed an equation relating the composition of mixtures to their melting points or, more precisely, to the termination of fusion, T^f. This equation allows the calculation of the liquidus curve of D or L of the binary diagram (Fig. 2). For the case of enantiomer mixtures, the Schröder–Van Laar equation may be written as follows:

$$\ln x = \frac{\Delta H_A^f}{R}\left(\frac{1}{T_A^f} - \frac{1}{T^f}\right) - \frac{C^l - C_A^s}{R}\left(\ln\frac{T_A^f}{T^f} + 1 - \frac{T_A^f}{T^f}\right) \qquad (1)$$

In this equation, x is the mole fraction of the more abundant enantiomer ($0.5 \leqslant x \leqslant 1$) of a mixture whose melting terminates at T^f (in degrees K); ΔH_A^f (in $\text{cal}\cdot\text{mol}^{-1}$) and T_A^f (also in degrees K) are the enthalpy of fusion and melting point of the pure enantiomers, respectively; C^l and C_A^s (in $\text{cal}\cdot\text{mol}^{-1}\cdot\text{K}^{-1}$) are the specific heats of the liquid and of the solid enantiomer, respectively, while R is the gas constant ($R = 1.9869\ \text{cal}\cdot\text{mol}^{-1}\cdot\text{K}^{-1}$). Calculation of eq. (1) with $0 < x < 0.5$ furnishes parts *Ed* or *El* of the liquidus curves (see Fig. 2) which describe metastable equilibria occurring when one enantiomer remains supercooled.

The validity of eq. (1) rests upon the following conditions:

1. Immiscibility of the enantiomers in the solid state.
2. Ideality of the enantiomer mixtures in the liquid state.
3. The constancy of the difference $C^l - C_A^s$ as a function of temperature.

As we shall see later (p. 52), the second term of eq. (1) containing the specific heats is generally negligible relative to the first term. Equation (1) is virtually always employed in its simplified form:

$$\ln x = \frac{\Delta H_A^f}{R}\left(\frac{1}{T_A^f} - \frac{1}{T^f}\right) \qquad (2)$$

The validity and accuracy of eqs. (1) and (2) can be demonstrated by a number of tests which we describe in the following section.

2.2.4 The Ideality of Enantiomer Mixtures in the Liquid State

Of the three postulates upon which the validity of the Schröder–Van Laar equation rests, it is the second which warrants close examination. An enantiomer mixture would be said to behave ideally in the liquid state if its enthalpy of mixing were equal to zero and its entropy of mixing were $\Delta S_l^m = -R(x_D \ln x_D + x_L \ln x_L)$ per mole (that is, for the racemate, where $x_D = x_L = 0.5$, $\Delta S_l^m = R \ln 2 = 1.38\ \mathrm{cal \cdot mol^{-1} \cdot K^{-1}}$).

Does theory require that liquid mixtures of enantiomers behave in an ideal manner? The answer to this question is no. As a matter of fact, it is entirely logical to suppose that the energy of intermolecular homochiral interactions ($d \cdots d$ or $l \cdots l$) in the liquid state or in solution must be different from that of the corresponding heterochiral interactions ($d \cdots l$). Consequently, the evolution or absorption of heat should be anticipated when two pure enantiomers are mixed either in solution or in the liquid state in absence of solvent. An exotherm or endotherm would result according to whether heterochiral interactions are stronger or weaker than homochiral ones.

Various investigators who have attempted to analyze this "chiral discrimination" phenomenon from a theoretical point of view have estimated the enthalpies of mixing to be between several calories and tens of calories per mole.[230–236] The few experimental data available (Table 1) actually lie between 0.5 and 50 cal · mol⁻¹.[223,224,237]

Table 1 Enthalpies of mixing of enantiomers in the liquid state or in solution[a]

	Solvent	ΔH_l^m (cal · mol^{-1})	Refs.
Mixing of liquid enantiomers			
2-*p*-Nitrophenylbutane	None	+ 0.45 ± 0.05	224
2-Octanol	None	+ 3.1 ± 0.3	224
α-Methylbenzylamine	None	+ 2.3 ± 0.2	224
Mixing of enantiomers in solution			
α-Methyl-α-ethylsuccinic acid	$CHCl_3$	− 33.5 ± 2.4	237
α-Methyl-α-isopropylsuccinic acid	$CHCl_3$	− 50.2 ± 4.8	237
Tartaric acid	H_2O	+ 0.47 ± 0.06	223
Threonine	H_2O	− 1.31	223

[a] Determinations carried out at 25°C (refs. 224 and 237) and 25.6°C (ref. 223).

These enthalpies of mixing often represent very much less than 1% of the enthalpies of fusion that figure in the Schröder–Van Laar equation (5 to 10 kcal · mol^{-1}). It is evidently because the energy differences between homochiral and heterochiral interactions are very small relative to the enthalpies of fusion that statements are occasionally made to the effect that mixtures of enantiomers behave ideally in the liquid state.

In addition to these direct measurements of enthalpies of mixing, which are difficult to carry out, there are several simple ways of demonstrating the practical validity of the ideality hypothesis: (a) The enthalpy of fusion of a pure enantiomer and that of the corresponding conglomerate can be measured, following which one may show that the entropy of mixing of the pure enantiomers is indeed equal to $R \ln 2$. (b) The enthalpy of fusion of an enantiomer may be accurately calculated from the experimental phase diagram. (c) It is possible to determine the phase diagram and, in particular, to calculate the melting point of the conglomerate from calorimetric measurement of the enthalpy of fusion of one enantiomer. Let us examine each of these possibilities in detail.

(a) Enthalpies of fusion of enantiomers and of their conglomerate; entropy of mixing of enantiomers in the liquid state

It is now possible to determine enthalpies of fusion with milligram quantities of substance and often with samples of less than 1 mg. This has been made possible by differential microcalorimeters which are now commercially available and are found in many laboratories (see Section 2.1). Table 2 gives a number of thermodynamic data concerning conglomerates and the corresponding enantiomers. The entropies of fusion (not listed in the table) may be calculated from the enthalpies of fusion

Table 2 Melting points and enthalpies of fusion of enantiomers and the corresponding conglomerate. Entropy of mixing of enantiomers in the liquid state[a]

	T_A^f	ΔH_A^f	T_R^f	ΔH_R^f	"$R \ln 2$" eq. (4)	T_R^f eq. (2)
o-Chloromandelic acid	392.5	5.9	358.5	4.8	1.29	359.5
p-Bromophenylhydracrylic acid	398	8.5	371	6.9	1.40	374
p-Chlorophenylhydracrylic acid	385	7.1	357	6.7	1.40	358
m-Fluorophenylhydracrylic acid	311	5.8	290	4.9	1.24	290
Phenylhydracrylic acid	391	7.8	366	7.1	1.30	366
3-Hydroxy-3-phenylpivalic acid	431	9.5	407	8.9	1.26	406
1,2-Dichloroacenaphthene	375	5.1	339	4.9	1.41	340.5
2-Naphthoxypropionamide	475	9.1	445	9.0	1.28	443
Hydrobenzoin	420.5	8.2	393	7.5	1.30	393
α-Methyl-4-methoxydesoxybenzoin	353	6.3	326	5.2	1.35	328
Methylphenylnaphthylfluorosilane	340.5	5.6	312	5.4	1.47	314
Anisylidenecamphor	399.5	7.2	371.5	6.3	1.27	371
					⟨1.33⟩	

[a] Temperatures in K; enthalpies in kcal · mol^{-1}; $R \ln 2$ in cal · mol^{-1} · K^{-1}. From ref. 238 and M. Leclercq, A. Collet, and J. Jacques, Tetrahedron, 1976, **32**, 821. Reproduced by permission of Pergamon Press Ltd.

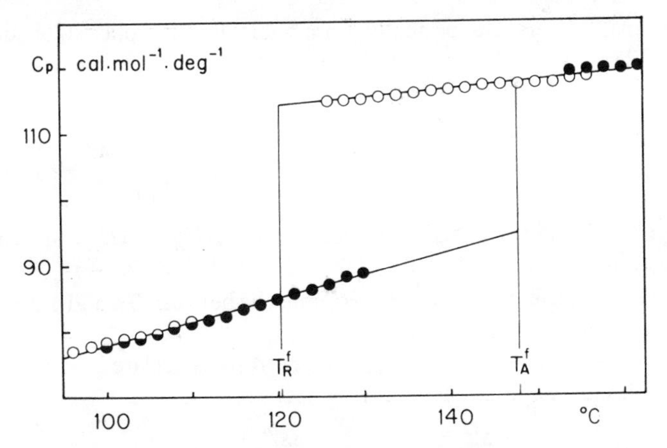

Figure 3 Heat capacities of solid and liquid hydrobenzoin. White circles: racemate; black circles: enantiomer.[238]

and the melting points by means of the equation $\Delta S^f = \Delta H^f/T^f$. It is evident from the data that the enthalpies of fusion of the enantiomers are always higher than those of the corresponding conglomerate, but the differences are sometimes small; we shall see later why this is so.

Very few data concerning heat capacities of racemates and the corresponding enantiomers in the solid and in the liquid states are available. An example is given in Fig. 3, which shows the effect of temperature on the specific heat of *threo*-hydrobenzoin, measured on the conglomerate and on the pure enantiomer. Note that the specific heat of the solid is smaller than that of the melt, as is true generally. Moreover, the specific heats of the solid enantiomer and racemate are identical ($C_A^s = C_R^s$) as is expected for a conglomerate. There is no evidence of any difference between the liquid racemate and the corresponding optically active melt ($C_A^l = C_R^l$) to a precision of the order of 1%. And, finally, the difference $C^l - C^s$ is practically constant between T_A^f and T_R^f.

We are now in a position to compare the experimental data of Table 2 with those to be expected for an ideal system. Let us consider a pair of enantiomers in the solid state, D_s and L_s, melting at temperature T_A^f to yield liquid phases D_l and L_l, respectively. The solid conglomerate DL_s melts at T_R^f ($T_R^f < T_A^f$) yielding the racemic liquid DL_l. The heat capacities of the solids are represented by C^s ($C_A^s = C_R^s = C^s$) and those of the liquids by C^l ($C_A^l = C_R^l = C^l$). The thermodynamic cycle shown below is constructed for the system just described; in this cycle, ΔH_l^m and ΔH_s^m represent the enthalpies of mixing of the enantiomers in the solid and liquid states, respectively.

$$\begin{array}{ccccccc} D_s + L_s & \xrightarrow{\Delta H_A^f} & D_l + L_l & \xrightarrow{\Delta H_l^m} & DL_l & & (\text{at } T_A^f) \\ \uparrow & & C^s(T_A^f - T_R^f) & C^l(T_R^f - T_A^f) & \downarrow & & \\ D_s + L_s & \xleftarrow[-\Delta H_s^m]{} & DL_s & \xleftarrow[-\Delta H_R^f]{} & DL_l & & (\text{at } T_R^f) \end{array}$$

If the enantiomers are perfectly immiscible in the solid state and if their mixture is ideal in the liquid state, then ΔH_s^m and ΔH_l^m equal zero. We may then write

$$\begin{aligned} \Delta H_s^m &= 0 = C^s(T_A^f - T_R^f) + \Delta H_A^f + C^l(T_R^f - T_A^f) - \Delta H_R^f \\ \Delta H_A^f - \Delta H_R^f &= (C^l - C^s)(T_A^f - T_R^f) \end{aligned} \tag{3}$$

Since $C^l > C^s$ and $T_A^f > T_R^f$, it follows that $\Delta H_A^f > \Delta H_R^f$. Moreover, since $C^l - C^s$ has a magnitude of 20 to 40 cal · mol^{-1} · K^{-1} and $T_A^f - T_R^f$ is of the order of 20 to 30 K, the difference $\Delta H_A^f - \Delta H_R^f$ must lie between 0.4 and 1.2 kcal · mol^{-1}. This is in fact what is observed (Table 2).

Let us now examine the changes in entropy associated with the thermodynamic cycle shown above:

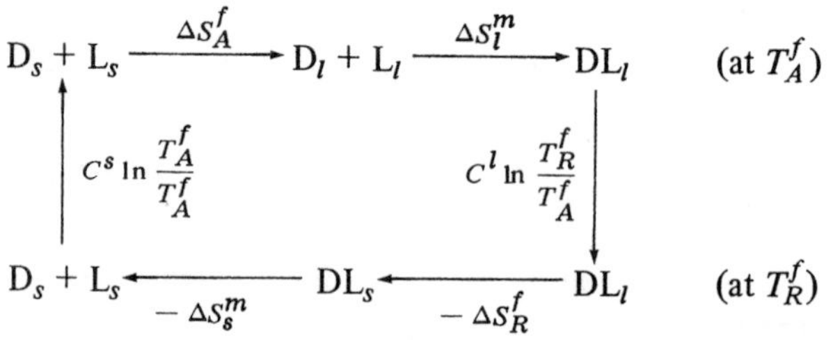

$$\begin{array}{ccccccc} D_s + L_s & \xrightarrow{\Delta S_A^f} & D_l + L_l & \xrightarrow{\Delta S_l^m} & DL_l & & (\text{at } T_A^f) \\ \uparrow C^s \ln \dfrac{T_A^f}{T_A^f} & & & & \downarrow C^l \ln \dfrac{T_R^f}{T_A^f} & & \\ D_s + L_s & \xleftarrow[-\Delta S_s^m]{} & DL_s & \xleftarrow[-\Delta S_R^f]{} & DL_l & & (\text{at } T_R^f) \end{array}$$

In this cycle, or elsewhere for that matter, a conglomerate DL_s exists in the form of two separate phases, D_s and L_s, which are present as a mechanical mixture without implication as to homogeneity.* Consequently, no change in entropy need be associated with this particular "mixing process" ($\Delta S_s^m = 0$). Of course, this is not the case for the mixing of the liquid enantiomers, where we expect that the corresponding entropy change, ΔS_l^m, must equal $R \ln 2 = 1.38$ cal · mol^{-1} · K^{-1} if the system behaves ideally.

Our analysis of the cycle permits us to write the following:

$$\begin{aligned} \Delta S_s^m &= 0 = C^s \ln \frac{T_A^f}{T_R^f} + \Delta S_A^f + \Delta S_l^m + C^l \ln \frac{T_R^f}{T_A^f} - \Delta S_R^f \\ -\Delta S_l^m &= \frac{\Delta H_A^f}{T_A^f} - \frac{\Delta H_R^f}{T_R^f} - \frac{\Delta H_A^f - \Delta H_R^f}{T_A^f - T_R^f} \ln \frac{T_A^f}{T_R^f} \end{aligned} \tag{4}$$

Equation (4) allows us to calculate the entropy of mixing of the liquid enantiomers to yield the liquid racemate from the enthalpies of fusion and the melting points, which can be accurately measured by microcalorimetry. The values of ΔS_l^m thus calculated are given in the next to last column of Table 2. They are quite close to the value expected from theory.

* "Thermodynamically speaking, the eutectic is a mixture of two phases, not a separate phase, and must always be so considered when calculating degree(s) of freedom. . . ." (From ref. 219, p. 136.)

(b) *Calculation of the enthalpy of fusion from the phase diagram*

In order to obtain a value of the enthalpy of fusion ΔH_A^f with fair precision, it is necessary to start from a phase diagram that itself has been established with great care. In fact, this is not always the case for diagrams found in the chemical literature.

Nonetheless, it is possible to find suitable examples such as the value of the enthalpy of fusion of 3-hydroxy-3-phenylpivalic acid calculated from the data of Matell[225] with the aid of the simplified Schröder–Van Laar equation [eq. (2)]. The results are given in Table 3. The enthalpy of fusion thus calculated (ΔH_A^f = 9.9 kcal · mol^{-1}) is close to that measured directly by differential scanning microcalorimetry (9.5 kcal · mol^{-1}).[238]

Table 3 Enthalpy of fusion ΔH_A^f of active 3-hydroxy-3-phenylpivalic acid calculated from the experimental binary diagram[a]

% (+)-Enantiomer	x^b	T^f	ΔH_A^f (calcd)	$T^{f\,c}$
0	1	158		
17.2	0.828	152	11.46	150.8
34.6	0.654	142.5	9.76	142.1
45.8	0.542	136	9.76	135.4
50	0.50	134	10.07	132.6
51.8	0.518	135	10.00	133.8
54.0	0.540	136	9.82	135.3
59.1	0.591	139	9.77	138.5
66.5	0.665	137.5	7.00	142.7
76.2	0.762	148	9.81	147.7
83.3	0.833	151.5	10.23	151.0
91.5	0.915	155	10.86	154.6
100	1	158		
			⟨9.9⟩	

[a] Temperature in °C; enthalpy of fusion in kcal · mol^{-1}. Data from ref. 225.
[b] Mole fraction of the more abundant enantiomer ($0.5 \leqslant x \leqslant 1$).
[c] T^f calculated from the experimental value $\Delta H_A^f = 9.5$ kcal · mol^{-1} by means of eq. (2).

(c) *Calculation of the phase diagram from the enthalpies of fusion and the melting point of an enantiomer*

Since our interest, and indeed our goal, is the understanding of resolutions and since we hope to make it easier to carry out resolutions, a more important focus than the analysis of Section **(b)** is its inverse. That is, we need to concern ourselves more with the *a priori* determination of the phase diagram from accessible experimental data (i.e., enthalpies of fusion and melting points) than with the inverse.

Let us once again consider the preceding example, that of 3-hydroxy-3-phenylpivalic acid. From the measured value of the enthalpy of fusion (9.5 kcal · mol^{-1}), melting points have been calculated for the compositions corre-

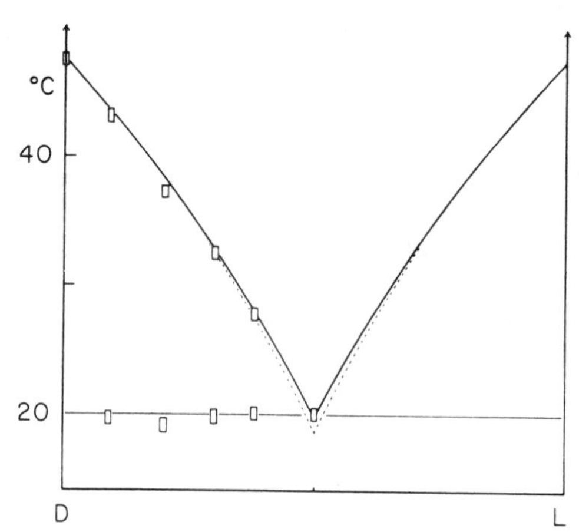

Figure 4 Melting point diagram of hydrobenzoin (see text).[238]

sponding to the mixtures employed by Matell (Table 3) on the basis of the simplified eq. (2). We see first of all that the calculated values do not always coincide with those observed by Matell. Nonetheless, the agreement is good (the difference is less than 2°) for the majority of cases, particularly if we take into account the fact that the measurements involved visual observation of the final melting temperature. It is possible, however, that the differences between the observed melting points and those calculated by means of eq. (2), which are slightly lower, are not just a consequence of systematic experimental error. They may arise as a consequence of the nature of the equation employed and of the approximation which tends to neglect the effect of heat capacities.

To assess this possibility, that is, to probe the approximation built into the simplified eq. (2), we have carefully examined the binary phase diagram for hydrobenzoin. For this substance the enthalpy of fusion ΔH_A^f as well as the heat capacities in both the solid and liquid states have been measured by differential microcalorimetry[238] (Table 2 and Fig. 3). In Fig. 4, we compare the phase diagram calculated from the simplified eq. (2) (solid line) with that derived from the original eq. (1) (dotted line). The difference between the two calculated melting points for the eutectic (T_R^f) is of the order of 1.5°. For enantiomeric purities larger than 50% ($x > 0.75$), the difference between the two curves becomes clearly smaller than the experimental error.

Several experimental points (from microcalorimetry) are indicated on the diagram of Fig. 4; the agreement with the calculated liquidus curves is remarkably good. However, we observe that the measured melting point for the racemate (120 ± 0.5°C) is closer to that found on the curve obtained by means of the simplified equation than that derived from the complete equation (119.8 and 118.6°C, respectively). This result may be indicative either of a slight deviation from ideality, or of the existence of a small region of partial miscibility in the solid state (solid solution) at the extremities of the diagram.

Finally, we have recalculated by means of eq. (2) the racemate melting points of a series of compounds from the enthalpies of fusion and melting points of the corresponding enantiomers. The results are assembled in the last column of Table 2. The agreement between calculated and measured values is very good indeed.

The above analysis taken as a whole demonstrates the validity of the Schröder–Van Laar equation, particularly in its simplified form [eq. (2)]. This circumstance largely facilitates the quantitative description of phenomena that take place during resolutions. Moreover, the above analysis lays the groundwork for an examination of a calorimetric method for determining enantiomeric purities, which is taken up in Section 2.7.

2.2.5 Inventory of Enantiomer Mixtures That Exist as Conglomerates

In this section, we have assembled as complete a list as possible of those racemates that are likely to exist as conglomerates. These are the compounds that may be resolved by direct crystallization according to methods to be described in Chapter 4 of this monograph. Several such inventories have already appeared in print, the earliest being that of Ebel[226] in 1933, then those of Delépine[227] (1935), Klabunowski[228] (1960), and Secor[229] (1962). A much more complete inventory, carried out by Collet, Brienne, and Jacques in a systematic search for new cases of spontaneous resolutions and which includes the older cases, was published in 1972.[1] An updated list[2] was completed in 1976. The lists which follow (Tables 4 through 6) and which contain nearly 250 systems believed to exist as conglomerates include some cases identified in 1977–1979.

Table 4 consists of compounds with accessible melting points, that is, for the most part, covalent organic compounds. Table 5 includes amino acid salts and their derivatives or precursors which permit them to be resolved by direct crystallization on an industrial scale. Some of these amino acids (in underivatized form) or their covalent derivatives which appear in Table 4 are referred to Table 5. Consultation of Table 5, therefore, gives access to all known possible instances of resolution of synthetic amino acids by direct crystallization. Table 6 is concerned with various other organic salts and organometallic complexes.

The validity of these tables is dependent upon the accurate identification of conglomerate behavior for each listed compound. The criteria employed to identify such behavior differ from case to case, and more than one criterion is applied for some of the entries. We list here the eight criteria (recognition tests) employed, together with the abbreviations which identify them in the tables:

- S — Observation of a spontaneous resolution
- SN — Examination, in a nematic phase, of the solution of a crystal isolated from the racemate
- BD — Examination of the binary phase diagram of the system
- TD — Examination of the solubility diagram of the enantiomer mixture (ternary phase diagram)
- E — Resolution by entrainment
- ‡ — Resolution by crystallization in an optically active solvent

X Determination of the crystal space group by X-ray methods

IR Comparison of the infrared spectra of the enantiomers and of the racemate in the crystal state, in some cases (f) after melting and resolidification

A detailed description of each of these criteria follows.

(a) Spontaneous resolution (S)

One of the most obvious and easily applied criteria is the direct observation of a spontaneous resolution. Provided that we can obtain crystals of sufficient size (e.g., 1 mg) from a racemic solution, all that one needs to do is to measure the rotatory power of one of these dissolved in a small amount of suitable solvent. The crystal should be separated from its neighbors as much as possible. Use of a photoelectric polarimeter of high sensitivity (reproducibility ± 0.005°) is indicated as well as analysis at several wavelengths.

The utility of this resolution test is limited only by the tendency of certain racemates to form twinned conglomerate crystals under some conditions. In such crystals, homochiral lattice fragments of both chiralities are so closely intermixed that individual crystals of conglomerates exhibit low or no rotation when dissolved. See, for example, the detailed crystallographic observations on *threo*-phenylglyceric acid by V. M. Goldschmidt cited by Riiber.[41,242]

(b) Solution in a nematic phase of a crystal isolated from a racemate (SN)

In a certain sense, this procedure constitutes a variant of the preceding process inasmuch as it has to do with the indirect measurement of what is equivalent to optical rotation. Moreover, this process is a much more sensitive one. It is based upon the observation that an optically active substance dissolved in a nematic liquid crystal converts the latter into a cholesteric liquid crystal. The experimental technique employed permits the recognition of optical activity in a crystal weighing as little as 0.001 mg or even less.[127] The method does require some experience in the recognition of the properties of cholesteric mesophases. As the adjoining photographs (Fig. 5) make clear, the effect, which is seen on the heating stage of a polarizing microscope, is not an especially subtle one.

The test is carried out as follows: A mesomorphic compound (ca. 1 mg) is placed on a microscope slide, covered with a cover slip, and brought to a temperature at which the nematic phase is present. Compounds such as MBBA (N-*p*-methoxybenzylidene-*p*-butylaniline, nématic range 21 to 46°C) or "Merck Nematic Phase V" (a eutectic mixture of azoxy compounds, nematic range −5 to 75°C) may be used for convenience. Both are commercially available and have the advantage of being nematic at ambient temperature.

A monocrystal around 0.1 mm long is isolated from the racemate with a needle and is brought in contact with the nematic phase. The sample is heated slightly in order to dissolve the crystal – or at least part of it – and an isotropic liquid phase appears in the part of the nematic phase where the crystal dissolves. The sample is allowed to cool, whereupon the mesomorphic phase reappears. If the dissolved sample is optically active, then the phase which forms in the vicinity of the dissolved crystal is cholesteric. Since an optically active crystal is by definition found only in racemates that have conglomerate character, the appearance of a cholesteric phase is clear evidence for a conglomerate. Cholesteric character is in turn evidenced by a variety of specific microscopic characteristics the most common of which are (1) colors associated with rotatory power, that is, these colors are unchanged when the sample is rotated without rotating the analyzer; (2) periodic striations; (3) myelin forms; (4) black spiral crosses, which may be observed during the transformation of isotropic liquid to mesomorphic phase which takes place upon cooling.

It is essential that the compound under examination be soluble in the mesomorphic phase and that it not react with it. These condition are almost always met when a covalent organic compound is added to MBBA or Phase V. Contrariwise, the method is not likely to be employed in the case of organic salts or of organometallic complexes that are not soluble in these lipophilic solvents.

Other mesomorphic substances whose nematic phases exist at, or above, room temperature may be employed. Care must be taken to protect the microscope optics if heating is required.

The clear presence of cholesteric character may be taken as proof of the optical activity of the dissolved substance. Hence, qualitatively this method permits one to detect optical activity in an extremely small sample – whether taken from a racemate or not – which might be insufficient for conventional polarimetric measurements of optical rotation.

On the other hand, the absence of cholesteric characteristics should not be construed as absolute proof of the absence of optical activity. Some globular compounds, such as camphor, while optically active, do not induce sufficiently obvious cholesteric characteristics so as to be recognizable through the simple visual test described.[1] Moreover, when one isolates a crystal from a conglomerate, it may happen that the optical purity of the isolated particle is too low for optical activity to manifest itself in the manner described. The occurrence of twinning of (+) and (–) crystals may conceivably lead to such difficulties.

(c) Examination of the binary phase diagram of the enantiomer mixture (BD)

If samples of the two enantiomers are available, or of one enantiomer and of the racemate, the binary phase diagram can easily be determined (see Section 2.1). Systems whose phase diagrams exhibit only an eutectic at a mole ratio of 0.5 exhibit conglomerate behavior by definition.

(d) Examination of the solubility diagram of the enantiomer mixture (TD)

As we shall see (Chapter 3), the ternary phase diagram of the two enantiomers in the presence of a solvent – the solubility diagram – has the same form as the corresponding binary phase diagram and thus allows one to recognize the nature of the racemate. It is a fact that more effort is required to construct solubility diagrams than binary phase diagrams. Few conglomerates have, in fact, been identified through this recognition test. However, for infusible substances or substances that melt with decomposition, the only phase diagrams accessible are, in fact, solubility diagrams.

A simpler test related to this criterion is the following: When a pure enantiomer

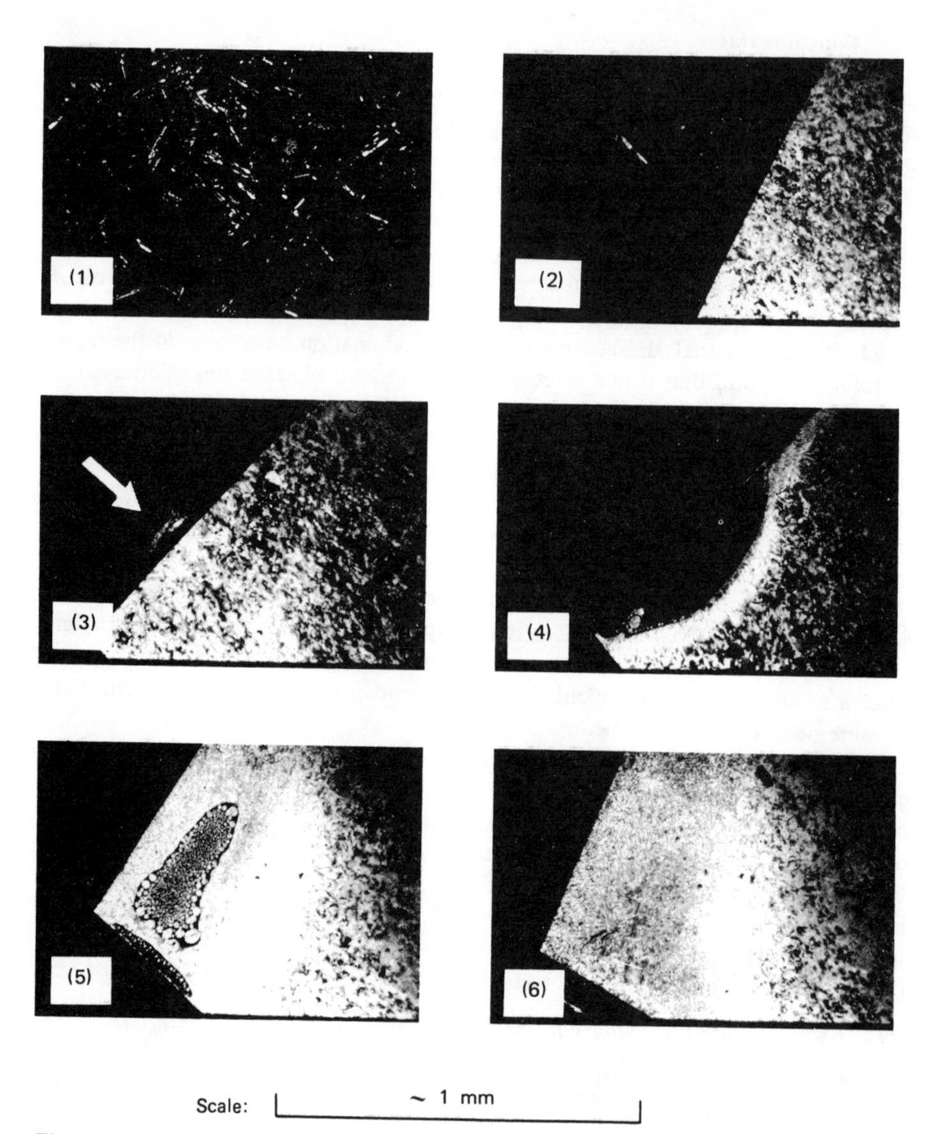

Figure 5 Microdiagnostic recognition test (SN) for conglomerates. (1) The racemate (p. 69, formula 113) consists of small needles ~0.01 to 0.1 mm in length. (2) A single crystal is isolated from the racemate and placed in the vicinity of the nematic phase (Merck Nematic Phase V). (3) The crystal (arrow) is placed in contact with the nematic phase. (4) When the preparation is heated, the isotropic liquid phase appears (black under crossed Nicols) and the racemate crystal dissolves. (5), (6) Cooling and formation of the new mesomorphic phase. (7), (8) Enlargement of (5) and (6) clearly showing the cholesteric characteristics: essentially these are colors associated with rotatory power (the colors are unchanged when the hot stage is rotated) and *periodic striations.* (9), (10) Other figures frequently observed in cholesteric phases: *myelin forms.*

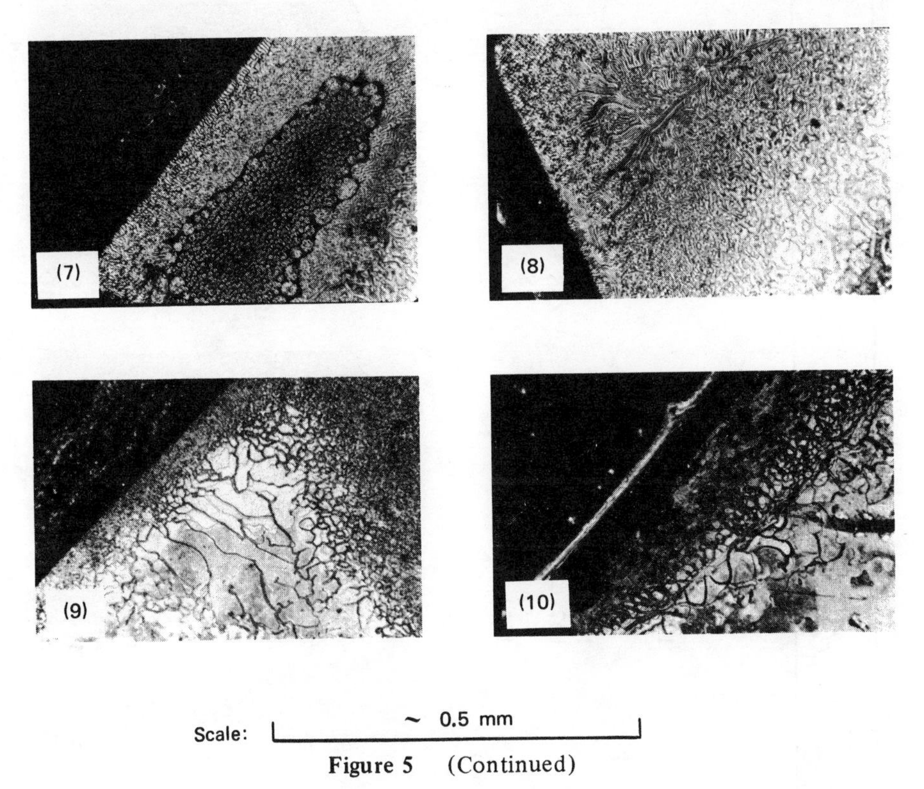

Figure 5 (Continued)

is obtained through recrystallization of a partially resolved mixture of low optical purity, it raises the likelihood that the racemate exhibits conglomerate behavior. This presumption must nevertheless be treated with caution (see entry 64 of Table 4 and Chapter 3).

(e) Resolution by entrainment (E)

The ability to carry out a resolution by entrainment (Chapter 4) implies that the solubility diagram of the compound in question be that of a conglomerate at least under the experimental conditions employed (solvent, temperature). The conglomerate could even be a metastable one.

(f) Crystallization of a racemate in an optically active solvent to yield a pure enantiomer (‡)

The success of this operation implies that the racemate is a conglomerate. We elaborate this point in Section 4.4.

(g) Determination of the crystal space group by X-ray methods (X)

Crystals isolated from a racemate are themselves enantiomorphous when the latter is a conglomerate. The space group to which the crystals belong is therefore devoid of inverse symmetry elements, something which can generally be ascertained by X-ray methods (see Chapter 1).

Table 4 Inventory of organic conglomerates

Entry No.	Name	T_A^f (°C)	T_R^f (°C)	Recogn. Test	Refs.
	C_1				
1	Methanesulfinic acid	–	–	X	3
	C_4				
2	2-Chloro-3-hydroxysuccinic acid	175	145	BD	4
3	*trans*-2,3-Dibromo-1,4-dioxane	–	74	X	5
4	4-Methylazetidinone	27	−12	E	240
5	*threo*-2,3-Dibromo-1,4-butanediol	115	90	≠	1, 6
6	Trichlorfon	107	75	X, S, E	7, 8
7	Asparagine	–	–	S	9
8	Threonine	255d	230d	X, E	10, 11, 12
9	2-Amino-4-sulfobutyric acid (homocysteic acid)	–	–	E	13
10	Threitol	88	72	IR	1, 14, 15
	C_5				
11	4-Vinylazetidinone	27	−1	E	240
12	3, 4-Dimethylazetidinone	61	24	E	240
13	2-Methylaspartic acid	240	233 167	S	16
14	Glutamic acid	–	–	IR, S, E	17
	C_6				
15	1*a*, 2*a*, 4*e*, 5*e*-Tetrabromocyclohexane	–	185	X	2, 18
16	1*a*, 2*a*, 4*e*, 5*e*-Tetrachlorocyclohexane	–	174	X	2, 19
17	5-Methyl-5-vinylhydantoin	166	129	IR	20
18	Serine anhydride	248	224	IR	21
19	Gulonolactone	180	160	S	22
20	Dilactyldiamide	208	184	BD	23
21	*threo*-2,5-Hexanediol	53	25	BD, S	24

$CH_3-S(=O)-OH$

1

$HO-CH-CO_2H$ / $Cl-CH-CO_2H$

2

O, O, Br, Br

3

CH_3, HN, O

4

Br, CH_2OH / Br, CH_2OH

5

$Cl_3C-CH(OH)-P(=O)(OCH_3)-OCH_3$

6

$H_2NCO-CH_2-CH(NH_2)-COOH$

7

$CH_3-CH(OH)-CH(NH_2)-CO_2H$

8

$CH_2(SO_3H)-CH_2-CH_2-CH(NH_2)-CO_2H$

9

HO, CH_2OH / HO, CH_2OH

10

HN, O

11

CH_3, CH_3, HN, O

12

$CH_3-C(NH_2)(CO_2H)-CH_2-CO_2H$

13

$HO_2C-CH_2CH_2CH(NH_2)-CO_2H$

14

X, X, X, X

15,16

CH_3, O, HN, NH, O

17

H, CO-NH, CH_2OH, C, C, $HOCH_2$, NH-CO, H

18

$HOCH_2-CH-CH(OH)-CH(OH)-CH(OH)-C=O$ (ring O)

19

$CH_3-CH-CONH_2$ / O / $CH_3-CH-CONH_2$

20

$CH_3-CH(OH)-CH_2-CH_2-CH(OH)-CH_3$

21

Table 4 Continued

Entry No.	Name	T_A^f (°C)	T_R^f (°C)	Recogn. Test	Refs.
	C_7				
22	2-Methoxy-2-(2-thienyl)acetic acid	86	61	BD	25
23	2-Methoxy-2-(3-thienyl)acetic acid	86	60	BD	25
24	N-chloroacetylproline	120	88	E	26, 27
25	N-acetylproline	–	–	E	27
26	N-acetylglutamic acid	199	182	E	28
27	Methyl mannoside	191	166	S	29
	C_8				
28	*o*-Chloromandelonitrile	–	47	SN	30
29	*o*-Chloromandelic acid	120	86	S, BD, IR	30
30	*trans*-4,5-Dicyanocyclohexene	–	123	SN	1
31	2-Phenylglycine	d	d	E	31
32	*O*-Acetyl pantolactone	–	–	IR, BD, E	32
33	2,5-Dimethyladipic acid	104	75	IR	33
34	*N*-Acetylleucine	192	164	E, BD	34, 35
35	*N*-Acetylisoleucine	151	119	E	34
36	3,3′-Iminodibutyric acid	180	160	E	36
	C_9				
37	2-Amino-5-phenyloxazolin-4-one	271	–	IR	37
38	3-(*p*-Bromophenyl)hydracrylic acid	125	98	IR, SN, BD	38
39	3-(*m*-Chlorophenyl)hydracrylic acid	95	67	BD	2, 38
40	3-(*p*-Chlorophenyl)hydracrylic acid	112	84	SN, BD	38
41	3-(*m*-Fluorophenyl)hydracrylic acid	38	17	BD	38
42	1-Indanol	71	38 } 50 }	IR, BD	39
43	3-Phenylhydracrylic acid	116	93	S	38, 40
44	3-Phenyllactic acid	125	96 } 93 }	IR	2
45	*threo*-3-Phenylglyceric acid	166	141	S	41, 42

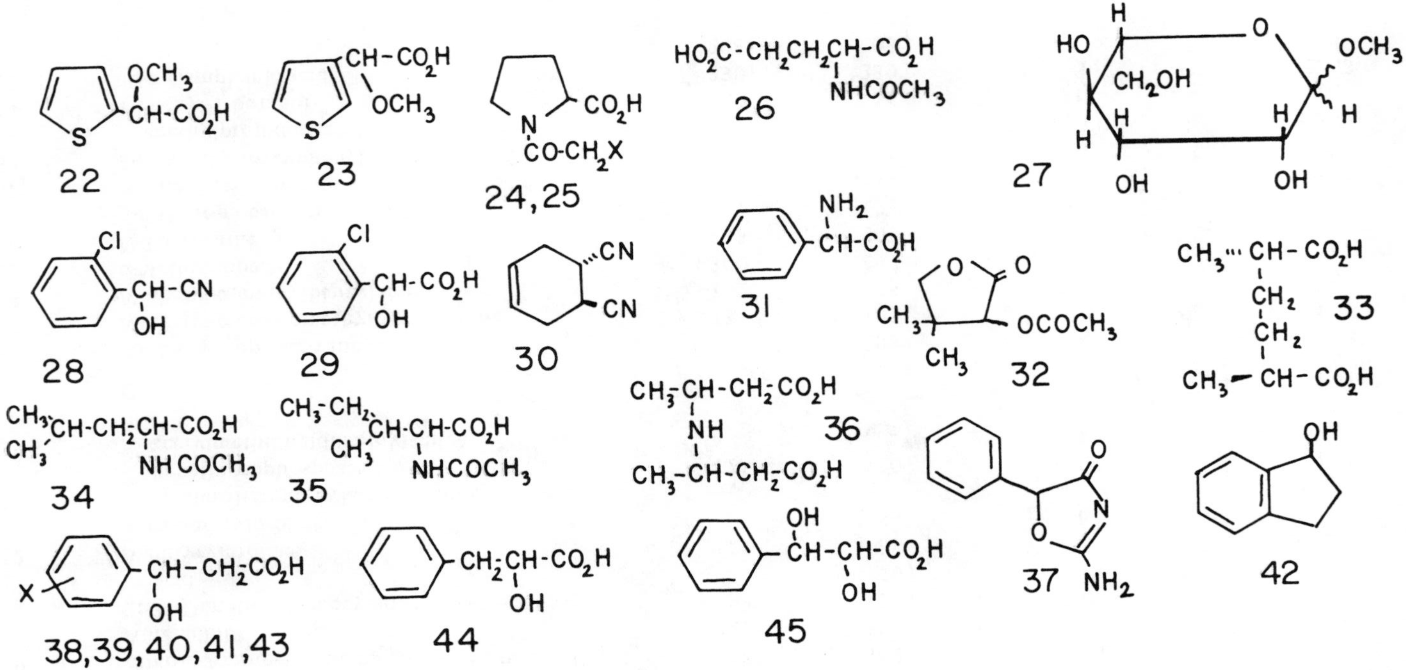

22: S, OCH_3, $CH\text{-}CO_2H$

23: S, $CH\text{-}CO_2H$, OCH_3

24,25: N, CO_2H, $CO\text{-}CH_2X$

26: $HO_2C\text{-}CH_2CH_2CH\text{-}CO_2H$, $NHCOCH_3$

27: HO, H, O, OCH_3, CH_2OH, H, H, H, H, OH, OH

28: Cl, $CH\text{-}CN$, OH

29: Cl, $CH\text{-}CO_2H$, OH

30: CN, CN

31: NH_2, $CH\text{-}CO_2H$

32: O, O, CH_3, CH_3, $OCOCH_3$

33: $CH_3\text{-}CH\text{-}CO_2H$, CH_2, CH_2, $CH_3\text{-}CH\text{-}CO_2H$

34: CH_3, CH_3, $CH\text{-}CH_2\text{-}CH\text{-}CO_2H$, $NHCOCH_3$

35: $CH_3\text{-}CH_2$, CH_3, $CH\text{-}CH\text{-}CO_2H$, $NHCOCH_3$

36: $CH_3\text{-}CH\text{-}CH_2\text{-}CO_2H$, NH, $CH_3\text{-}CH\text{-}CH_2\text{-}CO_2H$

37: O, N, O, NH_2

42: OH

38,39,40,41,43: X, $CH\text{-}CH_2\text{-}CO_2H$, OH

44: $CH_2\text{-}CH\text{-}CO_2H$, OH

45: OH, $CH\text{-}CH\text{-}CO_2H$, OH

Table 4 Continued

Entry No.	Name	T_A^f (°C)	T_R^f (°C)	Recogn. Test	Refs.
46	6-Hydroxy-5-iodobicyclo[2.2.2]octane-2-carboxylic acid, lactone	116	86	IR	43
47	*m*-Tyrosine	–	–	X	44
48	3-(3,4-Dihydroxyphenyl)alanine	–	–	E	45
49	*threo*-2-Amino-1-*p*-nitrophenyl-1,3-propanediol	164	142	E, BD	46, 47
50	Adrenaline	221	191	E	48, 49
51	2*t*,3*c*-Dimethyl-4-oxocyclohexane-1*r*-carboxylic acid	49	25 20	BD	50
52	*N*-Butyrylproline	115	88	E	26, 27
53	*N*-Isobutyrylproline	–	86	E	26, 27
54	(γ-Picolino)(trimethylamino)chloroborane, hexafluorophosphate	126	120	S	51
55	*N*-Acetylmethionyldimethylamide	–	–	X	52
	C_{10}				
56	3,4-Dehydroproline anhydride	–	208	X	53
57	*N*-Acetyl-α-chloromethylbenzylamine	134	105	IR	54
58	*N*-Acetyl-α-methylbenzylamine	102	75	IR	1, 55
59	α-Methyldopa	300	297	E	1, 56
60	*N*-Tosylserine	233	205	BD	57
61	α-π-Dibromocamphor	157	122	BD	58
62	α-Thio-*S*-carbomethoxyphenylacetic acid	–	–	BD	59
63	3-(3,4-Dihydroxyphenyl)-2-hydrazino-2-methylpropionic acid	205	–	E, TD	60
64	1-Phenyl-1-butanol	50	16	TD	1, 61
65	*cis*-π-Camphanic acid	230 208	230 208 170	S	1, 62, 63

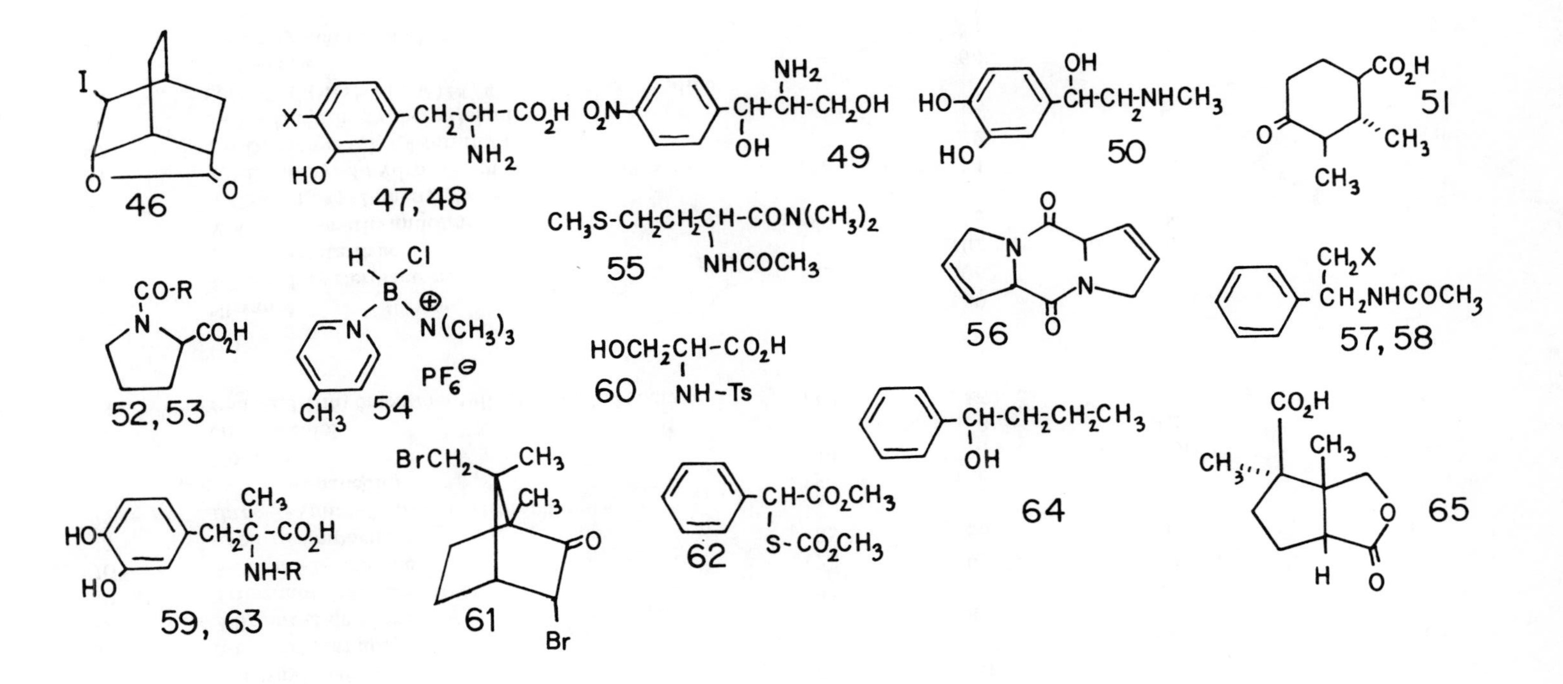
46
47, 48
49
50
51
52, 53
54
55
56
57, 58
60
64
59, 63
61
62
65
CH3S-CH2CH2CH-CON(CH3)2
NHCOCH3
HOCH2CH-CO2H
NH-Ts

Table 4 Continued

Entry No.	Name	T_A^f (°C)	T_R^f (°C)	Recogn. Test	Refs.
66	Dimethyl diacetyltartrate	103	80	BD	1, 64
67	α-Bromocamphor	74	39	BD	1, 65, 66
68	3-Bromo-2-decalone	121	89	X	67, 68
69	Tribromotetrahydrocarvone	92	62	IR	1, 15, 69
70	3-Chloro-2-decalone	122	92	X	67, 68
71	ω-Nitrocamphene	85	66	E	70
72	*threo*-2-Amino-1-(*p*-methylthiophenyl)-1,3-propanediol	152	126	E	71
73	Noludar	84	74	S, BD	1, 72, 73
74	*cis*-2-Decalol	38	18	BD?	74
75	*trans*-2,3-Dihydroxymethylbicyclo[2.2.2]octane	118	86	IR	2
	C_{11}				
76	Nirvanol	–	199	S, BD	75
77	4-Phenyl-4-valerolactone	123	106	E	76
78	2-Phenylglutaric acid	131	102	BD	77
79	*N*-Acetyl-2-methylindoline	89	56	SN	78
80	*N*-Carbethoxy-2-phenylglycine	–	–	E	79
81	2,2-Dimethyl-3-hydroxy-3-phenylpropionic acid	158	134	BD	80
82	2,6-Dimethyl-1,2,3,4-tetrahydroquinoline	53	32	S	1, 15, 81
83	3-(3,4-Dimethoxyphenyl)alanine	–	–	E	82
84	Spiro-2,2′-(norbornane)(4′,4′-dimethyloxazolidine-3′-oxyle)	–	68	X	83
85	*N*-Formylneomenthylamine	118	86	BD	1, 15, 84, 85, 86

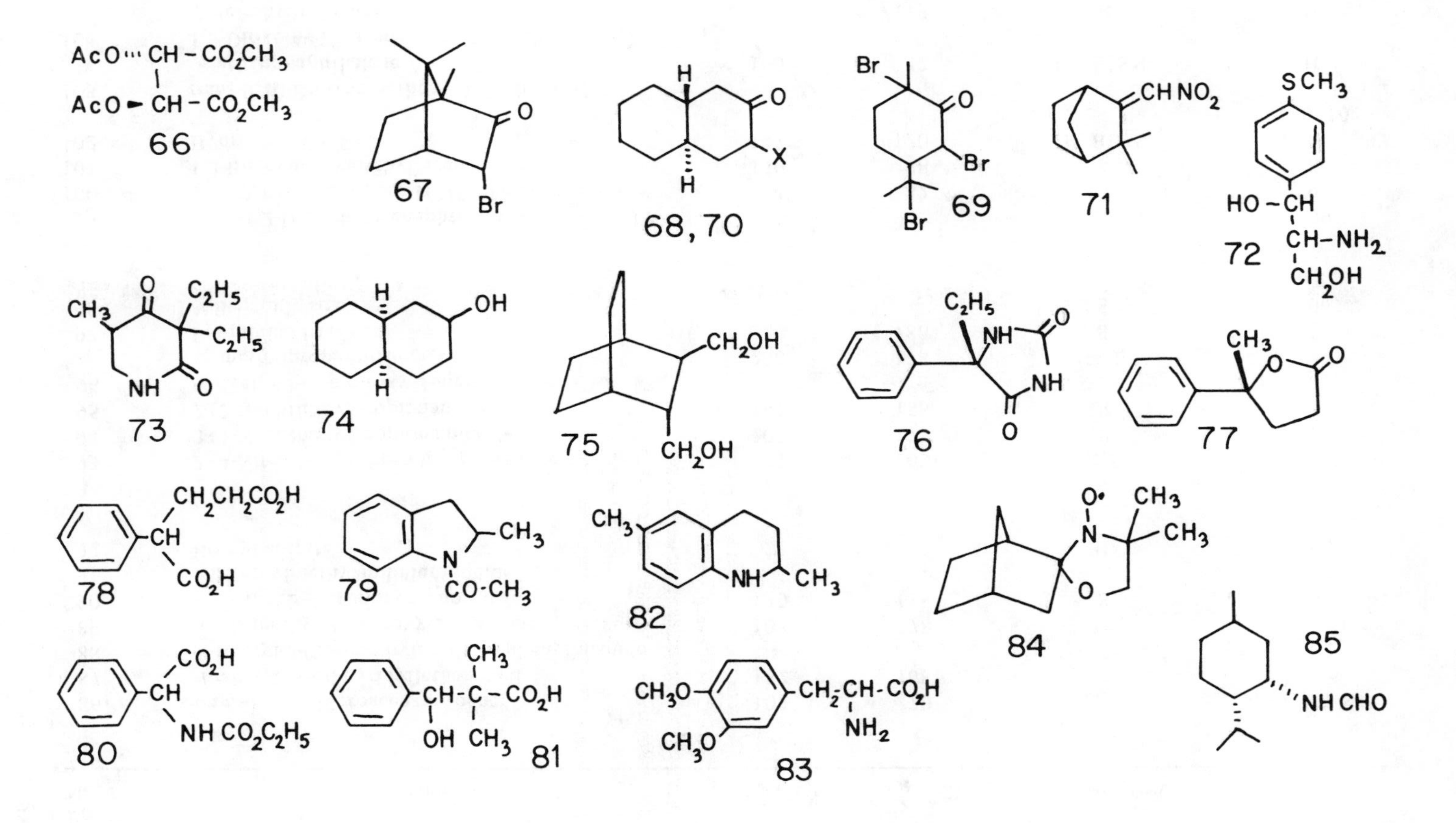
AcO··· CH - CO2CH3
AcO - CH - CO2CH3
66
O
Br
67
H
O
X
H
68, 70
Br
O
Br
Br
69
CH NO2
71
SCH3
HO - CH
CH-NH2
CH2OH
72
O
C2H5
CH3
C2H5
NH
O
73
H
OH
H
74
CH2OH
CH2OH
75
C2H5
NH
O
NH
O
76
CH3
O
O
77
CH2CH2CO2H
CH
CO2H
78
CH3
N
CO-CH3
79
CH3
NH
CH3
82
O•
CH3
N
CH3
O
84
CO2H
CH
NH CO2C2H5
80
CH3
CH-C-CO2H
OH CH3
81
CH3O
CH3O
CH2-CH-CO2H
NH2
83
85
NH CHO

Table 4 Continued

Entry No.	Name	T_A^f (°C)	T_R^f (°C)	Recogn. Test	Refs.
	C_{12}				
86	*trans*-1,2-Dichloroacenaphthene	102	70	IR, X, SN	1, 87, 88
87	*threo*-1,2-Bis(4-pyridyl)ethanediol	182	168	‡	1, 6
88	*N*-Acetyl-3-(3,4-methylenedioxyphenyl)alanine	–	–	E	82
89	3,3-Dimethyl-2-hydroxy-1-tetralone	108	78	IR	1, 90
90	4,4′-Azobis-4-cyanopentanoic acid	130	118	S	91
91	Allylethylmethylanilinium iodide	–	–	S	97
92	Bornyl acetate	27	7	BD	92
	C_{13}				
93	2-(4-Nitro-1-naphthoxy)propionamide	228	203	IR(f)	1, 93
94	2-(1-Naphthoxy)propionamide	202	172	IR	1, 93
95	2-(2-Naphthoxy)propionamide	187	158	IR	1, 93
96	*N*-Acetyl-3-(4-hydroxy-3-methoxyphenyl)-2-methylpropionitrile	206	178	IR, E	94
97	Epiquinide triacetate	220	188	S	95
98	*N*-Benzoyllysine	–	–	S	96
	C_{14}				
99	*trans*-1,2-Dimethylacenaphthene			X	89
100	1,9,10,10*a*-Tetrahydro-3(2*H*)-phenanthrenone	114	82	IR	98
101	1,3-Bis(hydroxymethyl)acenaphthene	130	98	IR	89
102	Hydrobenzoin	147.5	120	IR, BD, S	1, 15, 84, 99, 100, 101
103	*trans*-9,10-Dicarbomethoxy-1,4-dihydro-1,4-ethanonaphthalene	100	72	IR, E, SN	102
104	1,5-Dioxo-4-(2′-carboxyethyl)-7a-ethyl-5,6,7,7a-tetrahydroindane	148	119	IR	103

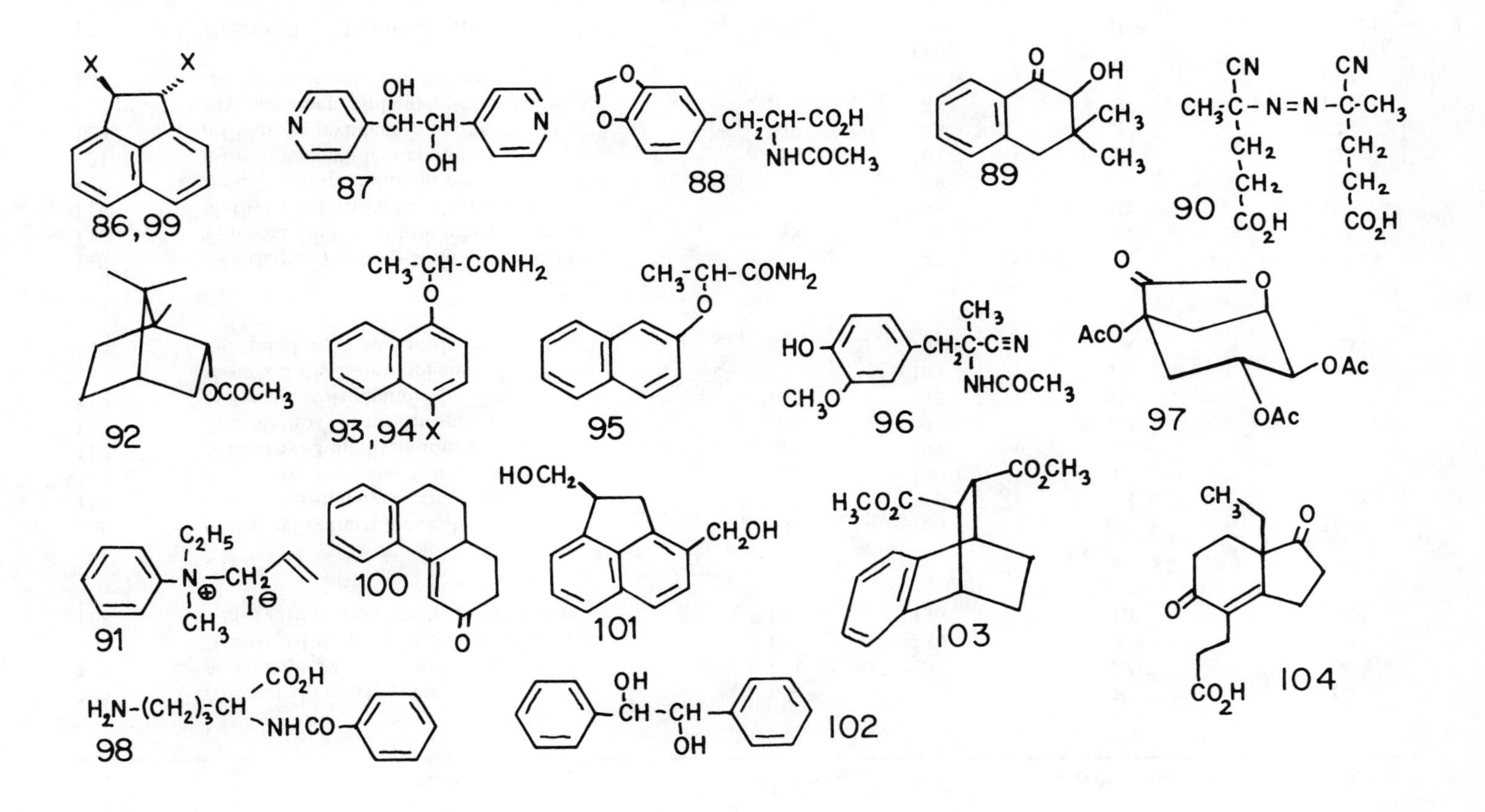
86,99
87
88
89
90
92
93,94 X
95
96
97
91
100
101
103
104
98
102

Table 4 Continued

Entry No.	Name	T_A^f (°C)	T_R^f (°C)	Recogn. Test	Refs.
	C_{15}				
105	*N*-Benzoylphenylglycine	–	–	E	104
106	*N*-Benzoyl-2-amino-2-phenylethanol	181	153	IR	54
107	*threo*-1,3-Dihydroxy-1,3-diphenylpropane	153	130	S	105
108	17β-Hydroxy-des-*A*-androst-9-en-5-one	170	136	IR	2, 106
	C_{16}				
109	β-Benzoylhydratropic acid	182	154	BD	107
110	*threo*-2,3-Diphenylsuccinic acid	214	–	IR	1, 108
111	*N*-Benzoyl-3,4-dichloroamphetamine	170	140	E	109
112	*N*-Benzoyl-2-methylindoline	119	92	S	78
113	2-(*p*-Methoxyphenyl)propiophenone	80	60	S, BD, E	1, 15, 35, 110
114	*O*-Benzylatrolactamide	142	113	BD	112
115	*threo*-2,3-Bis(2-aminophenyl)-2,3-butanediol	205	183	IR	111
116	2-Octanol, hydrogen phthalate	75	55	IR	113
	C_{17}				
117	2-(2-Carboxybenzyl)-2-chloro-1-indanone	179	147	IR	114, 115
118	Fluoromethyl-2-naphthylphenylsilane	68	41	IR	116
119	Methyl-1-naphthylphenylgermane	72	48	BD	117
120	*trans*-2,3-Tetralindiol monotosylate	–	88	SN	15
121	*p*-Bromobenzylidenecamphor	134	101	S	2
122	*p*-Chlorobenzylidenecamphor	109	78	S	2
123	Erythrose benzylphenylhydrazone	104	83	S	118
124	*trans*-2-Decalonoxime benzoate	135	102 110	S	119
125	Menthyl 3,5-dinitrobenzoate	154	130	IR	120, 121

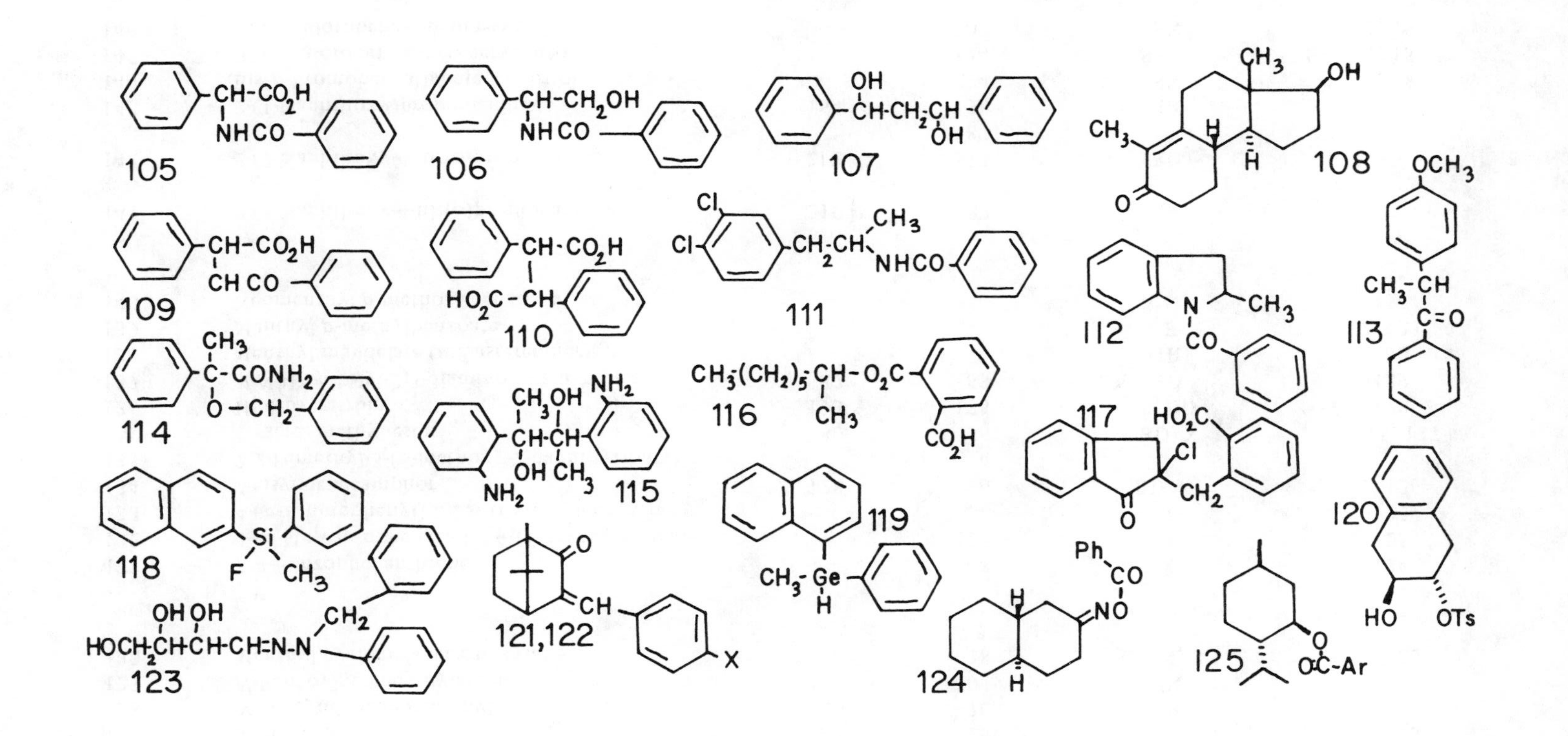

Table 4 Continued

Entry No.	Name	T_A^f (°C)	T_R^f (°C)	Recogn. Test	Refs.
126	1-Benzylamino-2-nitrosopinene	144	123	S	122
127	Menthyl benzoate	55	25	E, SN, BD	2, 123
128	*N*-Salicylideneneomenthylamine	100	70	IR	85
129	*N*-Benzoylneomenthylamine	121	102	IR	85
130	Menthyl cyclohexanecarboxylate	48	28	E, SN, BD	2, 123
	C_{18}				
131	3:4-Benzophenanthrene	–	68	X	124
132	4-(*p*-Mercaptophenyl)-2,2,4-trimethylchromane	–	135	X	125
133	4-(*p*-Aminophenyl)-2,2,4-trimethylchromane	–	–	X	125
134	Anisylidenecamphor	128	96	BD, IR	1, 126
135	2,2-Dimethyl-3-(6-methoxy-2-naphthyl)butyric acid, methyl ester	80	56	BD, X	1, 15, 127
136	Hydroveratroin	193	175	S, BD	128
137	*N,N'*-Di-*p*-tolyl-2,3-diaminobutane	87	63	BD	129
138	Menthyl mandelate (*n* diastereomer)			IR	130
139	Menthyl *p*-methylbenzoate	–	–	E	2, 123
140	Neomenthyl *p*-methoxybenzoate	–	–	E	131
	C_{19}				
141	2-(1-Naphthoxy-4-nitro)propionanilide	210} 221}	187	IR	93
142	2-(2-Naphthoxy-4-nitro)propionanilide	212	173} 187}	IR(f)	93
143	2-(1-Naphthoxy)propionanilide	164	138	IR	93
144	Bis(*p*-bromobenzal)pentaerythritol	–	224	SN	1, 15
145	Bis(*o*-chlorobenzal)pentaerythritol	–	144	SN, S	1, 15
146	Bis(*p*-chlorobenzal)pentaerythritol	–	202	SN	1, 15

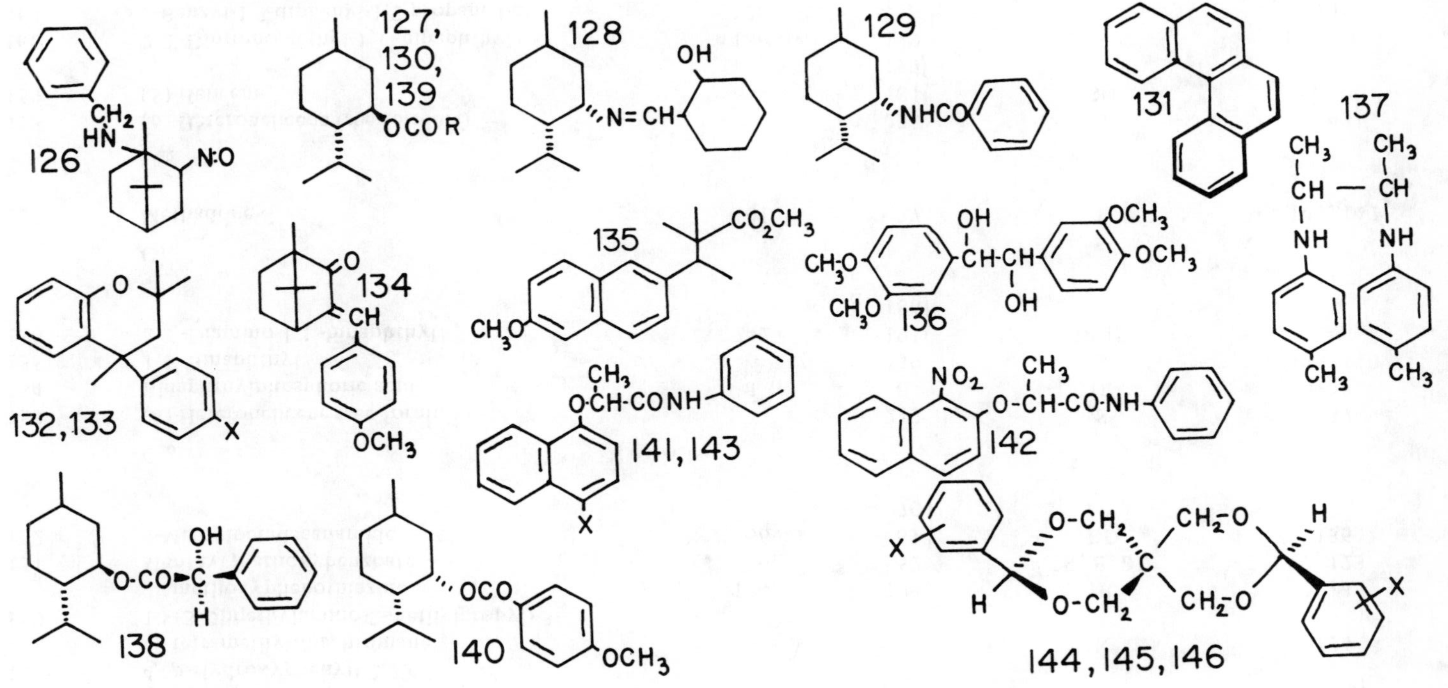
126
127, 130, 139
128
129
131
137
132,133
134
135
136
141,143
142
138
140
144,145,146

Table 4 Continued

Entry No.	Name	T_A^f (°C)	T_R^f (°C)	Recogn. Test	Refs.
147	(See formula)	135	110	IR	114
148	Dibenzalpentaerythritol	189	160	S, SN	1, 15, 132, 133
149	4-(*p*-Hydroxyphenyl)-2,2,4,7-tetramethylthiachromane	–	–	X	134
150	10-(3-Dimethylamino-2-methylpropyl)-3-methoxyphenothiazine	124	103	IR	1, 135
151	Menthyl *p*-ethoxybenzoate	76	52	IR, E, BD	2, 123
152	9-Methyloctadecanamide	66	61, 70	BD	136
	C_{20}				
153	(6)-Heterohelicene (see formula)	–	212	S	137
154	Binaphthylphosphoric acid	d	d	IR	138
155	1,1′-Binaphthyl	–	159	S	139, 140
156	2,2′-Diamino-1,1′-binaphthyl	242	191, 186	IR (f)	93
	C_{21}				
157	Methadone	101	77	E	141, 142
	C_{22}				
158	(6)-Heterohelicene (see formula)	–	225	S	143
159	(5)-Helicene	–	181, 167	SN	1, 15
160	2,2′-Dibromomethyl-1,1′-binaphthyl	186	149	S	1, 15, 144
161	2-Benzyl-1,3-diphenyl-1,2-propanediol	136	116	IR	145
162	(See formula)	72	50	IR	146

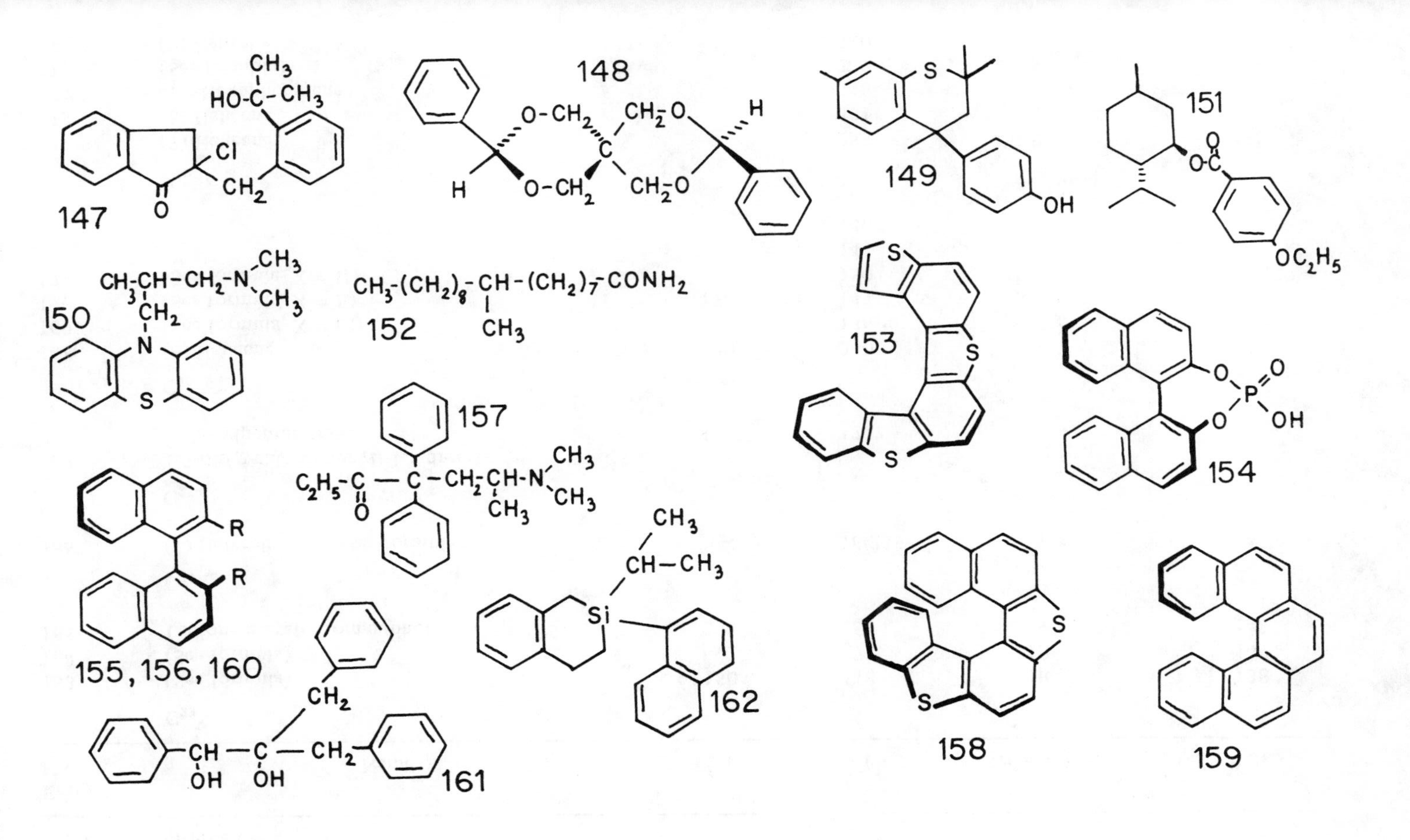
147
148
149
151
150
152
CH3-(CH2)8-CH-(CH2)7-CONH2
CH3
153
154
157
155, 156, 160
162
161
158
159

Table 4 Continued

Entry No.	Name	T_A^f (°C)	T_R^f (°C)	Recogn. Test	Refs.
	C_{23}				
163	(See formula)	150	118	IR	138
164	(See formula)	92	72	IR	147
165	Lactone acetate from euphol	290	250	S	148
	C_{24}				
166	(7)-Heterohelicene (see formula)	–	216-236	S	143
	C_{25}				
167	1,5-Bis(*p*-chlorophenyl)-1,5-di-*tert*-butylpentatetraene	159?	172?	S	1, 149
	C_{26}				
168	(6)-Helicene	270	240	S	150
169	(See formula, X = Cl)	–	136	S, SN	1, 15, 127
170	(See formula, X = F)	170	147	IR, SN	1, 15, 127
171	(See formula, X = H)	178	150 144 140	IR, SN	1, 15, 127
	C_{30}–C_{38}				
172	(7)-Helicene	–	255	S	151
173	(8)-Helicene	–	331	S	151
174	*dl-o*-Hexaphenylene	–	346	S	2, 152
175	(See formula)	–	209	S	153
176	(9)-Helicene	–	360	S	151

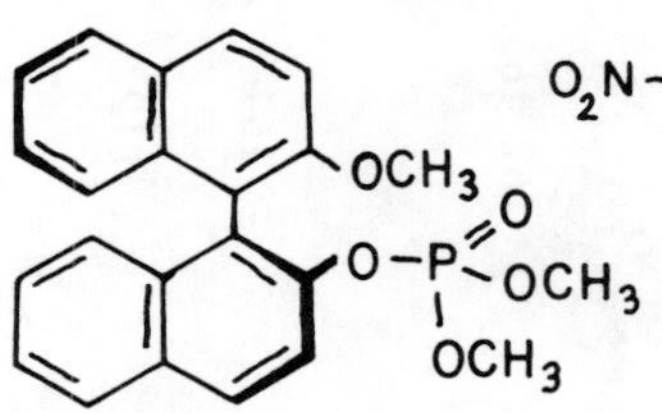

163

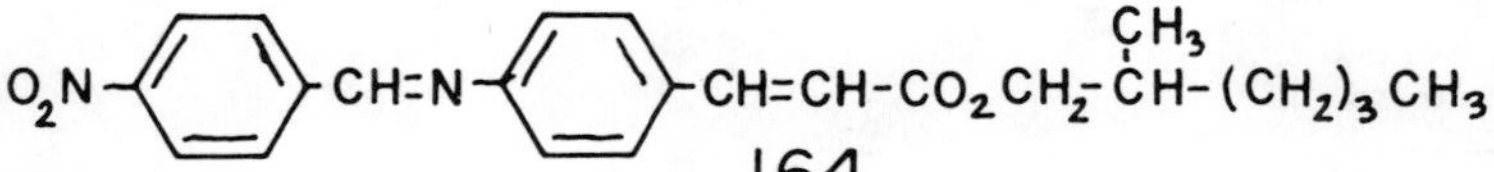

$O_2N-C_6H_4-CH{=}N-C_6H_4-CH{=}CH-CO_2CH_2-CH(CH_3)-(CH_2)_3CH_3$

164

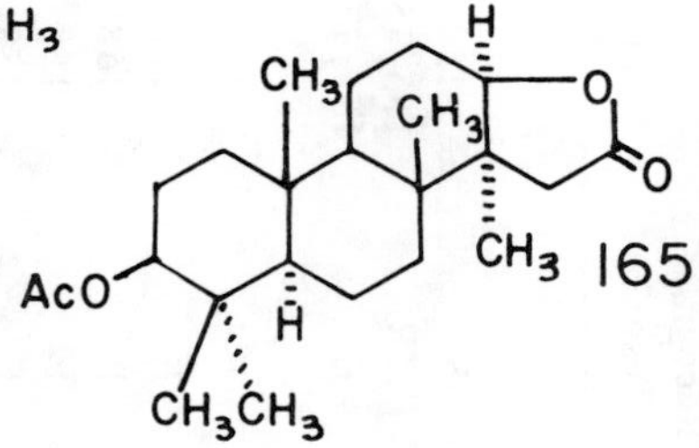

165

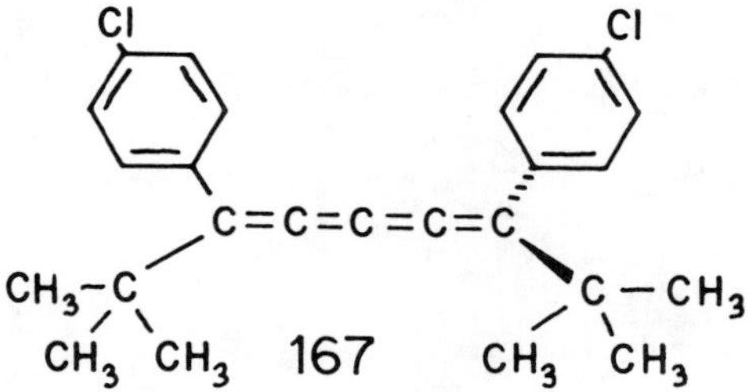

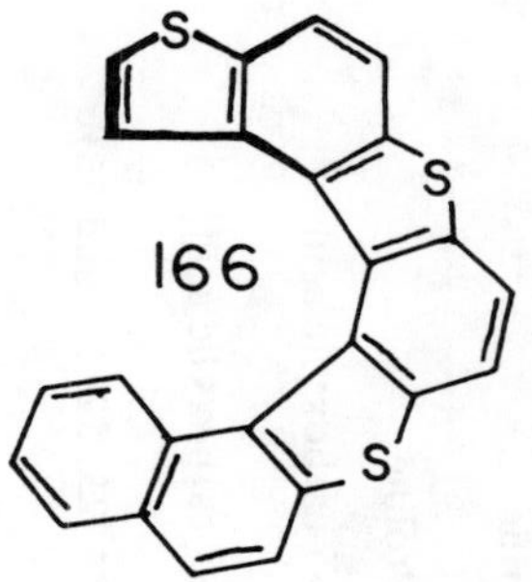

166

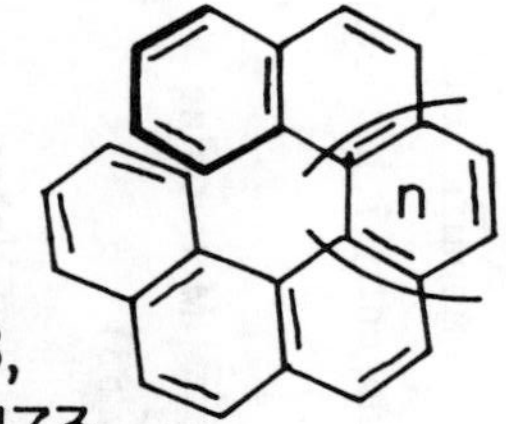

167

168, 172, 173, 176

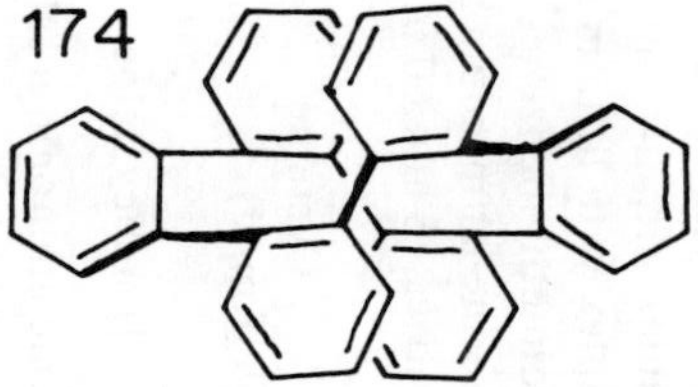

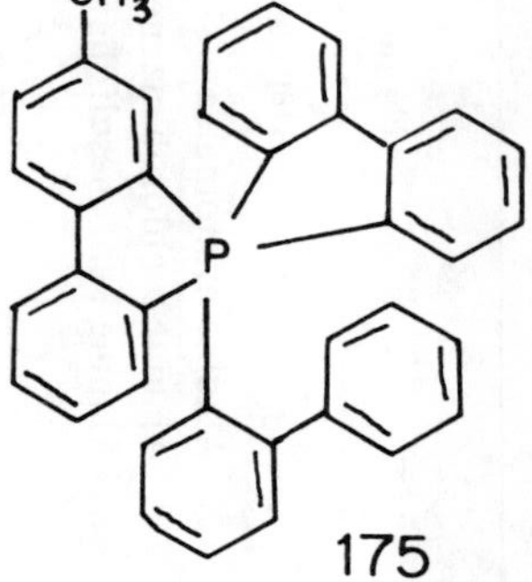

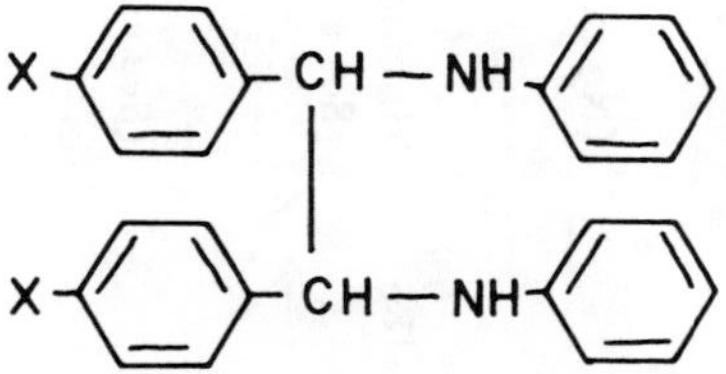

169, 170, 171

174

175

Table 5 Salts and complexes involving amino acids and their derivatives

Entry No.	Name	Recogn. Test	Refs.
	Alanine		
177	Alanine monomaleate	S	154
178	Alanine *p*-chlorobenzenesulfonate	E, IR	155, 156
179	Alanine benzenesulfonate	E	155, 157
	Serine		
180	Serine complex with lithium oxalate	E	158
181	Serine complex with lithium chloride	E	158
182	Serine complex with lithium iodide	E	158
183	Serine complex with lithium nitrate	E	158
184	Serine 2,4-dimethylbenzenesulfonate	E	159
185	N-Benzoylserine ammonium salt	E	160
	(See also entries 18 and 60 in Table 4)		
	Cysteine		
	(No example found)		
	Homocysteine		
186	Homocysteine ammonium salt	E	161
	(See also entry 9)		
	Threonine		
	(See entry 8)		
	Aspartic acid		
	(Asparagine, entry 7; homoaspartic acid, entry 13)		
	Methionine		
	(See entry 55)		
	Valine		
187	Valine hydrochloride	E	162, 163
	Proline		
188	2-Oxopyrrolidine-5-carboxylic acid, lithium salt	E	164
189	2-Oxopyrrolidine-5-carboxylic acid, zinc salt	E	1, 165
	(See also entries 24, 25, 52, 53, and 56)		

Table 5 Continued

Entry No.	Name	Recogn. Test	Refs.
	Glutamic acid		
190	Glutamic acid ammonium salt	E	1, 166
191	Glutamic acid magnesium salt	E	167
192	Glutamic acid zinc salt	E	168
193	Glutamic acid hydrochloride	E	169
	(See also entries 14 and 26)		
	Leucine		
194	Leucine benzenesulfonate	E, IR	155, 156
	(See entry 34)		
	Isoleucine		
	(See entry 35)		
	Lysine		
195	Lysine *p*-aminobenzenesulfonate	E, IR	155, 156, 170
196	Lysine 3,5-dinitrobenzoate	E	171
197	Lysine 1-chloro-4-naphthalenesulfonate	E	171
198	Lysine anthraquinone-2-sulfonate	E	171
199	α-Amino-ε-caprolactam hydrobromide	E	183
200	α-Amino-ε-caprolactam nickel chloride complex	E	184, 216
201	α-Amino-ε-caprolactam 2-amino-1-naphthalenesulfonate	E	185
202	α-Amino-ε-caprolactam 2-naphthalenesulfonate	E	185
	(See also entry 98)		
	Histidine		
203	Histidine hydrochloride	E, TD	1, 172
	Arginine		
	(No example found)		
	Phenylglycine		
204	Phenylglycine benzenesulfonate	E	173
205	Phenylglycine camphorsulfonate (*p* salt)	E	174
206	*N*-acetylphenylglycine ammonium salt	E	175
207	*p*-Hydroxyphenylglycine *p*-toluenesulfonate	E	176
208	*p*-Hydroxyphenylglycine *m*-xylenesulfonate	E	176
	(See also entries 31, 80, and 105)		

Table 5 Continued

Entry No.	Name	Recogn. Test	Refs.
	Phenylalanine		
209	Methyl phenylalaninate hydrogen sulfate	E	177
	Tyrosine		
	(No example found for *p*-tyrosine; *m*-tyrosine, see entry 47)		
	Dopa		
210	3-(3,4-Dihydroxyphenyl)alanine, 2-naphthol-6-sulfonate	E, IR	155, 156
211	*N*-Acetyl-3-(3,4-methylenedioxyphenyl)-alanine, ammonium salt	E	178, 179
212	*N*-Acetyl-3-(3,4-methylenedioxyphenyl)-alanine, di-*n*-butylammonium salt	E, IR, TD	179
	(See also entries 48, 83, and 88)		
	Methyldopa		
213	3-(3,4-Methylenedioxyphenyl)-2-methylalanine *p*-hydroxybenzene-sulfonate	E, IR	155, 156
214	*N*-Acetyl-3-(3,4-methylenedioxyphenyl)-2-methylalanine hydrazinium salt	E, IR, TD	179
	(See also entries 59, 63, and 95)		
	Tryptophan		
215	Tryptophan benzenesulfonate	E, IR	155, 156
216	Tryptophan *p*-hydroxybenzenesulfonate	E	180
217	*N*-Formyltryptophan ammonium salt	E	181
218	*N*-Acetyltryptophan ammonium salt	E	181, 182
219	*N*-Propionyltryptophan ammonium salt	E	181
	Miscellaneous		
220	3-Amino-4-hydroxy-*N*-benzylvaleric acid, lactone, hydrochloride	X	186

Table 6 Miscellaneous salts and complexes forming conglomerates

Entry No.	Name	Recogn. Test	Refs.
	Organic salts		
221	Ammonium zinc lactate	TD	1, 187, 188
222	Ammonium dimolybdomalate	S	1, 189
223	Sodium ammonium tartrate	S, TD	1, 190
224	Potassium tartrate	TD	1, 191
225	Potassium sodium tartrate	TD	1, 192
226	Rubidium tartrate	TD	1, 193
227	Ammonium hydrogen malate	TD	1, 194, 195
228	Silver 2-methylbutyrate	TD	196
229	Potassium methylvalerate	TD	197
230	Ammonium pantoate	E	198
231	Spiroheptanediamine bishydrobromide	X	1, 199
232	Spiroheptanediamine bishydrochloride	X	1, 199
233	α-Methylbenzylamine hydrogen sulfate	IR, TD	1, 200, 201
234	*N*-Methylamphetamine hydrochloride	BD	202
235	Pumiliotoxine C hydrochloride	X	203
236	2-(*p*-Nitrobenzoxy)-1-(2-pyridinium)-propane *p*-nitrobenzoate	S, E	2, 204
	Organometallic complexes		
237	Cobalt(3−) trioxalato, tripotassium	TD	1, 205
238	Rhodium(3−) trioxalato, tripotassium	TD	1, 206, 207
239	Cobalt(1+) dinitrobis(ethylenediamine) *cis*, bromide	TD, E, IR	1, 15, 208
240	Cobalt(1+) dinitrobis(ethylenediamine) *cis*, chloride	TD, E, IR	1, 15, 209
241	Cobalt(1+) tri(cyclopentyldiamine), perchlorate	TD	210
242	Cobalt(1+) oxalatobis(ethylenediamine), bromide	S, E	209, 211
243	Nickel(1+) aquobis(ethylenediamine)-tetrafluoroborato, tetrafluoroborate	X	212
244	Palladium dichloro(*O*-methyl-*N*-allylthiocarbamato)	X	213
245	Molybdenum π-cyclopentadienyl-*trans*-iododicarbonyl(trimethylphosphite)	X	214
246	Molybdenum π-(methylcyclopentadienyl)-*trans*-iododicarbonyl(trimethylphosphite)	X	214
247	Zinc bis(benzothiazoline)	X	215
	Inorganic complexes		
248	Ammonium tris(pentasulfido)platinate(IV)	S	241

(h) Comparison of the infrared spectra of enantiomers and of the corresponding racemate in the solid state (IR)

All properties of crystals of conglomerates which do not depend on chirality are identical for the enantiomers and the racemate (densities, specific heats, refractive indexes, etc.). This identity does not obtain with racemic compounds or pseudoracemates. As a consequence it is possible to devise recognition tests that take advantage of the difference. In current laboratory practice, the comparison of infrared spectra of racemates and their corresponding enantiomers in the crystal state (in Nujol suspension or as KBr pellets) constitutes an excellent test of the nature of the racemate. When the latter exists as a racemic compound, the two spectra are generally clearly different (see Section 1.2.4, Fig. 11); in the case of conglomerates, they are perfectly superposable.

When polymorphism exists (discussed in Section 2.5), the conglomerate form may manifest itself only after melting and resolidification of a sample obtained through crystallization in a solvent. Cases of this type are identified in the inventory with the symbol f.

The retrieval of new instances – or even of old cases – of spontaneous resolutions described in the literature, that is, identifying conglomerates, is a hit-or-miss proposition. Such behavior is often described only in experimental sections of articles, sometimes incidentally or imprecisely. This is true as well of relative solubility data, densities, crystal habits, and so on, of racemates and resolved substances. Thus, there is no guarantee that an inventory of well-described cases is exhaustive.

To offset somewhat the element of chance in the search associated with the inventory that follows, we have chosen to employ a method which provides a first screening and depends only on the availability of melting point data: those of optically active substances and of the corresponding racemates, being values that are commonly found in articles or tables. The enthalpy of fusion of organic compounds varies within a very wide range; for methane it is 0.2 kcal · mol^{-1}, while for tristearin the value reaches 50 kcal · mol^{-1}. However, for most ordinary organic substances, ΔH^f generally falls within the 5 to 10 kcal · mol^{-1} range. If the upper value of this range of the enthalpy of fusion is substituted in the Schröder–Van Laar equation [eq. (2), Section 2.2.3], the probable maximum melting point T_R^f of the conglomerate corresponding to an enantiomer melting at T_A^f may be calculated. It may be evident that a racemic compound is not excluded if the experimental melting point is below the calculated, but the existence of a conglomerate melting above the calculated temperature is highly improbable. Thus, from a knowledge only of the melting points of a racemate and of the enantiomers, one knows whether it is reasonable to proceed with one of the recognition tests for a conglomerate.

In practice, it is convenient to apply the screening process just described with the aid of the nomograph given in Fig. 6, which gives the $T_A^f - T_R^f$ difference as a function of enantiomer melting point for enthalpies of fusion in the range of 5 to 10 kcal · mol^{-1}. This difference in melting points lies typically in the range of 25 to 35°, but lower or higher values may occasionally be found.

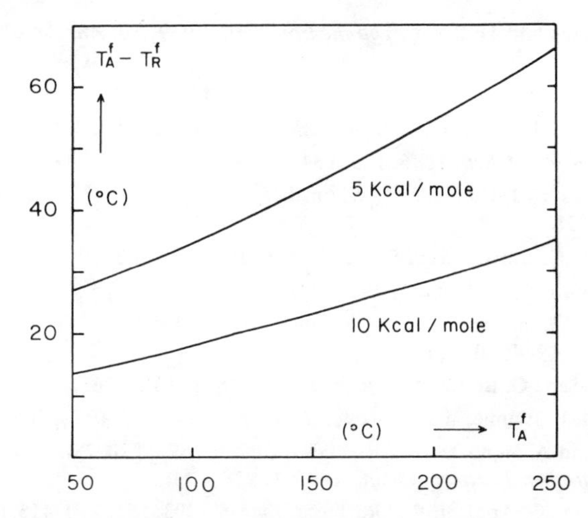

Figure 6 Nomograph giving the lowering in melting point expected for a conglomerate with respect to the melting point of the corresponding enantiomers.

To assess the frequency with which chiral compounds exist in the form of conglomerates under ordinary conditions, we have carried out a relatively simple analysis based upon 1308 chiral crystalline compounds listed in 11 volumes of the Beilstein Handbook. In this sample we found 126 compounds whose enantiomer melting point is at least 20°C greater than that of the corresponding racemate; with such a difference, the probability of having a conglomerate in hand is quite good. The validity of this finding was assessed by preparing and reexamining 32 of these 126 candidates. Of the 32 compounds, 21 are actually conglomerates. From this analysis, which is admittedly qualitative and not very precise, it would appear that for the general run of organic compounds (not including salts) the frequency of conglomerates may be estimated to be in the range 5 to 10%. This analysis is not meant to suggest, however, that it may not be possible to increase the probability observed in the arbitrary sample chosen.

Recent studies have led Jacques, Leclercq, and Brienne to conclude[243] that the frequency of conglomerate formation among salts appears to be two to three times greater than that observed among covalent compounds. This result stems from an examination of ca. 500 salts derived from racemic acids or bases combined with achiral (or racemic) basic or acidic reagents.

REFERENCES 2.2

1 A. Collet, M. J. Brienne, and J. Jacques, *Bull. Soc. Chim. Fr.*, 1972, 127.
2 A. Collet, M. J. Brienne, and J. Jacques, *Bull. Soc. Chim. Fr.*, 1977, 494.
3 F. Wudl, D. A. Lightner, and D. J. Cram, *J. Am. Chem. Soc.*, 1967, **89**, 4099.
4 J. Timmermans, P. Machtelinckx, and J. Mokry, *Bull. Soc. Chim. Belg.*, 1951, **60**, 424.
5 C. Altona, C. Knobler, and C. Romers, *Rec. Trav. Chim.*, 1963, **82**, 1089.

6 A. Lüttringhaus and D. Berrer, *Tetrahedron Lett.*, 1959, 10. *Angew. Chem.*, 1958, **70**, 439.

7 E. Höhne and K. H. Lohs, *Z. Naturforsch.*, 1969, **244**, 1071.

8 M. J. Brienne and J. Jacques, *C.R. Acad. Sci. Ser. C*, 1975, **280**, 291.

9 A. Piutti, *C.R. Acad. Sci.*, 1886, **103**, 134.

10 D. P. Shoemaker, J. Donohue, V. Shomaker, and R. B. Corey, *J. Am. Chem. Soc.*, 1950, **72**, 2328.

11 L. Velluz and G. Amiard, *Bull. Soc. Chim. Fr.*, 1953, **20**, 903.

12 G. Amiard, *Bull. Soc. Chim. Fr.*, 1956, 447.

13 M. Hara, T. Akashi, and M. Ono, Japanese Patent 1409 (1964) (to Ajinomoto Co.). *Chem. Abstr.*, 1964, **60**, 12104h.

14 L. Maquenne and G. Bertrand, *C. R. Acad. Sci.*, 1901, **132**, 1565.

15 A. Collet, M. J. Brienne, and J. Jacques, *Bull. Soc. Chim. Fr.*, 1972, 336.

16 W. Körner and A. Menozzi, *Atti R. Accad. Lincei*, 1893, [5], **2**, II, 368. *Ber. Ref.*, 1894, **27**, 121. *Jahresber. Fortschr. Chem.*, 1893, **1**, 978.

17 A. Menozzi and G. Appiani, *Atti R. Accad. Lincei*, 1893, [5], **2**, II, 415. *Chem. Zentralbl.*, 1894, I, **65**, 463.

18 E. Halmoy and O. Hassel, *J. Am. Chem. Soc.*, 1939, **61**, 1601.

19 O. Hassel and E. Wang Lund, *Acta Crystallogr.*, 1949, **2**, 309.

20 N. Takamura, S. Terashima, K. Achiwa, and S. Yamada, *Chem. Pharm. Bull.*, 1967, **15**, 1776.

21 H. Brockmann and H. Musso, *Chem. Ber.*, 1956, **89**, 241.

22 E. Fischer and R. S. Curtiss, *Ber.*, 1892, **25**, 1027.

23 P. Vieles, *Ann. Chim.* (Paris), 1935, **3**, 147. *C. R. Acad. Sci.*, 1934, **198**, 2102.

24 K. Serk-Hanssen, S. Ställberg-Stenhagen, and S. Stenhagen, *Ark. Kemi*, 1953, **5**, 203.

25 T. Raznikiewicz, *Acta Chem. Scand.*, 1962, **16**, 1097.

26 British Patent, 1,150,851 (1969), (to Tanabe Seiyaku Co.). *Chem. Abstr.*, 1969, **71**, 50523j.

27 C. Hongo, M. Shibazaki, S. Yamada, and I. Chibata, *J. Agric. Food Chem.*, 1976, **24**, 903.

28 T. Akashi, *Nippon Kagaku Zasshi*, 1962, **83**, 532. *Chem. Abstr.*, 1963, **59**, 4029e.

29 E. Fischer and L. Beensch, *Ber.*, 1896, **24**, 2927.

30 A. Collet and J. Jacques, *Bull. Soc. Chim. Fr.*, 1973, 3330.

31 J. Santhanam, British Patent, 1,210,495 (1967), (to Struthers Scientific and Int. Corp.). *Chem. Abstr.*, 1971, **74**, 42618r.

32 M. Inagaki, I. Kuniyoshi, and S. Nabeta, *Yakugaku Zasshi*, 1976, **96**, 71; *Chem. Abstr.*, 1976, **84**, 135388.

33 I. Hedlund, *Ark. Kemi*, 1955, **8**, 93.

34 M. Shibazaki, Japanese Patent 1576 (1965) (to Tanabe Seiyaku Co.). *Chem. Abstr.*, 1965, **62**, 13233g.

35 C. Fouquey and J. Jacques, *Bull. Soc. Chim. Fr.*, 1966, 165.

36 H. Scheibler, *Ber.*, 1912, **45**, 2272.

37 H. Najer, R. Giudicelli, J. Menin, and J. Loiseau, *Bull. Soc. Chim. Fr.*, 1964, 47.

38 A. Collet and J. Jacques, *Bull. Soc. Chim. Fr.*, 1972, 3857.

39 W. Hückel and F. Mössner, *Liebigs Ann. Chem.*, 1960, **637**, 57.

40 H. Wieland, W. Koschara, E. Dane, J. Renz, W. Schwarze, and W. Linde, *Liebigs Ann. Chem.*, 1939, **540**, 156.

41 C. N. Riiber, *Ber.*, 1915, **48**, 823.

42 A. Collet, *Bull. Soc. Chim. Fr.*, 1975, 215.

43 D. Varech and J. Jacques, *Tetrahedron*, 1972, **28**, 5671.

44 A. Byrkjedal, A. Mostad, and C. Rømming, *Acta Chem. Scand.*, 1974, **B28**, 760.

45 K. Vogler and H. Baumgartner, *Helv. Chim. Acta*, 1952, **35**, 1777.

46 L. Velluz, G. Amiard, and R. Joly, *Bull. Soc. Chim. Fr.*, 1953, 342.

47 T. Ishiguro, M. Yagyu, S. Takanashi, and J. Mitsui, *Nippon Yakugaku Zasshi*, 1960, **80**, 733. *Chem. Abstr.*, 1960, **54**, 24501f.

48 E. Darmois, *C. R. Acad. Sci.*, 1953, **237**, 124.

49 E. Calzavara, French Patent, 763,374 (1934) (to Manufacture de Produits Chimiques Purs). *Chem. Abstr.*, 1934, **28**, 5472.

50 C. Fouquey, L. Lacombe, J. Jacques, and G. Azadian, *Bull. Soc. Chim. Fr.*, 1976, 469.

51 G. E. Ryschkewitsch and J. M. Garrett, *J. Am. Chem. Soc.*, 1967, **89**, 4240.

52 A. Aubry, M. Marraud, J. Protas, and J. Neel, *C. R. Acad. Sci. Ser. C*, 1971, **273**, 959. M. Marraud, personal communication.

53 I. L. Karle, H. C. J. Ottenheym, and B. Witkop, *J. Am. Chem. Soc.*, 1974, **96**, 539.

54 H. Reihlen, L. Knöpfle, and W. Sapper, *Liebigs Ann. Chem.*, 1934, **534**, 247.

55 F. Nerdel and H. Liebig, *Liebigs Ann. Chem.*, 1959, **621**, 42.

56 Dutch Patent Appl. 6,514,950 (1966), (to Merck & Co.). *Chem. Abstr.*, 1966, **65**, 14557b.

57 T. Perlotto and M. Vignolo, *Farmaco* (Pavia), 1966, **21**, 30.

58 E. J. Corey, S. W. Chow, and R. A. Scherrer, *J. Am. Chem. Soc.*, 1957, **79**, 5773.

59 A. Fredga, *Bull. Soc. Chim. Fr.*, 1973, 173.

60 S. Karady, M. G. Ly, S. H. Pines, and M. Sletzinger, *J. Org. Chem.*, 1971, **36**, 1946.

61 J. Kenyon and S. M. Partridge, *J. Chem. Soc.*, 1936, 128.

62 F. S. Kipping and W. J. Pope, *J. Chem. Soc.*, 1897, **71**, 984.

63 M. J. Brienne and J. Jacques, *Tetrahedron*, 1970, **26**, 5087.

64 A. Findlay and A. N. Campbell, *J. Chem. Soc.*, 1928, 1768.

65 M. Padoa and G. Rotondi, *Atti R. Accad. Lincei*, 1912, **21**, II, 626. *Gazz. Chim. Ital.*, 1915, **45**, I, 51.

66 J. Timmermans, *Bull. Soc. Chim. Belg.*, 1930, **39**, 239.

67 L. Mion, A. Casadevall, and E. Casadevall, *Bull. Soc. Chim. Fr.*, 1968, 2950.

68 J. Lapasset and J. Falgueirettes, *C. R. Acad. Sci. Ser. C*, 1969, **268**, 1097.

69 O. Wallach, *Liebigs Ann. Chem.* 1895, **286**, 141.

70 (a) G. Jacob, G. Ourisson, and A. Rassat, *Bull. Soc. Chim. Fr.*, 1959, 1374. (b) P. Lipp, *Liebigs Ann. Chem.* 1911, **382**, 296.

71 L. M. Long, U.S. Patent, 2,767,213 (1966), (to Parke, Davis and Co.). *Chem. Abstr.*, 1957, **51**, 7414.

72 K. Vogler and M. Kofler, *Helv. Chim. Acta*, 1956, **39**, 1387.

73 M. Kuhnert-Brandstätter, K. Schleich, and K. Vogler, *Monatsh. Chem.*, 1970, **101**, 1817.

74 W. Hückel and C. Kühn, *Ber.*, 1937, **70**, 2479.

75 W. T. Cave, U.S. Patent, 2,942,004 (1960), (to Monsanto Chemical Co.). *Chem. Abstr.*, 1960, **54**, 22684d.

76 J. A. Reid and E. E. Turner, *J. Chem. Soc.*, 1951, 3219.

77 L. Westmann, *Ark. Kemi*, 1959, **14**, 115.

78 W. J. Pope and G. Clarke, Jr., *J. Chem. Soc.*, 1904, **85**, 1330.

79 K. Ohata, H. Fukumi, H. Ishiwata, and M. Yajima, Japanese Kokai 76 52,156 (1976). *Chem. Abstr.*, 1976, **85**, 159702.

80 M. Matell, *Ark. Kemi,* 1949, **1**, 455.

81 W. J. Pope and E. M. Rich, *J. Chem. Soc.,* 1899, **75**, 1093.

82 British Patent, 1,241,405 (1971), (to Ajinomoto Co.). *Chem. Abstr.,* 1971, **75**, 118 607e.

83 M. Moutin, A. Rassat, D. Bordeaux, and J. Lajzerowicz-Bonneteau, *J. Mol. Struct.,* 1976, **31**, 275.

84 J. Read and C. C. Steele, *J. Chem. Soc.,* 1927, 910.

85 J. Read, A. M. R. Cook, and M. I. Shannon, *J. Chem. Soc.,* 1926, 2223.

86 J. Read and G. J. Robertson, *J. Chem. Soc.,* 1926, 2209.

87 S. J. Cristol, F. R. Stermitz, and P. S. Ramey, *J. Am. Chem. Soc.,* 1956, 78, 4939.

88 M. T. Le Bihan and M. C. Perucaud, *Acta Crystallogr.,* 1972, **B28**, 629.

89 J. Canceill, J. Jacques, and M. C. Perucaud-Brianso, *Bull. Soc. Chim. Fr.,* 1974, 2833.

90 J. Bedin, Thesis, Paris, 1971.

91 C. G. Overberger and D. A. Labianca, *J. Org. Chem.,* 1970, **35**, 1763.

92 W. J. Considine, *J. Org. Chem.,* 1960, **25**, 671.

93 J. Jacques, C. Fouquey, J. Gabard, and W. Douglas, *C. R. Acad. Sci.,* 1967, **265**, 260.

94 D. F. Reinhold, R. A. Firestone, W. A. Gaines, J. M. Chemerda, and M. Sletzinger, *J. Org. Chem.,* 1968, **33**, 1209.

95 J. Corse and R. E. Lundin, *J. Org. Chem.,* 1970, **35**, 1904.

96 K. Kotera and Y. Sato, *Tanabe Seyaku Kenkyu Nempo,* 1959, **4**, 77. *Chem. Abstr.,* 1960, **54**, 320e.

97 E. Havinga, *Biochim. Biophys. Acta,* 1954, **13**, 171.

98 E. Touboul and G. Dana, *Bull. Soc. Chim. Fr.,* 1974, 2269.

99 E. Erlenmeyer, Jr., *Ber.,* 1897, **30**, 1531.

100 J. Read, I. G. M. Campbell, and T. V. Barker, *J. Chem. Soc.,* 1929, 2305.

101 F. Eisenlohr and L. Hill, *Ber.,* 1937, **70**, 942.

102 M. J. Brienne and J. Jacques, *Bull. Soc. Chim. Fr.,* 1974, 2647.

103 French Patent, 1,476,509 (1965), (to Roussel-UCLAF); Neth. Appl., 6,414,755; *Chem. Abstr.,* 1965, **63**, 18223f.

104 G. Bison, P. Jansen, and R. Schuebel, French Patent, 2,163,740 (1973), *Chem. Abstr.,* 1974, **80**, 71096.

105 J. Dale, *J. Chem. Soc.,* 1961, 910.

106 G. Saucy, R. Borer, and A. Fürst, *Helv. Chim. Acta,* 1971, **54**, 2034.

107 C. L. Bickel and A. T. Peaslee, Jr., *J. Am. Chem. Soc.,* 1948, **70**, 1790.

108 H. Wren and C. J. Still, *J. Chem. Soc.,* 1915, **107**, 444.

109 Dutch Patent Appl., 6,514,950 (1966), (to Merck & Co.). *Chem. Abstr.,* 1966, **65**, 14557b.

110 Mme. Bruzau, *C. R. Acad. Sci.,* 1933, **196**, 122. *Ann. Chim.,* (Paris) 1934, **1**, 321.

111 A. D. Thomsen and H. Lund, *Acta Chem. Scand.,* 1969, **23**, 3582.

112 W. A. Bonner and R. A. Grimm, *J. Org. Chem.,* 1967, **32**, 3022.

113 J. Kenyon, *J. Chem. Soc.,* 1922, **121**, 2540.

114 M. J. Luche, S. Bory, M. Dvolaitzky, R. Lett, and A. Marquet, *Bull. Soc. Chim. Fr.,* 1970, 2564.

115 H. Leuchs, *Ber.,* 1915, **48**, 1015.

116 L. H. Sommer, C. L. Frye, G. A. Parker, and K. W. Michael, *J. Am. Chem. Soc.,* 1964, **86**, 3271.

117 A. J. Jean, Thèse 3ème Cycle, Paris, 1970.

118 O. Ruff, *Ber.*, 1901, **34**, 1362.

119 W. Hückel and M. Sachs, *Liebigs Ann. Chem.*, 1932, **498**, 166.

120 H. Böhme, K. Van Emster, and M. Warmbier, *Arch. Pharm.*, 1960, **293**, 711.

121 J. B. Cohen and H. P. Armes, *J. Chem. Soc.*, 1906, **89**, 1479.

122 M. Delépine, R. Alquier, and F. Lange, *Bull. Soc. Chim. Fr.*, 1934, **1**, 1250.

123 J. Fleischer, K. Bauer, and R. Hopp, German Offen., 2,109,456 (1972), (to Haarmann & Reimer); *Chem. Abstr.*, 1972, 77, 152393h.

124 F. Herbstein and G. M. J. Schmidt, *J. Chem. Soc.*, 1954, 3302.

125 A. D. U. Hardy, D. D. MacNicol, J. J. McKendrick, and D. R. Wilson, *Tetrahedron Lett.*, 1975, 4711.

126 J. Minguin and E. G. de Bollemont, *C. R. Acad. Sci.*, 1901, **132**, 1573.

127 J. P. Penot, J. Jacques, and J. Billard, *Tetrahedron Lett.*, 1968, 4013.

128 J. Grimshaw and J. S. Ramsey, *J. Chem. Soc.*, 1966 (C), 653.

129 G. T. Morgan, W. J. Hickinbottom, and T. V. Barker, *Proc. Roy. Soc. London*, 1926, **110A**, 502.

130 M. J. Brienne, unpublished results.

131 J. Fleischer, R. Hopp, H. J. Kaminsky, K. Bauer, H. Mack, and M. Köpsel, *Int. Congr. Essent. Oils*, 1974, 142. *Chem. Abstr.*, 1976, **84**, 105791.

132 J. Böeseken and B. B. C. Felix, *Ber.*, 1928, **61**, 787.

133 H. J. Backer and H. B. J. Schurink, *Proc. K. Ned. Akad. Wet.*, 1928, **31**, 370.

134 A. D. U. Hardy, J. J. McKendrick, and D. D. MacNicol, *J. Chem. Soc. Chem. Comm.*, 1974, 972.

135 R. M. Jacob and J. G. Robert, U.S. Patent, 2,837,518 (1958), (to Rhône-Poulenc). *Chem. Abstr.*, 1958, **52**, 16382d.

136 B. Hallgren, *Ark. Kemi*, 1956, 9, 389.

137 M. B. Groen, Thesis, Groningen, 1970.

138 J. Jacques, C. Fouquey, and R. Viterbo, *Tetrahedron Lett.*, 1971, 4617.

139 R. E. Pincock and K. R. Wilson, *J. Am. Chem. Soc.*, 1971, **93**, 1291.

140 K. R. Wilson and R. E. Pincock, *J. Am. Chem. Soc.*, 1975, **97**, 1474.

141 H. E. Zaugg, *J. Am. Chem. Soc.*, 1955, **77**, 2910.

142 H. E. Zaugg, U.S. Patent, 2,983,757 (1961), (to Abbott Laboratories). *Chem. Abstr.*, 1961, **55**, 16491b.

143 H. Wynberg and M. B. Groen, *J. Am. Chem. Soc.*, 1968, **90**, 5339.

144 M. Hall and E. E. Turner, *J. Chem. Soc.*, 1955, 1242.

145 R. Roger and A. McKenzie, *Ber.*, 1929, **62**, 272.

146 J. P. Massé, Thesis, Poitiers, 1969.

147 M. Leclercq, J. Billard, and J. Jacques, *Mol. Cryst.*, 1969, **8**, 367.

148 R. E. Ireland, S. W. Baldwin, D. J. Dawson, M. I. Dawson, J. E. Dolfini, J. Newbould, W. S. Johnson, M. Brown, R. J. Crawford, P. F. Hudrlik, G. H. Rasmussen, and K. K. Schmiegel, *J. Am. Chem. Soc.*, 1970, **92**, 5743.

149 M. Nakagawa, K. Shingu, and K. Naemura, *Tetrahedron Lett.*, 1961, 802.

150 M. S. Newman, R. S. Darlak, and L. Tsai, *J. Am. Chem. Soc.*, 1967, **89**, 6191.

151 R. H. Martin, M. Flammang-Barbieux, J. P. Cosyn, and M. Gelbcke, *Tetrahedron Lett.*, 1968, 3507.

152 G. Wittig and K. D. Rümpler, *Liebigs Ann. Chem.*, 1971, **751**, 1.

153 D. Helwinkel, *Chem. Ber.*, 1966, **99**, 3642.

154 S. Asai, H. Tazuke, and H. Kageyama, British Patent, 1,345,113 (1974). *Chem. Abstr.*, 1974, **80**, 133825.

155 S. Yamada, M. Yamamoto, and I. Chibata, *Chem. Ind.* (London), 1973, 528.

156 S. Yamada, M. Yamamoto, and I. Chibata, *J. Org. Chem.*, 1973, **38**, 4408.

157 I. Chibata, S. Yamada, M. Yamamoto, and M. Wada, *Experientia*, 1968, **24**, 638.

158 T. Nakamura and Y. Murayama, Japanese Patent 71 35,248 (1971) (to Ajinomoto Co.); *Chem. Abstr.*, 1972, **76**, 14923.

159 I. Chibata, S. Yamada, and M. Yamamoto, German Offen., 1,950,018 (1970), (to Tanabe Seiyaku Co.); *Chem. Abstr.*, 1970, **73**, 15251p.

160 Dutch Patent Appl., 6,500,316 (1965), (to Tanabe Seiyaku Co.); *Chem. Abstr.*, 1966, **64**, 2159a.

161 M. Hara, T. Akashi, and M. Ono, Japanese Patent 1409 (1964) (to Ajinomoto Co.); *Chem. Abstr.*, 1964, **60**, 12104h.

162 S. Tatsumi, S. Haruichiro, and H. Ono, Japanese Patent 18,470 (1962) (to Ajinomoto Co.); *Chem. Abstr.*, 1963, **59**, 11659c.

163 S. Tatsumi, I. Sasaji, and H. Ono, British Patent, 969,128 (1964), (to Ajinomoto Co.); *Chem. Abstr.*, 1965, **62**, 13232g.

164 M. Iida, H. Tazuke, and H. Kageyama, German Offen., 2,211,252 (1972), (to Ajinomoto Co.); *Chem. Abstr.*, 1972, **77**, 164456.

165 M. Dazai, N. Mizoguchi, and K. Ito, Japanese Patent 68 27,859 (1968) (to Ajinomoto Co.); *Chem. Abstr.*, 1969, **70**, 57639q.

166 See, *inter alia*, British Patent, 833,823 (1960), (to International Minerals and Chemical Corp.). *Chem. Abstr.*, 1961, **55**, 407h.

167 Y. Kawamura, K. Takenouchi, Y. Sakata, and T. Houkawa, Japanese Patent, 1963, 5266 (to Ajinomoto Co.). *Chem. Abstr.*, 1963, **59**, 11660c.

168 T. Ogawa and I. Komori, Japanese Patent 9022 (1956) (to Ajinomoto Co.). *Chem. Abstr.*, 1958, **52**, 11905d.

169 F. Kögl, H. Erxleben, and G. J. Vanveersen, *Z. Phys. Chem.*, 1943, **277**, 260.

170 W. K. Van der Linden and G. H. Suverkropp, German Offen., 1,949,585 (1970), (to Stamicarbon N.V.). *Chem. Abstr.*, 1970, **73**, 15252q.

171 N. Sato, T. Uzuki, K. Toi, and T. Akashi, *Agric. Biol. Chem.* (Tokyo), 1969, **33**, 1107. *Chem. Abstr.*, 1969, **71**, 124865w.

172 R. Duschinsky, *Chem. Ind.* (London), 1934, 10. *Festschrift Emil Barell,* F. Reinhardt Verlag, Basel, 1936, pp. 375–393.

173 French Demande, 2,226,376 (1974), (to Asahi Chemical Industry Co.). *Chem. Abstr.*, 1975, **83**, 10852g.

174 T. Watanabe, S. Hayashi, S. Ouchi, and S. Senoo, Japanese Kokai 73 78,137 (1973). *Chem. Abstr.*, 1974, **80**, 71099.

175 I. Chibata, S. Yamada and M. Yamamoto, German Offen., 2,014,874 (to Tanabe Seiyaku Co.). *Chem. Abstr.*, 1970, **73**, 130785c.

176 T. Shirai, Y. Tashiro, and S. Aoki, German Offen., 2,501,957 (1975), (to Nippon Kayaku Co.). *Chem. Abstr.*, 1975, **83**, 179614s.

177 T. Uzuki, M. Yuda, and K. Toi, Japanese Kokai 73 75,540 (1973). *Chem. Abstr.*, 1974, **80**, 71102.

178 I. Senhata, S. Yamada, and M. Yamamoto, Japanese Patent 75 22,547 (1975) (to Tanabe Seiyaku Co.). *Chem. Abstr.*, 1976, **84**, 31497f.

179 S. Yamada, M. Yamamoto, and I. Chibata, *J. Org. Chem.*, 1975, **40**, 3360.

180 I. Chibata, S. Yamada, M. Yamamoto and S. Sanematsu, German Patent, 2,348,616 (1974). *Chem. Abstr.*, 1974, **81**, 4268.

181 French Patent, 1,302,248 (1962), (to Ajinomoto Co.). *Chem. Abstr.*, 1963, **58**, 12672g.

182 I. Sasaharu, H. Ono, and J. Kato, Japanese Patent 6183 (1963) (to Ajinomoto Co.). *Chem. Abstr.*, 1963, **59**, 11661g.

183 T. Takeshita, Y. Ono, and H. Watase, Japanese Kokai 74 41,388 (1974). *Chem. Abstr.*, 1974, **81**, 151568.

184 A. M. Kubanek, S. Sifniades, and R. Fuhrmann, U.S. Patent, 3,824,231 (1974). *Chem. Abstr.*, 1974, **81**, 135484.

185 British Patent, 1,192,097 (1970), (to Sanyo Chem. Ind.). *Chem. Abstr.*, **73**, 44932s.

186 P. T. Cheng, C. H. Koo, I. P. Mellor, S. C. Nyburg, and J. M. Young, *Acta Crystallogr.*, 1970, **B26**, 1339.

187 T. Purdie, *J. Chem. Soc.*, 1893, **63**, 1143.

188 T. Purdie and J. W. Walker, *J. Chem. Soc.*, 1895, **67**, 616.

189 E. Darmois and J. Perin, *C. R. Acad. Sci.*, 1923, **176**, 391. *Bull. Soc. Chim. Fr.*, 1924, **35**, 353.

190 J. H. Van't Hoff, H. Goldschmidt, and W. P. Tomssen, *Z. Phys. Chem.*, 1895, **17**, 49.

191 J. H. Van't Hoff and W. Muller, *Ber.*, 1899, **32**, 857.

192 J. H. Van't Hoff and H. Goldschmidt, *Z. Phys. Chem.*, 1895, **17**, 505.

193 J. H. Van't Hoff and W. Muller, *Ber.*, 1898, **31**, 2206.

194 F. B. Kenrick, *Ber.*, 1897, **30**, 1749.

195 J. H. Van't Hoff and H. M. Dawson, *Ber.*, 1898, **31**, 528.

196 W. Marckwald, *Ber.*, 1899, **32**, 1089.

197 C. Neuberg, *Biokhimiya*, 1937, **2**, 383. *Chem. Abstr.*, 1937, **31**, 5323.

198 French Patent, 1,522,111 (1968), (to Fuji Chem. Ind. Co.). *Chem. Abstr.*, 1969, **71**, 12566t.

199 S. E. Janson and W. J. Pope, *Proc. Roy. Soc. London*, 1936, **A154**, 53.

200 A. Ault, *Org. Synth.* 1969, **49**, 93.

201 W. Markwald and R. Meth, *Ber.*, 1905, **38**, 801.

202 M. Kuhnert-Brandstätter, W. Heindl, and R. Linder, *Mikrochim. Acta*, 1976, **1**, 363.

203 T. Ibuka, Y. Inubushi, I. Saji, K. Tanaka, and M. Masaki, *Tetrahedron Lett.*, 1975, 323.

204 D. Butruille, G. Fodor, and G. A. Cooke, 50ème Congrès de l'Institut de Chimie du Canada, Toronto, June 1967.

205 F. M. Jaeger and W. Thomas, *Proc. K. Ned. Akad. Wet.* 1919, **21**, 693. *Rec. Trav. Chim.*, 1919, **38**, 250.

206 F. M. Jaeger, *Proc. K. Ned. Akad. Wet.*, 1917, **20**, 264.

207 A. Werner, *Ber.*, 1914, **47**, 1954.

208 F. P. Dwyer and F. L. Garvan, *Inorg. Synth.*, **6**, 195.

209 A. Werner, *Ber.*, 1914, **47**, 2171.

210 J. M. Jaeger and H. B. Blumendal, *Z. Anorg. Allg. Chem.*, 1928, **175**, 211.

211 D. G. Brewer and K. T. Kan, *Can. J. Chem.*, 1971, **49**, 965.

212 A. A. G. Tomlinson, M. Bonamica, G. Dessy, V. Fares, and L. Scaramuzza, *J. Chem. Soc., Dalton*, 1972, 1671. A. A. G. Tomlinson, private communication.

213 P. Porta, *J. Chem. Soc.*, 1971 (A), 1217.

214 A. D. U. Hardy and G. A. Sim, *J. Chem. Soc., Dalton*, 1972, 1900.

215 L. F. Lindoy, D. M. Busch, and V. Goedken, *J. Chem. Soc. Chem. Comm.*, 1972, 683.

216 S. Sifniades, W. J. Boyle, Jr., and J. F. Van Peppen, *J. Am. Chem. Soc.*, 1976, **98**, 3738.

217 I. Prigogine and R. Defay, *Thermodynamique Chimique*, Desoer, Liège, Belgium, 1950. English translation, *Chemical Thermodynamics*, 4th ed. Longmans, London, 1967.

218 J. E. Ricci, *The Phase Rule and Heteregeneous Equilibrium*, D. Van Nostrand, New York, 1951.

219 A. Findlay, *Phase Rule*, 9th ed., revised by A. N. Campbell and N. O. Smith, Dover, New York, 1951.

220 I. Schröder, *Z. Phys. Chem.*, 1893, **11**, 449.

221 J. J. Van Laar, *Arch. Neerl.* 1903, II, 8, 264.

222 H. Le Chatelier, *C. R. Acad. Sci.*, 1894, **118**, 638.

223 S. Takagi, R. Fujiskiro, and R. Amaya, *Chem. Comm.*, 1968, 480.

224 J. P. Guetté, D. Boucherot, and A. Horeau, *Tetrahedron Lett.*, 1973, 465.

225 M. Matell, *Ark. Kemi,* 1949, **1**, 455.

226 F. Ebel, "Die Isomerieerscheinungen," in *Stereochemie*, Vol. 1, K. Freudenberg, Ed., F. Deuticke, Leipzig, 1933, p. 564ff.

227 M. Delépine, in *Traité de Chimie Organique*, Vol. 1, V. Grignard, Ed., Masson, Paris, 1935, p. 935ff.

228 J. I. Klabunowski, *Asymmetrische Synthese*, WEB Deutscher Verlag der Wissenschaften, Berlin, 1963, p. 175ff.

229 R. M. Secor, *Chem. Rev.*, 1963, **63**, 297.

230 K. Amaya, *Bull. Chem. Soc. Jpn*, 1961, **34**, 1689.

231 C. Mavroyannis and M. J. Stephen, *Mol. Phys.*, 1962, 5, 629.

232 D. P. Craig, E. A. Power, and T. Thirunamachandran, *Proc. Roy. Soc. London,* 1971, **A322**, 165.

233 D. P. Craig and P. E. Shipper, *Proc. Roy. Soc. London*, 1975, **A342**, 19.

234 D. P. Craig and D. P. Mellor, *Top. Curr. Chem.*, 1976, **63**, 1.

235 S. F. Mason, *Ann. Rep. Progr. Chem. Sect. A,* 1976, 53.

236 R. Kuroda, S. F. Mason, C. D. Rodger, and R. H. Seal, *Chem. Phys. Lett.* 1978, **57**, 1.

237 A. Horeau and J. P. Guetté, *Tetrahedron*, 1974, **30**, 1923.

238 M. J. Brienne, A. Collet, C. Fouquey, J. Gabard, J. Jacques and M. Leclercq, unpublished results.

239 M. Leclercq, A. Collet, and J. Jacques, *Tetrahedron*, 1976, **32**, 821.

240 H. Jensen, German Offen. 1,807,495 (1970) (to Farberke Hoechst A.-G.). *Chem. Abstr.*, 1970, **73**, 77222p.

241 R. D. Gillard and F. L. Wimmer, *J. Chem. Soc. Chem. Comm.*, 1978, 936.

242 For a crystallographic description of twinning of enantiomorphous crystals, see M. van Meerssche and J. Feneau-Dupont, *Introduction à la Cristallographie et à la Chimie Structurale*, Vander, Paris, 1973, pp. 405–417.

243 J. Jacques, M. Leclercq, and M. J. Brienne, *Tetrahedron*, 1981, in press.

2.3 RACEMIC COMPOUNDS

The second type of racemate identified in Chapter 1 is racemic compound, which is characterized by a crystal form in which the two enantiomers coexist in the same unit cell. Mixtures of enantiomers forming racemic compounds yield phase diagrams which were predicted by Roozeboom[1] just before the turn of the century.

2.3.1 Phase Diagrams

A relatively large number of phase diagrams of enantiomer mixtures exhibiting racemic compound formation have been experimentally determined and published.

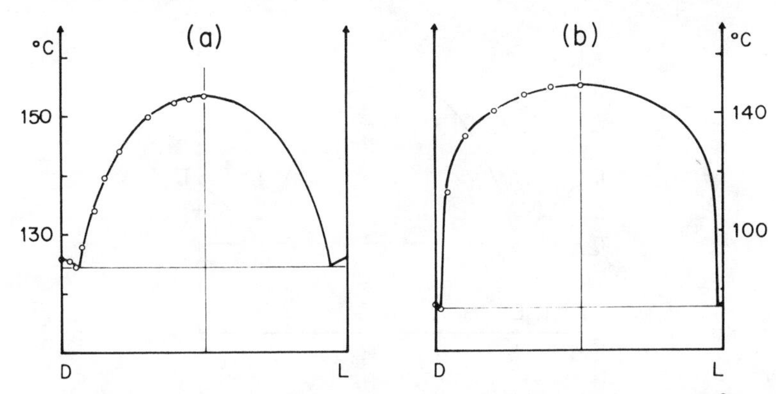

Figure 1 Melting point diagrams of α-(1-naphthoxy)propionic acid[2] (a) and α-(1-naphthyl)propionic acid[3] (b). Experimental points are indicated by circles. The liquidus curves shown were calculated by means of the Prigogine–Defay equation (Section 2.3.2).

In particular, we are indebted for this to A. Fredga and his associates at the University of Uppsala. The shape of these diagrams can vary within rather large limits depending upon whether the racemic compound melting point is greater, lower, or conceivably equal to that of the enantiomers.

When the racemic compound melts at a much higher temperature than the enantiomers do, the eutectic sometimes cannot be discerned on the phase diagram when it is very close to the edges of the diagram (Fig. 1). At the other extreme it is the very existence of the racemic compound that may be difficult to detect (Fig. 2). This is particularly true when its melting point is lower than that of the conglomerate, which may be calculated by means of the equation of Schröder–Van Laar (see Section 2.2.3 and Fig. 2*b*).

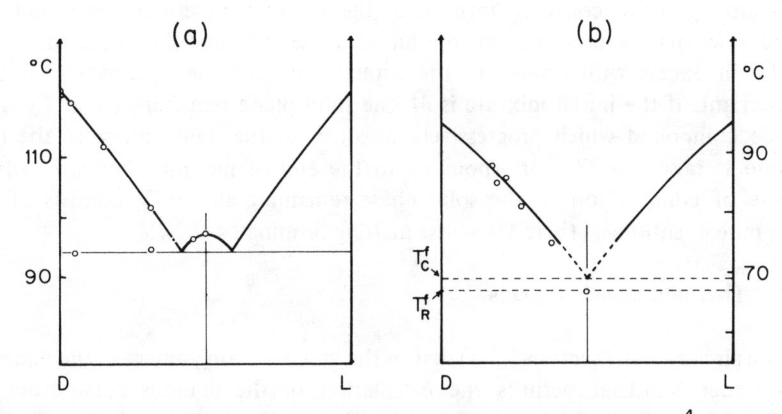

Figure 2 Melting point diagrams of (a) *m*-fluoromandelic acid,[4] and (b) *m*-chlorophenylhydracrylic acid.[5] Experimental points are indicated by circles. Liquidus curves are calculated by means of Schröder–Van Laar and Prigogine–Defay equations.[19] In (b), T_R^f is the melting point of the racemic compound and T_c^f is the calculated melting point for a conglomerate.

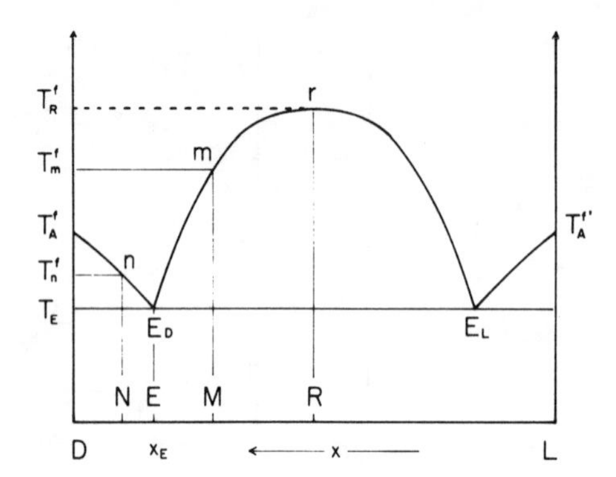

Figure 3 Common melting point diagram of enantiomers forming a racemic compound.

Aside from exceptional cases, diagrams having shapes as different as those of Figs. 1 and 2*a* as well as intermediate cases essentially have identical properties which can be analyzed by means of a diagram such as Fig. 3.

In this figure, T_R^f represents the racemic compound melting point while T_A^f (or $T_A^{f'}$) is that of the pure optically active substance. The eutectic E_D consisting of x_L mole of D and $1 - x_L$ mole of L, is actually a mixture of $2x_E - 1$ mole of *crystalline* D and of $2(1 - x_E)$ mole of *racemic compound.*

This mixture of melting point T_E melts as if it were a pure compound. Let us consider, in a general way, a solid mixture (i.e., at T lower than T_E) which consists of $2x - 1$ mole of enantiomer crystals and of $2(1 - x)$ mole of racemic compound crystals.* This system is monovariant and its temperature may be raised without a change in state, at least up to T_E. At this temperature, the mixture begins to melt and the system becomes invariant in view of the coexistence of three phases; the temperature remains constant until all of the eutectic present in the sample has melted. Two phases then remain: the liquid phase and the solid phase which was initially in excess with respect to the eutectic composition. The system is then monovariant. If the initial mixture is M, the solid phase remaining above T_E is the racemic compound which progressively dissolves in the liquid phase as the temperature is raised to T_m^f corresponding to the end of melting. Similarly, with a mixture of composition N, the solid phase remaining above T_E consists of one of the pure enantiomers (here D) whose melting terminates at T_n^f.

2.3.2 Thermodynamic Aspects

We have already seen (Section 2.2.4) that in the case of a conglomerate, the equation of Schröder–Van Laar permits the calculation of the liquidus curve from the melting point and the enthalpy of fusion of a pure enantiomer. In the case of a racemic compound, the same equation may be applied to the calculation of that

* We do not repeat here in any detail the type of phase diagram analysis we developed for the conglomerate case (Section 2.2).

part of the liquidus found between the pure enantiomers and the corresponding eutectic, that is, curve $T_A^f E_D$ (and $T_A^{f'} E_L$) in Fig. 3 (the enantiomer branch).

The $E_D r E_L$ part of the curve (below which the solid phase consists of pure racemic compound) may be calculated by means of eq (1):

$$\ln 4x(1-x) = \frac{2\Delta H_R^f}{R}\left(\frac{1}{T_R^f} - \frac{1}{T^f}\right) \tag{1}$$

in which x represents the mole fraction of one of the enantiomers in the mixture whose melting point (end of fusion) is T^f (in degrees K); T_R^f (also in degrees K) and ΔH_R^f ($\text{cal}\cdot\text{mol}^{-1}$) are, respectively, the melting point and the enthalpy of fusion of the racemic compound; and R ($1.9869\ \text{cal}\cdot\text{mol}^{-1}\cdot\text{K}^{-1}$) is the gas constant. This equation, which does not incorporate specific heats of the constituent solids and liquids, is formally comparable to the simplified Schröder–Van Laar equation.

Equation (1), which is applicable to all binary systems in which a crystalline addition compound is formed containing *one* molecule of each constituent, is due to Prigogine and Defay[6] (see also refs. 7 and 8). The conditions under which it is valid are the same as those which obtain for the equation of Schröder–Van Laar (Section 2.2.3). Various tests are available to ensure that these conditions are well satisfied, for example, (a) the enthalpy of fusion of the racemic compound may be calculated from the experimental phase diagram; and (b) the liquidus curve may be calculated from the heat of fusion and melting point of the racemic compound, both of which are accessible from microcalorimetry.

(a) Enthalpy of fusion from experimental phase diagram

Prigogine and Defay, and later Mauser, tested eq. (1) by calculating the enthalpies of fusion of racemic compounds from melting point phase diagrams found in the literature.[6,7] If the experimental values of $1/T^f$ are plotted against $\ln x(1-x)$, a straight line is obtained whose slope gives the enthalpy of fusion ΔH_R^f. An example of such a plot derived from the melting point diagram shown in Fig. 1*a* is given in Fig. 4. The experimental data of Fredga and Matell[2] are tabulated in Table 1. A variant of the process described above consists in directly calculating the value of ΔH_R^f corresponding to each experimental point. The values of ΔH_R^f so calculated for the example of Fig. 1*a* are also given in Table 1. The mean value of ΔH_R^f thus found ($8.1 \pm 0.4\ \text{kcal}\cdot\text{mol}^{-1}$) is in very good agreement with that derived from the plot described by Fig. 4 ($8.0\ \text{kcal}\cdot\text{mol}^{-1}$).*

* In its original and most general form, the equation of Prigogine–Defay may be written as follows:

$$\ln 4x(1-x) = \frac{\Delta H_{AB}^f}{R}\left(\frac{1}{T_{AB}^f} - \frac{1}{T^f}\right)$$

In this equation, ΔH_{AB}^f corresponds to the melting of "1 mole" of addition compound AB, containing 1 mole of each constituent. Since the enthalpy of fusion of a racemic compound is defined on the basis of $\frac{1}{2}$ mole of (+) and $\frac{1}{2}$ mole of (–) enantiomer, $\Delta H_{AB}^f = 2\Delta H_R^f$, whence the coefficient 2 of eq. (1).

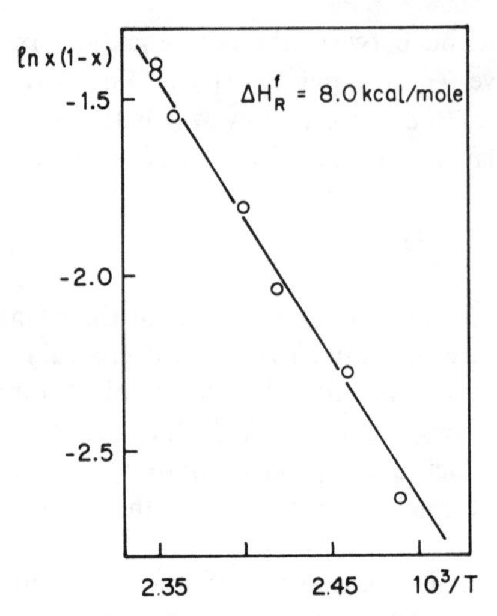

Figure 4 Test of the Prigogine–Defay equation. Graph of $\ln x(1-x)$ vs $1/T^f$ for α-(1-naphthoxy)propionic acid (compare Table 1 with Fig. 1*a*).

(b) Racemic compound phase diagram from its enthalpy of fusion

Equation (1) is of particular interest in making possible the *a priori* determination of the phase diagram of mixtures of enantiomers which form a racemic compound from a knowledge of the melting point and enthalpy of fusion of the latter. Since these values are easily accessible through microcalorimetry it is a simple matter to trace the phase diagram. For verification of the calculated diagram, determination of the melting points (showing beginning and termination of fusion) of just a few

Table 1 Test of the Prigogine–Defay eq. (1) for α-(1-naphthoxy)propionic acid

Experimental Points[a]		Calculations			
x	T^f (°C)	ΔH^f_R (kcal · mol^{-1})	T^f (°C)[b]	$\ln x(1-x)$	$1000/T^f$ ($10^3 \cdot$K^{-1})
0.500	153.5				
0.550	153	3.63	153.3	−1.40	2.35
0.602	152.5	7.67	152.5	−1.43	2.35
0.697	150	8.64	149.8	−1.55	2.36
0.794	144	7.90	144.2	−1.81	2.40
0.846	139.5	8.14	139.4	−2.04	2.42
0.885	134	7.95	134.3	−2.28	2.46
0.923	128	8.39	127.1	−2.64	2.49
		⟨8.1⟩			

[a] Experimental points are those of Fredga and Matell.[2]
[b] T^f calculated by means of eq. (1) with a mean value of $\Delta H^f_R = 8.1$ kcal · mol^{-1}

enantiomer mixtures of known composition should suffice to show that these coincide with the calculated liquidus curve.

2.3.3 Comparison of the Enthalpies of Fusion and Specific Heats of Racemic Compounds with Those of the Crystalline Enantiomers

We have already seen that the enthalpies of fusion of racemic compounds and those of their enantiomer constituents may be determined from the phase diagram of their mixtures. However, it is much more convenient to determine these data directly by microcalorimetry. Numerous determinations of this type have been carried out. Table 2 brings together a consistent set of data determined[9] by a single group of experimenters at the Collège de France (Paris). The assembled data do not reveal the fact that the phenomenon of polymorphism complicates the measurements in some cases (see Section 2.5). Note that, with few exceptions, the enthalpies of fusion range between 5 and 10 kcal · mol^{-1}.

Comparative heat capacity data for enantiomers and their racemic compound are rare and of recent origin.[9] In the solid state, the specific heats C_A^s and C_R^s differ significantly and vary appreciably with temperature. On the other hand, there is no difference (± 0.5%) between the heat capacities of the two species in the liquid state, a result which is consistent with the hypothesis of ideality of liquid enantiomer mixtures (Section 2.2.4). These data are illustrated in Figs. 5 and 6; they were obtained with the Perkin–Elmer DSC 2 microcalorimeter.

2.3.4 The Problem of Racemic Compound Stability

The concept of racemic compound stability and the analysis of the various factors that determine it have been discussed by several authors. An excellent review on

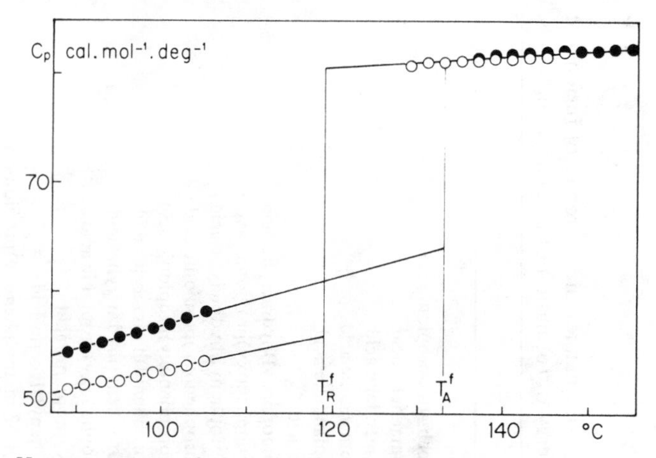

Figure 5 Heat capacities of solid and liquid mandelic acid between 90 and 155°C. Open circles for the racemate; solid circles for the enantiomer. M. Leclercq, A. Collet, and J. Jacques, *Tetrahedron*, 1976, **32**, 821. Reproduced by permission of Pergamon Press Ltd.

Table 2 Enthalpies and entropies of fusion and melting points of optically active compounds and the corresponding racemic compounds.[a] Free energy of formation of racemic compounds.[b,c]

Name	T_R^f	ΔH_R^f	ΔS_R^f	T_A^f	ΔH_A^f	ΔS_A^f	ΔG^ϕ
Malic acid	402	6.6	16.4	376	5.5	14.6	–0.95
Methylenebisthiopropionic acid	429	9.4	21.9	355	5.0	14.1	–2.1
p-Chloromandelic acid	394	6.5	16.5	394	5.5	14.0	–0.5
o-Fluoromandelic acid	390	7.2	18.5	363	5.0	13.8	–1.0
m-Fluoromandelic acid	370	5.9	15.9	394	5.8	14.7	–0.15
p-Fluoromandelic acid	403	7.0	17.4	426	7.3	17.1	–0.2
Mandelic acid	392	6.1	15.6	406	6.3	15.5	–0.3
3-(*m*-Bromophenyl)hydracrylic acid	349	6.4	18.3	350	5.7	16.3	–0.5
2-(*p*-Bromophenoxy)propionic acid	385	7.6	19.7	380	6.6	17.4	–0.6
3-(*m*-Chlorophenyl)hydracrylic acid	340	5.7	16.8	368	6.7	18.2	0
2-(*o*-Chlorophenoxy)propionic acid	388	7.7	19.8	369	6.4	17.3	–0.9
2-(*m*-Chlorophenoxy)propionic acid	386	7.9	20.5	367.5	7.1	19.3	–0.9
3-(*o*-Fluorophenyl)hydracrylic acid	342	6.5	19.0	348	5.4	15.5	–0.4
3-(*p*-Fluorophenyl)hydracrylic acid	362	6.6	18.2	381	7.4	19.7	–0.1
2-(*p*-Nitrophenoxy)propionic acid	411.5	7.7	18.7	362	5.0	13.8	–1.4
2-Phenoxypropionic acid	388	7.9	20.4	359	5.4	15.0	–1.1
erythro-Phenylglyceric acid	395	7.5	19.0	371.5	5.6	15.0	–0.95
2-(2-Chloro-3-methylphenoxy)propionic acid	391.5	7.3	18.6	359.5	5.3	14.7	–1.1
3-Hydroxy-3-phenylbutyric acid	330	4.7	14.2	357	5.4	15.1	0
Dimethyl *O,O'*-diacetyltartrate	357.5	6.6	18.5	377.5	6.5	17.2	–0.1
3-Hydroxy-3-phenylvaleric acid	394	8.4	21.3	379	7.4	19.5	–0.8
1,2-Dibromoacenaphthene	397	6.0	15.1	416	6.3	15.1	–0.3

Table 2 Continued

Name	T_R^f	ΔH_R^f	ΔS_R^f	T_A^f	ΔH_A^f	ΔS_A^f	ΔG^ϕ
2-(1-Nitro-2-naphthoxy)propionamide	431	7.0	16.2	461.5	7.3	15.8	−0.1
2-(1-Naphthyl)propionic acid	422.5	7.3	17.3	342	3.4	9.9	−1.7
1,5-Dichloro-9,10-dihydro-9,10-ethanoanthracene	424	6.6	15.6	353.5	3.0	8.5	−1.6
N,N'-Bis-α-methylbenzylthiourea	410.5	7.9	18.0	471.5	8.6	18.1	0
Benzylidenecamphor	350.5	5.5	15.5	371	5.6	15.2	−0.2
1,5-Dichloro-11,12-di(hydroxymethyl)-9,10-dihydro-9,10-ethanoanthracene *trans* (exo)	519.5	12.8	25.1	527	12.8	24.3	−0.5
1,5-Dichloro-11,12-di(hydroxymethyl)-9,10-dihydro-9,10-ethanoanthracene *trans* (endo)	441	7.9	17.9	435	4.7	10.8	−0.7
11,12-Di(iodomethyl)-9,10-dihydro-9,10-ethanoanthracene	468.5	8.0	17.1	491	8.4	17.1	−0.3
11,12-Di(hydroxymethyl)-9,10-dihydro-9,10-ethanoanthracene	474.5	9.85	20.7	405.5	5.7	14.1	−2.0
1,5-Dichloro-9,10-dihydro-9,10-ethano-11,12-dicarbomethoxyanthracene (exo)	465	8.8	18.9	424	5.55	13.0	−1.4
1,5-Dichloro-9,10-dihydro-9,10-ethano-11,12-dicarbomethoxyanthracene (endo)	436	6.3	14.4	427	5.9	13.8	−0.7
Dimethyl *O,O'*-dibenzoyltartrate	422.5	11.7	27.7	409	11.0	26.9	−0.9
9,10-Dimethyl-9,10-dihydro-9,10-ethano-11,12-dicarbomethoxyanthracene *trans*	465	9.8	21.1	393	4.5	11.3	−2.1
2,3,3-Triphenylvaleric acid	480	8.9	18.5	441.5	6.4	14.5	−1.3

[a] Enthalpies in kcal · mol^{-1}; entropies in cal · mol^{-1} · K^{-1}; temperatures in K.
[b] The free energy of formation of racemic compounds was calculated for each example at whichever of the two temperatures is the lower – that of the racemic compound melting point or that of the enantiomer melting point.
[c] M. Leclercq, A. Collet, and J. Jacques, *Tetrahedron*, 1976, **32**, 821. Reproduced by permission of Pergamon Press Ltd.

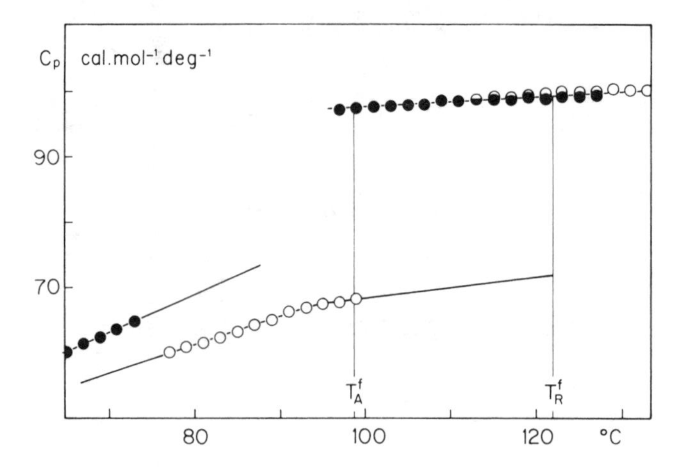

Figure 6 Heat capacities of *erythro*-phenylglyceric acid between 65 and 130°C. Open circles for the racemate; solid circles for the enantiomer. M. Leclercq, A. Collet, and J. Jacques, *Tetrahedron*, 1976, **32**, 821. Reproduced by permission of Pergamon Press Ltd.

this topic has been given in Petterson,[10] who has also proposed a definition of the stability of a racemic compound from a consideration of phase diagrams of enantiomer mixtures. This can be done with reference to a variety of criteria.

The first criterion takes into account the length of the racemic compound region as it is limited by the two eutectics. This length is represented by $E_D E_L$ of Fig. 3. The more the eutectic composition approaches a mole fraction of 0.5, the more unstable is the racemic compound. The limiting case is that in which the racemic compound no longer exists (when the eutectic corresponds to the conglomerate).

Another criterion may be derived from the difference in the melting points of the racemic compound and of the eutectics ($T_R^f - T_E$) or from the difference in those of the racemic compound and of the enantiomers ($T_R^f - T_A^f$). While these methods of assessment of stability generally correlate in the same sense for a given racemic compound, their application is not totally unambiguous.*

Finally, Petterson found it more satisfactory to relate the maximal height of the racemic compound, that is, $T_R^f - T_E$, to the difference between enantiomer and eutectic melting points, $T_A^f - T_E$. This relationship is given by the ratio i of these two differences: $i = (T_R^f - T_E)/(T_A^f - T_E)$. It was suggested that a ratio i less than 0.5 would indicate a small tendency to form a racemic compound, i between 0.5 and 1.5 a moderate tendency, and i larger than 1.5 a large tendency. These i values correspond roughly to racemic compound relative areas of less than 50%, from 50 to 75%, or of more than 75%, respectively, of each phase diagram. It must

* For example, hydratropamide is a racemic compound which melts ca. 1.5° above the eutectics and 7° below the enantiomers. This might suggest a low tendency to form a racemic compound. Contrariwise, the racemic domain $E_D E_L$ takes up almost half of the phase diagram, which suggests that the racemic compound is relatively stable.

be said that while these indicators are not without merit, they are arbitrary in character.

In fact, to define the stability of a racemic compound rigorously, it must be done through the free energy difference ΔG^{ϕ}, which corresponds to the "reaction" between the crystalline D and L species that gives rise to the crystalline racemic compound:[9]

$$\text{D crystal} + \text{L crystal} \rightarrow \text{racemic compound}$$

This free energy of formation can also be expressed as a function of the changes in enthalpy and entropy that accompany the preceding reaction, $\Delta G^{\phi} = \Delta H^{\phi} - T\Delta S^{\phi}$. The magnitudes of these thermodynamic parameters cannot be directly measured, but they can be determined from the data accessible through calorimetry, that is, the enthalpies and entropies of fusion and melting points as well as the heat capacities of the different species present.

The symbols employed in the following discussion are as follows: T, temperature in degrees K; T^f, T^v, melting and boiling point of a pure compound; ΔH^f, ΔH^v, ΔH^s, enthalpies of fusion, vaporization, and sublimation; ΔS^f, ΔS^s, entropies of fusion and sublimation; C^s, C^l, C^v, heat capacities of the solid, of the liquid, and of the vapor (at constant pressure); ΔH^{ϕ}, ΔS^{ϕ}, ΔG^{ϕ}, enthalpy, entropy and free energy of formation of a racemic compound from crystalline enantiomers. The subscripts A and R characterize the data for enantiomers and for the racemic compound, respectively.

As we have already seen, mixtures of two enantiomers in the liquid state can be treated as ideal systems, that is, the enthalpy of mixing is zero and the entropy of mixing is equal to $R(x_{\mathrm{D}} \ln x_{\mathrm{D}} + x_{\mathrm{L}} \ln x_{\mathrm{L}})$, that is, $R \ln 2$ for a racemate (see Section 2.2.4).

(a) The enthalpy of formation ΔH^{ϕ}

The enthalpy of formation of a racemic compound from its crystalline enantiomers, ΔH^{ϕ}, corresponds to the difference in enthalpies of sublimation of the crystalline optically active species and of the racemic compound. The enthalpy of sublimation of a crystal, at temperature T, is given by eq. (1):

$$\Delta H_T^s = \Delta H_{T^f}^f + \Delta H_{T^v}^v + \int_T^{T^f} C^s\, dT + \int_{T^f}^{T^v} C^l\, dT + \int_{T^v}^{T} C^v\, dt \qquad (1)$$

Since the enthalpies of vaporization as well as the heat capacities in the liquid and gaseous states of the enantiomers and of their mixtures are indistinguishable, when eq. (1) is applied to the racemate and to the enantiomer, the desired enthalpy of formation may be obtained as follows:

$$\Delta H_T^{\phi} = (\Delta H_A^s)_T - (\Delta H_R^s)_T$$

$$= \Delta H_A^f - \Delta H_R^f + \int_T^{T_A^f} C_A^s\, dT - \int_T^{T_R^f} C_R^s\, dT + \int_{T_R^f}^{T_A^f} C^l\, dT \qquad (2)$$

When $T_A^f < T_R^f$, this equation becomes

$$\Delta H_T^\phi = \Delta H_A^f - \Delta H_R^f + \int_{T_A^f}^{T_R^f} (C^l - C_R^s)\, dT + \int_T^{T_A^f} (C_A^s - C_R^s)\, dT \qquad (3a)$$

And when $T_R^f < T_A^f$,

$$\Delta H_T^\phi = \Delta H_A^f - \Delta H_R^f + \int_{T_A^f}^{T_R^f} (C^l - C_A^s)\, dT + \int_T^{T_R^f} (C_A^s - C_R^s)\, dT \qquad (3b)$$

Inasmuch as $C^l - C_{A\,(\text{or}\,R)}^s$ is virtually constant between T_A^f and T_R^f, eqs. (3a) and (3b) finally reduce to

$$\Delta H_{T_A^f}^\phi = \Delta H_A^f - \Delta H_R^f + (C^l - C_R^s)(T_R^f - T_A^f) \qquad (4a)$$

$$\Delta H_{T_R^f}^\phi = \Delta H_A^f - \Delta H_R^f + (C^l - C_A^s)(T_R^f - T_A^f) \qquad (4b)$$

In these equations, ΔH^ϕ is calculated at the melting point of whichever species [eq. (4a) enantiomer; (4b) racemic compound] melts lower, that is, at the maximum temperature at which the "reaction" D crystal + L crystal → racemic compound may occur.

Table 3 summarizes the enthalpies of formation of several compounds calculated with eqs. (4a) or (4b). In most of the cases, the enthalpy of formation is negative, thus indicating that the formation of the racemic compound from its enantiomers in the solid state is exothermic. A positive enthalpy of formation (endothermic reaction) seems to be less common.

(b) The entropy of formation ΔS^ϕ

The entropy of formation at temperature T, ΔS_T^ϕ, is given by the difference between the entropies of sublimation of the optically active and racemic species to which must be added the entropy of mixing, $R \ln 2$:

$$\Delta S_T^\phi = (\Delta S_A^s)_T - (\Delta S_R^s)_T + R \ln 2$$

Table 3 Enthalpies and entropies of formation of racemic compounds calculated at the melting point of whichever species (racemic compound or enantiomer) melts lower.[a,b]

Name	$C^l - C_{A\,\text{or}\,R}^s$	$\Delta H_{T_{A\,\text{or}\,R}^f}^\phi$	$\Delta S_{T_{A\,\text{or}\,R}^f}^\phi$
Mandelic acid	20	−0.3	+0.2
p-Fluorophenylhydracrylic acid	25	+0.3	+1.3
erythro-Phenylglyceric acid	28	−1.2	−0.8
α-(1-Naphthyl)propionic acid	20	−2.3	−1.7
m-Fluoromandelic acid	25	−0.7	−1.4

[a] Heat capacities in cal · mol^{-1} · K^{-1}; ΔH^ϕ in kcal · mol^{-1}; ΔS^ϕ in kcal · mol^{-1} · K^{-1}.
[b] M. Leclercq, A. Collet, and J. Jacques, *Tetrahedron*, 1976, 32, 821. Reproduced by permission of Pergamon Press Ltd.

Reasoning just as we did in the preceding section, when $T_R^f > T_A^f$, we have

$$\Delta S^{\phi}_{T_A^f} = \Delta S_A^f - \Delta S_R^f + R \ln 2 + (C^l - C_R^s) \ln \frac{T_R^f}{T_A^f} \quad (5a)$$

And when $T_R^f < T_A^f$,

$$\Delta S^{\phi}_{T_R^f} = \Delta S_A^f - \Delta S_R^f + R \ln 2 + (C^l - C_A^s) \ln \frac{T_R^f}{T_A^f} \quad (5b)$$

Table 3 lists values of the entropies of formation of the same compounds for which enthalpies of formation were calculated as above.

(c) The free energy of formation ΔG^{ϕ}. The stability of racemic compounds

The free energy of formation can be obtained from the previously defined enthalpies and entropies of formation. When $T_A^f < T_R^f$, the free energy of formation at T_A^f is obtained from eqs. (4a) and (5a):

$$\Delta G^{\phi}_{T_A^f} = \Delta H_R^f \left(\frac{T_A^f}{T_R^f} - 1\right) - T_A^f R \ln 2 + (C^l - C_R^s)\left(T_R^f - T_A^f - T_A^f \ln \frac{T_R^f}{T_A^f}\right) \quad (6a)$$

When $T_R^f < T_A^f$, eqs. (4b) and (5b) yield

$$\Delta G^{\phi}_{T_R^f} = \Delta H_A^f \left(1 - \frac{T_R^f}{T_A^f}\right) - T_R^f R \ln 2 + (C^l - C_A^s)\left(T_R^f - T_A^f - T_R^f \ln \frac{T_R^f}{T_A^f}\right) \quad (6b)$$

In the case of mandelic acid, for example, by means of eq. (6b) one obtains the following from the heat capacity difference (Table 3) and from the enthalpies of fusion and melting points (Table 2):

$$\Delta G^{\phi}_{T_R^f} = 0.21 - 0.54 - 0.05 = -0.33 \text{ kcal} \cdot \text{mol}^{-1}$$

For *p*-fluorophenylhydracrylic acid, eq. (6b) yields

$$\Delta G^{\phi}_{T_R^f} = 0.37 - 0.50 - 0.01 = -0.14 \text{ kcal} \cdot \text{mol}^{-1}$$

and for α-(1-naphthyl)propionic acid, eq. (6a) yields

$$\Delta G^{\phi}_{T_A^f} = -1.39 - 0.45 + 0.16 = -1.70 \text{ kcal} \cdot \text{mol}^{-1}$$

In fact, the contribution of the term containing the heat capacities in eqs. (6) is always a small one, effectively negligible. Even in the last example given, where $T_R^f - T_A^f$ attains the exceptionally large value of 80°, this term accounts for only 10% of ΔG^{ϕ}.

Finally, the last column of Table 2 gives the values of the enthalpies of formation of the racemic compounds studied, neglecting the term that takes into account the heat capacities. These values of ΔG^{ϕ} have been plotted as a function of the difference $T_R^f - T_A^f$ in Fig. 7; it becomes evident then that the stability of a racemic compound is approximately proportional to the difference between the melting point of the racemate and that of the enantiomers. We see that the definition of racemic compound stability in terms of free energy of formation approaches

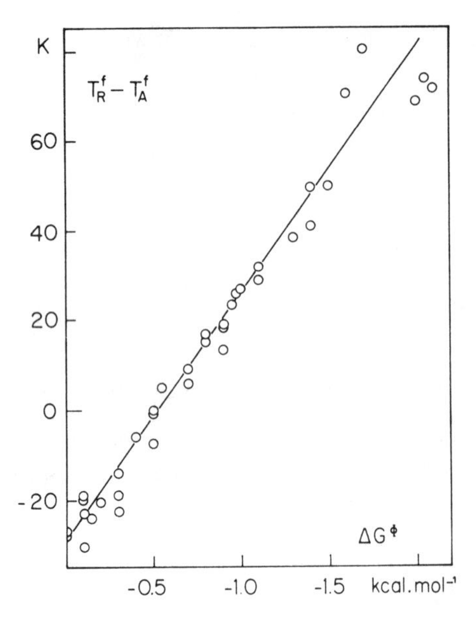

Figure 7 Variation in the free energy of formation of racemic compounds from their crystalline enantiomers as a function of the difference between the melting point of the racemic compound and that of the enantiomers ($T_R^f - T_A^f$). M. Leclercq, A. Collet, and J. Jacques, *Tetrahedron*, 1976, **32,** 821. Reproduced by permission of Pergamon Press Ltd.

and, indeed, justifies the definition earlier proposed purely on intuitive grounds such as, in particular, the definition based on the difference $T_R^f - T_A^f$.

This apparent linearity is, in fact, easily explained. Table 2 shows that the entropies of fusion of the compounds studied approach the mean value $\Delta S_m^f = 17\ \text{cal} \cdot \text{mol}^{-1} \cdot \text{K}^{-1}$.* Since the melting points of these substances are not far from $T_m = 400$ K, on the basis of eqs. (6a) and (6b) one can write the following approximate linear relationship:

$$\Delta G^\phi \simeq \Delta S_m^f(T_R^f - T_A^f) - T_m R \ln 2 \qquad (7)$$

2.3.5 Quasi-Racemates

The discovery of those special types of addition compounds known as quasi-racemates was announced by Pasteur in the same paper as that describing the preparation of the alkaloid salts of tartaric acid.[12] The first examples described were those of compounds formed by ammonium hydrogen (+)-tartrate and ammonium hydrogen (−)-malate, and by (+)-tartramide and (−)-malamide.

* This constant should be compared to the one Walden calculated for the entropy of fusion of some achiral compounds.[11] "Walden's constant" (not valid for spherical or long-chain compounds) corresponds for the case of melting to what Trouton's rule expresses for the case of vaporization, namely, the entropy of vaporization is a constant ($\simeq 21\ \text{cal} \cdot \text{mol}^{-1} \cdot \text{K}^{-1}$).

It was not until the 1920s that Delépine[13] and Timmermans[14] drew attention to these types of compounds that arise from two substances which are structurally similar and configurationally quasi-enantiomeric. No precise definition of the bounds of similarity has thus far been attempted. Yet compound formation analogous to that of racemic compounds often takes place; it is these which are termed quasi-racemates. The true importance of these compounds was recognized even later through the work of Fredga and his school,[15] who showed that their use permitted the diagnosis of stereochemical relationships and, in particular, to facilitate the determination of absolute configuration at a time when application of the X-ray method was difficult and consequently rarely used.

(a) Phase diagrams and thermodynamic aspects

The construction of phase diagrams between pairs of "quasi-enantiomeric" compounds constitutes the method of choice for demonstrating the existence of quasi-racemates which have only rarely been isolated as such or revealed by infrared spectroscopy or through X-ray measurements. It is to Fredga and his collaborators that we owe both the determination of hundreds of such phase diagrams and the establishment of their utility. The shapes of these diagrams can vary over rather wide limits depending on the stability of the quasi-racemate formed (in the sense in which we have discussed the stability of racemic compounds in Section 2.3.4).

The detailed analysis of quasi-racemate phase diagrams is analogous to that carried out earlier for the case of racemic compounds (Section 2.3.2). The equation of Prigogine–Defay is applicable to all binary systems in which an addition compound of stoichiometry 1:1 is formed. This equation allows one to calculate the part of the liquidus curve that corresponds to the quasi-racemate from the enthalpy of fusion and the melting point of the latter; or inversely, one may calculate the enthalpy of fusion of the quasi-racemate from the experimental melting point diagram of mixtures of quasi-enantiomers. For example, Brienne and Jacques have calculated the phase diagram for the quasi-racemate formed between the anthracene diols **1** and **2** (Fig. 8).[16] The data necessary for the calculation are given in Table 4.

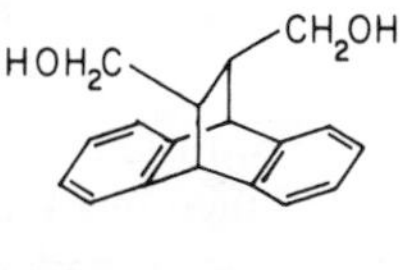

(+) 1 (+) 2

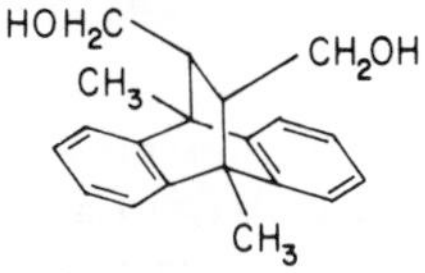

(1) **(2)**

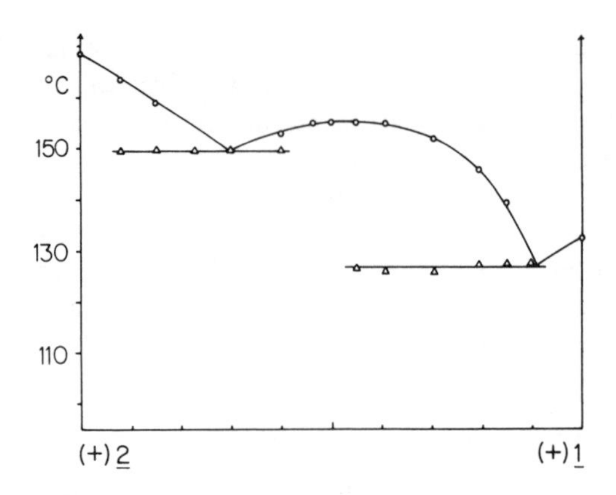

Figure 8 Melting point diagram of quasi-enantiomeric diols (+)-**1** and (+)-**2** with formation of a quasi-racemate. The liquidus curves of the pure substances are calculated by means of the equation of Schröder–Van Laar and that of the quasi-racemate by means of the Prigogine–Defay equation, utilizing the data of Table 4. Experimental points are indicated by triangles (for beginning of fusion) and circles (for termination of fusion). M. J. Brienne and J. Jacques, *Bull. Soc. Chim. Fr.*, 1974, 2647. Reproduced by permission of the Société Chimique de France.

(b) Crystal structure of quasi-racemates

The spatial arrangement of quasi-enantiomeric molecules constituting a quasi-racemate represents an interesting problem which was first examined by Karle and Karle[17] in 1966. The principal conclusion of the crystal structure analysis of the addition compound formed from equal parts of (+)-*m*-methoxyphenoxy-propionic acid and (−)-*m*-bromophenoxypropionic acid is that the component molecules arrange themselves about a pseudocentre of symmetry (Fig. 9) exactly as they do about a true center of symmetry in most racemic compounds.

Table 4 Melting points and enthalpies of fusion of the quasi-enantiomeric diols (+)-**1** and (+)-**2** and of their quasi-racemate[a]

	mp (°C)	ΔH^f (kcal·mol^{-1})
(+)-**1**	133	6.0
(+)-**2**	168.5	7.0
Quasi-racemate	155.5	7.25[b]

[a] From ref. 16.
[b] One mole of quasi-racemate consists of 0.5 mole of each constituent.

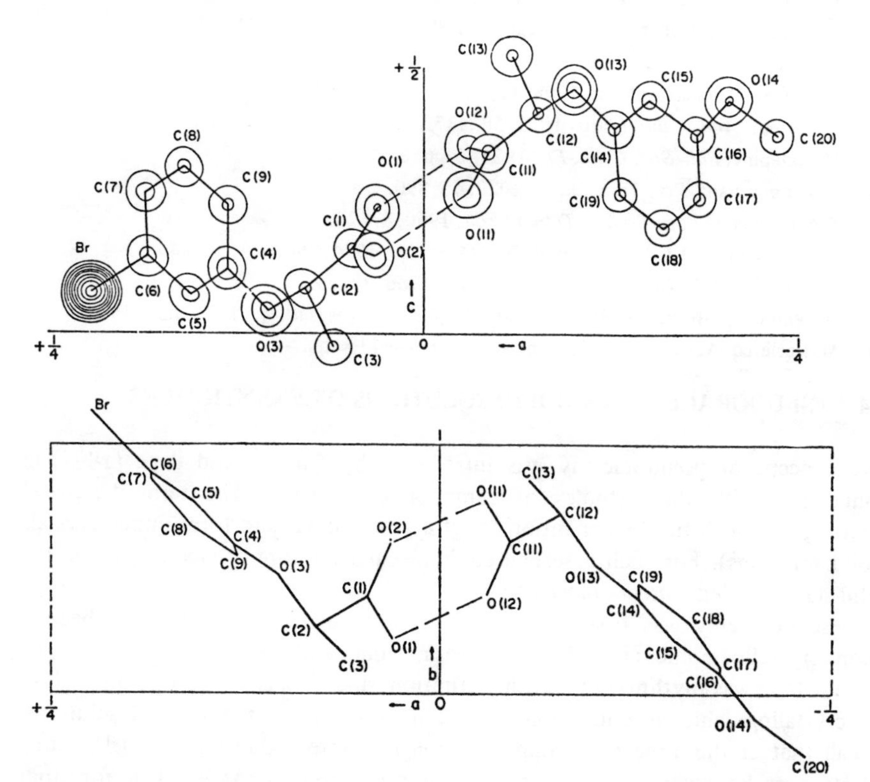

Figure 9 Crystal structure of the quasi-racemate formed from (+)-*m*-methoxy- and (−)-*m*-bromophenoxypropionic acids. The quasi-enantiomeric molecules form a dimer about a pseudocenter of symmetry. From ref. 18 by I. L. Karle and J. Karle, *J. Am. Chem. Soc.*, 1966, **88**, 24. Reproduced by permission of the publisher. Copyright 1966, American Chemical Society.

Whuler et al. determined the crystal structures of the racemic Werner complexes (±)-$Co(en)_3Cl_3 \cdot 3H_2O$ and (±)-$Cr(en)_3Cl_3 \cdot 3H_2O$ and subsequently that of the quasi-racemate formed between the (+)-cobalt and the (−)-chromium salts.[18] There too the analogy between racemic compound and quasi-racemate holds.

References 2.3

1 H. W. B. Roozeboom, *Z. Phys. Chem.*, 1899, **28**, 494.

2 A. Fredga and M. Matell, *Ark. Kemi*, 1951, **3**, 429.

3 B. Sjöberg, *Ark. Kemi*, 1957, **11**, 439.

4 A. Collet and J. Jacques, *Bull. Soc. Chim. Fr.*, 1973, 3330.

5 A. Collet and J. Jacques, *Bull. Soc. Chim. Fr.*, 1972, 3857.

6 I. Prigogine and R. Defay, *Thermodynamique Chimique*, Desoer, Liège, Belgium, 1950. English translation, *Chemical Thermodynamics*, 4th ed., Longmans, London, 1967.

7 H. Mauser, *Chem. Ber.*, 1957, **90**, 299. *Ibid*, p. 307.

8 R. P. Rastogi, *J. Chem. Educ.*, 1964, **41**, 443.
9 M. Leclercq, A. Collet, and J. Jacques, *Tetrahedron*, 1976, **32**, 821.
10 K. Petterson, *Ark. Kemi*, 1956, **10**, 297.
11 P. Walden, *Z. Elektrochem.*, 1908, **14**, 713.
12 L. Pasteur, *Ann. Chim.* (Paris) 1853, **38**, 437.
13 M. Delépine, *Bull. Soc. Chim. Fr.*, 1921, **29**, 656.
14 J. Timmermans, *Rec. Trav. Chim.*, 1929, **46**, 890.
15 See, for example, A. Fredga, *Tetrahedron*, 1960, 8, 126.
16 M. J. Brienne and J. Jacques, *Bull. Soc. Chim. Fr.*, 1974, 2647.
17 I. L. Karle and J. Karle, *J. Am. Chem. Soc.*., 1966, 88, 24.
18 A. Whuler, C. Brouty, P. Spinat, and P. Herpin, *Acta Crystallogr.*, 1976, **B32**, 194.
19 M. Leclercq, A. Collet, and J. Jacques, unpublished results.

2.4 PSEUDORACEMATES. SOLID SOLUTIONS OF ENANTIOMERS

The concept of pseudoracemy was introduced by Kipping and Pope (1897) in connection with their studies of camphor derivatives.[1] The term designates mixtures in which the two enantiomers of a given compound form mixed crystals (solid solutions). For such cases, an equimolecular mixture of enantiomers in the solid state is called a *pseudoracemate*.

Pseudoracemy was first proposed as a consequence of observations bearing upon crystallographic analogies between racemic and resolved compounds. At first the term was anything but clear in definition, since it was not easy to distinguish on crystallographic grounds alone between conglomerates and solid solutions. Recall that at this time no examples of conglomerates exhibiting normal melting points were known in the domain of organic compounds. As was true for other racemic forms, solid solutions of enantiomers were not well characterized until Roozeboom defined them in terms of the phase diagrams of enantiomer mixtures.[2]

The experimental behavior of systems of enantiomers forming solid solutions at all concentrations is summarized in Fig. 1. The three classic cases identified by Roozeboom as type I, type II, and type III obtain where the two constituents are enantiomers.[2,3]

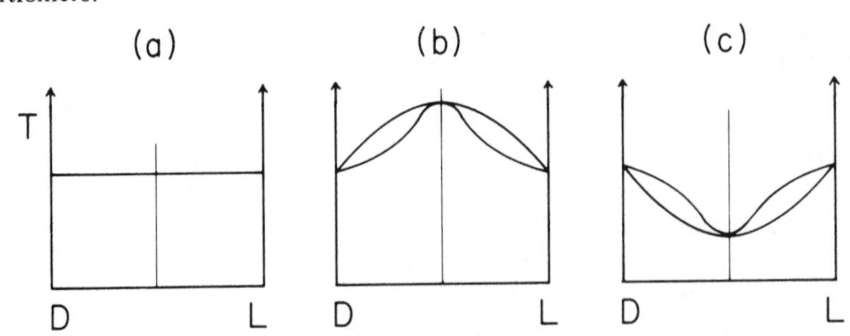

Figure 1 Melting point phase diagrams for enantiomer mixtures forming solid solutions over the entire concentration range: (a) Roozeboom Type I (ideal solid solution); (b) Roozeboom Type II (with maximum melting point); (c) Roozeboom Type III (with minimum melting point).

In the first case, mixtures of (+) and (−) enantiomers in all proportions melt at the same temperature as the pure enantiomers. In the second case, the phase diagram exhibits a maximum melting point for the racemate, and in the third case, a minimum melting point.

The number of cases of pseudoracemy described in the literature is not very large. Relatively complete lists of pseudoracemates are given below for the three types of behavior that characterize these systems (Tables 1, 2, and 3).

2.4.1 Ideal Solid Solutions

Table 1 is an inventory of compounds whose enantiomer mixtures melt at a constant temperature over the entire range of concentrations. These systems may be said to behave ideally. That is, the enthalpy of mixing in the solid state is zero, and the entropy of mixing is equal to $R \ln (x_D \ln x_D + x_L \ln x_L)$, that is, $R \ln 2$ for the pseudoracemate itself. In fact, phase diagrams have not been determined for all compounds in Table 1. We shall see later that there are good arguments for keeping them in the list, nonetheless.

The compounds in Table 1 may be divided into three categories. In the first two, the compounds are polymorphic; that is, prior to melting into isotropic liquids they exhibit "mesomorphic" properties either of liquid crystals or of plastic crystals within well-defined temperatures ranges. The third category consists of compounds which, while not exhibiting either of the two preceding types of behavior, nonetheless give rise to ideal solid solutions of enantiomers.

(a) Plastic crystals[5]

Timmermans was the first to draw attention (in 1935) to plastic crystals which, above a certain transition temperature, undergo a change in resistance upon crushing or extrusion.[4] However, these specific mechanical properties constitute but one of their characteristics. With respect to their optical properties, plastic crystals generally belong to the cubic system. They are therefore isotropic and, if composed of a resolved substance, may exhibit optical rotation. Probably the simplest way of recognizing plastic crystals is the microscopic examination of their behavior upon heating and cooling (for instance, they exhibit typical growth figures while crystallizing). X-Ray analysis of plastic crystals reveals the mobility of individual molecules within the crystal lattice; this mobility is manifested by a local dynamic disorder. Rotation of the molecules nonetheless maintains the positions of their centers of gravity within the lattice. Below the transition temperature leading to the appearance of the plastic crystal phase, this local dynamic disorder is frozen.

According to Aston,[5b] the temperature of the transition leading to the rotation in a plastic crystal is best determined as part of a series of heat capacity measurements. The appearance of this mobility can also be demonstrated through the measurement of changes in dielectric constant as a function of temperature.[6,43] Fig. 2 shows some examples of curves so obtained.

And finally, from the thermodynamic viewpoint, since some molecular disorganization has already taken place below the melting point of the crystal, the

Table 1 Solid solutions of enantiomers (identical racemate and enantiomer melting points)

	Name	mp (°C)[a]	Refs.
C_6	Caproic amide (3-methylpentanamide)	126	11
C_8	2,2,5,5-Tetramethyl-3-hydroxypyrrolidine-1-oxyl	126 (D)	12
C_9	Camphenylone (2,2-dimethylnorbornan-3-one)	38–39 (P)	13
C_{10}	*trans*-π-Camphanic acid	164	1
	trans-Camphotricarboxylic anhydride	154	1
	π-Bromocamphoric anhydride	156	1
	ω-Bromocamphoric anhydride	216	1
	Camphoquinone	202 (P)	6
	Camphoric anhydride	224 (D, P)	14
	π-Bromocamphor	93	1
	π-Chlorocamphor	–	1
	Camphorimide	249 (P)	6
	3-Nitrocamphor	102 (P)	6
	Camphene	46.5 (D, P)	14, 15
	Bornylene	113 (P)	13
	2,6-Dibromocamphane	169 (P)	13
	2,10-Dibromocamphane	91	44
	2,6-Dichlorocamphane	173 (P)	13
	Camphor	178 (D, P)	14
	Thiocamphor	139 (P)	6
	Bornyl bromide	95 (P)	6, 13
	Isobornyl bromide	136 (P)	13
	Bornyl chloride	132 (P)	13
	Camphoroxime	119 (D, P)	9, 18
	Camphane	156 (P)	13
	Isocamphane	65 (P)	13
	Borneol	208 (D, P)	14
	Isoborneol	212 (P)	41, 43
	Bornylamine	163 (P)	13
C_{11}	1,2,3,4-Tetrahydro-2-naphthoic acid	97–100	41, 46
	3-Cyanocamphor	129 (P)	6
C_{12}	2-Amyl alcohol phenylurethane	31 (D)	11
C_{13}	2-Amyl 1-(3-nitrophthalate)	116	11
	2-Amyl 2-(3-nitrophthalate)	155	11
C_{18}	4′-(2-Methylbutyloxy)biphenyl-4-carboxylic acid	239 (S, N)	8
	Bornyl hydrogen phthalate	163 (D)	14, 15
C_{20}	2-Naphthyl β-camphorsulfonate	100	15
C_{22}	2-Methylbutyl 4-(4-methoxybenzylidene)aminocinnamate	47 (S, N)	8
	Bornyl oxalate	106	17
C_{23}	4′-(2-Benzylpropyloxy)biphenyl-1-carboxylic acid	205 (S, N)	8
	2-Methylbutyl 4-(4-ethoxybenzylidene)aminocinnamate	40 (S, N)	8
C_{25}	2-Methylhexyl 4-(4-ethoxybenzylidene)aminocinnamate	106 (S, N)	8

[a] For explanation of D, N, P, and S, see text.

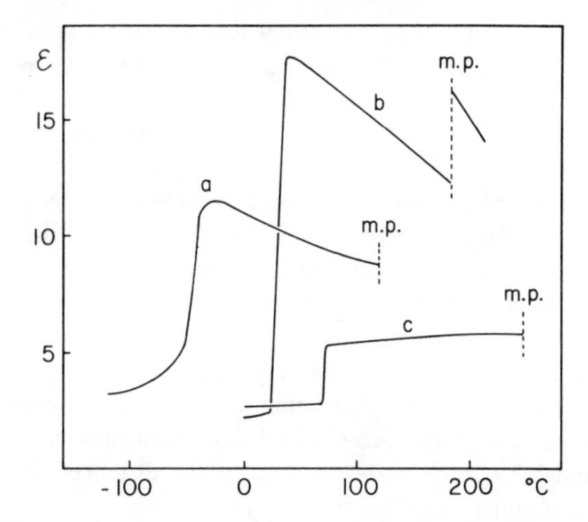

Figure 2 Dielectric constant of: (a) *dl*-thiocamphor; (b) *dl*-2,3-camphanedione; (c) camphoric imide. Adapted from A. H. White and W. S. Bishop, *J. Am. Chem. Soc.*, 1940, **62**, 8 by permission of the publisher. Copyright 1940, American Chemical Society.

passage to the liquid state is accompanied by only a relatively small increase in disorder. That is, the entropy of fusion of plastic crystals is generally only of the order of several $cal \cdot mol^{-1} \cdot K^{-1}$ (less than 5). This is to be compared to the common entropy of fusion of organic compounds, which has a mean value of $17\ cal \cdot mol^{-1} \cdot K^{-1}$ (Section 2.3.4). The properties described are most often associated with a globular molecular structure. Some examples are camphane or adamantane derivatives, molecules which can adopt a variety of orientations without upsetting the crystalline network that contains them. As Aston puts is,[5a]

> When the molecule is almost spherical it can rotate in the crystal without much motion of the center of gravity to aid it. Thus the molecule has as much, or almost as much, rotational entropy in the plastic crystal as in the liquid and the entropy of fusion is close to R ($= 2\ cal \cdot mol^{-1} \cdot K^{-1}$), the communal entropy.

While it is evident that these properties are not unique to chiral molecules which are the subject of this survey, nonetheless the molecular mobility described above contributes to the blurring of differences between the two enantiomers and, thus, favor their perfect miscibility.

(b) Liquid crystals[7]

Three main types of mesomorphic phases constitute the so-called liquid crystalline state; they are called *smectic*, *nematic*, and *cholesteric*. All three are characterized by greater molecular freedom relative to that which obtains in ordinary crystals. Here, too, this additional freedom manifests itself above a certain transition temperature.

In the case of smectic phases, the molecules, which are generally rather long, acquire the ability to spin about their longitudinal axes even while maintaining sufficient organization to keep themselves aligned parallel to one another (with allowance being made for thermal motion) and arranged in layers perpendicular to the major molecular axis (Fig. 3*a*).

In the case of nematic phases, the degree of molecular freedom is even greater since the layered organization is missing. Only the parallelism of the major molecular axes remains (Fig. 3*b*). When the molecules possessing this type of property are chiral, then one finds that when the racemate is nematic the enantiomers are cholesteric.[8] The latter mesophase involves layers of molecules whose longitudinal axes are parallel to one another. These layers approximate parallel planes. The directions of the major axes in neighboring planes subtend a constant angle such that the aggregate of the planes describes a helix which may be defined by a sense (right or left) and by a pitch corresponding to the distance between two planes having the same molecular orientations (Fig. 3*c*).

Without attempting to understand how the chirality of each molecule determines the helical organization and the pitch of the cholesteric phase formed, it is easy to see that enantiomeric molecules of this type, where the presence of asymmetric centers may not radically perturb the relatively elongated molecular shapes, may not appear too different from one another and hence would be more easily miscible.

Finally, we specify that the inverse situation, namely, the existence of plastic or liquid crystal phases melting at the same temperature irrespective to their optical purity, does not necessarily require that the enantiomers be miscible in the solid state. While there is a dearth of experimental data on this point, one can easily imagine that such systems can form racemic compounds as well as conglomerates at temperatures lower than the transition crystal mesomorphic phase. Only X-ray crystallographic studies or the determination of solubility diagrams can give information on this point. The case of camphoroxime, studied first by Adriani[9] and subsequently by Gabard and Jacques,[10] illustrates this type of problem (see Section 3.4).

(c) Static disorder in the crystal

While systems illustrating the two preceding cases exhibit miscibility properties associated with both static and dynamic disorder, a third kind of ideal solid solution may be identified which is rarer than the others. In the crystals of the latter, the enantiomers are statistically interchangeable at each molecular site of the lattice, that is, there is positional static disorder.[12] Unlike the other two cases, some structural feature such as hydrogen bonding has eliminated the possibility of dynamic disorder.

(d) Inventory of ideal or quasi-ideal solid solutions of enantiomers

We have listed in Table 1 all cases of ideal or quasi-ideal solid solutions of enantiomers known to us. However, we must point out that all the compounds of Table 1 have not been studied to the same extent. Hence, not all cases can be rigorously

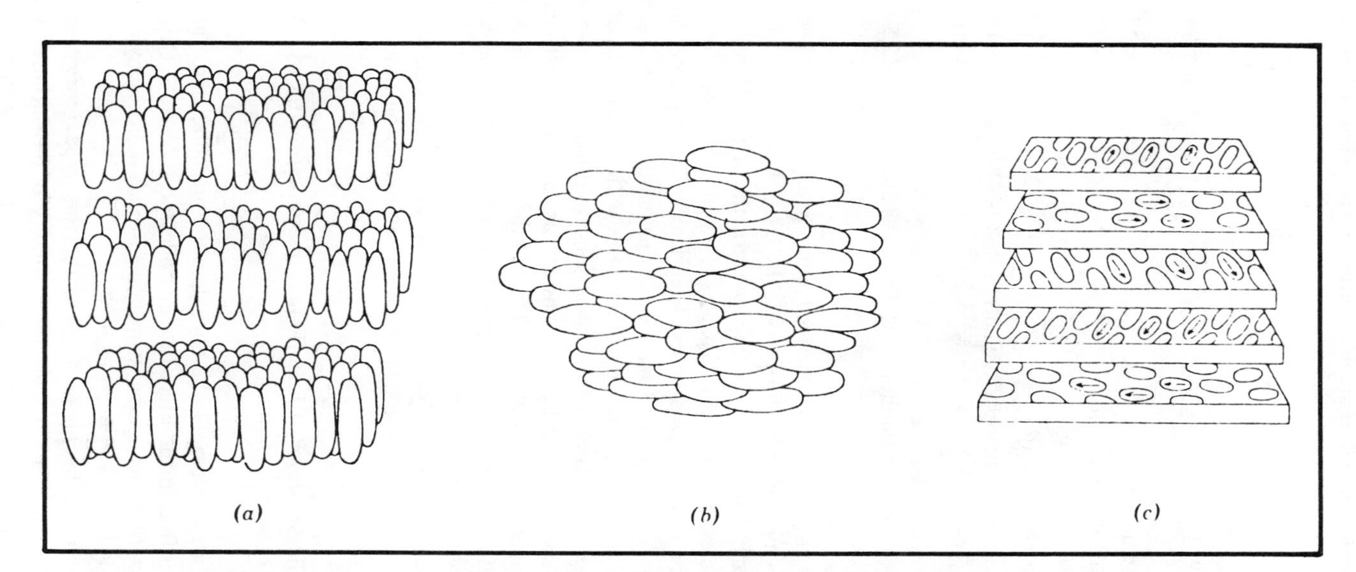

Figure 3 Mesomorphic phases: (a) smectic; (b) nematic; (c) cholesteric. R. Steinstrasser and L. Pohl, *Angew. Chem.*, 1973, **85**, 706; *Angew. Chem. Int. Ed.*, 1973, **12**, 617. Reproduced by permission of Verlag Chemie GMBH.

Table 2 Inventory of solid solutions of enantiomers (maximum melting point foı the pseudoracemate)

	Name	T_R^f (°C)	T_A^f (°C)	Refs.
C_8	*threo*-2,3-Diacetoxybutane	43	28	22
C_{10}	Carvoxime	91.5	72	9, 18
	Dihydrocarvoxime	113–114	88–89	9, 18
C_{11}	2-Thenyl-2-thienylacetic acid	92.5–94	79–81	23
	2-(*p*-Trifluoromethylphenyl)butyric acid	45–46	36	41, 45
C_{13}	Bornyl hydrogen malonate	72	66 (D)	24
C_{14}	2-Amino-2′-nitro-6,6′-dimethylbiphenyl	124–126	108–109	25
C_{17}	Benzoylcarvoxime	105	96 (D)	9, 18
C_{18}	*o*-Ethoxyphenylaminocamphor	59	53	26
	m-Ethoxyphenylaminocamphor	120.5	82.5	26

classed according to the categories described immediately above. Where we have information on their mesomorphic properties, the letter P so indicates for a plastic phase (or for dynamic disorder within the crystal); the letters N and S similarly indicate nematic/cholesteric or smectic phases. Few are the cases for which the phase diagram for the enantiomer pairs has been experimentally determined; even in these cases the diagrams are based on the melting points of a very small number of enantiomer mixtures. We have labeled these cases with the letter D. Sample diagrams illustrating the three categories of ideal solid solution are shown in Fig. 4.

For some of the compounds, we consider as sufficient justification for their inclusion in Table 1 the following facts: (a) the melting points of the racemate and the enantiomers are identical; (b) their entropy of fusion is less than 5 cal · mol^{-1} · K^{-1}; or (c) that evidence of molecular motion in the crystal be available either through diffusion observed on the X-ray photographs in diffraction measurements or through dielectric dispersion measurements on racemates or/and on enantiomers.

2.4.2 Nonideal Solid Solutions of Enantiomers. Pseudoracemates with Maximum Melting Point

This type of solid solution appears to be rarer than the preceding one. The examples we have been able to find in the literature are collected in Table 2. Carvoxime has been subjected to very thorough studies with respect to crystallography and thermodynamics by Oonk et al.[19,21] and Baert et al.[20]

The establishment of the phase diagram for this type of system poses an experimental problem which arises from the difficulty of measuring accurately the melting point range of a given mixture. Visual melting point determinations tend to be uncertain, and even measurements by microcalorimetry are relatively imprecise due to the fact that true liquid/solid equilibria are practically impossible to establish

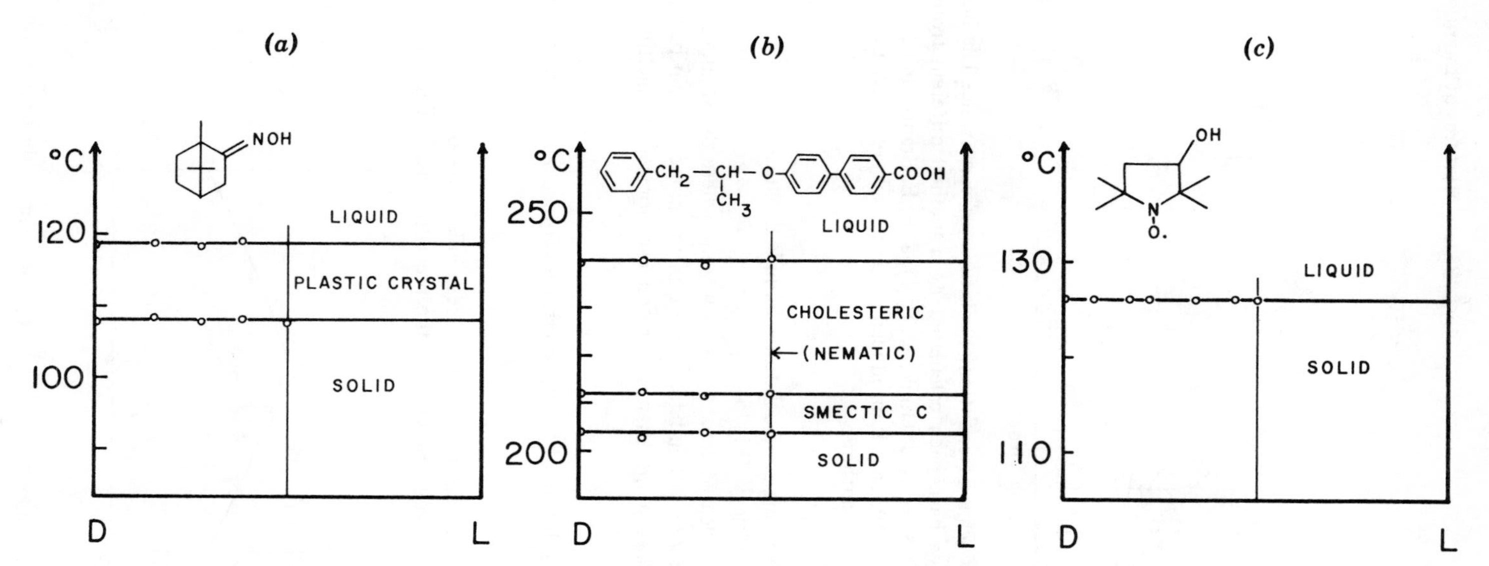

Figure 4 Ideal solid solutions of enantiomers; (a) existence of a plastic phase prior to melting; (b) liquid crystal phases – smectic *C* followed by cholesteric (nematic for the racemate) – prior to transformation into isotropic liquid; (c) solid solution with only static disorder.[12,41] B. Chion, J. Lajzerowicz, A. Collet, and J. Jacques, *Acta Crystallogr.*, 1976, **B32**, 339. Reproduced by permission of the publisher. 1976 Copyright © International Union of Crystallography.

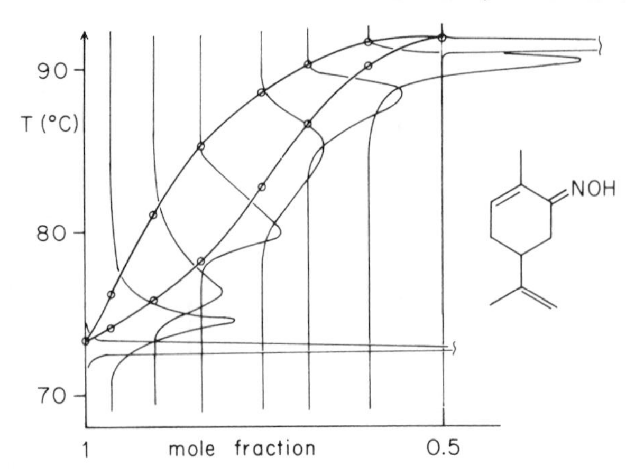

Figure 5 DSC melting scans of enantiomer mixtures of carvoxime. The liquidus and solidus curves shown have been calculated by a method different from that described in Section 2.4.5. H. A. J. Oonk, K. H. Tjoa, F. E. Brants, and J. Kroon, *Thermochim. Acta*, 1977, **19**, 161. Reproduced by permission of Elsevier Scientific Publishing Company and the authors.

in the case of systems forming solid solutions. Sample dsc traces for the carvoxime case illustrate these difficulties[21] (Fig. 5).

In the case of benzoylcarvoxime, the narrow zone where the solid and liquid phases coexist is even more difficult to estimate in the phase diagram because of the small difference between the enantiomer and pseudoracemate melting points (Fig. 6).

2.4.3 Nonideal Solid Solution of Enantiomers. Pseudoracemates with Minimum Melting Point

This last type of pseudoracemate, which corresponds to Roozeboom's type III case (Fig. 1), is even less well known than the preceding cases. The same experi-

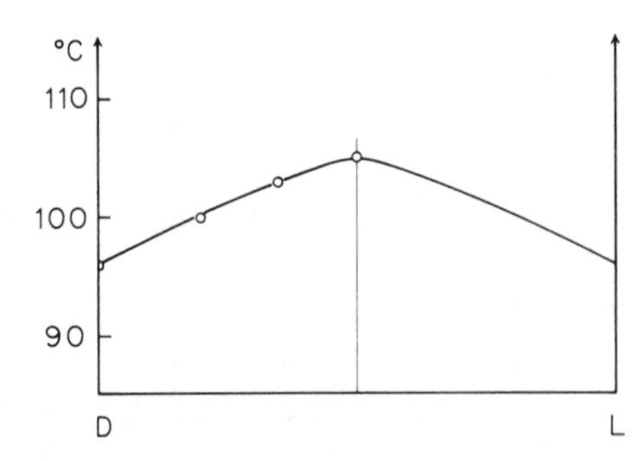

Figure 6 Melting point diagram for benzoylcarvoxime illustrating the near-coincidence of liquidus and solidus curves.[41]

Table 3 Inventory of solid solutions of enantiomers (minimum melting point for the pseudoracemate)

	Name	T_R^f (°C)	T_A^f (°C)	Refs.
C_6	2-Amyl carbamate	51.3	62.2	25
C_9	2,2,5,5-Tetramethylpyrrolidinyl-1-oxyl-3-carboxylic acid	200	204	30
C_{16}	*o*-Bromophenyliminocamphor	114	124.8	27
	o-Chlorophenyliminocamphor	113–114	123–124	28
	o-Chlorophenylaminocamphor	118	146–147	28
	Camphanodihydroquinoxaline	193	213	29
C_{20}	1,4-Di-*p*-tolyl-2,3-dimethylpiperazine	110	140	16
C_{24}	4′-(2-benzylpropyloxy)biphenyl-4-carboxylic acid, methyl ester	106	118	8
C_{26}	*p*-Phenylenebisiminocamphor	259.5	268.5	29

mental problems in the determination of phase diagrams mentioned earlier obtain here as well. *o*-Chlorophenyliminocamphor, whose phase diagram was determined either by visual measurements[28] or by differential scanning calorimetry, illustrates this type of pseudoracemy (Fig. 7). Other examples of such systems are listed in Table 3.

2.4.4 Thermodynamic Properties

Thermodynamic data on pseudoracemates (enthalpies of fusion and heat capacities) are relatively rare, with the exception of a number of compounds which form ideal solid solutions and which have been investigated by workers interested specifically in their plastic or liquid crystal properties.

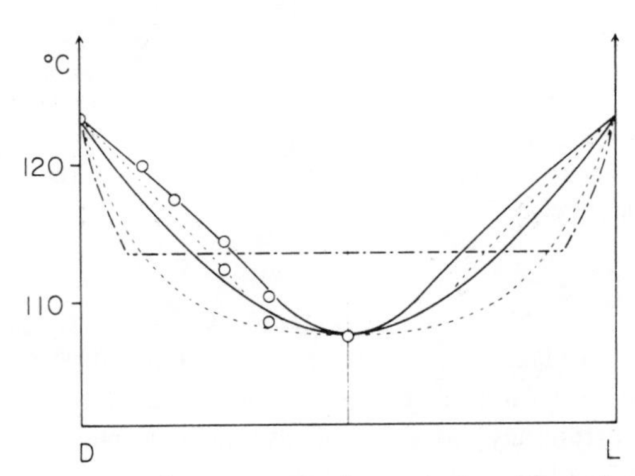

Figure 7 Melting point diagram of *d*- and *l*-*o*-chlorophenyliminocamphor: (—·—·—·—·—) visual measurements of Singh and Seth[28]; (– – – – – – –) calculated solidus and liquidus curves (see Section 2.4.5); circles and solid lines: experimental diagram obtained from dsc scans (ref. 41).

Table 4 Transition enthalpies, entropies, and temperatures of chiral substances forming plastic crystals and ideal solid solutions of enantiomers[a]

	Name		T^p (b)	T^p (c)	ΔH^p (c)	ΔS^p (c)
			Crystal → Plastic Crystal			
C_9	Camphenylone					
C_{10}	Camphoquinone	R	45–55	59	3.1	9.3
		A		66.5	3.5	10.4
	Camphoric anhydride		135		10.5	25.0
	Camphorimide		92–94			
	3-Nitrocamphor		40–49			
	Camphene	A	−124			
	Bornylene	A				
	2,6-Dibromocamphane					
	2,10-Dibromocamphane					
	2,6-Dichlorocamphane					
	Camphor	R, A	−31	−32	2.7	11.1
	Thiocamphor	R	−30 to −10			
	Bornyl bromide	R	−68 to −61			
	Isobornyl bromide					
	Bornyl chloride			−42	0.17	0.75
	Camphoroxime	A		112	3.5	9.1
		R		109	3.2	8.4
	Camphane					
	Isocamphane					
	Borneol		71–75	70	0.78	2.3
	Isoborneol	A	30–40			
		R				
	Bornylamine					
C_{11}	3-Cyanocamphor		110–116			

[a] Temperatures in °C, enthalpies in kcal · mol^{-1}, entropies in cal · mol^{-1} · K^{-1}. Data from (b) dielectric dispersion,[6] (c) microcalorimetric[41] and (d) cryoscopic measurements.[31] A and R refer to measurements carried out on optically active compound or on the pseudoracemate, respectively.

(a) Plastic crystals

The available data on plastic crystals have been assembled in Table 4. In this case the data more often than not are either for the racemate or for an enantiomer but not for both. Unfortunately, it is not always clear on which of these species the measurements were made.

The available data on enthalpies of fusion are mostly taken from the work of Pirsch and are derived from cryoscopic measurements.[31] It is evident that direct ΔH^f measurements based upon calorimetric traces[41] sometimes yield data

Table 4 Continued

	Name		Plastic Crystal → Liquid				
			T^f	ΔH^f		ΔS^f	
			(b)	(d)	(c)	(d)	(c)
C_9	Camphenylone		38–39	0.42		1.35	
C_{10}	Camphoquinone	R	196	1.61	2.1	3.41	4.5
		A					
	Camphoric anhydride		223	2.0		4.0	
	Camphorimide		249				
	3-Nitrocamphor		102				
	Camphene	A	44–46.5	0.88		2.74	
	Bornylene	A	113	1.20		3.12	
	2,6-Dibromocamphane		169	1.44		3.25	
	2,10-Dibromocamphane		91	0.68		1.87	
	2,6-Dichlorocamphane		173	1.47		3.29	
	Camphor	R, A	178	1.54	1.4	3.42	3.1
	Thiocamphor	R	139				
	Bornyl bromide	R	95	0.83		2.31	
	Isobornyl bromide		136	1.23		3.01	
	Bornyl chloride		(b) 132 (c) 126.5	1.20	1.0	2.97	2.5
	Camphoroxime	A	(c) 118		0.58		1.5
		R	(c) 118		0.58		1.5
	Camphane		156	1.71		3.99	
	Isocamphane		65	0.70		2.08	
	Borneol		206		1.6		3.4
	Isoborneol	A	212	1.54	1.8	3.1	3.8
		R	208		1.7		3.4
	Bornylamine		163	1.44		3.29	
C_{11}	3-Cyanocamphor		129				

that differ from the cryoscopic ones, but in general the agreement is rather good.

The temperatures of the indicated transitions have diverse origins. The melting points pose no problem. The crystal to plastic crystal transitions for the most part derive from measurements of dielectric dispersion.[6] Here, too, the calorimetric traces[41] and the earlier determinations of these transitions sometimes yield different values.

(b) Liquid crystals

Transition temperatures, enthalpies, and entropies for various mesomorphs which have been examined in both the racemic and optically active states are given in Table 5. Most of these values were obtained by the group of Jacques.[8,41]

Table 5 Transition enthalpies, entropies, and temperatures of chiral mesomorphic substances forming solid solutions of enantiomers[a]

No.		$K_1 \rightarrow K_2$			K → N or N*		
		T	ΔH	ΔS	*T*	ΔH	ΔS
I	A	238			239	4.6	9.0
	R				239	4.7	9.2
II	A	165	1.7	3.9	171	1.8	4.0
	R	167			171	2.8	6.3
V	A				205	4.7	9.8
	R				205.5	4.75	9.9
		K → S			$S_1 \rightarrow S_2$ (?)		
IV	A	44	3.7	11.7	60	0.28	0.8
	R	47	4.4	13.7	60	0.32	1.0
VI	A	38.5–40	3.6	11.5	75	0.36	1.0
	R	40	4.85	15.5	83	0.28	0.8
VIII	A	55	4.65	14.2			
	R	60	6.6	19.8			
IX	A	86	8.6	24.0			
	R	87	6.3	17.5			
		K → N or N*					
III	A	93.5	5.65	15.4			
	R	96.5	5.6	15.2			
VII	A	70	6.97	20.3			
	R	71					

[a] T, transition temperatures in °C; ΔH, ΔS, transition enthalpies and entropies, in kcal · mol^{-1} and cal · mol^{-1} · K^{-1}, respectively. K, Crystal; S, smectic; N, nematic; N*, cholesteric; L, liquid. A and R refer to measurements carried out on optically active compound or on the pseudoracemate, respectively.

I $CH_3-CH_2-CH(CH_3)-CH_2O$-⟨⟩-⟨⟩-CO_2H $C_{18}H_{20}O_3$

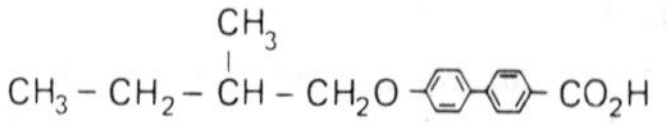

II $CH_3-CH_2-CH_2-CH_2-CH(CH_3)-CH_2O$-⟨⟩-⟨⟩-$CO_2H$ $C_{20}H_{24}O_3$

III $N\equiv C$-⟨⟩-$CH=N$-⟨⟩-$CH=CH-CO_2-CH_2-CH(CH_3)-CH_2CH_3$ $C_{22}H_{22}O_2N_2$

IV CH_3O-⟨⟩-$CH=N$-⟨⟩-$CH=CH-CO_2-CH_2-CH(CH_3)-CH_2CH_3$ $C_{22}H_{25}O_3N$

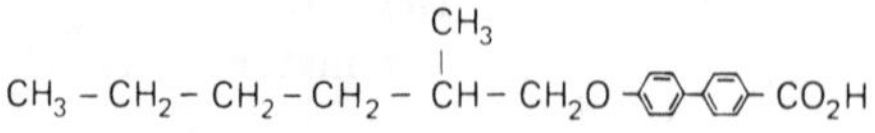

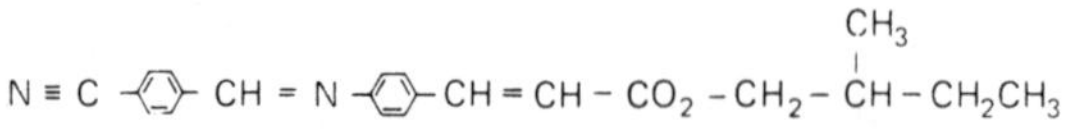

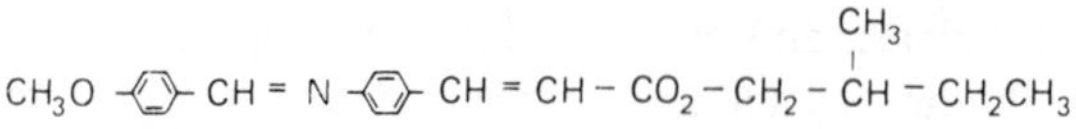

Table 5 Continued

No.		N_1 (N_1^*) → N_2 (N_2^*)			→ L		
		T	ΔH	ΔS	T	ΔH	ΔS
I	A				249	0.9	1.7
	R				249	0.9	1.7
II	A	215	1.1	2.2	229	0.9	1.8
	R	224	0.9	1.8	245	1.2	2.3
V	A	213	0.5	1.0	241	0.8	1.6
	R	213.5	0.5	1.0	241	0.9	1.7
		S → N or N*			→ L		
IV	A	82	0.46	1.3	102	0.06	0.2
	R	82	0.40	1.1	102	0.02	0.05
VI	A				117	1.03	2.6
	R				124	0.99	2.5
VIII	A	69.5	0.17	0.5	90.5	0.06	0.2
	R	70	0.17	0.5	92	0.11	0.3
IX	A				106	0.83	2.2
	R				99	0.6	1.6
					→ L		
III	A				108.5	0.06	0.2
	R				108	0.06	0.2
VII	A				97	0.08	0.2
	R				99	0.08	0.2

V $C_6H_5-CH_2-CH(CH_3)-CH_2O-C_6H_4-C_6H_4-CO_2H$ $C_{23}H_{12}O_3$

VI $CH_3CH_2O-C_6H_4-CH=N-C_6H_4-CH=CH-CO_2-CH_2-CH(CH_3)-CH_2CH_3$ $C_{23}H_{27}O_3N$

VII $N\equiv C-C_6H_4-CH=N-C_6H_4-CH=CH-CO_2-CH_2-CH(CH_3)-CH_2-CH_2-CH_2-CH_3$ $C_{24}H_{26}O_2N_2$

VIII $CH_3O-C_6H_4-CH=N-C_6H_4-CH=CH-CO_2-CH_2-CH(CH_3)-CH_2-CH_2-CH_2-CH_3$ $C_{26}H_{29}O_3N$

IX $CH_3-CH_2O-C_6H_4-CH=N-C_6H_4-CH=CH-CO_2-CH_2-CH(CH_3)-CH_2-CH_2-CH_2-CH_3$ $C_{27}H_{31}O_3N$

Table 6 Temperatures, enthalpies, and entropies of fusion of enantiomers forming solid solutions with maximum or minimum melting points[a]

	T_R^f (°C)	ΔH_R^f ($kcal \cdot mol^{-1}$)	ΔS_R^f ($cal \cdot mol^{-1} \cdot K^{-1}$)	T_A^f (°C)	ΔH_A^f ($kcal \cdot mol^{-1}$)	ΔS_A^f ($cal \cdot mol^{-1} \cdot K^{-1}$)
Carvoxime	92	5.3	14.5	73	4.1	11.9
o-Chlorophenyliminocamphor	107.5	6.8	17.9	123.5	6.2	15.6

[a] From ref. 41.

(c) Miscellaneous solid solutions

The few available data on solid solutions that do not belong to the preceding categories are given in Table 6. These are of relatively recent origin.[41]

2.4.5 Thermodynamic Aspects. *A Priori* Calculation of Pseudoracemate Phase Diagrams

Let us review first the composition of the phases at equilibrium as described in phase diagrams of solid solution with maxima or minima.

Consider, for example, the diagram shown in Fig. 8, which exhibits a maximum melting point. When a mixture of composition Y is gradually heated, it begins to melt at temperature T_s. At this temperature, the first traces of liquid have the composition X (or, as the diagram makes clear, points l_x and s_y located at the opposite ends of a tie line give the composition of the two phases at equilibrium at this temperature). At T_l, corresponding to the end of melting of the mixture, the last trace of solid remaining should have the composition Z which approaches that of the racemate. At the intermediate temperature T, the liquid should have the composition X_l and the solid the composition X_s, while the proportions of the two phases are given by application of the lever rule (see Section 2.2.2).

It may be easily deduced from the equilibrium diagram of Fig. 8 that the composition of the solid phase must change during the process of melting (from s_y to s_z), so as to keep the system at equilibrium for every temperature between T_s and T_l. As a matter of fact, this requirement is practically impossible to obtain, and for this reason the construction of true equilibrium phase diagrams of systems forming solid solutions is far from being obvious.

Just as it has been possible to calculate the phase diagram corresponding to enantiomer mixtures forming conglomerates or racemic compounds, so too one may calculate the theoretical diagram for the case of pseudoracemates with maximum or minimum melting point. However, here the problem is a bit different and more complicated – this by virtue of the nonideality of the solid solution even while the liquid mixture of enantiomers remains ideal. It is formally similar to that of the calculation of boiling point and condensation curves of fully miscible binary mixtures of liquids.[32] The equations obtained depend upon the hypothesis that one may make regarding the thermodynamic behavior of these solid solutions. Assuming as a first approximation that the solid solutions of enantiomers are regular,* the solidus and liquidus curves may be calculated from eqs. (1) and (2), which give, respectively, the composition of the solid x_s and that of the liquid x_l at equilibrium at temperature T:

$$x_s = \frac{e^{-A} - e^{-\lambda_1}}{e^{-\lambda_2} - e^{-\lambda_1}} \tag{1}$$

$$x_l = \frac{e^{\lambda_2} - e^{\lambda_1 + \lambda_2 + A}}{e^{\lambda_2} - e^{\lambda_1}} \tag{2}$$

* In the case of regular solutions, the activity coefficients γ follow a simple relationship of the form $RT \ln \gamma = \alpha x^2$, where α is a constant (cf. ref. 32).

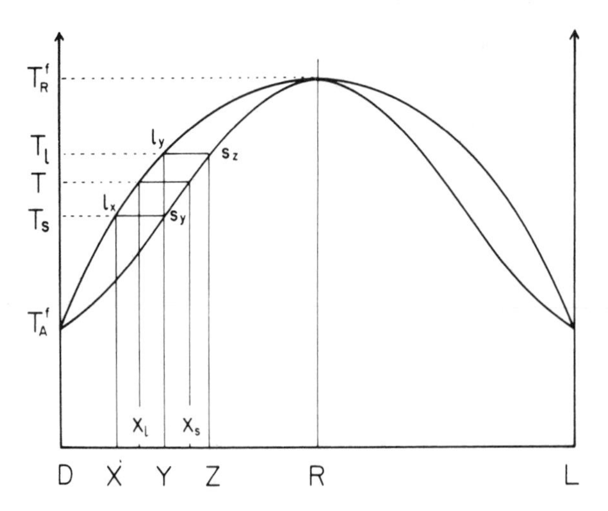

Figure 8 Melting point phase diagram for a nonideal solid solution of enantiomers.

In these equations,

$$A = \frac{\Delta H_A^f}{R}\left(\frac{1}{T_A^f} - \frac{1}{T^f}\right)$$

$$\lambda_1 = \frac{\alpha}{T^f}x^2$$

$$\lambda_2 = \frac{\alpha}{T^f}(1 - x^2)$$

with ΔH_A^f being equal to the enthalpy of fusion of a pure enantiomer melting at T_A^f, T_R^f being the melting point of the pseudoracemate. The normalization factor of the activity coefficients, α, is calculated by means of the relationship

$$\alpha = \frac{4\,\Delta H_A^f}{R}\left(\frac{T_A^f - T_R^f}{T_A^f}\right)$$

The sign of α is negative or positive according to whether one has a curve with a maximum or a minimum. In the case of an ideal solution, α is zero and a flat diagram results.

Unfortunately, there are insufficient data to test the hypothesis of *regularity* of solid solutions of enantiomers. And the difficulties already alluded to in obtaining precise experimental phase diagrams lend uncertainty to the comparison of experimental and calculated curves. By way of example, Fig. 7 shows the curves derived from eqs. (1) and (2) and the experimental ones obtained both by visual means and by differential scanning calorimetry in the case of *o*-chlorophenyliminocamphor.

2.4.6 Pseudoracemy and Its Relationship to Isomorphism

The existence of pseudoracemates leads us to consider in a more general way the problems associated with isomorphism and with related concepts to which we return later, particularly in connection with the behavior of diastereomeric salts.

It is well known that some mixtures of inorganic as well as organic compounds form mixed crystals or solid solutions (one also says that they cocrystallize, or syncrystallize, or that they are miscible in the solid state). A single and homogeneous crystalline phase may actually be formed from mixtures of two partners either in any proportion or only within a specified concentration range.[33] This miscibility may be demonstrated, as we have just seen, either in melting point phase diagrams or, as we shall see later (Section 3.4), in solubility diagrams in the presence of solvent.

Syncrystallization has been historically linked to the concept of isomorphism which describes the near identity of crystal forms, that is, forms that exhibit small angle differences and like parameters. These two concepts may nonetheless not be equated since compounds do exist, and in particular inorganic ones, which cristallographically are practically identical yet do not form mixed crystals.

The concept of isomorphism which refers to macroscopic properties may be associated with that of isosterism which deals with properties on a molecular scale. The analogies implicit in isosterism refer to geometric elements (molecular volumes, conformations, chirality, etc.), to electronic elements (number and type of electrons, charges, dipole moments, etc.), and even to thermodynamic elements (dilatation coefficients, specific heats, etc.).*

Isosterism is tolerant of some approximations. For example, a series of homologs with a very similar number of atoms may conceivably exhibit isosterism. The term isosterism may also be applied to groups of atoms. Let us also recall that the notion of isosterism was large developed in research on and rationalization of physiologic or pharmacologic activities.[38,40]

While isosterism usually implies the narrow relationship of crystals forms (i.e., isomorphism), exceptions to this may arise linked to the possibility that two different yet very similar molecules may prefer very different arrangements within a crystal. This is similar to what occurs in polymorphism, in which case the same molecule may assume different arrangements in the crystal. Similarly, isomorphism does not necessarily coincide with isosterism; some very different molecules may crystallize in nearly indiscernible forms, for example, the case of potassium nitrate compared to calcium carbonate. In spite of these diverse reservations, one can nevertheless state that in a general way, isosterism leads to isomorphism, which in turn *may* lead to syncrystallization. Conversely, the experimental observation of syncrystallization between two substances allows one to conclude with certainty that more or less perfect isosterism obtains for the pair. And finally, we believe that making a distinction between isomorphism and homeomorphism leads to the introduction of useless nuances.

* We use the term isosterism in the broad sense defined above, even though some authors may prefer to limit its use to electronic similarities, the way in which it was originally defined by Langmuir.[39] Nowadays, one uses the term isoelectronic for such analogies. We will avoid the term homeomerism utilized by Urbain[42] to deal with thermodynamic analogies at the molecular level. In our view, isosterism takes precedence over the notion of "molecular isomorphism", which we consider awkward.

Table 7 Lattice parameters of some solid solutions of enantiomers[a,b]

	1		**3**		**4**		**5**	
	A	R	A	R	A	R	A	R
a (Å)	9.95	10.02	8.09	8.04	10.25	9.90	9.24	9.52
b (Å)	6.68	6.67	10.10	10.12	11.68	11.84	10.20	9.76
c (Å)	13.94	13.97	12.40	12.49	8.56	8.52	8.12	8.03
α (deg)	–	–	–	–	–	–	90.91	91.37
β (deg)	–	–	–	–	103.07	100.4	99.11	98.02
γ (deg)	–	–	–	–	–	–	89.94	87.38
Space group	$P2_12_12_1$	$Cmcm$	$P2_12_12_1$	$Pnma$	$P2_1$	$P2_1/c$	$P1$	$P1$
Z	4	4	4	4	4	4	2	2
Density	1.133	1.119	1.221	1.216	1.098	1.115	1.18	1.19
Roozeboom type	I		III		II		II	

[a] Formulas **1** and **3** on p. 123, formulas **4** and **5** on p. 125.
[b] A, Enantiomer; R, pseudoracemate.

2.4.7 Crystalline Forms of Enantiomers That Cocrystallize

Among those elements which define isosterism, we have listed chirality or, otherwise stated, the identity of absolute configurations. In the case of solid solutions of enantiomers, there appears immediately a difficulty as a consequence of the contradiction between enantiomerism and isosterism. That is, enantiomeric molecules are not likely to be interchangeable in a crystal lattice (nor in a biologic receptor). Similarly, enantiomorphism and isomorphism seem to be incompatible; although the enantiomeric crystal structures are identical in distances and angles, they are not, by definition, superposable. It is probably for these reasons that solid solutions of enantiomers are so rare.

Very few crystallographic studies of such solid solutions have been carried out. When they do, in fact, form, it is a consequence of the near interchangeability of the molecules of opposite chirality; the lattice of a solid solution of any composition is thus similar to that of the pure enantiomers. Examination of Table 7 shows that effectively the cell parameters measured for the enantiomers and for their pseudoracemate are very similar, though not identical.

We have seen that several types of solid solutions of enantiomers may be characterized by means of their phase diagram (see Fig. 1). Another classification may be applied which is based upon their crystallographic behavior. It should be emphasized that these two criteria are not necessarily related, in the sense that a given type of phase diagram does not always correspond to a given structural type. The crystallographic classification is as follows:[30]

(a) Disordered type

This type results when one of the enantiomers may be replaced by the other in a totally random way, that is, independently of the occupancy of the neighboring sites and without evolution of heat of mixing. Such a solid solution is completely disordered. Statistically inverse symmetry elements are then generally found for the racemate. Such a solid solution corresponds to a very statistical disorder even at short distances. This situation arises, for example, for the nitroxide **1** whose

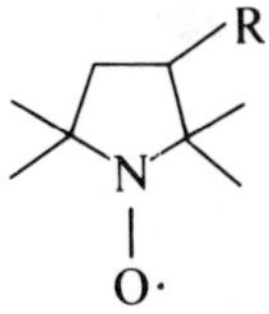

R

N

O·

1 R = OH
2 R = CH_2OH
3 R = COOH

crystal structure has been determined both for the active species and for the racemate[12] and for which we have already seen the "flat" binary phase diagram corresponding to an ideal solid solution (Fig. 4*c*).

Figure 9 shows the disposition of the molecules in the crystal structure of the pure (+)-**1** enantiomer. For the pseudoracemate, the packing is identical; however,

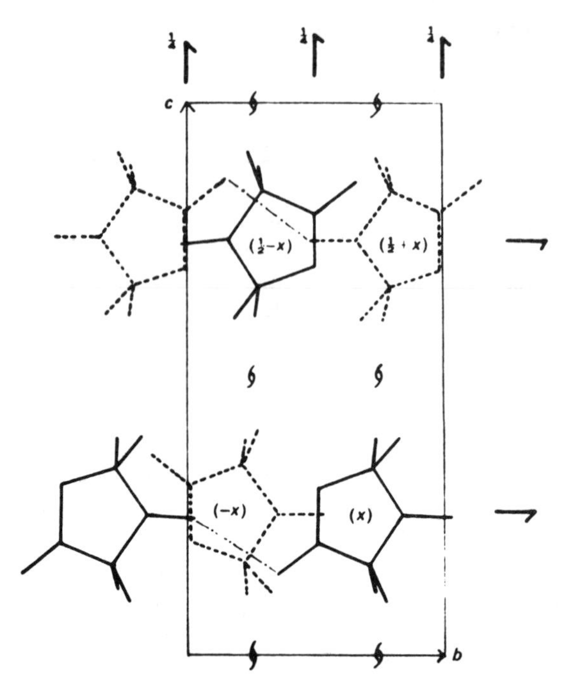

Figure 9 Projection of the structure of (+)-**1** on the *bc* plane. The molecular positions correspond to $x \sim 0.25$. Hydrogen bonds are marked (—·—·—·—·—). B. Chion, J. Lajzerowicz, A. Collet, and J. Jacques, *Acta Crystallogr.*, 1976, **B32**, 339. Reproduced by permission of the publisher. 1976 Copyright © International Union of Crystallography.

at each molecular site there is a 1/4 probability of finding a molecule in one of the arrangements labeled A_1, A_2, B_1, or B_2 corresponding to 2 mm symmetry elements of the *Cmcm* space group. Molecules in positions A_1 and A_2 are located by means of the binary axis and thus have the same chirality; the same is true for B_1 and B_2. The A_1 and B_1 (or A_2 and B_2) sites have a mirror image relationship. Figure 10 shows the four possible orientations of the molecules in the pseudoracemate.

(b) Short-range order type

A second type is obtained in the case where substitution of one enantiomer by the other takes place in a macroscopically random way yet is locally influenced by the occupancy of neighboring sites (e.g., a (−)-molecule has a greater chance of finding itself near a (+)-molecule rather than near another (−)-molecule). There must be an enthalpy of mixing which manifests itself by a change in the melting point as a function of composition. In spite of the macroscopic disorder, there exists short-range order. Statistically inverse symmetry elements may appear for the pseudoracemate. Since the melting point of the latter is different than that of the enantiomers, the phase diagram exhibits a maximum or a minimum for $x = 0.5$; the nitroxide **3** belongs to this category.[30,34]

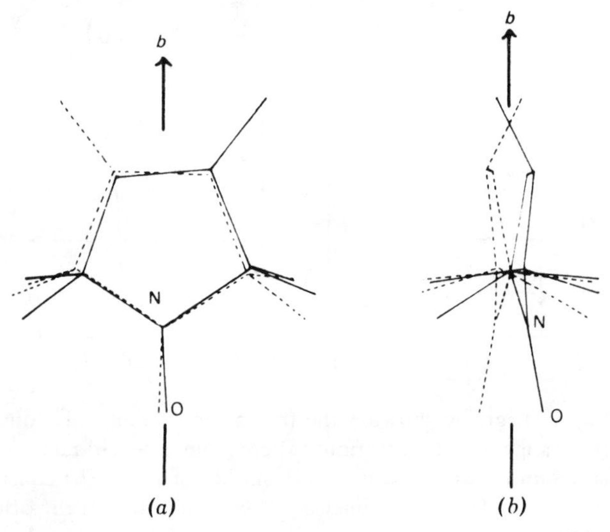

Figure 10 Projection of the four possible orientations of the (+) and (−) molecules of **1** in the pseudoracemate: (a) on plane *bc*; (b) on plane *ba*. B. Chion, J. Lajzerowicz, A. Collet, and J. Jacques, *Acta Crystallogr.*, 1976, **B32,** 339. Reproduced by permission of the publisher. 1976 Copyright © International Union of Crystallography.

(c) Nonstatistical inverse symmetry type

The two preceding cases may be described as continuous solid solution of two enantiomers. A third possibility exists, in which molecules of the pure enantiomer occupy two nonequivalent sites A and B in their crystal lattice. If the $(+)_A$- and $(+)_B$-molecules are substituted randomly by (−)-molecules, one recognizes one or the other of the two preceding categories **(a)** or **(b)**. But if, on the other hand, one substitutes (+)-molecules at one site only (e.g., B), then one obtains intermediate structures of type $(+)_A$, $(1-x)(+)_B$, $x(-)_B$ with (−)-molecules exclusively occupying site B. This necessarily leads to a racemate of type $(+)_A$, $(-)_B$ which is ordered and in which a real (nonstatistical) inverse symmetry element may appear which relates $(+)_A$ and $(-)_B$ in the lattice. The structure of the pseudoracemate is now equivalent to that of the usual racemic compound. One might consider this case to be that of a solid solution between a "racemic compound" and each enantiomer even though the binary phase diagram would not be different from those corresponding to case **(b)**. Carvoxime **4**[19,20] and benzoylcarvoxime **5**[35] are examples of this third possibility.

NOH
4

O
NO—C—
5

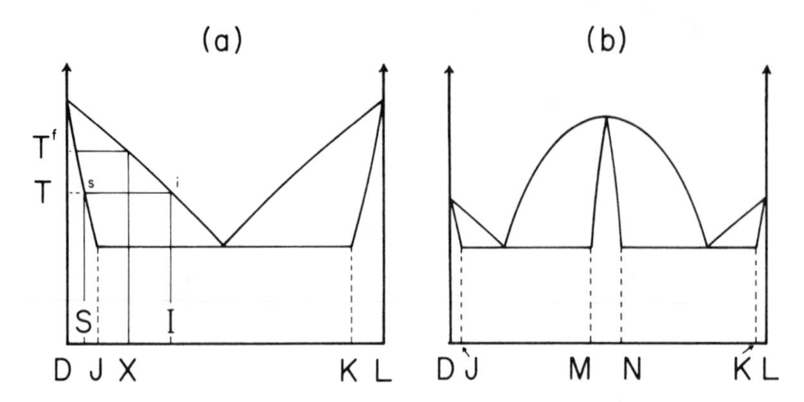

Figure 11 Binary diagrams showing the formation of solid solutions of enantiomers in a limited range of concentration: (a) conglomerate; (b) racemic compound. Samples whose composition lie between *D* and *J* (or *K* and *L*) consist of a solid solution of the less abundant enantiomer in the crystal lattice of the other. Between *M* and *N* there is incorporation in the racemic compound lattice of that enantiomer which is in excess.

Note that in all cases of racemic solid solutions, both ordered and nonordered, the real or statistical symmetry elements are superposed on the symmetry elements of the enantiomer space group. The enantiomer space group is then always a subgroup of that of the racemate, for example, $Cmcm \rightarrow P2_1 2_1 2_1$ or $P2_1/c \rightarrow P2_1$.

2.4.8 Partial Miscibility

We have already stated that, even when a racemate is a conglomerate or a racemic compound, an observable solid solution may exist between the several crystalline species but particularly in the vicinity of either pure enantiomer or of the racemic compound. Figure 11 illustrates two examples of phase diagrams which exhibit such partial solid solutions. In the first case (*a*), upon melting the mixture of composition *X*, the liquid composition is given by *I*. On the other hand, the solid present constitutes a single phase of composition *S*, and no longer pure enantiomer as before.

One way of determining the limits of the existence of the eutectic (or in another context, the limits of the solid solution range) is by means of Tammann curves, that is, the plotting of the *length of eutectic halts* as a function of the composition. Since the eutectic halt is proportional to the quantity of eutectic present in the mixture, a simple diagram allows one to extrapolate the composition at which the eutectic disappears.

The method of Tammann is actually no longer used in its original form since phase diagrams are now rarely derived from cooling curves. On the other hand, a method based upon an identical principle and which is quite rigorous consists of the measurement by microcalorimetry of the enthalpy of fusion corresponding to the melting of the eutectic present in mixtures of different compositions. By

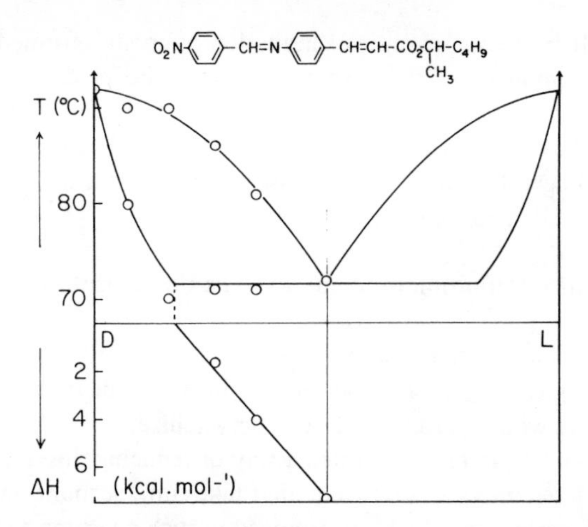

Figure 12 Construction of a Tammann diagram by differential microcalorimetry.[41]

plotting the values of these enthalpies of fusion, adjusted so as to correspond to equal quantities of mixture, as a function of mixture composition, a straight line is obtained the extrapolation of which to $\Delta H = 0$ gives the limit of partial miscibility. An example of such a plot is given in Fig. 12.

This method is impracticable when the eutectic present cannot be evidenced by thermal analytic methods. This arises, in particular, when the limit of partial miscibility is close to the eutectic composition. In the case of a conglomerate, this situation, which corresponds to a large domain of existence of solid solution, is illustrated in Fig. 13. In this example, analysis of melting point curves through microcalorimetry does not allow to distinguish between a continuous solid solution diagram (calculated diagram shown by dotted lines in Fig. 13) and a diagram with

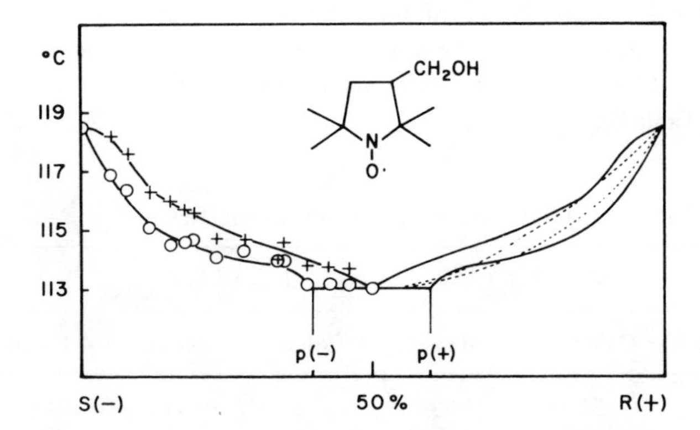

Figure 13 Solid solution of enantiomers limited to the 60 to 100% range of composition. The racemate is a conglomerate of solid solutions having the compositions $p(-)$ and $p(+)$. B. Chion, J. Lajzerowicz, D. Bordeaux, A. Collet, and J. Jacques, *J. Phys. Chem.*, 1978, **82,** 2682. Reproduced by permission of the publisher. Copyright 1978, American Chemical Society.

a real eutectic. It is the microscopic examination of crystals formed from a racemic solution which demonstrates the existence of two solid phases which are enantiomorphic. The monocrystals, when dissolved, exhibit an optical rotation which allows one to estimate the partial miscibility limit ($x = 0.4$ or 0.6). Structure determination by X-ray crystallography confirms this case as being one exhibiting conglomerate behavior with partial solid solution.[30]

2.4.9 Quantitative Definition of Similarity. The Coefficient of Isomorphism

It would be desirable to quantitatively describe the similarity, or isosterism, of two molecules. When greater than a certain minimum yet to be defined, such similarity between molecules would permit them to syncrystallize.

Kitaigorodskii has, in fact, suggested a way of reducing this degree of isosterism to numbers.[36] He employs a coefficient that takes into account molecular volumes common to two molecules which are arranged in such a fashion as to maximize the superposition of constituent groups and parts of the skeleton. Given two molecules A and B, one defines the volume which they may have in common, V_o, or the volume of overlap, and V_{no} the volume of nonoverlap. The coefficient of isomorphism ϵ is defined as follows:

$$\epsilon = 1 - \frac{V_{no}}{V_o}$$

According to Kitaigorodskii, no solid solution is likely to obtain when $\epsilon < 0.8$; a miscibility in all ranges of concentrations would imply $\epsilon \geqslant 0.9$.

This coefficient of isomorphism ϵ may be determined in two different ways. First of all, it can be obtained from experimental data on the geometry and the relative orientations of molecules derived from X-ray crystallography. In this case, the overlap and nonoverlap volumes are evaluated on the basis of the real positions of the molecules in the lattice of the solid solution. By way of example, Chion et al.[12,30] have calculated the ϵ coefficient in this way for the nitroxides **1** through **3** (Section 2.4.7). The experimental values found were in the range of 0.64 to 0.69 only (Table 8).

Table 8 Calculated and measured isomorphism coefficients[a]

	Mol. Vol. (Å^3)		V_{no} (Å^3)		ϵ	
Compound	Calcd.	Meas.	Calcd.	Meas.	Calcd.	Meas.
1	161	167	23	51	0.85	0.64
2	176	189	55	50	0.63	0.69
3	178	187	52	56	0.66	0.65

[a] V_o = Molecular volume − volume of R + volume of H; V_{no} = volume of R − volume of H (see formulas **1**, **2**, and **3**, Section 2.4.7).

Table 9 Calculated isomorphism coefficient for enantiomers forming ideal solid solutions

Name	ϵ	
trans-π-Camphanic acid	0.74	
π-Bromocamphoric anhydride	0.86	0.85
ω-Bromocamphoric anhydride	0.99	
Camphoric anhydride	0.85	0.99
π-Bromocamphor	0.83	
π-Chlorocamphor	0.86	
Camphene	0.85	
Camphor	0.97	
Camphoroxime	0.91	
Borneol	0.93	
2-Amyl alcohol, phenylurethane	0.89	
2-Amyl 1-(3-nitrophthalate)	0.91	
2-Amyl 2-(3-nitrophthalate)	0.91	
4′-(2-Methylbutyloxy)biphenyl-4-carboxylic acid	0.92	
2-Naphthyl β-camphorsulfonate	0.99	
2-Methylbutyl 4-(4-ethoxybenzylidene)aminocinnamate	0.94	

The coefficient may also be calculated *a priori* as Kitaigorodskii has done from the molecular volumes of the two substances which may be deduced from covalent and van der Waals radii. One first determines the volume of each molecule by adding up the increments proposed by Kitaigorodskii[36] or by Bondi.[37] To simplify the calculations even while maintaining their predictive value, one choses those conformations which best allow the maximum number of atoms of the molecules to coincide, thus allowing an overlap volume to be defined. The nonoverlap volume is given by the atoms that cannot be superposed. Table 9 lists coefficients calculated in this way for a number of compounds whose enantiomers cocrystallize. For several compounds of the table, two ϵ values are given which correspond to two different ways of superposing the (+)- and (−)-molecules. One can observe that most of the compounds considered here lead to a calculated $\epsilon > 0.8$. The corresponding experimental ϵ values are not available, however.

REFERENCES 2.4

1 F. S. Kipping and W. J. Pope, *J. Chem. Soc.*, 1897, **71**, 989.

2 H. W. B. Roozeboom, *Z. Phys. Chem.*, 1899, **28**, 494.

3 H. W. B. Roozeboom, *Z. Phys. Chem.*, 1891, **8**, 504.

4 J. Timmermans, *Bull. Soc. Chim. Belg.*, 1935, **44**, 17.

5 (a) J. G. Aston, *Physics and Chemistry of the Organic Solid State*, D. Fox, M. M. Labes, and A. Weissberger, Eds., Interscience, London, 1963, p. 543, (b) p. 557.

6 A. H. White and W. S. Bishop, *J. Am. Chem. Soc.*, 1940, **62**, 8.
7 R. Steinsträsser and L. Pohl, *Angew. Chem.*, 1973, **85**, 706; *Angew. Chem. Int. Ed.*, 1973, **12**, 617.
8 M. Leclercq, J. Billard, and J. Jacques, *Mol. Cryst. Liq. Cryst.*, 1969, **8**, 367.
9 J. H. Adriani, *Z. Phys. Chem.*, 1900, **33**, 453.
10 J. Jacques and J. Gabard, *Bull. Soc. Chim. Fr.*, 1972, 342.
11 W. Markwald and E. Nolda, *Ber.*, 1909, **42**, 1583.
12 B. Chion, J. Lajzerowicz, A. Collet, and J. Jacques, *Acta Crystallogr.*, 1976, **B32**, 339.
13 J. Pirsch, *Mikrochim. Acta*, 1956, **1**, 992.
14 J. D. M. Ross and I. C. Somerville, *J. Chem. Soc.*, 1926, 2770.
15 B. K. Singh and S. R. Sarma, *J. Ind. Chem. Soc.*, 1958, **35**, 49. *Chem. Abstr.*, 1959, **53**, 13197.
16 G. T. Morgan, W. J. Hickinbottom, and T. V. Barker, *Proc. Roy. Soc.* (London), 1925, **A110**, 502.
17 E. B. Abbot, A. McKenzie, and J. D. Ross, *Ber.*, 1938, **71**, 9.
18 H. Rheinbolt and M. Kircheisen, *J. Prakt. Chem.*, 1926, **113**, 351.
19 (a) H. A. P. Oonk and J. Kroon, *Acta Crystallogr.*, 1976, **B32**, 500. (b) J. Kroon, P. R. E. van Gurp, H. A. J. Oonk, F. Baert, and R. Fouret, *Acta Crystallogr.*, 1976, **B32**, 2561.
20 F. Baert and R. Fouret, *Cryst. Struct. Comm.*, 1975, **4**, 307.
21 H. A. J. Oonk, K. H. Tjoa, F. E. Brants, and J. Kroon, *Thermochim. Acta*, 1977, **19**, 161.
22 H. J. Lucas, F. W. Mitchell, Jr., and H. K. Garner, *J. Am. Chem. Soc.*, 1950, **72**, 2142.
23 K. Petterson, *Ark. Kemi*, 1954, **7**, 339.
24 E. B. Abbott, E. W. Christie and A. McKenzie, *Ber.*, 1938, **71**, 9.
25 J. T. Melillo and K. Mislow, *J. Org. Chem.*, 1965, **30**, 2149.
26 B. K. Singh and B. S. Saxena, *J. Ind. Chem. Soc.*, 1958, **35**, 103. *Chem. Abstr.*, 1959, **53**, 13197.
27 B. K. Singh and B. S. Saxena, *J. Sci. Ind. Res.*, 1959, **18B**, 479. *Chem. Abstr.*, 1960, **54**, 17451.
28 B. K. Singh and S. K. Seth, *J. Ind. Chem. Soc.*, 1956, **33**, 491. *Chem. Abstr.*, 1957, **51**, 6559.
29 B. K. Singh and B. S. Saxena, *J. Ind. Chem. Soc.*, 1960, **37**, 80. *Chem. Abstr.*, 1962, **56**, 1456.
30 B. Chion, J. Lajzerowicz, D. Bordeaux, A. Collet, and J. Jacques, *J. Phys. Chem.*, 1978, **82**, 2682.
31 J. Pirsch, *Monatsh. Chem.*, 1955, **86**, 216.
32 I. Prigogine and R. Defay, *Chemical Thermodynamics*, 4th ed., Longmans, London, 1967, p. 350.
33 G. Friedel, *Leçons de Cristallographie*, Berger-Levrault, Paris, 1926, p. 541ff.
34 S. S. Ament, J. B. Wetherington, J. W. Moncrief, K. Flohr, M. Mochizuki, and E. T. Kaiser, *J. Am. Chem. Soc.*, 1973, **95**, 7896.
35 F. Baert, Thesis, Lille, 1976.
36 A. I. Kitaigorodskii, *Organic Chemical Crystallography*, Consultants Bureau, New York, 1961, p. 230.
37 A. A. Bondi, *Physical Properties of Molecular Crystals, Liquids and Glasses*, Wiley, New York, 1968, p. 453.
38 A. Burger, *Medicinal Chemistry*, Vol. 1, Interscience, New York, 1951, pp. 36–50.
39 I. Langmuir, *J. Am. Chem. Soc.*, 1919, **41**, 868, 1543.

40 (a) H. Erlenmeyer and E. Berger, *Biochem. Z.*, 1932, **255**, 429. (b) H. Erlenmeyer and M. Leo, *Helv. Chim. Acta,* 1933, **16**, 892. (c) H. Erlenmeyer, *Z. Phys. Chem.*, 1934, **B27**, 404. (d) H. G. Grimm, M. Günther, and H. Tittus, *ibid.*, 1931, **14**, 169.

41 M. J. Brienne, A. Collet, C. Fouquey, J. Gabard, M. Leclercq, and J. Jacques, unpublished results.

42 G. Urbain, *Les Disciplines d'une Science*, G. Doin, Paris, 1921, p. 139ff.

43 W. A. Yager and S. O. Morgan, *J. Am. Chem. Soc.*, 1935, **57**, 2071; A. H. White and S. O. Morgan, *J. Am. Chem. Soc.*, 1935, **57**, 2078.

44 G. A. Wagner and W. Brickner, *Ber.*, 1899, **32**, 2302; see footnote p. 2303.

45 J. Capillon and J. P. Guetté, *Tetrahedron*, 1979, **35**, 1807; J. Capillon, Thesis, Paris, 1975.

46 A. Schoofs, J. P. Guetté, and A. Horeau, *Bull. Soc. Chim. Fr.*, 1976, 1215.

2.5 POLYMORPHISM IN BINARY SYSTEMS

The ability of an organic compound to crystallize in several forms is far more frequent than is generally supposed. Polymorphism is thus a phenomenon whose theoretical and practical importance is not negligible.

Considering only the latter point, polymorphism can aid – or befuddle – the characterization of organic compounds as a consequence of its influence on infrared spectra in the solid state or on melting points. It can play an important role on the outcome of recrystallizations through its influence on solubilities. And again, from the point of view of medicinal properties – specifically kinetics of drug activity – it can affect the rate of action of drugs introduced in the solid state into living organisms.[1]

But we are concerned here with polymorphism of enantiomers and racemates. It is of particular importance inasmuch as it sometimes permits the interconversion of the different types of racemates we have previously considered.

This section treats transformation between *unsolvated* crystalline substances that can be interpreted and illustrated by binary phase diagrams and for which thermodynamic data are relatively easily available. The equally important case of polymorphism associated with solvation of crystals is examined in Chapter 3.

2.5.1 Enantiotropy and Monotropy. Transition Temperature

Two types of polymorphism are known. We shall give a simplified description of them while assuming, as we have done earlier (Section 2.2.1), that no vapor phase is present and that the pressure is invariant. Under these conditions, a system consisting of one component ($C = 1$) existing in two crystalline forms ($\Phi = 2$) is invariant ($v = C - \Phi + 1 = 0$); this signifies that the two crystalline phases of the same compound can coexist in equilibrium only at a single temperature T_0. Examination of curves representing the free energy of the system as a function of temperature (Fig. 1) then allows one to distinguish two possibilities: enantiotropy,

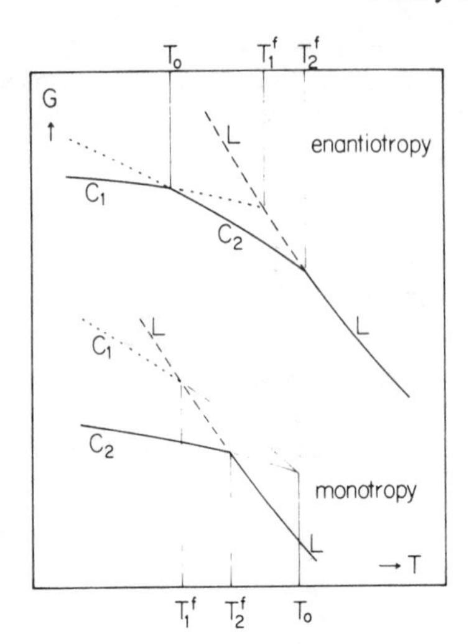

Figure 1 Simplified description of the enantiotropic and monotropic transformations at constant pressure: free energy of the system as a function of temperature. C_1 and C_2 are crystalline phases while L is the liquid phase; (———) stable (liquid or solid) phases; (– – – –) supercooled liquid; (- - - - - - - -) metastable crystal.

in which case T_0 is lower than the melting points of either of the two crystalline forms T_1^f and T_2^f. Each of the forms C_1 and C_2 is thus stable in its temperature range, and the passage from one form to the other can occur reversibly at T_0 (transition temperature). In the second case, called monotropy, T_0 is higher than the melting points of the two polymorphic forms (T_1^f and T_2^f). In this case, one of the forms (C_1 here) is always metastable relative to the other, and a transformation can take place in an irreversible manner only in the direction $C_1 \rightarrow C_2$, and this at any temperature lower than T_1^f. The transition temperature T_0 is then virtual, inasmuch as it occurs above the melting points. As a corollary, the metastable form C_1 is obtainable only upon cooling a substance in the molten state to a temperature lower than T_1^f.

While it is evident that melting is never subject to delay, delayed crystallization is a common occurrence. Such a phenomenon affects monotropic as well as enantiotropic systems. According to the so-called "law of successive reactions" of Ostwald, crystallization of a polymorphic substance from a solution or a melt yields crystals which at first are in the less stable form.[2] These crystals are subsequently transformed into the stable form either while standing in contact with the liquid (either solution or melt) or upon heating of the solid to an appropriate temperature. It is evidently this circumstance, that is, the ease of formation of metastable forms, which allows them to be detected and observed without particular difficulty (for instance, during the determination of their melting point). As a consequence of the slowness with which transformations take place in the solid state, crystals of a given metastable form can frequently be maintained virtually indefinitely.

Let us analyze the traces obtainable with a differential scanning calorimeter

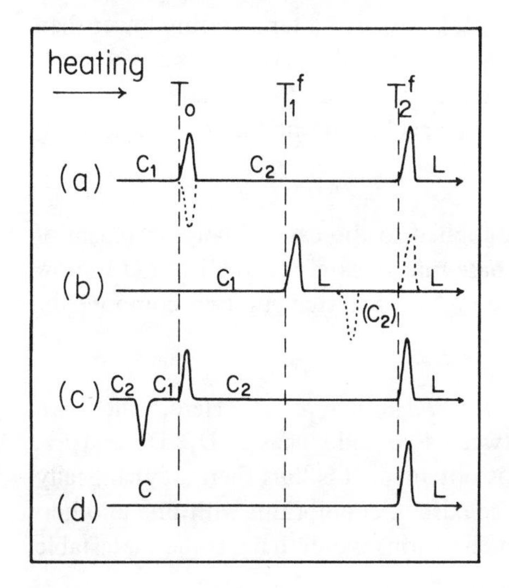

Figure 2 DSC traces: enantiotropy.

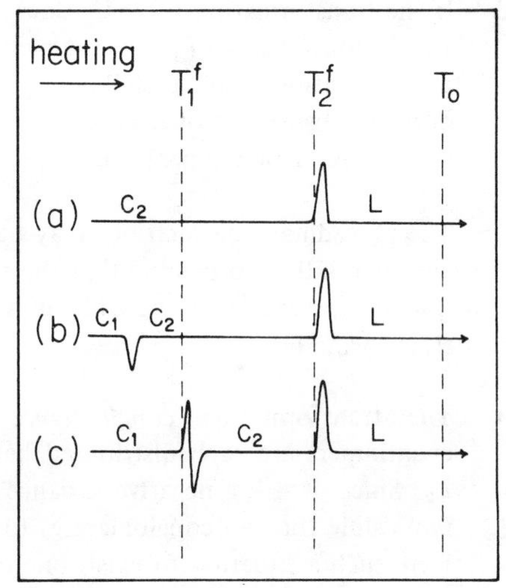

Figure 3 DSC traces: monotropy.

by reference to the diagrams of Figs. 1, 2, and 3. Consider first a case of enantiotropy (Fig. 2). Upon heating, different curves can be obtained according to the history of the sample, the speed of the crystal–crystal transformation, and the heating rate of the calorimeter.

1 When a sample of crystals C_1 is heated, the transition $C_1 \rightarrow C_2$ is observed at T_0 followed by the melting of C_2 at T_2^f. The transition $C_1 \rightarrow C_2$ may be either endothermic (solid line) or exothermic (dotted line).

2 When the sample is subjected to more rapid heating, there is insufficient time for the transition $C_1 \rightarrow C_2$ to occur; the melting of C_1 is observed at T_1^f (solid line), and this may be followed by the crystallization of C_2 and its melting at T_2^f (dotted line).

3 A crystal of C_2, which is unstable below T_0, is heated. Before T_0 is reached, the disequilibrium may suddenly end. While such an occurrence is unpredictable, when it does occur it is marked by an evolution of heat. The heating process then continues as in case **1** or case **2**.

4 Alternatively, if no event occurs below T_0, melting of C_2 takes place at its melting point T_2^f.

Let us next examine what takes place in the case of monotropy (Fig. 3).

1 When a crystal of C_2 (the stable form) is heated, the only thing observed is its melting at T_2^f.

2 If a crystal of C_1 is heated, then before reaching T_1^f it may be transformed into a crystal of C_2, which is more stable and which melts are T_2^f.

3 If the transformation $C_1 \rightarrow C_2$ does not take place at a temperature lower than T_1^f, C_1 must melt at T_1^f, but the melting may take place concomitantly with its transformation into crystalline C_2. This corresponds to the case where one observes absorption of heat followed by its evolution (melting/resolidification) both at the same temperature.

The preceding considerations may be applied to the case of polymorphism of enantiomer mixtures providing that the phase rule be carefully applied. Let us now examine the several cases possible in theory for a system of two components, namely, (+) and (−).

1 The transformation *conglomerate 1 ⇌ conglomerate 2*. Here, the two enantiomers are each distributed between two solid phases, D_1, D_2 and L_1, L_2. Since $\Phi = 4$, a negative variance is obtained. It is thus thermodynamically impossible for two conglomerates to coexist in equilibrium with one another. Were such a situation to exist, one of the forms would have to be metastable and the system necessarily monotropic.

2 The transformation *enantiomer 1 ⇌ enantiomer 2* in the presence of a nonpolymorphic racemic compound. Since the two components are distributed between three solid phases D_1, D_2 and R (or L_1, L_2 and R) the variance is zero. The three phases may thus coexist in equilibrium at a transition temperature T_0, and the system is enantiotropic or monotropic according to where T_0 lies relative to the melting points.

3 The transformation *racemic compound ⇌ conglomerate*. Here, two components are distributed between three solid phases, D, L, and R yielding a zero variance. The same conclusions may be drawn as in case 2.

4 The transformation *racemic compound 1 ⇌ racemic compound 2* in the presence of nonpolymorphic enantiomers. There are two ways of looking at this case both of which lead to the same conclusion. For a mixture of racemic composition, there is but one independent component (since the quantities of (+) and (−) are equal) and two solid phases, R_1 and R_2. The variance is thus equal to zero. Alternatively, for the case of partially resolved mixtures, there are two components and three phases, D, R_1, R_2 (or L, R_1, R_2). The variance is still zero, and the conclusion is identical to that of case 2.

5 The transformation *conglomerate ⇌ solid solution*. There are two components and three solid phases: D, L, and SS. The same conclusions may be drawn as in case 2.

6 The transformation *racemic compound ⇌ solid solution*. The argument is like that of case **4**: for the racemate, there is one independent component and there are two solid phases, R and SS; for any partially resolved mixture, on the other hand, there are two components and three solid phases, D, R, and SS (or L, R, and SS). The conclusions are as in 2.

7 The transformation *solid solution 1 ⇌ solid solution 2*. Here, there are two components and two solid phases, SS_1 and SS_2. The variance is 1. In contrast

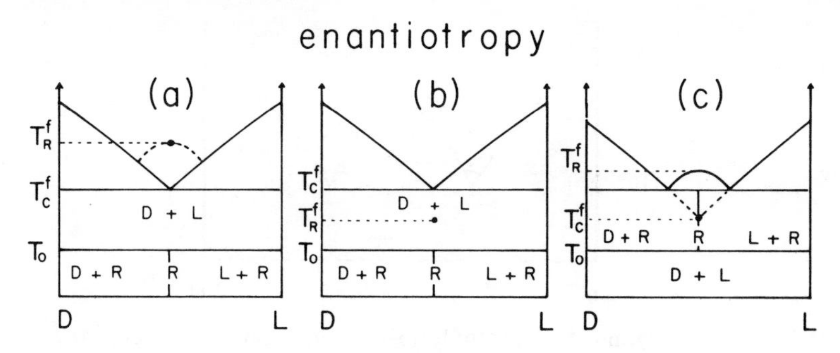

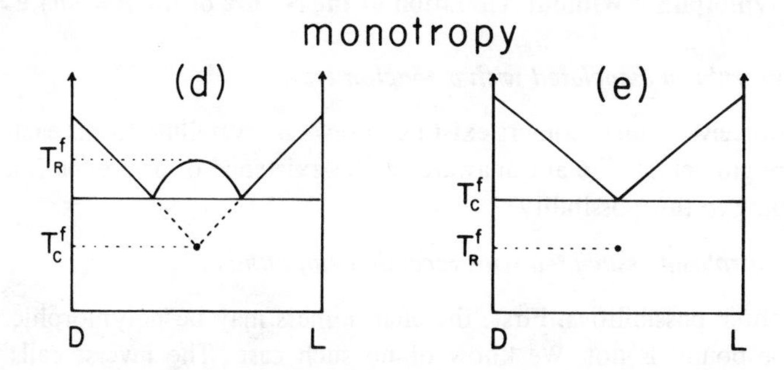

Figure 4 Phase diagrams which may correspond to polymorphism of type racemic compound ⇌ conglomerate: (a) The racemic compound is stable at low temperature and is transformed into a conglomerate at T_0; the racemic compound melts at T_R^f, but is liquidus curve (dotted curve) is observable only in the absence of stable thermodynamic equilibrium. (b) Case similar to (a), with $T_R^f < T_C$. (c) The conglomerate is stable at low temperature yielding a racemic compound above T_0; the dotted curve corresponds to a metastable equilibrium. (d) Monotropy with metastable conglomerate. (e) Monotropy with metastable racemic compound.

to preceding cases there thus exists a *range* of temperatures and of concentrations within which the two phases may coexist at equilibrium.

In practice, it is not always easy – if it is at all possible – to establish from experimental data such as dsc curves what type of polymorphism and what phase diagram one is dealing with. The similarity between some of the phase diagrams for the above cases is illustrated by those which may correspond to the transformation racemic compound ⇌ conglomerate (Fig. 4). Each of the other theoretical cases just cited corresponds to other phase diagrams which we feel it unnecessary to sketch. Interested readers will be able to construct them without difficulty.

We will now illustrate the preceding cases by describing in the following paragraphs some real systems that have been studied in more or less detail.

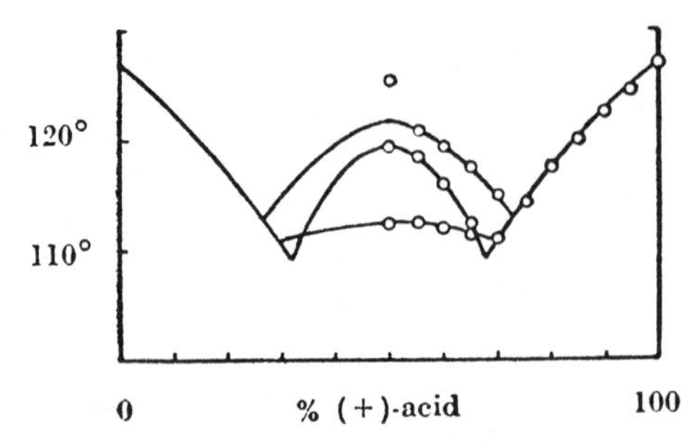

Figure 5 Phase diagram for 2-(ethylthio)succinic acid. M. Matell, *Ark. Kemi*, 1953, 5, 17. Reproduced by permission of the Royal Swedish Academy of Sciences.

2.5.2 Polymorphism Without Alteration of the Nature of the Racemate

(a) Polymorphism associated with a conglomerate

One may conceive of enantiomers existing in several crystalline forms each of which yields a conglomerate. We are unaware of the existence of any experimental data corresponding to this possibility.

(b) Polymorphism associated with racemic compounds

There are three possibilities. First, the enantiomers may be polymorphic while the racemic compound is not. We know of no such case. The inverse calls for polymorphism of the racemic compound only. Many examples of this type have been reported;[3,4,7] one of these is illustrated in Fig. 5. The mandelic acid system, in particular, has been thoroughly studied.[4] In the third instance, both the enantiomers and the racemic compound are polymorphic. Matell, for example, has described this type of polymorphism[3] (Fig. 6).

(c) Polymorphism associated with solid solutions

Carvoxime is one of the rare examples of this type which has been the subject of a relatively detailed study.[5,8] The pure enantiomers exist in two crystalline forms, mp 67 and 73°C each forming a solid solution between (+) and (−). Two pseudo-

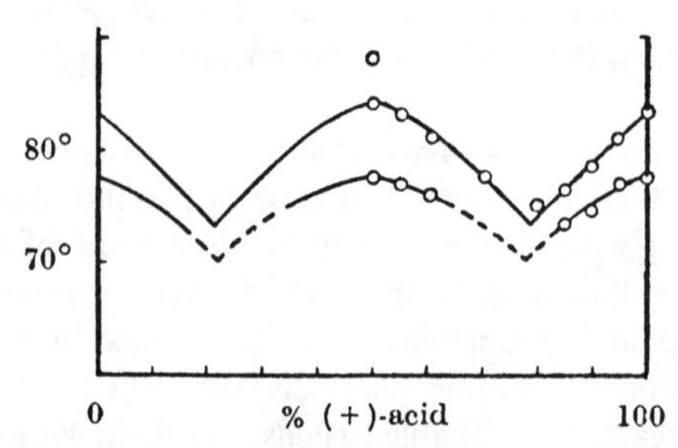

Figure 6 Phase diagram for *n*-hexylsuccinic acid. M. Matell, *Ark. Kemi*, 1953, 5, 17. Reproduced by permission of the Royal Swedish Academy of Sciences.

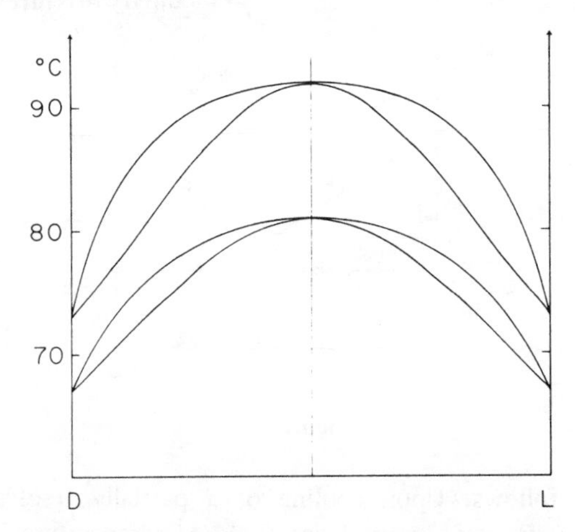

Figure 7 Polymorphism of the carvoxime system. The solidus and liquidus curves have been calculated by means of eqs. (1) and (2) of Section 2.4.5 and of the data given by Oonk[5]: Stable form, $T_A^f = 73°C$, $T_R^f = 92°C$; $\Delta H_A^f = 4.1\ \text{kcal} \cdot \text{mol}^{-1}$. Metastable form, $T_A^f = 67°C$, $T_R^f = 81°C$, $\Delta H_A^f = 3.3\ \text{kcal} \cdot \text{mol}^{-1}$.

racemates, mp 81 and 92°C, are thus obtained. The phase diagram is represented by Fig. 7. Two nicotine derivatives that manifest the same type of polymorphism have been described by Langhammer.[21]

2.5.3 Polymorphism with Alteration to the Nature of the Racemate

The type of polymorphism which engages our closest attention here involves the possible transformations between the various types of racemates. Such transformations have many consequences on the practice of resolution, such as, the control of enantiomeric purity, ease of crystallization, and so on.

It is once again Roozeboom[6] who was the first to systematically describe the different theoretical possibilities of polymorphism involving binary mixtures of enantiomers: (a) In the first case, a racemic compound exists only at low temperature and is transformed into a conglomerate, or, the inverse, a conglomerate is stable at a low temperature and is transformed into a racemic compound upon heating. (b) In the second case, a racemic compound exists at a higher temperature and is formed at the expense of a solid solution; the inverse is also possible. (c) Third, a conglomerate may be transformed into a solid solution, or the inverse transformation may take place.

Let us now look at these possibilities in some detail.

(a) Transformation of a racemic compound to a conglomerate

The racemic compound which exists at a low temperature is transformed into a conglomerate at a higher temperature. The anilide of 2-(1-nitro-2-naphthoxy)-propionic acid, whose phase diagram is shown in Fig. 8, is such a system. It may

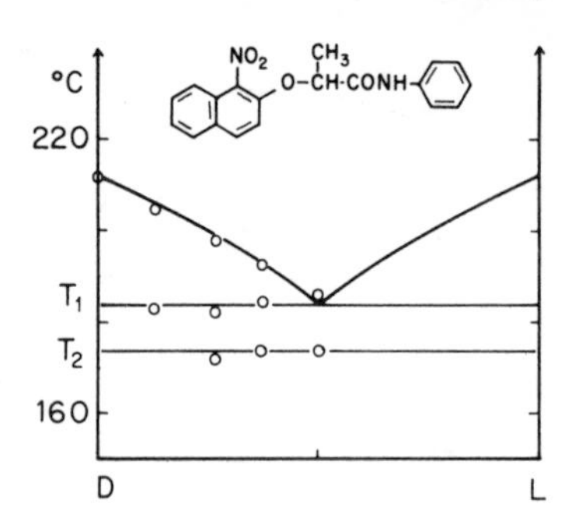

Figure 8.

be interpreted as follows: Upon cooling of a partially resolved mixture, the enantiomer present in excess crystallizes first. At temperature T_1, a crystalline mixture of the two enantiomers precipitates which is transformed into a mixture of one enantiomer and a racemic compound at a temperature below T_2. When the different racemic forms are examined by infrared spectroscopy (as a mull in Nujol or as a KBr pellet), one observes that the racemate formed below T_2 has a spectrum that is different from that of either enantiomer, while the cristalline form that melts at T_1 has a spectrum identical to that of the enantiomers.

Another system that corresponds to this case is that of binaphthyl, which is discussed in Section 2.5.4.

(b) Transformation of a conglomerate to a racemic compound

The conglomerate that exists at low temperature is transformed into a racemic compound. An example of this type of racemate is 2,2′-diamino-1,1′-binaphthyl (Fig. 9). A sample crystallized from alcohol yields an infrared spectrum different from that of the enantiomers. However, after melting and resolidification of the sample, the spectrum of the racemate is observed to be identical to that of the enantiomers. In Fig. 9, T_1 is the eutectic melting point corresponding to the racemic compound form, while T_2 is the melting point for the conglomerate. The same type of diagram has been observed for α-bromocamphor.[9,10]

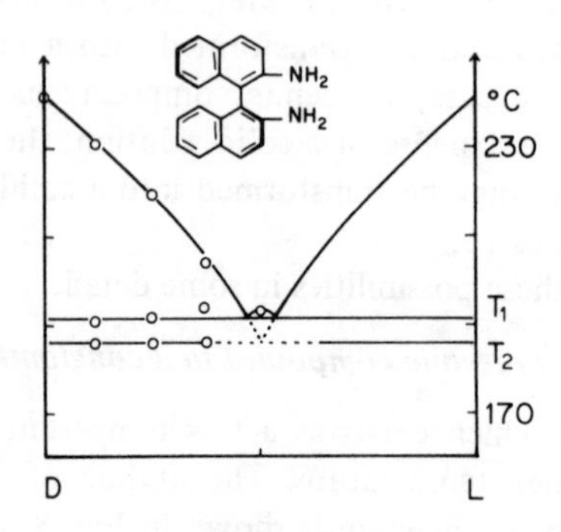

Figure 9.

(c) Transformations between a racemic compound and a solid solution

The racemic compound is transformed into a solid solution, or vice versa. The literature dealing with these systems (and, broadly speaking, much that deals with solid solutions of enantiomers) is particularly confusing principally as a consequence of the rareness of examples and of the difficulties attending the interpretation of experimental measurements dealing with them.

Camphoroxime has been described by Kipping and Pope[11] and later by Adriani[8] as an example of a racemic compound that is stable at a low temperature and which is transformed into a solid solution upon heating. The phase diagram given in the Adriani paper also indicates the existence of a solid solution ⇌ plastic crystal transition. Jacques and Gabard subsequently confirmed that the form stable at 25°C is a racemic compound.[12]

(d) Transformation between a conglomerate and a solid solution

The solid solution is transformed into a conglomerate, or vice versa. According to some observations of Kipping,[14] it would have seemed possible for *cis*-π-camphanic acid to be a conglomerate at low temperature and a solid solution at higher temperature. At the time, this interpretation was contested by Roozeboom,[6] who did not believe that conglomerate ⇌ solid solution polymorphism was possible.

The *cis*-π-camphanic acid case has since been reexamined,[10,13] and its phase diagram, which was determined by dsc, is shown in Fig. 10. The dextrorotatory acid exhibits a transition at 208°C and melts at 230°C. The racemate exhibits transitions at 170 and at 208°C and melts also at 230°C. Mixtures of 70 and 40% enantiomeric purity both exhibit the transition and melt at the same temperature as does the racemate. The phase diagram thus reduces to three horizontal lines: one at 170°C corresponding to the transformation conglomerate → solid solution; one at 208°C corresponding to the transformation solid solution → plastic crystal; and, of course one at 230°C corresponding to melting. In order to obtain the conglomerate form, it is necessary to follow the instructions of Kipping, that is, to carry out a slow crystallization by evaporating a methanol solution of the acid in the cold.

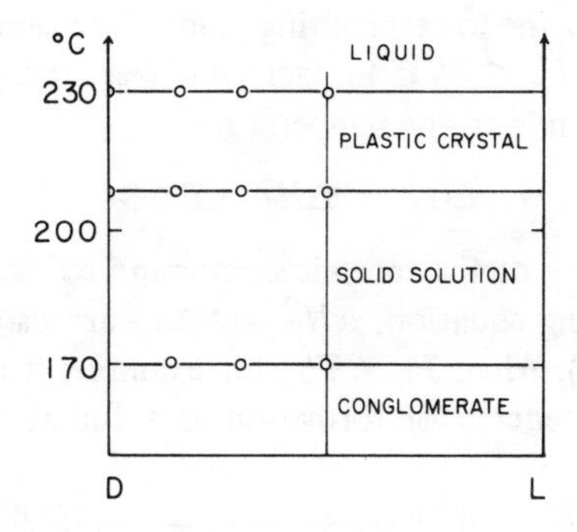

Figure 10 Phase diagram for *cis*-π-camphanic acid.

2.5.4 The Racemic Compound ⇌ Conglomerate Transformation

Among the problems associated with the phenomenon of polymorphism, that which deals with the possible transformation of racemic compounds into conglomerates, or the inverse, is certainly one of the most interesting, because of its practical as well as conceptual consequences.

We have already seen that conglomerates are few in number. Among them, some amino acids and their derivatives are prominent. Racemic compounds, on the other hand, constitute the overwhelming majority of equimolecular mixtures of enantiomers. Now, among the hypotheses proposed to explain the appearance of optical rotatory power in nature,[15] the one that proposes the spontaneous resolution of crystals of certain racemates is not the most unreasonable. It proposes that a single enantiomer separates from an inactive solution through crystallization without requiring the intervention of a physical agent able to induce an enantioselective synthesis of the original organic substance. Such a disequilibrium, while purely accidental, can ultimately be rendered irreversible, for example, as a consequence of the runoff of the mother liquors away from the crystalline deposit, or through chemical transformation of the enantiomer remaining in solution in the presence of a small quantity of an appropriate reagent. In the context of this hypothesis, it is evidently important to know whether *any* racemate can, at will and under appropriate experimental conditions, be found in the form either of a conglomerate or a racemic compound.

We now examine the extent to which it is possible to predict the existence of racemic compound ⇌ conglomerate transformations as a function of temperature and of pressure.

(a) *Polymorphism as a function of temperature*

The thermodynamic parameters permitting the measurement of racemic compound stability have been defined in Section 2.3.4. These are enthalpy, entropy, and free energy of formation, ΔH^{ϕ}, ΔS^{ϕ}, and ΔG^{ϕ}, corresponding to the solid phase "reaction"

$$\mathrm{D_{crystal}} + \mathrm{L_{crystal}} \rightarrow [\mathrm{DL}]_{\mathrm{crystal}}$$

We have calculated the values of these parameters at the melting point of the crystalline species having the lowest melting point (for example, at T_R^f for a racemic compound melting at $T_R^f < T_A^f$). In fact, the free energy of formation of the racemic compound depends on the temperature:

$$\Delta G_T^{\phi} = \Delta H_T^{\phi} - T\,\Delta S_T^{\phi}$$

and the range of stability of the racemic compound corresponds to negative values of ΔG_T^{ϕ}. In the preceding equation, ΔH_T^{ϕ} and ΔS_T^{ϕ} are also dependent on temperature (see Section 2.3.4). When $T_R^f < T_A^f$, for example, it is easy to show that the expression for the free energy of formation as a function of temperature is the following:[16]

$$\Delta G_T^{\phi} = \Delta H_{T_R^f}^{\phi} - T\,\Delta S_{T_R^f}^{\phi} + \int_T^{T_R^f} (C_A^s - C_R^s)\,dT - T \int_T^{T_R^f} (C_A^s - C_R^s)\frac{dT}{T} \qquad (1)$$

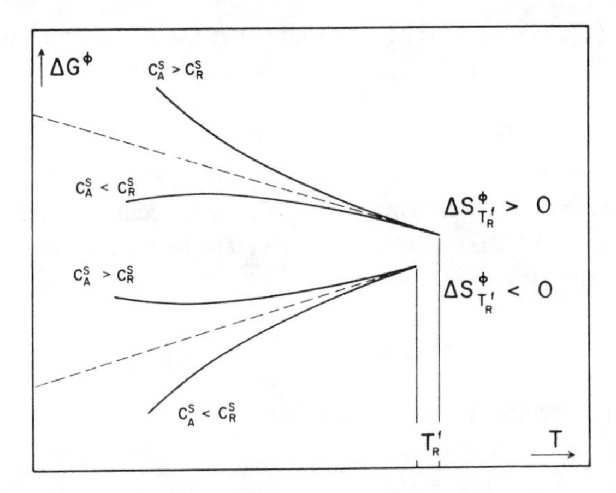

Figure 11 Variation of the free energy of formation of racemic compounds with temperature [eq. (2)]. The existence of a racemic compound–conglomerate transition implies that the curve ΔG^{ϕ} vs T (solid line) can cross the horizontal line $\Delta G^{\phi} = 0$ (which is not drawn). Dotted curves correspond to a variation of ΔG^{ϕ} vs T when $\alpha = 0$ [eq. (3)]. M. Leclercq, A. Collet, and J. Jacques, *Tetrahedron*, 1976, **32**, 821. Reproduced by permission of Pergamon Press Ltd.

[When $T_R^f < T_A^f$, the corresponding expression for ΔG_T^{ϕ} may be obtained simply by substituting T_A^f for T_R^f in eq. (1)]. In this equation, $\Delta H^{\phi}_{T_R^f}$ and $\Delta S^{\phi}_{T_R^f}$ are the enthalpy and entropy of formation of the racemic compound at its melting point, and C_A^s and C_R^s are the specific heats of the enantiomers and of the racemic compound, respectively, in the solid state.

In order to see how ΔG_T^{ϕ} varies with the temperature, and in particular to try to find possible racemic compound ⇌ conglomerate transitions ($\Delta G_T^{\phi} = 0$), it is important to have information about the function $(C_A^s - C_R^s)_T$. Studies carried out on a small number of cases (see Section 2.3.4) indicate that the difference in specific heats of the solids, $C_A^s - C_R^s$, can attain 5 to 10 cal · mol^{-1} · K^{-1} in the vicinity of the melting point, and that it decreases nearly linearly as the temperature drops. Granting that this difference tends to zero at low temperature, it would seem that, as a first approximation, $C_A^s - C_R^s = \alpha T$. Equation (1) then becomes

$$\Delta G_T^{\phi} = \Delta H^{\phi}_{T_R^f} - T\,\Delta S^{\phi}_{T_R^f} + \frac{\alpha}{2}(T_R^f - T)^2 \qquad (2)$$

To illustrate this point, we have drawn the shape (parabolic) of this relationship according to the signs of $\Delta S^{\phi}_{T_R^f}$ and $C_A^s - C_R^s$ (Fig. 11). To determine in which cases a racemic compound ⇌ conglomerate transition exists, it would suffice to determine under which conditions these curves cross the horizontal line $\Delta G^{\phi} = 0$. The reader is invited to consider on this figure the different situations possible in theory.

In practice, as a consequence of real experimental data, one must taken into account both the imprecision in $\Delta H^{\phi}_{T_R^f}$ and $\Delta S^{\phi}_{T_R^f}$ and the approximation implied in the function $(C_A^s - C_R^s) = \alpha T$. It is evident that the further one departs from the melting point T_R^f, the more the latter term becomes important. It follows that the part of the curves of Fig. 11 toward the lower temperatures is only of qualitative significance. As a consequence, the prediction of a transition far removed from the melting point, for example, $T_R^f - T > 100°$, is likely to be uncertain.

For this reason, the following discussion is limited to cases of transitions occurring near T_R^f. The ΔG_T^ϕ curves of Fig. 11 are then quite near to their tangent at T_R^f, which is none other than the line

$$\Delta G^\phi = \Delta H^\phi_{T_R^f} - T\,\Delta S^\phi_{T_R^f} \tag{3}$$

while the transition temperature approaches

$$T_0 \simeq \frac{\Delta H^\phi_{T_R^f}}{\Delta S^\phi_{T_R^f}} \tag{4}$$

It appears then that the existence of a racemic compound ⇌ conglomerate transition in the vicinity of the melting point implies that $\Delta H^\phi_{T_R^f}$ and $\Delta S^\phi_{T_R^f}$ *bear the same sign*.

Let us now examine the several possible cases in turn:

1 The most common case is that where $\Delta H^\phi_{T_R^f}$ and $\Delta S^\phi_{T_R^f}$ are both negative. Inasmuch as, in most instances, $\Delta G^\phi_{T_R^f}$ is negative, one has the situation corresponding to Fig. 12*a*: T_R^f is below T_0 and the racemic compound is always more stable than the conglomerate. The transition temperature is *virtual*. If in spite of this a conglomerate could be obtained, it would necessarily be metastable. In this event, since its melting point T_C^f is below T_R^f, one would observe an exothermic transition at $T \leqslant T_C^f$. The enthalpy of this transition corresponds to the (irreversible) enthalpy of formation of the racemic compound ΔH_T^ϕ from the conglomerate (typical monotropy).

An enantiotropic racemic compound ⇌ conglomerate transition can effectively be observed only if T_R^f is greater than T_0. Now, it is clear that in this situation (Fig. 12*b*) the free energy of formation of the racemic compound is *positive* from T_0 to T_R^f; in this temperature range, therefore, the racemic compound is metastable. However, the sluggishness with which transitions take place in the solid state may in some cases allow the observation of the racemic compound up to its melting point. An example of this has been described:[17] racemic binaphthyl exhibits two crystalline forms – the one having $T_R^f = 145°C$ is a racemic compound while the other with $T_C^f = 158°C$ is a conglomerate. The transition temperature above which the racemic compound becomes metastable was estimated experimentally to be 76°C. From the data given,[17] we may calculate $\Delta G^\phi_{T_R^f} \simeq +0.16\,\text{kcal}\cdot\text{mol}^{-1}$, $\Delta H^\phi_{T_R^f} \simeq -0.19\,\text{kcal}\cdot\text{mol}^{-1}$, and $\Delta S^\phi_{T_R^f} \simeq -5\,\text{cal}\cdot\text{mol}^{-1}\cdot\text{K}^{-1}$, which would place the transition in the

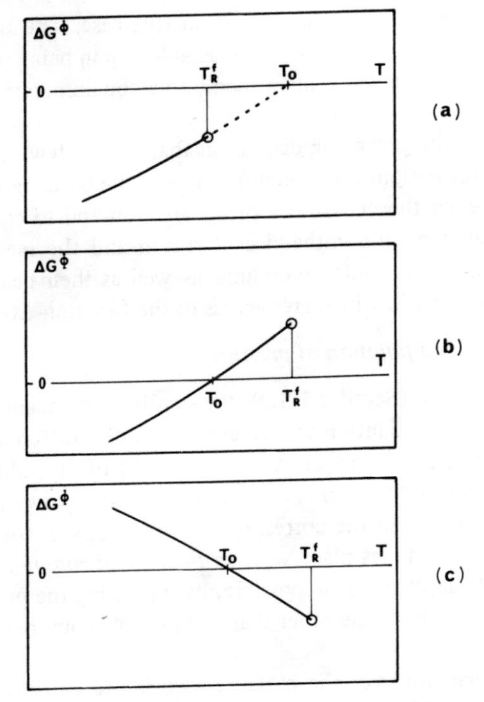

Figure 12 Racemic compound ⇌ conglomerate transitions in the vicinity of the racemate melting point. M. Leclercq, A. Collet, and J. Jacques, *Tetrahedron*, 1976, **32**, 821. Reproduced by permission of Pergamon Press Ltd.

vicinity of 100°C in fairly good agreement with fact.* The destabilization of this racemic compound with rising temperature is manifestly due to its large negative entropy of formation.

2 There remains the case where $\Delta H^{\phi}_{T^f_R}$ and $\Delta S^{\phi}_{T^f_R}$ are positive (Fig. 12*c*). Contrary to what we observed in the preceding case, the free energy of formation decreases as the temperature rises. Consequently, it is the conglomerate which is the stable form at low temperature. In fact, very few compounds belong to this category. Among them, one may cite *m*-chlorophenylhydracrylic acid,[16] for which $\Delta H^{\phi}_{T^f_R} = +0.56 \pm 0.13$ kcal · mol^{-1}, $\Delta S^{\phi}_{T^f_R} = +1.62 \pm 0.45$ cal · mol^{-1} · K^{-1}, and $C^s_A - C^s_R = 0$. From these data there may be calculated an enantiotropic transition temperature which is virtually equal to T^f_R ($T_0 = 72 \pm 20$°C, $T^f_R = 67$°C). The free energy of formation would be positive at all

* In this calculation, we have arbitrarily taken $C^l - C^s_A = 24$ cal · mol^{-1} · K^{-1} in the absence of experimental data. The values found can therefore only be approximate ones. Also note that since the enthalpy of fusion and the melting point of the pure enantiomers are inaccessible due to the rapid racemization of the product, these data must be replaced by the enthalpy of fusion and the melting point of the conglomerate. In the calculation of the entropy of formation, the term R ln 2 must then be omitted from eq. (5b) of Section 2.3.4.

temperatures below the melting point in this case. The racemic compound observed would be expected to be metastable. Upon being moderately heated, this racemic compound was transformed into a conglomerate.[16]

It follows from the preceding discussion that only a few racemic compounds may undergo transformations into conglomerates as the temperature is varied. Also, it is possible to predict the occurrence of a transition and to calculate its temperature, if one has at hand the enthalpies of fusion and the melting points of the enantiomers and of the racemic compound as well as their heat capacities in the liquid and solid states, all of which give access to the function $\Delta G^{\phi}(T)$.

(b) Polymorphism as a function of pressure

As we have already seen (Section 1.3.2), the densities of racemic compounds and those of the constituent enantiomers are generally quite different. In other words, the formation of racemic compounds from mixtures of crystalline enantiomers is accompanied by a change of volume, ΔV. Many cases are known in which a racemic compound is less dense than the corresponding enantiomers; the formation of the racemic compound then takes place with an *increase in volume*. In agreement with Le Chatelier's principle, it must be possible, by increasing the pressure, to promote the inverse reaction which is none other than the racemic compound $\rightleftharpoons$ conglomerate transformation.

The order of magnitude of the pressure required to carry out such a transformation can be estimated from the equation

$$\frac{\partial}{\partial P}(G) = V$$

If one allows that enantiomers and the corresponding racemic compound have identical compressibilities, one may write, as a first approximation,

$$\Delta P \simeq \Delta G/\Delta V$$

Given a free energy of formation of the order 0.1 to 0.5 kcal $\cdot$ mol^{-1} for racemic compounds at atmospheric pressure and a volume difference ($V_{\text{racemic}} - V_{\text{enantiomer}}$) of several cm$^3 \cdot$ mol^{-1} (see the list of densities, Section 1.3.2), one calculates a ΔP of 10^3 to 10^4 bars, which is a relatively easily accessible pressure.

It is thus not inconceivable that the frequency of spontaneous resolutions may be considerably increased under conditions of elevated pressure. This hypothesis has not, for the moment, been experimentally verified.

2.5.5 Polymorphism of Chiral Mesomorphs

Just as a given compound – whether chiral or not – may exist in the solid state in several crystalline forms, so some compounds in the liquid crystalline state may successively adopt a different organization upon being heated. This special kind of polymorphism may exist for racemates as well as for resolved substances; it is for this reason that a brief analysis is warranted here. Chiral mesomorphic substances generally form solid solutions of enantiomers in the crystalline state. This is discussed in Section 2.4.1, in which the reader will also find a brief description of the main structures known for mesomorphic phases.

The polymorphic transformations just alluded to are associated with and are made evident by changes in physical properties. They may be either monotropic or enantiotropic just as is the case for crystalline substances. In the literature, they are most often marked by transition temperatures. The existence of a mesophase may sometimes be detected by observation of its texture with a polarizing microscope. It may generally be identified by establishing its total miscibility with a reference mesophase. The structures which characterize the several known types of mesomorphic substances (nematic, cholesteric, smectic) have been determined, at least for some, by X-ray diffraction.

SCHEME 1

Achiral compound or racemate		$S_E\ S_G\ S_B\ S_F\ S_C\ S_D\ S_A$	N	
	Crystal	Smectic	Nematic Cholesteric	Liquid
Resolved compound		$S_E\ S_G\ S_B^*\ S_F\ S_C^*\ S_D\ S_A$	N*	
——— Increasing temperature ———→				

Scheme 1 summarizes the succession of smectic (S), nematic (N), or cholesteric (N*) phases that may be observed with increasing temperature for achiral or racemic mesomorphic substances on the one hand and for resolved substances on the other. The scheme includes the different known smectic phases conventionally distinguished from one another by subscript letters (A, B, . . .), where the alphabetic order only serves to recall the chronology of their discovery.

For resolved molecules, the asterisk indicates that the mesomorphic phase itself possesses a chiral structure. For example, N* is synonymous with cholesteric. Polymorphic liquid crystals naturally do not generally exhibit *all* the phases shown in the scheme. Only the order of appearance is well defined.

As we have already seen (Section 2.4.1), smectic phases have a layered structure. There are structural variants here too, of which only the most common designated by S_A, S_B, and S_C will be mentioned.

While the average direction of the principal molecular axes are perpendicular to the layers in smectic A, molecules of smectic C are inclined with respect to each layer. Smectic B has additional order: the molecules are arranged in a hexagonal lattice. There are two types of smectic B according to whether the molecules are or are not perpendicular to the layers. In the tilted smectic B and in smectic C, the superposition of layers may give rise to a helical (chiral) structure reminiscent of that of cholesteric phases. This arrangement does not exist in the case of nontilted smectic B or smectic A, even when formed of resolved molecules.

Numerous mesomorphs, chiral as well as achiral, exhibit an especially rich polymorphism. Compound **1** is one of many examples which could be cited; upon heating, it exhibits successively four mesomorphic phases:[18]

T	82	113	162	205	212 °C
Crystal	S_E	S_B	S_A	N	Liquid

C_6H_5–C_6H_4–CH=N–C_6H_4–CH=CH–CO_2–CH_2–CH(CH_3)–CH_3 **1**

Compounds **2** and **3**, which are also chiral mesomorphs, similarly, give rise to polymorphism in the mesomorphic state:[19,20]

T	205	213	241 °C
Crystal	S_C	N	Liquid

C_6H_5–CH_2–CH(CH_3)–CH_2–O–C_6H_4–C_6H_4–CO_2H **2**

T	144	160	208	224 °C
Crystal	S_B	S_C	N	Liquid

C_2H_5–CH(CH_3)–$CH_2CH_2CH_2$–O–C_6H_4–N=CH–C_6H_4–CH=N–C_6H_4–O–$CH_2CH_2CH_2$–CH(CH_3)–C_2H_5 **3**

REFERENCES 2.5

1 K. Münzel, *Progr. Drug Res.*, 1966, **10**, 204.

2 A. Findlay, *The Phase Rule*, 9th ed., revised by A. N. Campbell and N. O Smith, Dover, New York, 1951, p. 54.

3 M. Matell, *Ark. Kemi,* 1953, **5**, 17.

4 M. Kuhnert-Brandstätter and R. Ulmer, *Mikrochim. Acta*, 1974, 927.

5 H. A. J. Oonk, K. H. Tjoa, F. E. Brandt, and J. Kroon, *Thermochim. Acta*, 1977, **19**, 161.

6 H. W. B. Roozeboom, *Z. Phys. Chem.*, 1899, **28**, 494.

7 A. Collet and J. Jacques, *Bull. Soc. Chim. Fr.*, 1973, 3330.

8 J. H. Adriani, *Z. Phys. Chem.*, 1900, **33**, 453.

9 P. Padoa and G. Rotondi, *Atti R. Accad. Lincei,* 1912, **21**, 11, 626.

10 A. Collet, M. J. Brienne, and J. Jacques, *Bull. Soc. Chim. Fr.*, 1972, 127.

11 F. S. Kipping and W. J. Pope, *J. Chem. Soc.*, 1897, **71**, 989.

12 J. Jacques and J. Gabard, *Bull. Soc. Chim. Fr.*, 1972, 342.

13 M. J. Brienne and J. Jacques, *Tetrahedron*, 1970, **26**, 5087.
14 F. S. Kipping and W. J. Pope, *J. Chem. Soc.*, 1897, **71**, 962.
15 See, for example, G. Natta and M. Farina, *Molécules en Trois Dimensions*, Masson, Paris, Chapter 9.
16 M. Leclercq, A. Collet, and J. Jacques, *Tetrahedron*, 1976, **32**, 821.
17 K. R. Wilson and R. E. Pincock, *J. Am. Chem. Soc.*, 1975, **97**, 1474.
18 D. Coates, K. J. Harrison, and G. W. Gray, *Mol. Cryst.*, 1973, **22**, 99.
19 M. Leclercq, J. Billard, and J. Jacques, *Mol Cryst.*, 1969, **8**, 367.
20 D. Coates and G. W. Gray, *Mol. Cryst. Lett.*, 1976, **34**, 1.
21 L. Langhammer, *Arch. Pharm.*, 1975, **308**, 933.

2.6 ANOMALOUS RACEMATES

In all cases examined up to this point, crystals of racemic compounds by definition contain the (+) and (−) species in a 1–1 ratio. However, some instances are known in which the two enantiomers may form molecular combinations whose stoichiometry is different.

Bergmann and Lissitzin were among the first to study and report an "anomalous" racemate.[1] They showed that 4-benzoylamino-3-hydroxybutyric acid, **1**, previously described by Tomita and Sendju,[2] may exist in three forms: (a) a racemic compound which crystallizes in anhydrous form from water; (b) the enantiomers, mp 114°C (anhydrous) and 80–81°C (monohydrate), $[\alpha]_D \pm 22°$ (0.5*M* aqueous NaOH); and (c) two addition compounds, D_2L and L_2D, $[\alpha]_D \pm 7°$. Unfortunately, no phase diagrams for this system are available. The specific rotation observed is consistent with the stoichiometry 2–1 of the addition compounds.

C_6H_5-CONH-CH_2-CH(OH)-CH_2CO_2H **1**

HO_2C-CH(CH_3)-CH_2-CH(CH_3)-CO_2H **2**

A second anomalous addition compound has been described by Fredga,[3] namely, that formed by 2,4-dimethylglutaric acid, **2**. The phase diagram has been determined (Fig. 1). Because of the huge difference in melting point between the enantiomer and the racemic compound, the latter is highly stable. The phase diagram itself yields little information on the composition of the anomalous addition compounds which are evidenced by the beginning of a slight break in the liquidus curve. A clearer indication of the existence of the stoichiometry of this combination is furnished by the phase diagrams resulting from mixtures of (+)-**2** and (−)-dilactic acid,[4] **3** (Fig. 2*a*), and from (+)-**2** and (−)-2-methylglutaric acid,[3] **4** (Fig. 2*b*). In these two cases, the formation of a quasi-racemate (ratio 1–1) and of an addition compound (ratio 1–3) confirms the tendency of **2** to form anomalous addition compounds.

HO_2C-CH(CH_3)-O-CH(CH_3)-CO_2H **3**

HO_2C-CH(CH_3)-CH_2-CH_2-CO_2H **4**

HO_2C-CH(OH)-CH_2-CO_2H **5**

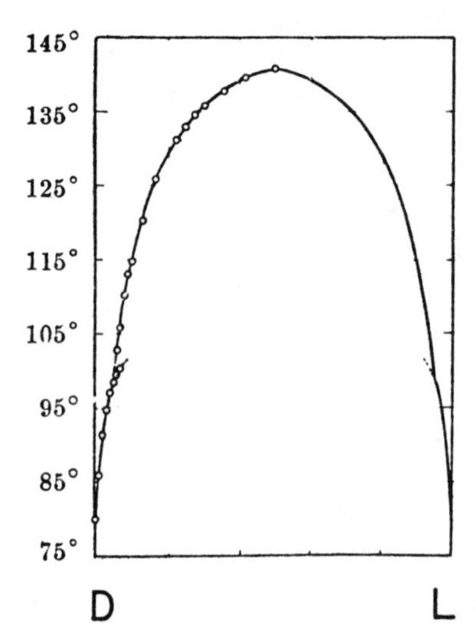

Figure 1 Phase diagram for 2,4-dimethylglutaric acid. Adapted from A. Fredga, *Ark. Kemi, Mineral. Geol.*, 1947, **24A**, No. 32 by permission of the Royal Swedish Academy of Sciences.

Another case is that of malic acid, **5**, which has been studied by examination of its phase diagram and of powder X-ray diagrams.[5] The latter reveals the formation of a molecular combination in which the enantiomers are present in a 1–3 ratio.

Albano et al.[7] have described four examples of anomalous racemates of a type different from those already cited; these are organometallic complexes of type $MX(PPh_3)_3$. The compounds $CuCl(PPh_3)_3$, $CuBr(PPh_3)_3$, $Pt(CO)(PPh_3)_3$, and

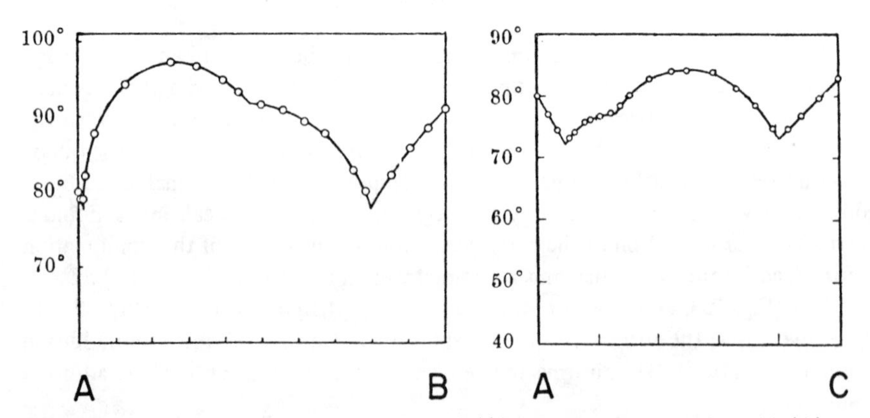

Figure 2 Anomalous quasi-racemates of (+)-2,4-dimethylglutaric acid (A), with (−)-dilactic acid (B), and with (+)-2-methylglutaric acid (C). Adapted from A. Fredga, *Ark. Kemi, Mineral. Geol.*, 1940, **14B**, No. 12 and 1947, **24A**, No. 32 by permission of the Royal Swedish Academy of Sciences.

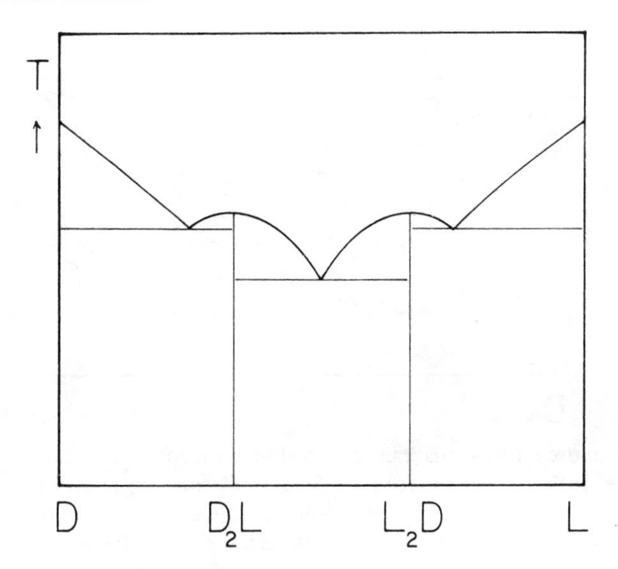

Figure 3 Proposed phase diagram for a conglomerate of addition compounds of D and L in a 2–1 ratio.

$Ir(NO)(PPh_3)_3$ are isomorphs that crystallize in space group *P*3. X-Ray crystallographic structure determination carried out on crystals obtained from a racemic solution shows that they contain the two enantiomers in a 2–1 ratio. The racemate is thus a conglomerate of D_2L and L_2D crystals; the authors call this "unbalanced packing of chiral molecules." Each crystal when dissolved would be expected to have one third the rotatory power of one of the pure enantiomers. The expected phase diagram corresponding to such a system is shown in Fig. 3.

The sedative and hypnotic agent 3,3-diethyl-5-methyl-2,4-piperidinedione (methyprylon, or Noludar®), **6**, provides an example, fortunately quite rare, of the complexity possible for anomalous racemates. This complexity is responsible for the difficulties occasionally encountered in the determination of the phase diagrams.

H N O H C_2H_5 CH_3 C_2H_5 O

6

The two enantiomers of this compound are dimorphic. One form is stable, mp 84.5°C; the other form, mp 79.5°C, is unstable. According to Vogler,[6] each stable form yields a solid solution with the unstable enantiomer which exhibits a minimum melting point at 73.3°C. The two spindle-shaped curves thus obtained intersect each other to yield a eutectic (racemic) corresponding to a conglomerate of mixed crystals (Fig. 4). Crystallization of racemic Noludar rarely yields rigorously inactive crystals. The study by Vogler et al. is unusual also in that the system was

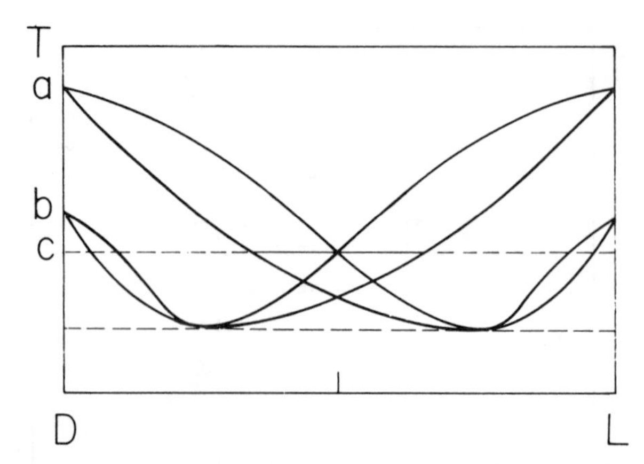

Figure 4 Proposed phase diagram for Noludar (according to ref. 6): (a) melting point of the enantiomer (stable form) = 84.5°C; (b) melting point of the enantiomer (unstable form) = 79.5°C. (c) melting point of the eutectic = 75.6°C. M. Kuhnert-Brandstätter, K. Schleich, and K. Vogler, *Monatsh. Chem.*, 1970, **101**, 1817. Reproduced by permission of Springer-Verlag.

subjected to careful analysis by dsc, by infrared spectroscopy, as well as by application of the Kofler contact method (Section 2.1.5). In fact, the equilibria associated with such a system are clearly difficult to understand. In our opinion, other explanations of the data cannot be precluded.

The most recent anomalous racemate described is that of tetramisole, **7**, a compound exhibiting antihelminthic behavior.[8] The melting point phase diagram points to formation of a relatively stable 3–1 complex between the enantiomers in addition to the normal 1–1 racemic compound.[8,9]

C_6H_5 N N S

7

REFERENCES 2.6

1 M. Bergmann and M. Lissitzin, *Ber.*, 1930, **63**, 310.

2 M. Tomita and Y. Sendju, *Z. Physiol. Chem.*, 1927, **169**, 263.

3 A. Fredga, *Ark. Kemi, Min. Geol.*, 1947, **24A**, No. 32.

4 A. Fredga, *Ark. Kemi, Min. Geol.*, 1940, **14B**, No. 12.

5 M. Andersson, A. Fredga, and B. Jerslev, *Acta Chem. Scand.*, 1966, **20**, 1060.

6 M. Kuhnert-Brandstätter, K. Schleich, and K. Vogler, *Monatsh. Chem.*, 1970, **101**, 1817.

7 (a) V. G. Albano, P. L. Bellon, and M. Sansoni, *J. Chem. Soc. D.*, 1969, 899. (b) V. G. Albano, G. M. Basso Ricci, and P. L. Bellon, *Inorg. Chem.*, 1969, **8**, 2109. (c) V. G. Albano, P. L. Bellon, and M. Sansoni, *J. Chem. Soc.*, 1971 (A), 2420.

8 L. Toke, M. Acs, E. Fogassy, F. Faigl, S. Gál, and J. Sztatisz, *Acta Chim. Acad. Sci. Hung.*, 1979, **102**, 59.

9 E. Fogassy, M. Acs, J. Felmeri, and Z. Aracs, *Period. Polytech., Chem. Eng.*, 1976, **20**, 247. *Chem. Abstr.*, 1980, **87**, 68233e.

2.7 CALORIMETRIC DETERMINATION OF ENANTIOMERIC PURITY

It is evident that phase diagrams are indispensable for the description of properties of enantiomers and their mixtures. These types of diagrams also provide a relatively simple solution to the determination of enantiomeric purity of crystallized enantiomer mixtures.

Differential scanning calorimetry happily complements the other known methods for the determination of enantiomeric purity,[1,2] such as methods based upon nmr spectroscopy, chromatography, enzymatic reactions, and isotope dilution techniques, all of which are described in Section 7.5. While the dsc procedure does have some limitations – just as its competitors have – in favorable cases it is very competitive with the other methods.[3]

The dsc method consists of two variants:

1 A *direct* method which may be employed if one has a sample of the racemate. The method requires the reconstitution of the phase diagram taking advantage of all the information the dsc traces can furnish from the melting of the pure racemate and from that of a partially resolved mixture of unknown enantiomeric purity. This method is applicable to samples of enantiomeric purities that are neither too large nor too small and for which the eutectic melting point is not too close to that of the termination of fusion.

2 An *indirect* method, based upon analysis of the melting scan according to the general procedure applicable to nearly pure crystalline substances,[4–7] which in turn can furnish enantiomeric purity (p) information of fair precision. Thus, it is applicable either to crystalline samples of high p or to nearly pure racemic compound, that is, samples of low p.

2.7.1 Principles and Limitations of the Direct Method

(a) Conglomerates

The recording of a dsc trace of the melting of a partially resolved conglomerate (Fig. 1) yields a curve in which the eutectic appears as an isolated peak whose area is directly proportional to the heat necessary to melt the racemate present in the mixture. From a knowledge of the enthalpy of fusion of the pure racemate and the total weight of the sample, one easily calculates the percentage of eutectic present which is equal to $100(1-p)$.

This analysis naturally is valid only if the peaks (that of the eutectic and that of the enantiomer, see Fig. 1) are well separated and if the area of the peak is measured with precision. Typically, the measurements are significant when the eutectic constitutes from 10 to 70% of the sample, that is, from 30 to 90% p. Within this range the error is of the order of ± 4%. The determination of the termination of fusion for the sample is carried out during the same recording of the dsc trace. When the p of the sample studied is not too low, it may be possible to estimate the actual liquidus curve of the phase diagram and to extrapolate the latter to the melting point of the pure enantiomer (dotted line, Fig. 1*b*).

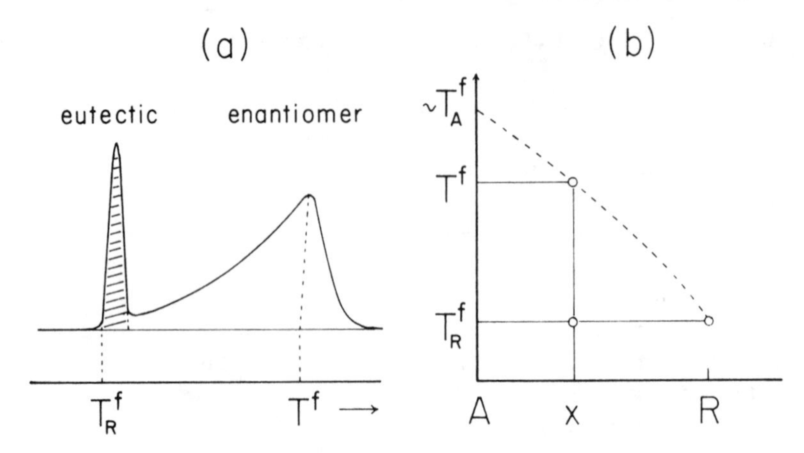

Figure 1 Determination by the direct method of enantiomeric purity in the case of a conglomerate: (a) The area of the eutectic peak is proportional to the quantity of racemate present in the sample. (b) The liquidus curve (dotted line) may be estimated from a knowledge of the enantiomeric composition (x) and from the temperature of termination of fusion (T^f).

(b) Racemic compounds

In this case (Fig. 2), one measures the temperature corresponding to the termination of fusion, T^f, of the unknown mixture and also the melting point and the enthalpy of fusion of the racemic compound. The introduction of these values into the Prigogine–Defay equation (Section 2.3.2) directly furnishes the composition x of the mixture. The method requires that the T^f measured be on the racemic compound branch of the liquidus curve. This is equivalent to saying that the enantio-

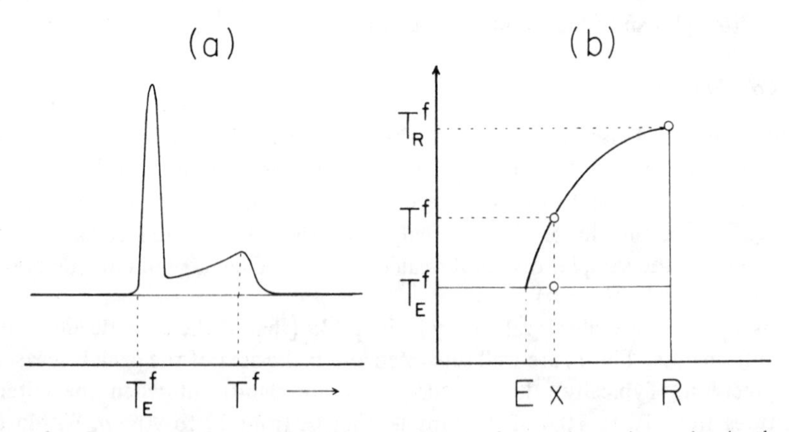

Figure 2 Determination by the direct method of enantiomeric purity in the case of a racemic compound: (a) T^f and T_E^f are read off the dsc trace of the melting of the unknown sample. (b) The value of T^f taken to the calculated liquidus curve allows one to directly read the enantiomeric composition (x); from T_E^f the composition of the eutectic (E) can be read off the phase diagram.

meric purity of the sample must be less than that corresponding to the eutectic. To ascertain on which side of the eutectic the sample composition is, a small quantity of racemic compound is added to a sample of the unknown mixture and the dsc curve is redetermined; the termination of fusion T^f is *raised* on the racemic compound branch while it is *lowered* on the enantiomer branch.

It is not difficult to assess the lack of precision of this method for cases of low enantiomeric purity. In effect, in the vicinity of the racemic compound, the lowering of the melting point, $T_R^f - T^f$, is very small. Since T_R^f and T^f are measured in two different scans, it is virtually impossible to determine with precision a lowering of less than 2°. Such a temperature lowering may correspond to enantiomeric purities in the 20 to 40% range, for example.

It is well to remember that virtually the entire phase diagram can be constructed from T_R^f, ΔH_R^f, and the dsc scan of the melting of a single unknown mixture (Fig. 2). The data on the racemic compound effectively allow one to calculate its liquidus curve, while the dsc scan of the unknown mixture yields the eutectic melting point. One can thus very rapidly determine the composition of the latter, which, as we shall see, is important in ensuring the success of the resolution of the racemate.

2.7.2 Theory of the Indirect Method

When the direct method is not applicable, recourse may be had to the determination of purity of nearly pure substances by calorimetry.[4-7] This method allows one to measure, from the dsc melting scan, the lowering of the melting point, $T_A^f - T^f$ for the enantiomer and $T_R^f - T^f$ for the nearly pure racemic compound. *Precise* values of T^f having thus been obtained, the equations of Schröder–Van Laar or of Prigogine–Defay may be employed to determine the sample composition x and thus enantiomeric purity.[3]

There are two regions in the phase diagram in which the indirect method may be applied. These are the shaded zones of Fig. 3. The enantiomer branch and the racemic compound branch require slightly different treatments, which are now examined in some detail.

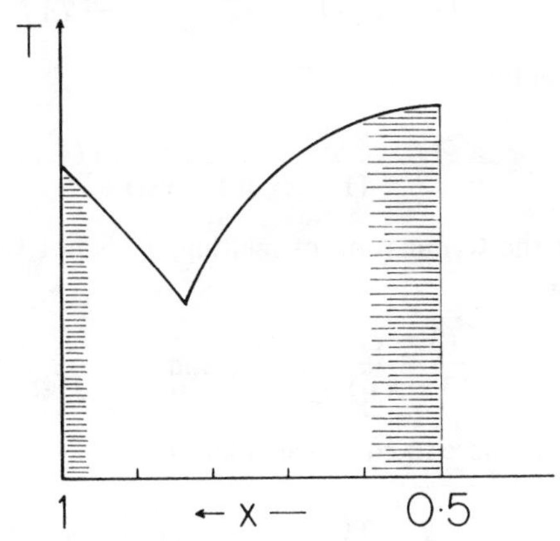

Figure 3.

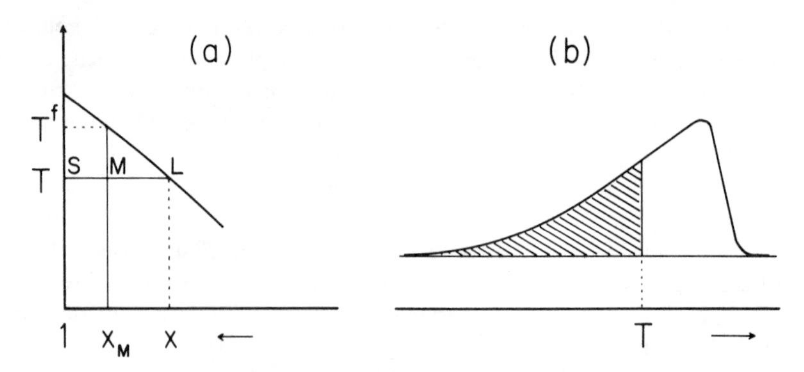

Figure 4 Principle of the indirect method (enantiomer branch).

(a) Enantiomer branch

The following analysis is applicable to conglomerates or to racemic compounds of high enantiomeric purity. During the melting of a sample of composition x_M, at a given temperature T, the liquid has the composition x (Fig. 4*a*), while after melting is complete, at T^f, it has the composition x_M of the initial mixture. Expansion of the logarithmic function of the equation of Schröder–Van Laar in its simplified form [eq. (2) of Section 2.2.3] up to its second term, when applied to x and to x_M, gives

$$(x-1)\frac{(3-x)}{2} = \frac{\Delta H_A^f}{R}\left(\frac{1}{T_A^f}-\frac{1}{T}\right)$$

$$(x_M-1)\frac{(3-x_M)}{2} = \frac{\Delta H_A^f}{R}\left(\frac{1}{T_A^f}-\frac{1}{T^f}\right)$$

Equating the two expressions, one obtains

$$\frac{(1-x)}{(1-x_M)}\frac{(3-x)}{(3-x_M)} = \frac{T-T_A^f T^f}{T^f-T_A^f T}$$

Rearranging, one obtains

$$T = T_A^f - \frac{(1-x)(3-x)}{(1-x_M)(3-x_M)}(T_A^f-T^f)\frac{T}{T^f} \tag{1}$$

In the vicinity of the termination of melting, T^f is not very different from T and x is near x_M. Then

$$\frac{(3-x)}{(3-x_M)} \sim 1 \qquad \text{and} \qquad \frac{T}{T^f} \sim 1$$

Equation (1) then reduces to eq. (2) as follows:

$$T \sim T_A^f - \frac{(1-x)}{(1-x_M)}(T_A^f-T^f) \tag{2}$$

This equation reveals the fraction F of sample melted at temperature T, which is experimentally measurable (see Fig. 4):

$$F = \frac{(1 - x_M)}{(1 - x)} = \frac{q}{Q}$$

Equation (2) may then be written as follows:

$$T \sim T_A^f - \frac{1}{F}(T_A^f - T^f) \tag{3}$$

and a plot of T versus $1/F$ will yield a straight line whose slope gives $(T_A^f - T^f) = \Delta T$.

The principle of operation is summarized in Fig. 4, while the practical application of the method is detailed in the following section. For a mixture of composition x_M (Fig. 4*a*), the fraction of compound melted at temperature T is given by the lever rule: $F = SM/SL = (1 - x_M)/(1 - x)$. In the dsc fusion scan of the same sample (Fig. 4*b*), the total area corresponds to the enthalpy of fusion of the substance as a whole (Q), while the hatched area represents the enthalpy of fusion of the fraction melted at temperature T. Thus, $F = q/Q$.

The validity of equation (3) rests upon the following main requirements: (i) the impurity (i.e., the minor enantiomer in the case of a conglomerate, or, the racemic compound) must form an eutectic with the nearly pure substance examined, and must not cocrystallize with the latter; (ii) the analysis of the dsc melting scan must be carried out in the vicinity of the termination of fusion.

(b) Racemic compound branch

The following analysis applies to racemic compounds of low enantiomeric purity.

Expansion of the equation of Prigogine–Defay [eq. (1) of Section 2.3.2] up to the second term of $\ln 4x(1 - x)$, and identical reasoning to that of the preceding paragraphs leads to the following relationships in which $y = x - 0.5$ has replaced x (Fig. 5):

$$\frac{\ln 4x(1 - x)}{\ln 4x_M(1 - x_M)} = \frac{\ln(1 - 4y^2)}{\ln(1 - 4y_M^2)} \sim \frac{y^2(1 + 2y^2)}{y_M^2(1 + 2y_M^2)} = \frac{(T_R^f - T)T^f}{(T_R^f - T^f)T} \tag{4}$$

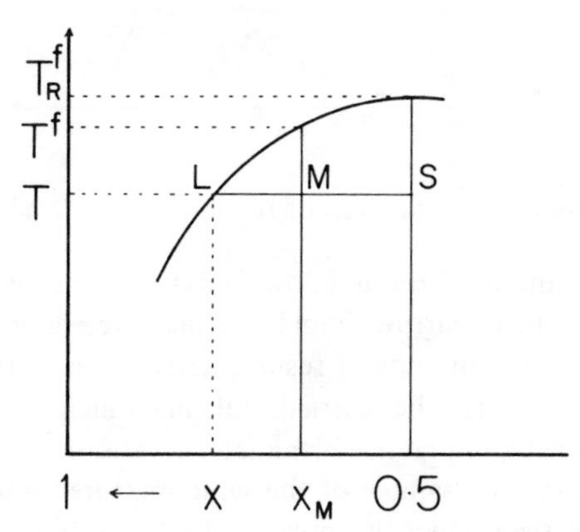

Figure 5 For a mixture of composition x_M, the fraction of sample melted at temperature T is given by $F = SM/SL = (x_M - 0.5)/(x - 0.5) = Y_M/Y$.

Since $y^2/y_M^2 = 1/F^2$ (Fig. 5), and $T^f/T \sim 1$, eq. (4) finally reduces to

$$T \sim T_R^f - \frac{1}{F^2}(T_R^f - T^f) \tag{5}$$

As before, the dsc scan is analyzed in the vicinity of the termination of fusion (y near to y_M), and a plot of T versus $1/F^2$ yields a straight line whose slope gives $(T_R^f - T) = \Delta T$.

The validity of eq. (5) depends upon the same requirements as those previously mentioned for eq. (3), that is, *inter alia*, the nonformation of a solid solution between the impurity (i.e., in this case the major enantiomer) and the racemic compound crystals.

2.7.3 Application of the Indirect Method

In practice, it is not usually possible to construct usable plots of T versus $1/F$ or T versus $1/F^2$ from the dsc fusion scans as described in Fig. 4. In the region of enantiomeric purity in which the indirect method is applied, the eutectic is often undetectable. The melting curve rises so slowly that measurement of the total area Q and of the partial areas q_i with precision becomes quite difficult. By the same token, it also becomes difficult to locate the temperatures T on the fusion scans. The following procedure, illustrated in Fig. 6, avoids these difficulties.

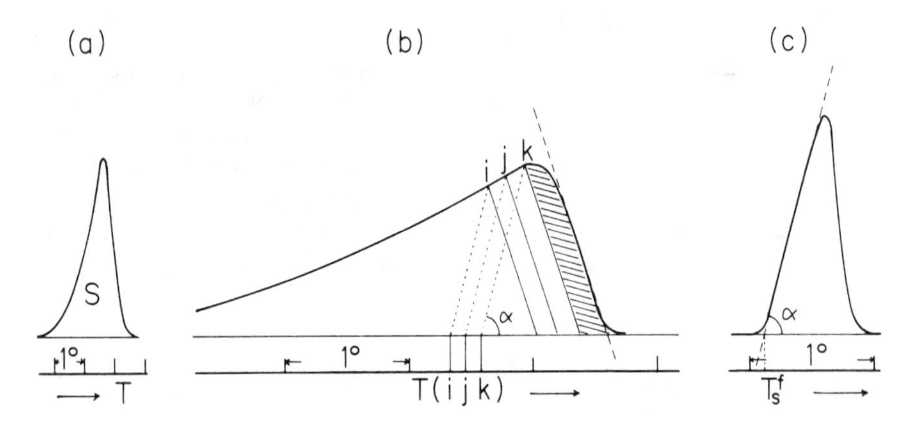

Figure 6 Application of the indirect method.

First (Fig. 6*a*), a sample of the unknown mixture is melted under conditions such that the fusion peak is narrow (rapid heating rate, slow paper speed). The area of this peak, S, gives the enthalpy of fusion, ΔH_m^f, of the mixture. Incidentally, while area measurements may be carried out in principle mechanically (disc integrator) or electronically, in practice planimetry gives satisfactory results.

Second (Fig. 6*b*), another sample of the same mixture is melted in such a way as to obtain a melting scan which is spread out (slow rate of heating, fast paper speed). The total area of the spread-out melting curve, Q, is *calculated* from the enthalpy of fusion, ΔH_m^f, previously measured. The measurements of q toward the end of the melting process (near the top of the curve) are given by difference from the planimetric area determinations corresponding to the *nonmelted* substance. One obtains $Q - q$, and since Q is known, the value of q follows (e.g., in Fig. 6*b*, the hatched area corresponds to $Q - q_k$).

The main experimental problem is the determination of the temperatures T ($i, j, k, \ldots$) corresponding to the arbitrary points (i, j, k, . . .) of the fusion curve. The method used is to transfer the angle α made by the leading edge of the peak obtained in a dsc melting scan of a pure organic standard such as naphthalene (Fig. 6*c*) determined under identical conditions to the trace of the unknown. This angle incorporates the finite time necessary for melting to occur. As a matter of fact, the temperature of the pure sample in the process of melting remains constant and equal to T_S^f along the leading edge of the peak. While the use of the same angle α for different samples may appear unjustified, the method works nonetheless.

From the foregoing, the fraction of substance melted at temperature T is given by $F = q/Q$. Then the plotting of T versus Q/q, or of T versus Q^2/q^2 as described earlier, furnishes a straight line whose slope yields the value of $\Delta T = T_A^f - T^f$ or $\Delta T = T_R^f - T^f$.

Application of the Schröder–Van Laar equation or of the Prigogine–Defay equation requires two additional data likely to be unknown: the melting point and the enthalpy of fusion of the *pure* substance, that is, T_A^f or T_R^f, and ΔH_A^f or ΔH_R^f, respectively.

The melting point of the pure substance may be taken as T_A^f (or T_R^f) $= T^f + \Delta T$, where T^f is the temperature of termination of fusion, which corresponds to the top of the melting curve (Fig. 6*b*). We emphasize that any error in the *absolute* value of T^f so determined has little effect on the precision of the method; what matters only is the value of ΔT.

On the other hand, the value of ΔH_m^f obtained from the area of the narrow melting scan (Fig. 6*a*) gives a good approximation of the enthalpy of fusion of the pure substance.

The mole fraction x_M of the more abundant enantiomer in the unknown sample is calculated as follows for the enantiomer branch and for the racemic compound branch, respectively:

$$\ln x_M = \frac{\Delta H_m^f}{R}\left(\frac{1}{T^f + \Delta T} - \frac{1}{T^f}\right)$$

$$\ln 4x_M(1 - x_M) = \frac{2\Delta H_m^f}{R}\left(\frac{1}{T^f + \Delta T} - \frac{1}{T^f}\right)$$

A detailed error analysis of the dsc determination of enantiomeric purity has been carried out by Fouquey and Leclercq.[3b] Some of the errors are systematic, such as the imprecision of the area measurements or the use of approximations in the equations. Others depend upon the nature of the substances analyzed such as the formation of partial solid solutions or the presence of chemical impurities. Table 1 gives some examples of determinations carried out by these authors.

Table 1 Determination of enantiomeric purity by differential scanning calorimetry[a]

Compound		True Value		Direct Method		Indirect Method	
		P (%)	x	P (%)	δP (%)	P (%)	δP(%)
(a) *Enantiomer Branch of the Phase Diagram*							
2-(*p*-Methoxyphenyl)propiophenone	($x_E = 0.5$)	83.0	0.915	84.0	+1.0	84.5	+1.5
		58.6	0.793	61.4	+2.8	57.8	−0.8
		41.6	0.705	45.2	+3.6	–	–
		26.6	0.633	27.4	+0.8	–	–
Mandelic acid	($x_E = 0.7$)	94.0	0.97	–	–	93.8	−0.2
		85.4	0.927	–	–	86.9	+1.5
		70.6	0.853	–	–	73.0	+2.4
		62.0	0.81	–	–	57.4	−4.6
(b) *Racemic Compound Branch of the Phase Diagram*							
2-(1-Naphthyl)propionic acid	($x_E = 0.99$)	20.0	0.6	28.4	+8.4	17.8	−2.2
		40.0	0.7	35.4	−4.6	36.0	−4.0
O, O'-Dibenzoyldimethyltartrate	($x_E = 0.90$)	19.6	0.598	26.4	+6.8	21.2	+1.6
		30.0	0.650	29.8	−0.2	29.6	−0.4
		40.0	0.70	42.8	+2.8	42.0	+2.0
2-(*p*-Bromophenoxy)propionic acid	($x_E = 0.77$)	9.6	0.548	19.8	+10.2	16.4	+6.8
		32.8	0.644	39.6	+6.8	35.6	+2.8
2-(*m*-Methylphenoxy)propionic acid	($x_E = 0.96$)	6.0	0.53	–	–	10.4	+4.4
		18.0	0.59	–	–	23.0	+5.0
2-(*o*-Chlorophenoxy)propionic acid	($x_E = 0.90$)	5.2	0.526	–	–	4.9	−0.3
		30.4	0.652	25.2	−5.2	29.2	−1.2
		40.6	0.702	43.0	+2.4	41.8	+1.2

[a] P = Enantiomeric purity; $\delta P = P(\text{found}) - P(\text{true})$; x, x_E are, respectively, the mole fraction of the enantiomer present in excess in the sample and in the eutectic. C. Fouquey and J. Jacques, *Tetrahedron,* 1970, **26**, 5637. Reproduced by permission of Pergamon Press Ltd.

The determination of enantiomeric purity by dsc is broadly applicable and is not even limited to organic compounds that are crystalline at room temperature. Skell et al. have applied it successfully to liquids such as 2,3-dibromobutane (racemic compound mp − 37°C).[8]

REFERENCES 2.7

1 M. Raban and K. Mislow, *Topics Stereochem.*, 1967, **2**, 199.

2 (a) J. Campbell, *Aldrichim. Acta*, 1972, **5**, 29. (b) E. Sullivan, *Topics Stereochem.*, 1978, **10**, 287. (c) E. Gil-Av and D. Nurok, *Adv. Chromatogr.*, 1975, **10**, 99. (d) H. Furukawa, E. Sakakibara, A. Kamei, and K. Ito, *Chem. Pharm. Bull.*, 1975, **23**, 1625. (e) E. Gil-Av, *J. Mol. Evol.*, 1975, **6**, 131. (f) M. D. McCreary, D. W. Lewis, D. L. Wernick, and G. M. Whitesides, *J. Am. Chem. Soc.*, 1974, **96**, 1038.

3 (a) C. Fouquey and J. Jacques, Tetrahedron, 1967, **23**, 4009. (b) C. Fouquey and M. Leclercq, *Tetrahedron*, 1970, **26**, 5637.

4 J. H. Badley, *J. Phys. Chem.*, 1959, **63**, 1991.

5 A. P. Gray, *Instrum. News* (Perkin Elmer Corp.), 1966, **16** (3), 9.

6 R. Reubke and J. A. Mollica, *J. Pharm. Sci.*, 1967, **56**, 822.

7 M. E. Brown, *J. Chem. Educ.*, 1979, **56**, 310.

8 P. S. Skell, R. R. Pavlis, D. C. Lewis, and K. J. Shea, *J. Am. Chem. Soc.*, 1973, **95**, 6735.

2.8 SOLID–VAPOR EQUILIBRIA. SUBLIMATION OF ENANTIOMER MIXTURES

2.8.1 Phase Diagrams for Sublimation

The thermodynamic questions raised when a solid phase is converted to a gaseous phase resemble those of the solid/liquid conversion in the sense that the sublimation phase diagrams for enantiomer mixtures have the same form as the corresponding fusion diagrams. However, contrary to what we have done up to this point, one must naturally consider the *pressure* as an additional variable when the phase rule is applied: $v = C - \Phi + 2$.

For example, let us consider the sublimation of a conglomerate in which x_M is the mole fraction of the more abundant enantiomer. Such a system consists of, at most, two solid phases (D_s and L_s) and one vapor phase. Since the vapor pressure of each solid enantiomer is identical to that of the other and is independent of the quantities of the crystals of each species – only their presence is required – the vapor phase is necessarily *racemic* no matter what the enantiomeric purity of the solid mixture. Since three phases are present in the system, the latter is monovariant. That is, only P or T are variable since the compositions of the solid and vapor phases are fixed.

Let us first examine what obtains at *constant pressure*, P_0. Here, the three phases may coexist within a given temperature *range*. For example, at temperature T_1 (Fig. 1), the system of composition x_M consists of a mixture (D_s, L_s) represented by C and of a racemic vapor V. The limit of this range is given as that temper-

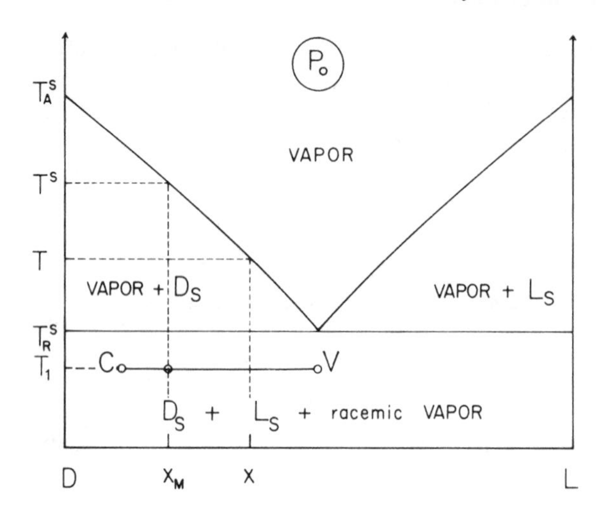

Figure 1 Sublimation of a conglomerate at constant pressure (P_0).

ature at which, at pressure P_0, all of the racemate [that is, $2(1-x_M)$] present in the initial mixture x_M has passed into the gas phase. In the phase diagram, given in Fig. 1, the temperature limit is shown as T_R^s, the sublimation temperature of the racemate at pressure P_0. Above T_R^s, there remains one solid phase, the enantiomer present in excess (D here), and a vapor phase whose composition is now variable. The variance has increased by one unit ($v = 2$). Since P_0 is constant, there is but one degree of freedom.

We now have a relationship between the two remaining variables, the temperature T and the vapor composition x. It is possible to show[1] that the relation is none other than the Schröder–Van Laar equation:

$$(\ln x)_{P_0} = \frac{\Delta H_A^s}{R}\left(\frac{1}{T_A^s} - \frac{1}{T^s}\right) \tag{1}$$

In this equation, T_A^s is the temperature at which the pure enantiomer crystals are at equilibrium with their vapor at pressure P_0 (sublimation temperature at P_0); ΔH_A^s is the enthalpy of sublimation of the pure enantiomer at temperature T_A^s and pressure P_0; and x is the mole fraction of the more abundant enantiomer in the vapor at equilibrium with the solid at temperature T^s.

Finally, if the temperature is further increased, all of the solid disappears at T^s to give a vapor phase of composition x_M.

A similar analysis allows us to construct a sublimation diagram at *constant temperature*. This is illustrated in Fig. 2. A mixture x_M at a pressure P_1 greater than P_R^s (the vapor pressure of the racemate at T_0) consists of the two solid enantiomers (C) in the presence of the racemic vapor (V). When the pressure is lowered to P, there remain crystals of the major enantiomer (D_s) at equilibrium with vapor x. At P^s, the last crystal of D disappears and a vapor phase of composition x_M remains.

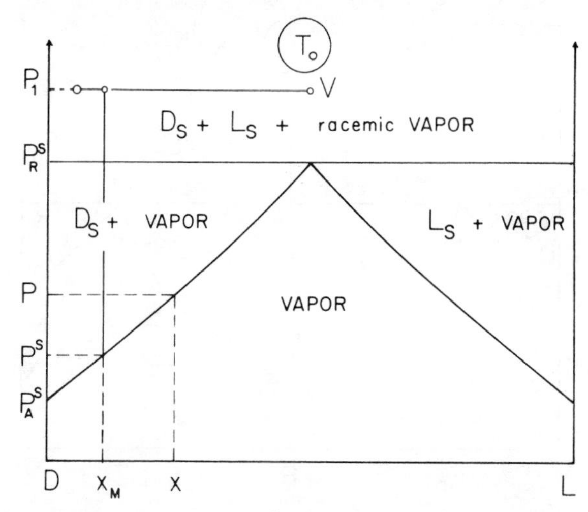

Figure 2 Sublimation of a conglomerate at constant temperature (T_0).

The case of enantiomer mixtures forming a racemic compound can be described analogously. The phase diagrams for sublimation at constant pressure and at constant temperature are shown in Figs. 3 and 4. Since the eutectic composition is related to the enthalpies of sublimation, ΔH_A^s and ΔH_R^s, which are different from the corresponding enthalpies of fusion, there is no reason to expect the eutectics in the sublimation and fusion diagrams to have *precisely* the same compositions. By neglecting the influence of specific heats, we may write $\Delta H^s \sim \Delta H^f + \Delta H^v$, where ΔH^f and ΔH^v are the enthalpies of fusion and vaporization, respectively (Section 2.3.4). Since ΔH_A^v and ΔH_R^v have practically identical values, $\Delta H_A^s - \Delta H_R^s \sim \Delta H_A^f - \Delta H_R^f$. Consequently, the vapor eutectic and the liquid eutectic should be found in the same region of these diagrams.

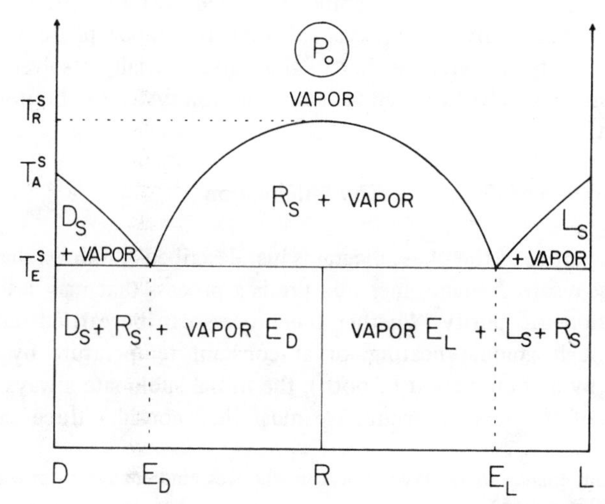

Figure 3 Sublimation of a racemic compound at constant pressure (P_0).

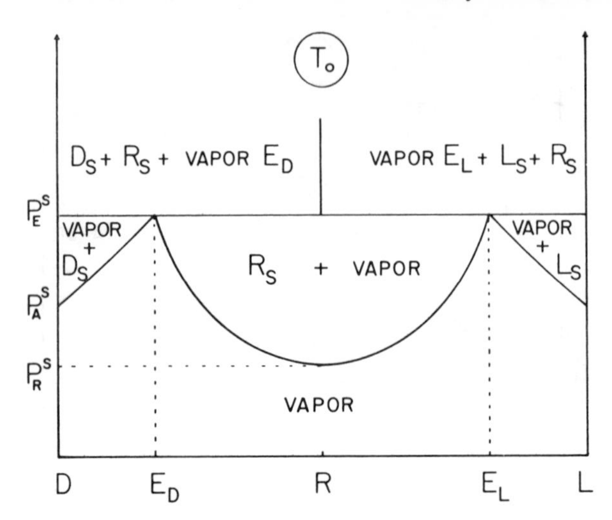

Figure 4 Sublimation of a racemic compound at constant temperature (T_0).

It is important to remember that the sublimation phase diagrams described in Figs. 1 through 4 hold only for those cases in which the sublimation temperatures of optically active and racemic substances are lower than their respective melting points. For most organic compounds, this condition is not met at atmospheric pressure. However, it can generally be met by reducing the pressure. Only a few organic compounds, generally globular and odoriferous ones, may be sublimed at atmospheric pressure.

The sublimation of solid solutions of enantiomers may be analyzed in similar fashion. We shall limit ourselves to the following conclusions: (1) In an ideal solid solution, the solid and vapor phases have identical compositions whatever may be the original composition of the sample sublimed. (2) In a solid solution with maximum melting point, it is expected that the vapor phase would have an enantiomeric purity higher than that of the initial partially resolved solid sample. Inverse behavior would be expected for solid solution systems exhibiting a minimum melting point.

2.8.2 Separation of Enantiomers by Sublimation*

From the properties of the phase diagrams just described it follows that sublimation of a partially resolved enantiomer mixture is a process that may lead to a change in the enantiomeric purity. Whether the process can be carried out at constant pressure through gradual heating, or at constant temperature by lowering the pressure (or by a combination of both), the initial sublimate always possesses the composition of the vapor eutectic. We must then consider three cases according

* We extend our thanks to Dr. David L. Garin who was kind enough to provide a number of references for this section.

to the nature of the racemate and the enantiomeric purity of the mixture being examined.

1 Partially resolved conglomerates: The fraction which sublimes first is *racemic*. The enantiomeric purity of the nonsublimed solid is thus increased. Above temperature T_R^s (Fig. 1) or below pressure P_R^s (Fig. 2), the solid remaining is the pure enantiomer.

2 Racemic compound with enantiomeric purity less than that of the eutectic in the sublimation phase diagram: The part of the mixture which first sublimes has a greater enantiomeric purity than that of the substance taken; the non-sublimed residue is enriched in racemate.

3 The inverse case, that is, racemic compound whose enantiomeric purity is greater than that of the eutectic: That which first sublimes is less pure than the initial mixture, while the enantiomeric purity of the residue rises and tends toward the pure enantiomer.

The literature reveals but few reports of the sublimation of enantiomer mixtures. The first observation appears to be that of Pracejus,[2] who found that the first fractions of enantiomerically enriched *N*-phthalyl-β-phenylalanyl esters obtained upon vacuum sublimation were of higher optical activity than the last and inferred that this was not simply a consequence of the removal of an impurity. She then demonstrated that the optically active form was more volatile than the racemate.

Zahorsky and Musso[3] later demonstrated through isotope labeling of one enantiomer of each of several compounds (3,3′-diiodobimesityl, mandelic acid, and α-methylbenzylamine) that its preferential vaporization relative to the racemate may be monitored by observing the molecular ion in the electron ionization mass spectrum as a function of time. The same effect has been observed on dimethyl tartrate by Fales and Wright by means of chemical ionization mass spectrometry.[4]

While no physicochemical studies of solid/vapor systems in thermodynamic equilibrium have been reported, the results of two fractional sublimation studies confirm and accurately illustrate the theoretical treatment earlier developed.

The first such experiments are those of Kwart and Hoster,[5] who studied the fractional sublimation of α-ethylbenzylphenyl sulfide whose racemate melts at 39–42°C while the optically active substance melts at 32–35°C, suggesting that one is dealing with a racemic compound.

1 A sample of α-ethylbenzylphenyl sulfide, $[\alpha]_D^{25} = +5.37°$(ethanol), was heated at 48°C for 12 hours. At this temperature, the sample was completely molten and the "sublimate" (actually the distillate) had the same optical activity as the residue. Nothing surprising in this result (see Section 2.9).

2 A mixture of ca. 12% optical purity was sublimed at 35°C for 12 hours. The sublimate (12.5% by weight) had $[\alpha]_D^{25} = +128.1°$(ethanol) corresponding to 73% optical purity. This result corresponds to the theoretical case (2).

Garin and his associates have carried out a more systematic study of the sublimation of mandelic acid of variable enantiomeric purity.[6] Their most inter-

Table 1 Fractional sublimation of mandelic acid[a]

Sample	Weight (mg)	$[\alpha]^{23}_{365}$ (deg)	Enantiomeric Purity (%)
	Experiment 1		
Starting material	202.1	+118	20.7
Fraction 1	43.7	+212	37.2
Fraction 2	26.1	+179	31.5
Fraction 3	33.3	+144	25.2
Fraction 4	32.8	+ 91	16.0
Fraction 5	64.1	+ 27	4.7
	200.0 (99% recovery)		
	Experiment 2		
Starting material	163.5	+344	60.2
Fraction 1	53.7	+300	52.5
Fraction 2	53.9	+354	62.0
Fraction 3	31.5	+366	64.1
Fraction 4	16.4	+424	74.3
	155.5 (95% recovery)		

[a] Fractions collected are those of sublimate. All rotations measured in 95% ethanol (1 to 2%). D. L. Garin, D. J. C. Greco, and L. Kelley, *J. Org. Chem.*, 1977, **42**, 1249. Reproduced by permission of the publisher. Copyright 1977, American Chemical Society.

esting results are given in Table 1. The samples (100 to 200 mg each) were sublimed at a pressure of 10 torr and at a temperature 10° below that of the melting point of the mixture. Every 2 to 4 hours, the sublimed fraction was collected and its optical activity measured. The eutectic of the *melting point* phase diagram of mandelic acid corresponds to an enantiomeric purity of ca. 50%. The vapor eutectic is expected to lie in the same range of composition.

In the first experiment (Table 1), the composition of the initial sample was intermediate between the eutectic and the racemate. In the second, the initial enantiomeric purity was greater than that of the eutectic. It is evident that the sublimation took place as expected in both cases [theoretical cases (2) and (3), respectively].

In favorable cases in which the eutectic is not too close to the racemic compound, sublimation may be method of choice to obtain a sample of high enantiomeric purity from a small quantity of substance of very low enantiomeric purity. The practicability of such a process was demonstrated by Garin[6]; from 101 mg mandelic acid of 1.2% enantiomeric purity, 6 mg initial sublimate having 20%

enantiomeric purity was collected. This sublimate fraction contained virtually all of the enantiomeric excess present in the initial sample.

Similarly, Kwart[5] underscores the large advantage that sublimation possesses over fractional crystallization to obtain α-ethylbenzylphenyl sulfide of high enantiomeric purity from poorly resolved mixtures. Undoubtedly, more such experiments will be carried out at least on a laboratory scale by those needing small samples of highly enriched substances now that the utility of this procedure has been demonstrated.

REFERENCES 2.8

1 I. Prigogine and R. Defay, *Chemical Thermodynamics*, 4th ed., Longmans, 1967, p. 247.
2 G. Pracejus, *Liebigs Ann. Chem.*, 1959, **622**, 10.
3 U.-I. Zahorsky and H. Musso, *Chem. Ber.*, 1973, **106**, 3608.
4 H. M. Fales and G. J. Wright, *J. Am. Chem. Soc.*, 1977, **99**, 2339.
5 H. Kwart and D. P. Hoster, *J. Org. Chem.*, 1967, **32**, 1867.
6 D. L. Garin, D. J. C. Greco, and L. Kelley, *J. Org. Chem.*, 1977, **42**, 1249.

2.9 LIQUID–VAPOR EQUILIBRIA. DISTILLATION OF ENANTIOMER MIXTURES

In Section 2.2.4 we have seen that enantiomer mixtures in the liquid state behave in a quasi-ideal fashion. The vapor pressure p_M of such a mixture having composition x_D, x_L at a given temperature is given by

$$p_M = p_A x_D + p_A x_L = p_A \tag{1}$$

where p_A is the vapor pressure of a pure enantiomer at the same temperature.[1]

Alternatively, at a given pressure the boiling points of the enantiomers, of the racemate, and of all partially resolved mixtures are equal; the binary phase diagram for distillation (T^v, x) reduces to a horizontal line. One may conclude from this that, contrary to crystallization or sublimation, distillation of a partially resolved mixture is an operation that cannot lead to a modification of the enantiomeric purity.

Aside from these theoretical considerations, there are actually some experimental results in the literature which attempt to make a case for differences in volatility between racemic and enantiomeric liquids. Thus, in 1912 Groh[2] believed that he had evidence for very small differences in vapor pressure and heat of vaporization between racemic and (+)-methyl tartrate. More recently, McGinn[3] measured a substantial difference (2.5°C) between the boiling points of racemic and optically active 2-octanols. In spite of this difference, the author was unable to effect any enantiomeric enrichment through distillation of the partially resolved alcohol.

On the other hand, Dupont and Desalbres[4] claimed to have separated racemic from levorotatory pinene by distillation, while Nerdel and Diepers[5] reported the possibility of altering the enantiomeric purity of partially resolved 2-(*p*-nitrophenyl)-butane also by distillation.

The illusory or at very least overestimated character of these observations has been exposed by the work of Horeau and Guetté.[6,7] It is a fact that departures from ideality of liquid enantiomer mixtures are extremely small and rarely detectable. If the vapor phase is ideal (which is more than likely), the difference in the heats of vaporization ΔH_A^v and ΔH_R^v is none other than the heat of mixing of the liquid enantiomers ΔH_l^m. Since, according to Trouton's rule, the entropy of vaporization of most unassociated organic compounds is constant ($\Delta S^v = \Delta H^v/T^v \simeq 20$ cal · mol^{-1} · K^{-1}),[9] the difference ΔT^v in the boiling points of a pure enantiomer, T_A^v, and that of its racemate, T_R^v, may be estimated:

$$\Delta T^v \simeq \frac{\Delta H_A^v - \Delta H_R^v}{20} = \frac{\Delta H_l^m}{20} \tag{2}$$

For 2-octanol, $\Delta H_l^m = 3.1$ cal · mol^{-1} as measured by calorimetry[6,7] (see Section 2.2.4, Table 1). This leads to a ΔT^v value of the order of 0.15°C. In fact, contrary to the results of McGinn,[3] Horeau and Guetté did not observe a significant difference in the boiling points of the racemate and the enantiomers. Similarly, Ambrose and Sprake concluded on the basis of very careful measurements[8] that T_A^v and T_R^v of the analogous 2-butanol are identical (within the limits of experimental error).

For 2-(*p*-nitrophenyl)butane[5] ($\Delta H_l^m = 0.45$ cal · mol^{-1}), the expected difference ΔT^v is less than 0.03°C, which makes it difficult to envisage a separation or even enrichment by distillation. The positive result reported by Nerdel and Diepers corresponds only to the trivial separation of the *ortho*- and *para*-nitro isomers.

REFERENCES 2.9

1 H. Mauser, *Chem. Ber.*, 1957, **90**, 299.
2 J. Groh, *Ber.*, 1912, **45**, 1441.
3 C. J. McGinn, *J. Phys. Chem.*, 1961, **65**, 1896.
4 G. Dupont and L. Desalbres, *C. R. Acad. Sci.*, 1923, **176**, 1881.
5 F. Nerdel and W. Diepers, *Tetrahedron Lett.*, 1962, 783.
6 J. P. Guetté, D. Boucherot, and A. Horeau, *Tetrahedron Lett.*, 1973, 465.
7 A. Horeau and J. P. Guetté, *Tetrahedron,* 1974, **30**, 1293.
8 D. Ambrose and C. H. S. Sprake, *J. Chem. Soc.*, (A) 1971, 1261.
9 See, for example, S. Glasstone, *Textbook of Physical Chemistry*, 2nd ed., Van Nostrand, New York, 1946, p. 457.

CHAPTER 3

Solution Properties of Enantiomers and Their Mixtures

At the turn of the century, numerous studies were carried out on the relationship between racemate solubility and that of the constituent enantiomers. Subsequently, it would seem that such inquiries were virtually ignored by physical and physical-organic chemists. Certainly, organic chemists have not particularly been concerned with quantitative considerations of solubility of enantiomers and their racemates even when carrying out resolutions involving differences in solubility between stereoisomers.

A knowledge of the rules that govern the solubility of the three fundamental categories of enantiomer mixtures described in the preceding chapters is not a matter of theoretical interest only. We shall see that, among other possibilities, recrystallization of a partially resolved compound can lead either to the racemate or to the pure enantiomer, according to the initial enantiomeric purity and to the type of racemate it forms. The ignorance of this possibility can be a source of error in a mechanistic study (where isolation of a crystalline product might lead one to conclude the occurrence either of racemization or of retention of chiral integrity) or, more commonly, in the determination of enantiomeric yield in an asymmetric synthesis. Knowledge and use of solubility properties of enantiomer mixtures allows one, moreover, to carry out resolutions in a more rational manner than usually obtainable.

3.1 GRAPHIC REPRESENTATION OF TERNARY SYSTEMS

3.1.1 Quantitative Definitions of Concentration and Solubility

Concentrations, or solubilities, may be expressed in many ways. The weight of substance dissolved may be related to the weight or to the volume of solvent or to those of the solution. One can also focus upon the number of moles of dissolved substances and relate this to the weight, the volume, or the number of moles of

Table 1 Ways of Expressing Concentrations

Definitions		Example
M	Molecular weight of the solute	$M = 200$
W	Weight of solute (in g)	$W = 22.22\,\text{g}$
M'	Molecular weight of the solvent	$M' = 50$
W'	Weight of solvent (in g)	$W' = 100\,\text{g}$
d'	Density of the solvent	$d' = 0.9$
d	Density of the solution	$d = 1.2$

Concentration		Results with Above Data
A (g/100 g solvent)	$= \dfrac{100W}{W'}$	$A = 22.22$
B (g/100 ml solvent)	$= Ad'$	$B = 20.00$
γ (g/100 g solution)[a]	$= \dfrac{100A}{100 + A}$	$\gamma = 18.18$
D (g/100 ml solution)	$= \gamma d$	$D = 21.82$
m (molality: mole/1000 g solvent)	$= \dfrac{10A}{M}$	$m = 1.11$
$\mathcal{M}$ (molarity: mole · liter^{-1} solution)	$= \dfrac{10D}{M}$	$\mathcal{M} = 1.09$
x (mole fraction)[a]	$= \dfrac{WM'}{WM' + W'M}$	$x = 0.053$

[a] γ and x are the expressions we use most often in this book.

solvent. The seven expressions of concentration most often used today are given and illustrated in Table 1.

We emphasize that these modes of expressing concentration must not be confused, particularly in the representation of the solubility diagrams which we describe in the following discussion.

As in the previous chapter, we submit that phase diagrams are convenient ways of representing and summarizing data for interpretive and analytic purposes as well as for predictive ones. Although ternary diagrams, in particular, may appear to many organic chemists to be archaic and/or out of place in a work directed as much to synthetic chemists as to those inclined to quantitative studies, with some introduction, careful description and explanation, we hope quickly to dispel any feelings of strangeness that such diagrams may engender.

3.1.2 Representation of Ternary Systems. Generalizations

A mixture of enantiomers, D and L, in the presence of solvent (S) constitutes a ternary system. While in a binary system variations in temperature and concentration can be represented in a single plane, for a ternary system the same data require three dimensions. The most common way of representing this is by means of a triangular prism whose vertical axes describe the temperature of the system (Fig. 1). In the case of an enantiomer mixture in the presence of solvent, one of the faces of the prism constitutes the *binary* phase diagram (D, L) which has been employed earlier in the book. The other two faces are those of the binary diagrams (D, S) and (L, S) involving the solvent.

We grant that it is inconvenient to represent and to describe a three-dimensional diagram on a sheet of paper. Hence, we limit ourselves to two dimensions by dealing with a two-dimensional slice of the prism perpendicular to the vertical axes. One thus obtains a triangle (*isothermal* slice) which represents the solubility diagram at temperature T_0. Such an isothermal slice is usually drawn as an equilateral triangle or as a right triangle. Other representations are occasionally employed. The different possibilities together with their advantages, disadvantages, and inconveniences are described in the paragraphs which follow.

3.1.3 Triangular Phase Diagrams

Triangular phase diagrams make possible the representation of the *relative* concentrations of three substances. The composition of each of these must then be stated

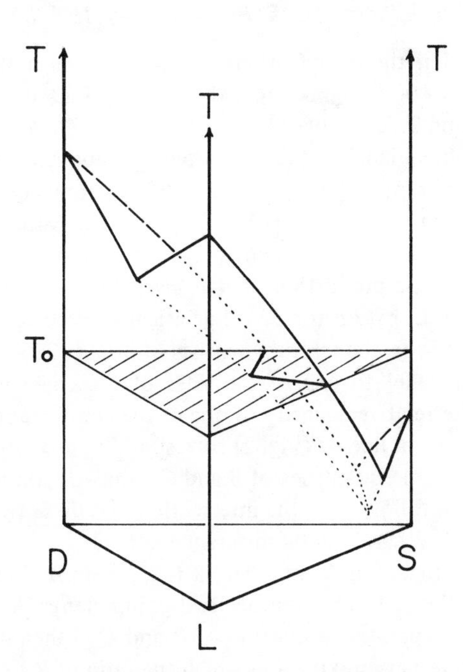

Figure 1 Three-dimensional representation of a ternary system of two enantiomers in a solvent *S*. One of the faces of the prism (at left) corresponds to the binary diagram of D and L (here a conglomerate). Shaded area: isothermal section representing the solubility diagram at temperature T_0.

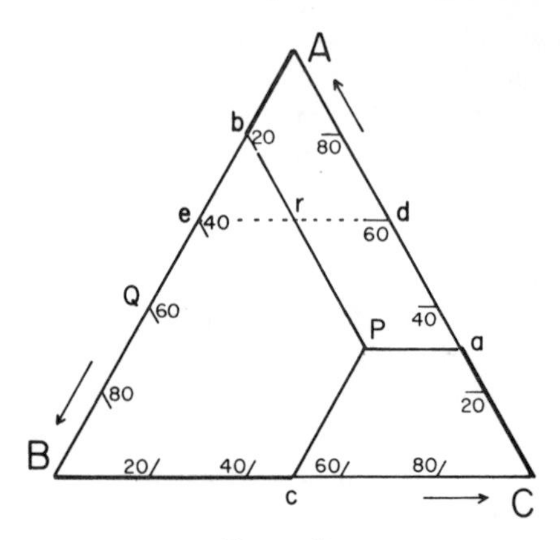

Figure 2.

with reference to the whole ternary system (as mole fractions or weight fractions). Either an equilateral (Fig. 2) or a right triangle (Fig. 4) form may be used.

In the equilateral triangle, the coordinates of a given point P (Fig. 2) may be defined by the lines Pa, Pb, and Pc which are drawn parallel to the sides. The geometry now requires that

$$Pa + Pb + Pc = AB = BC = CD = Ab + Bc + Ca$$

Let AB be equal to 1 if the compositions are expressed in mole fractions; alternatively, AB is equal to 100 g of mixture if the compositions are given as percentages (units x or γ in Section 3.1.1, Table 1).

Each vertex of the triangle represents a pure component, A, B, or C. A point located on a side represents a binary mixture. For example, Q (Fig. 2) represents a mixture of A and B in which $x_A = 0.4$ and $x_B = 0.6$ or, alternatively, 40%A and 60%B. In the interior of the triangle, any point represents a ternary mixture. Thus, for mixture P (Fig. 2) the proportion of A is given by Ca (or Pc), that of B by Ab (or Pa), and that of C by Bc (or Pb). Specifically, P corresponds to $x_A = 0.3$, $x_B = 0.2$, and $x_C = 0.5$; or, the mixture consists of 30%A, 20%B, and 50%C.*

A line drawn parallel to one of the sides of the triangle represents a composition in which the relative quantity of one of the components remains constant. For example, all points on line ed (Fig. 2) correspond to mixtures containing 60%A (or $x_A = 0.6$) and variable quantities of B and C. Similarly, on lines Pb all mixtures contain 20%B. It then follows that the intersection r of these two lines establishes a composition of 60%A, 20%B and, by difference, 20%C.

A straight line originating from a vertex (e.g., from A, Fig. 3) corresponds to mixtures for which the ratio of the other two components (B and C) remains constant. Point r on BC represents a mixture of B and C in the ratio Cr/Br. All points on Ar correspond to ternary mixtures in which the ratio of B to C is given by Cr/Br. Alternatively, line Ar describes the *dilution* of initial mixture r by pure A.

* The mole fractions (x) and weight percent (γ) expressions are not implied to be equivalent. The examples above imply alternative definitions as well as concentration units.

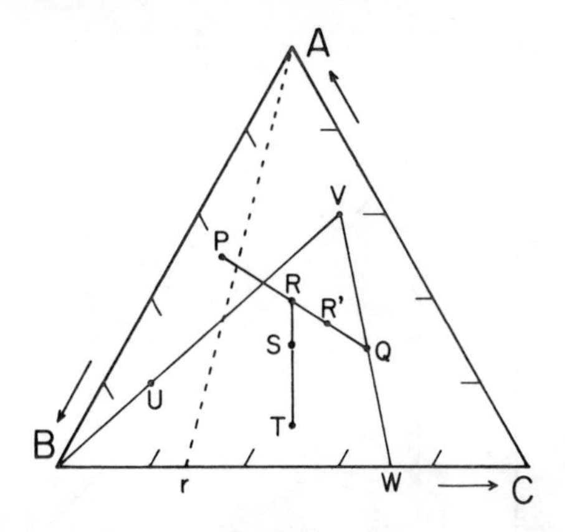

Figure 3.

If a system of composition $P(x_A, x_B, x_C)$ is mixed with another of composition $Q(x'_A, x'_B, x'_C)$, the resulting mixture has a composition R found on the line PQ (Fig. 3). If p and q are the masses of P and Q, respectively, the point on the diagram which represents mixture R (of mass $p + q$) is given by $RP/RQ = q/p$. The coordinates of R may also be calculated by means of the following expression:

$$x_i^R = \frac{px_i + qx'_i}{p + q} \qquad \text{with } i = A, B, C$$

Point R of Fig. 3 (40%A, 30%B, and 30%C) corresponds to a mixture of equal parts of P and Q ($q/p = 1$). If three parts Q are mixed with one part P, point R' then represents a final mixture having a composition 35%A, 25%B, and 40%C.

Through a similar exercise, one can find the point representing a mixture of three systems with compositions P, Q, and T and the corresponding mass p, q, and t, respectively. Point S, for example, corresponds to a mixture of equal parts of P, Q, and T.

Conversely, a mixture of known composition can also be "decomposed" into two (or 3 . . . n) mixtures whose respective compositions and weights may be determined by means of the diagram. Thus, mixture U (consisting of 70%B, 20%A, and 10%C) may be constituted from two parts of pure B and one part of mixture V (which consists of 10%B, 60%A, and 30%C).

Lines such as BUV or PRQ are called *tie lines*. As we shall see (Section 3.1.6), they are essential for the construction of experimental solubility diagrams since they connect the different parts (or phases) of a heterogeneous system. In Fig. 3, for example, let V represent a saturated solution of B and C in solvent and U the overall composition of the system. The tie line VUB indicates that the solid phase consists of pure B. Similarly, if the composition of the system was Q, the solid in equilibrium with saturated solution V would have a composition W (assuming that the solid was not solvated).

When one of the components of the mixture plays a special role, the solvent, for instance, then the equilateral triangle is sometimes replaced by a right triangle

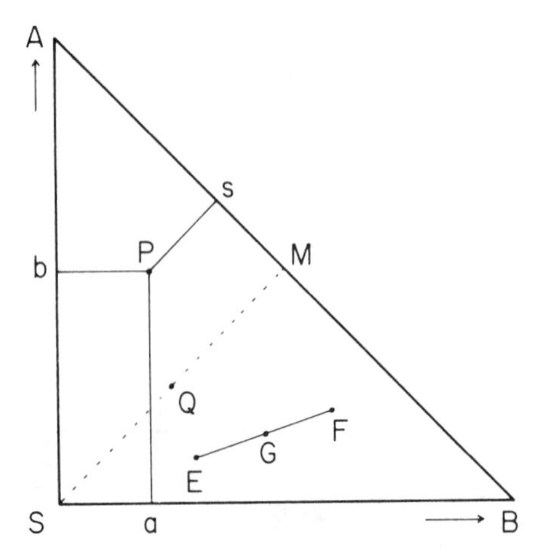

Figure 4 Right triangular phase diagram.

which has the advantage of not requiring special graph paper for its construction. As before, each apex (Fig. 4) represents a pure component, with the solvent S being placed at the right angle. The composition of a mixture P is given by lines perpendicular to the three sides. The amounts of A and B are Pa and Pb, respectively; that of the solvent is $2Ps/\sqrt{2}$. The latter value is more easily given by difference ($x_S = 1 - x_A - x_B$). In the example of Fig. 4, P has a composition corresponding to 20%B, 50%A, and therefore 30% solvent.

Everything said about the properties of equilateral triangles remains valid for the right triangles. For example, a mixture composed of equal parts of two systems, E and F, is represented by point G found in the middle of the tie line EF. Point Q, located in the middle of SM, is made up of equal quantities of solvent and mixture M, which itself is composed of equal parts of B and A.

3.1.4 The Enlargement of Triangular Phase Diagrams

Whether equilateral or right, triangular diagrams such as we have just described are inconvenient to use when one of the components largely predominates, for example, when one is dealing with solutes that are scarcely soluble. In such cases, the solubility curves are removed toward one of the apexes of the triangle and the area representing saturated solutions is quite small. There are two simple ways of overcoming the difficulties of reading such diagrams. In the first, it suffices to enlarge the part of interest in the diagram as shown in Fig. 5 for equilateral triangles (the same holds true for right triangles). This does not require a change in the way in which compositions are expressed; the latter are still related to the whole system (mole fraction or g per 100 g mixture). Note that points A and B, which correspond to pure components, are outside the limits of the enlarged diagram but remain at finite distance from S.

The second way, which is often used, consists of relating the quantities of components A and B to a given quantity of solvent (e.g., in grams A or B per liter

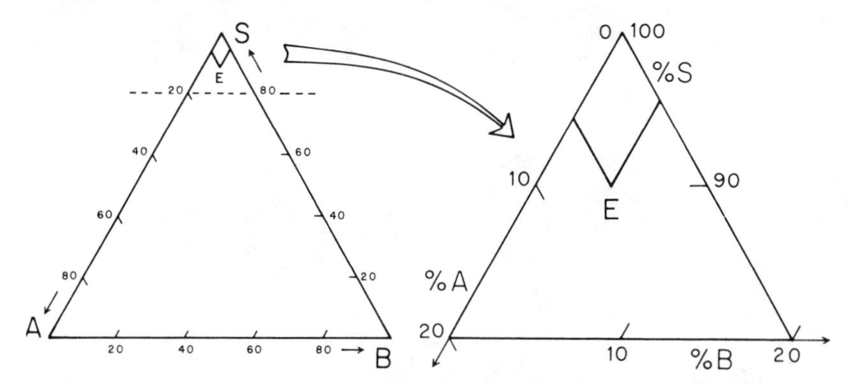

Figure 5 Enlargement of the ternary phase diagram – first method.

solvent). One commonly employs a right triangle, as shown in Fig. 6, but with the difference that the points representing pure components A and B are now removed to infinity (Fig. 6*b*).

3.1.5 Other Representations of Solubility Diagrams

Figure 7 illustrates another way in which solubilities may be represented diagrammatically as a function of composition. The abscissa of the graph gives the relative composition of the binary mixture A,B (where $A + B = 1$ mole, or 100 g) while the ordinate gives the quantity of solvent per mole or per 100 g of $A + B$. This representation actually is derived from a triangular diagram in which apex S is projected to infinity.

Lastly, Fig. 8 illustrates a way of representing solubility which, unlike all of the above, does not derive from a section of the solubility prism of Fig. 1. The quantity of dissolved substance, C, is plotted on rectangular coordinates as a function of the total quantity of substance, Q, all in a given amount of solvent. C and Q may, for example, be given as weight or volume of solvent.

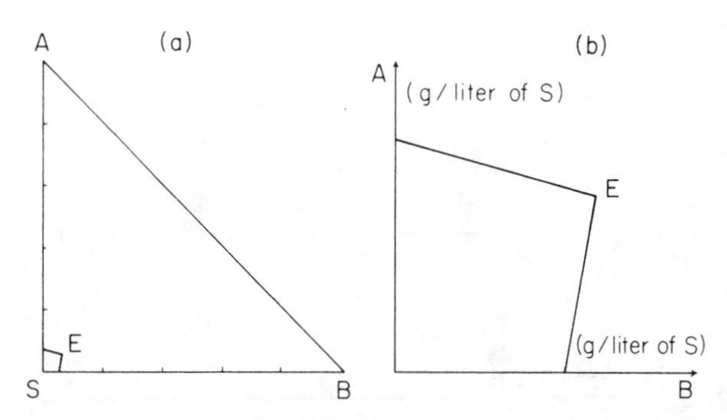

Figure 6 Enlargement of the ternary phase diagram – second method: (a) Original diagram; the concentration refers to the whole system (mole fraction or weight percent of the mixture). (b) Enlarged diagram; the concentrations of A and B are related to a given amount of S.

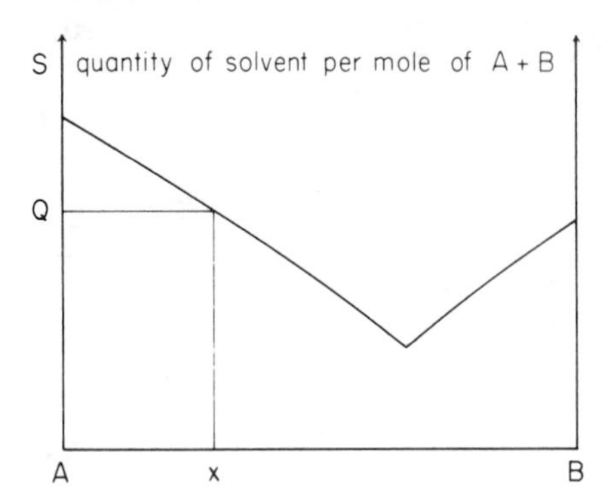

Figure 7 Nontriangular representation of a ternary diagram. Q is the quantity of solvent necessary to dissolve 1 mol of mixture $A + B$ of composition x.

In this representation, the diagonal line *Sd* corresponds to systems in which all of the substance is in solution ($C = Q$). If the substance is pure, when saturation has been attained (A) the quantity dissolved (C) no longer varies, no matter how much excess solid may be present ($Q - C$); a horizontal line (*Ap*) obtains. If the solute is a mixture of two substances that gives rise to an eutectic, then the solubility line will exhibit two breaks. The first, B, indicates the point beyond which the component in excess relative to the composition of the eutectic no longer dissolves. The second, D, followed by horizontal line *De*, corresponds to the solubility of the eutectic (S_E).

This last way of representing solubility is actually much less clear and is less easy to use than the others. Nevertheless, in Section 3.6 (determination of enantiomeric purity from solubility data) we see examples of studies in which this representation has been used.

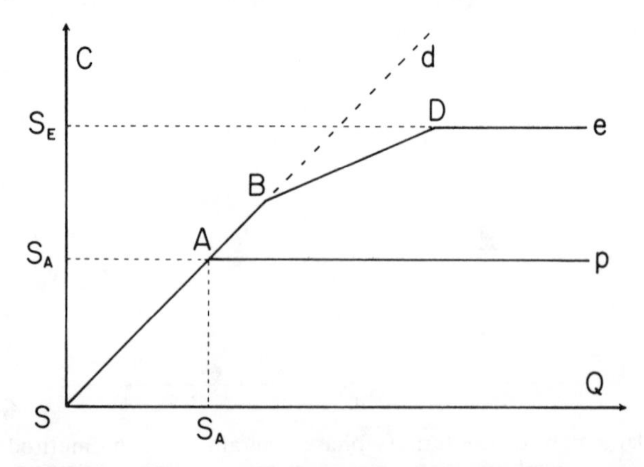

Figure 8 Solubility curves; quantity of solute dissolved (C) as a function of total amount taken (Q); *SAp* is the solubility curve of a pure substance and *SBDe* that of a binary mixture forming a eutectic.

3.1.6 Construction of Solubility Diagrams

A solubility diagram is the graphic representation of a heterogeneous system consisting of a liquid phase, the saturated solution, in equilibrium with one or more solid phases. For ternary mixtures consisting of two enantiomers and one solvent, the construction of the diagram requires the the following steps all carried out at constant temperature: (a) determination of the solubility curve which represents the composition of the saturated solutions as a function of the total composition of the system; and (b) determination of the number and nature of the solid phases in equilibrium with the solution. The experimental results must then be reported in an adequate scheme. For the purposes of this discussion we have adopted the equilateral triangle whose properties were given in detail in Section 3.1.3 as being most useful and compact.

The determination of solubility of enantiomer mixtures does not involve any unusual technical problems. On the contrary, optical rotation data facilitate the determination of the compositions. Several mixtures of enantiomers and racemic compound (or of the two enantiomers) are stirred in a thermostatted bath ($T \pm 0.1$°C) in the presence of a chosen solvent. Use of well-stoppered flasks is essential, particularly if one employs the method of "algebraic extrapolation of tie lines,"[1] for which a knowledge of the overall composition of the system is needed (*vide infra*).

In addition to the pure enantiomer and the pure racemic compound, five or six mixtures covering the 0 to 100% range of enantiomeric purity in even increments generally suffice for the construction of the diagram. It is evident that at equilibrium undissolved solid must remain in contact with the solution.

The attainment of equilibrium, particularly if little solid is present, may require a long time. A minimum time period for this is 24 hours with constant stirring, and sometimes it is necessary to wait several days. It is also wise, as a precaution and to ensure the reproducibility of results, to repeat the determination at 24-hour intervals. The attainment of equilibrium may also be verified by approaching the equilibrium concentation from two directions: (1) crystallization of a supersaturated solution, and (2) dissolution of excess solid in pure solvent.

Equilibrium having been attained, the first step consists of determining the composition of the saturated solution. Most often, this analysis involves quite straightforward procedures or simply common sense.

When the solvents employed are volatilé, the determination of the quantity of solvent and of solute can be carried out by evaporation (rotary evaporator) of weighed quantities of solution in tared flasks. Generally, between 0.5 and 2 g of solution is required for this. In the case of aqueous solutions, which are often quite difficult to evaporate without accident, it is sometimes possible to titrate the total quantity of dissolved enantiomers by appropriate methods (acidimetry, halogen determination according to Charpentier–Volhardt, etc.). Refractometric analysis of the solution under study is often the method of choice following the establishment of a calibration curve based upon titrated solutions. The indexes of refraction are measured at a slightly higher temperature than that of the solubility diagram in

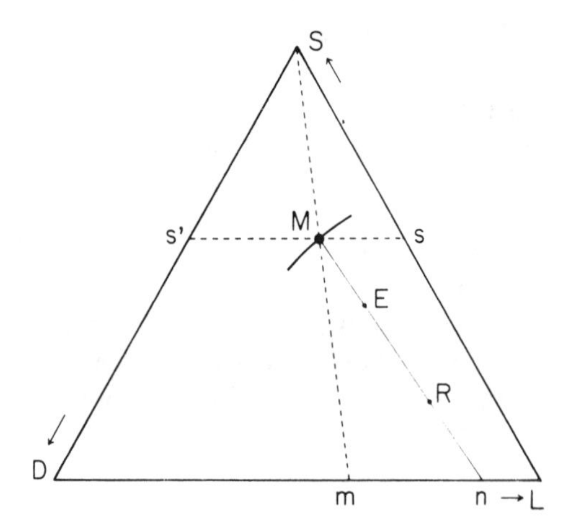

Figure 9 Construction of a triangular solubility diagram.

order to avoid crystallization during the measurement. The method requires only a drop of solution which often allows one to construct a diagram from less than 1 g of the enantiomers overall.

The preceding operations furnish the relative quantities of solvent and of the (D + L) mixture. The amount of solvent is indicated on the Fig. 9 diagram as segment *Ls*. The point which represents the composition of the solution studied must then be on the line *ss′*. The enantiomeric purity of the solute remains to be determined. This can be done through measurement of the optical rotation of the saturated solution whose concentration is known or of that of the solute itself following evaporation to dryness and dilution to a known volume of appropriate solvent. The composition of the (D + L) mixture is thus obtained. Point *M*, representing the saturated solution, is then found at the intersection of lines *ss′* and *Sm*.

The point *M* so determined is on the solubility curve. The composition of the solid in equilibrium with the solution can be obtained by two analytical methods both of which are based upon use of tie lines: the method of "algebraic extrapolation"[1] and the method of "wet residues."[2]

In the method of wet residues (method of rests, or method of Schreinemakers), once the composition of the saturated solution is known, one determines that of a sample of solid impregnated by mother liquor (the wet residue, or rest). This composition being *R*, one knows that that of the solid must lie on tie line *MR*, in the extension of the line beyond *R* (Fig. 9). In order that the method be as precise as possible *R* must be removed as far as possible from *M*. Hence, the solid sample must be as free from mother liquor as possible.

The method of algebraic extrapolation proposed by Hill and Ricci[1] is a variant of the method of wet residues. A point analogous to *R* is obtained if, instead of analyzing a sample of solid wet with mother liquor, one analyzes the whole system. This analysis is even unnecessary if one knows the quantities of the three components taken initially. Composition *E* thus being defined, we can draw the tie line

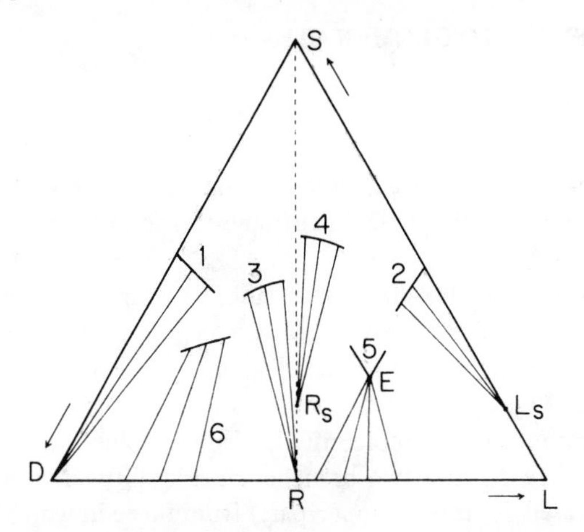

Figure 10 Tie lines associated with different systems: (1) a solid phase D (pure enantiomer) in the presence of mother liquor of variable composition; (2) a solid phase L*s* (solvated enantiomer) in the presence of mother liquor of variable composition; (3) a solid phase *R* (pure racemic compound); mother liquors of variable composition; (4) a solid phase *Rs* (solvated racemic compound); mother liquid of variable composition; (5) two solid phases, one enantiomer and the racemic compound (or, two enantiomers if *E* is on *SR*, i.e., for a conglomerate); mother liquor of fixed composition *E* (eutectic); (6) the tie lines do not converge; one solid phase is present (solid solution of D and L) in mother liquors of variable composition.

ME, which is in principle identical to *MR*. The latter method nonetheless is less precise than that of wet residues since point *E* is generally close to *M*.

The drawing of only *one* tie line is not very informative. It is indicative neither of the nature nor of the number of solid phases.* This information requires the examination of bundles of lines which correspond to systems of different compositions. The various situations encountered in the case of enantiomer mixtures are illustrated in Fig. 10. We return to this in greater detail in connection with the study of solubility diagrams that correspond to the different types of racemates.

REFERENCES 3.1

1 A. E. Hill and J. E. Ricci, *J. Am. Chem. Soc.*, 1931, **53**, 4305.
2 F. A. H. Schreinemakers, *Z. Phys. Chem.*, 1893, **11**, 75.

* On the other hand, where one knows that the crystals are *unsolvated*, point *n* in Fig. 9 gives the composition of the solid.

3.2 SOLUTIONS OF CONGLOMERATES

3.2.1 Theoretical Phase Diagrams and Experimental Properties

The only solid phases that may be present along with the solvent in a solution of the conglomerate are crystals of the enantiomers. The latter may be either pure or they may be solvated.

For a nonsolvated conglomerate, the solubility diagram at a given temperature is represented by the ternary isotherm shown in Fig. 1. It corresponds to a horizontal section of the three-dimensional diagram shown in the preceding section (Fig. 1 of Section 3.1.2).

In Fig. 1, points A and A' represent the (equal) solubilities of the pure enantiomers at temperature T_0. The two branches of solubility curve AEA' separate the domain of unsaturated solutions (upper part) from those in which one or two solid phases are in equilibrium with a saturated solution. The significance of the various zones of the diagram may be understood by consideration of the phase rule. At constant pressure, the variance of this ternary system is given by the expression

$$v = C - \Phi + 1 = 4 - \Phi$$

where C is the number of independent components (D, L, S) and Φ is the number of phases in the system (see Section 2.2.1). Given a solid mixture of the two enantiomers of composition M (Fig. 1*a*) to which a limited amount of solvent is added, one obtains a system P comprising three phases at equilibrium (solid D and L and a saturated solution) whose variance is 1. Consequently, *at constant temperature* the composition of the saturated solution in the presence of the two solid phases is fixed. In the isothermal diagram (Fig. 1*a*), it is represented by point E, which corresponds to the solubility of the racemate (eutectic) at T_0. Extension of the tie line EP leads to the composition N for the two solid phases in equilibrium with solution E. The area within which the three phases can coexist at equilibrium is triangle DEL in which all tie lines converge toward point E (Fig. 1*b*). When the temperature

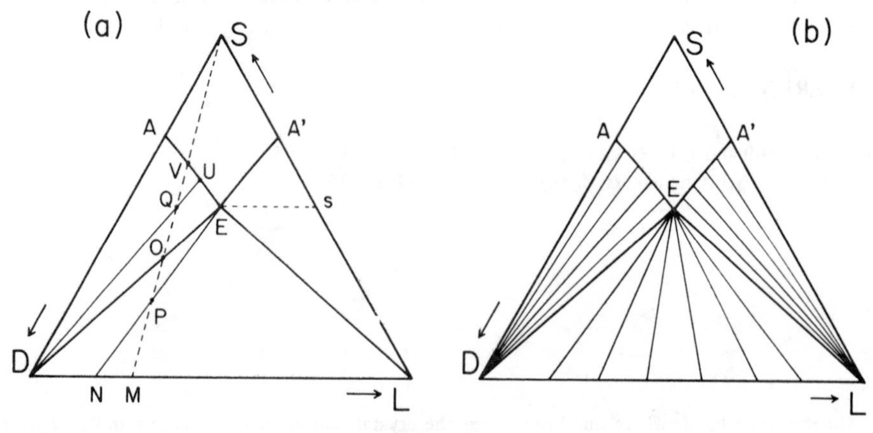

Figure 1.

is changed, the composition of the saturated solution is likewise changed. Point *E* is displaced along line *SC*. The relation between *T* and *E* is none other than the solubility curve of the racemate as a function of temperature.

Let us now return to temperature T_0. (Fig. 1*a*) and add to mixture *P* a quantity of solvent such that the new system be represented by *O*. All of the racemate contained in the initial solid mixture *M* has now passed into the saturated solution *E*. The only solid phase remaining is the enantiomer initially present in excess in *M* (enantiomer D in this case). The system now consists of only two phases and is consequently bivariant. At constant temperature *T* there still remains one degree of freedom. Indeed, further addition of solvent leads, between *O* and *V*, to systems consisting of pure solid (D) in the presence of saturated solutions whose compositions vary between *E* and *A* along the solubility curve. (The form of this solubility curve is discussed in Section 3.2.3.) System *Q*, for example, consists of pure solid D in a solution of composition *U*. The two-phase areas comprise triangles *ADE* and *A′LE* in which the tie lines converge toward the pure enantiomers (Fig. 1*b*).

Finally, from *V* on, the proportion of solvent becomes sufficient to completely dissolve the remaining solid. One enters the one-phase and trivariant domain of unsaturated solutions which is bound by *SAEA′*. (The three degrees of freedom of this monophasic system correspond to the possibility of simultaneously varying the temperature and the concentrations of two of the components, that of the third naturally being determined by choice of the other two.)

The quantitative analysis of these triangular diagrams utilizes the geometric properties which were reviewed in Section 3.1.3. A single example will be given here: 1 mole of system *P* (Fig. 1*a*) is distributed at equilibrium between *PN*/*NE* mole of solution of composition *E* and *PE*/*NE* mole of solid composition *N*. Moreover, the mole fraction of solvent in solution *E* is given by L*s*.

As illustration, consider the enantiomers of *N*-acetylleucine whose solubility

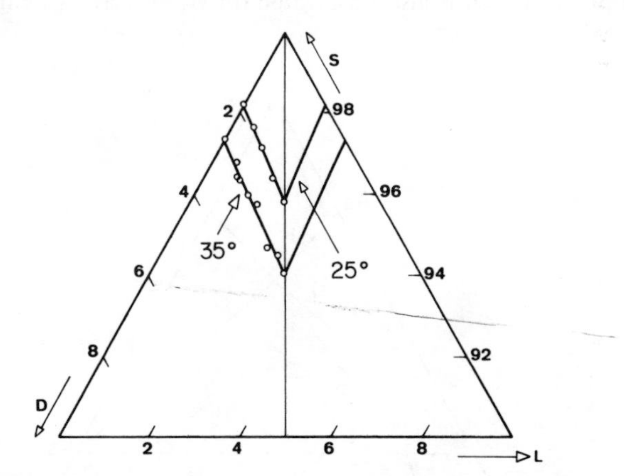

Figure 2 Solubility diagram of the (+)- and (−)-*N*-acetylleucines in acetone at 25 and 35°C. Only the upper part of the diagram is shown. Concentrations are in g/100 g mixture (γ unit). J. Jacques and J. Gabard, *Bull. Soc. Chim. Fr.*, 1972, 342. Reproduced by permission of the Société Chimique de France.

diagram was determined in acetone at 25° and 35°C (Fig. 2). This is a typical case of a conglomerate uncomplicated by polymorphism in the temperature range studied and forming no solvated forms with acetone. Note that the effect of temperature is simply the vertical displacement of the solubility curves which, in fact, are virtually straight lines.[1]

The behavior of solvated conglomerates is described by the isothermal solubility diagram of Fig. 3. A_s and A'_s represent the solubilities of the solvated enantiomers while D_s and L_s give the compositions of these solvates. Analysis of the SD_sL_s part of the diagram is identical to that given for the preceding case, except that the solid phases here are the solvated species D_s and L_s and not the pure compounds D and L. Thus, a system of composition P (Fig. 3*a*) at equilibrium consists of a mixture of two solid phases D_s and N_s of composition N_s (or N, if no account is taken of solvent molecules in the crystals) and of a racemic saturated solution of composition E.

Similarly, mixture Q at equilibrium deposits crystals of D_s in mother liquors of composition U. Below line D_sL_s the amount of solvent present in the system is no longer sufficient to solvate all of the solids. Thus, in this part of the diagram one has a mixture of solvated and nonsolvated crystals. An examination of the equilibria between the solid phases present, which from our point of view is only of marginal interest, is taken up in note 12.

Solvated crystals generally exist only within a given temperature range below or above a transition temperature. Desolvation of the crystals beyond this temperature (or through lack of solvent) can occasionally take place without change of racemate type (solvated conglomerate ⇌ desolvated conglomerate). Most often however, desolvation is accompanied by a change of racemate type (solvated conglomerate ⇌ solvated or unsolvated racemic compound).

Similarly, an unsolvated conglomerate can sometimes transform itself into a racemic compound which is also unsolvated (or vice versa), as we have already seen

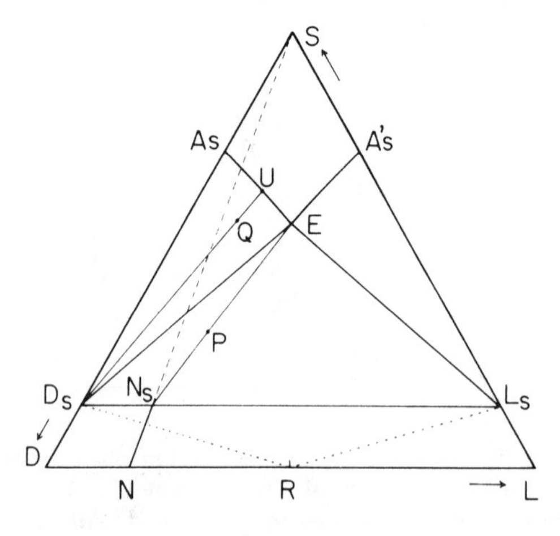

Figure 3.

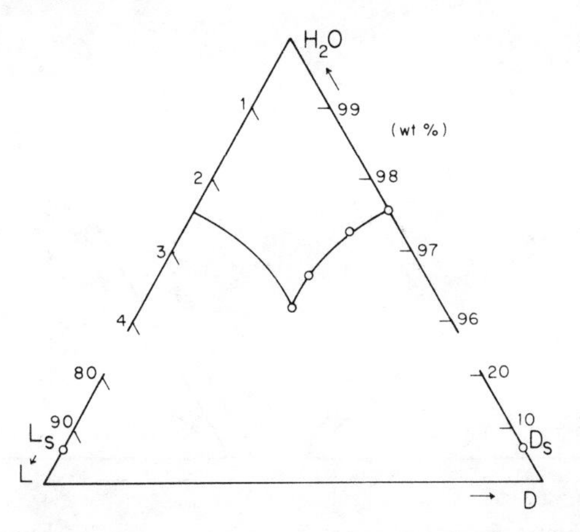

Figure 4 Solubility diagram of a solvated conglomerate. Adapted from Y. Shimura and K. Tsutsui, *Bull. Chem. Soc. Jpn.*, 1977, **50,** 145 by permission of the Chemical Society of Japan.

(Section 2.5). The consequences of these types of polymorphism on solubility diagrams are examined in Section 3.5.

Finally, the experimental solubility diagram of a solvated conglomerate[2] is shown in Fig. 4. The solid phases D_s and L_s are the monohydrates of the complexes (+)- and (−)-[Co(ox)(en)$_2$]Cl.

3.2.2 Solubility Rules for Partially Resolved Mixtures

It is particularly important, from a practical standpoint, to know the behavior of partially resolved mixtures of enantiomers upon recrystallization. The ternary phase diagrams just described allow us to follow the crystallization of partially resolved conglomerates in a very simple way, provided that the process be carried out under conditions of thermodynamic equilibrium. Let us examine, successively, isothermal recrystallization and recrystallization attended by concentration or cooling.

Isothermal recrystallization is illustrated in Fig. 5. Let us begin with a partially resolved conglomerate of composition P to which we add a definite quantity of solvent S. The point in the diagram which represents such a mixture is found on line PS. If the mixture is heated, the crystals dissolve. The system then is allowed to cool whereupon equilibrium is reestablished at T_0, at which temperature the solubility is represented by curves AEA'. There are two possibilities to consider. In the first, the amount of solvent is relatively small such that the mixture corresponds to point M_1 which is located within the bounds of area DEL. The crystals that deposit when equilibrium is reestablished have composition N, and the mother liquors are racemic. In the second case, a larger amount of solvent yields system M_3 which at equilibrium furnishes a precipitate of pure D crystals. Here, the mother liquors of composition s contain an excess of the D enantiomer. The maximum yield of enantiomer D sould be obtained from system M_2.

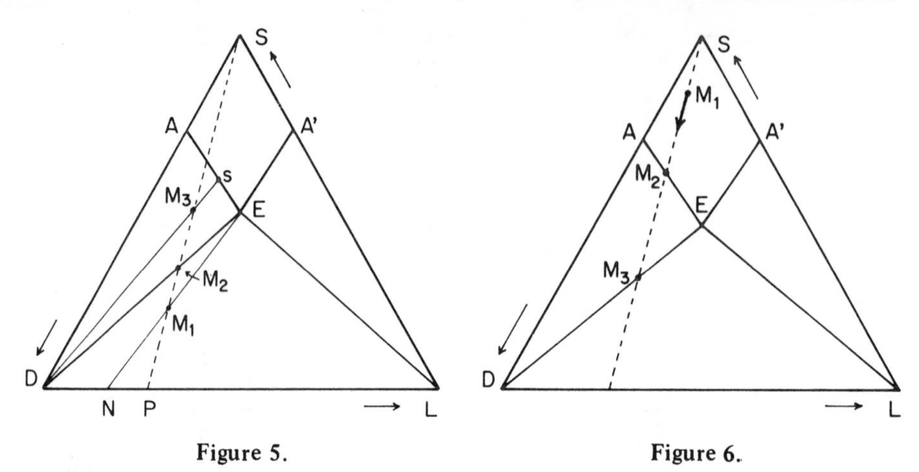

Figure 5. Figure 6.

Figure 6 describes recrystallization carried out by isothermal concentration of the solution or, alternatively, by cooling. At first, we consider system M_1, which represents an unsaturated solution. Upon evaporation of the solvent the system displaces downward along line SM_1. When it reaches the solubility curve at point M_2, crystals of pure D begin to deposit. This crystallization continues up to point M_3, where the two enantiomers deposit together from a solution which remains racemic.

The same sequence applies to a crystallization carried out by cooling of a solution, but in the latter case one has to consider that the solubility curves move upward toward point M_1, which does not vary.

It is noteworthy that, *in either case* (Fig. 5 or Fig. 6), recrystallization increases the enantiomeric purity of the crystals. With an adequate quantity of solvent one can always obtain the enantiomer originally present in excess pure in a single recrystallization. Moreover, upon concentration or cooling of solutions of conglomerates, the initial crystals are always those of the pure enantiomer.

We shall see later (Chapter 4) that these ternary phase diagrams may be used to describe recrystallizations performed *out of equilibrium*, which in turn makes possible resolution of conglomerates by entrainment.

3.2.3 Comparison of the Solubilities of Pure Enantiomers and Their Conglomerate*

It is generally recognized that in conglomerate systems the racemate is more soluble than the constituent enantiomers. This property is often invoked to explain enantiomeric enrichment obtained upon crystallization of a partially resolved mixture without reference to phase diagrams.

An old empirical rule – that of Meyerhoffer – deals with the quantitative aspect of this solubility difference. The so-called "double solubility" rule states that

* Adapted in part from A. Collet, M. J. Brienne, and J. Jacques, *Chem. Rev.*, 1980, **80**, 215 by permission of the editor. Copyright 1980, American Chemical Society.

a conglomerate has a solubility equal to the sum of the solubilities of the corresponding enantiomers.[9] Meyerhoffer rationalized this rule by considering vapor pressures, which is presumably legitimate since a gas may be thought of as a solution in a vacuum. The vapor pressure of an ideal binary mixture is, of course, equal to the sum of the vapor pressures of the constituents. Replacement of vacuum by an *inert* solvent and of vapor pressures by concentrations then leads immediately to the proposition that a conglomerate must be twice as soluble as each enantiomer, provided only that the solvent has no effect (associative or dissociative) upon the solute.

It is possible to explain this rule in a more precise way. Effectively, the Schröder–Van Laar equation (Section 2.2.3) indicates that in an ideal solution in equilibrium with crystals of a pure constituent at a given temperature, the mole fraction of this constituent in the liquid phase depends only on the enthalpy of fusion and on the melting point of the solute and not on the properties of the solvent, which may be either a single substance or a mixture (Fig. 7).

Let x_R and x_A be the solubilities (as mole fractions) of a conglomerate DL and of one of the pure enantiomers, respectively, at the same temperature. From the foregoing, the mole fraction of enantiomer L (or D) in the racemic saturated solution, which is equal to $x_R/2$, must also be equal to x_A. In the solubility relationship expressed in mole fractions, it follows that the solubility of the conglomerate is double that of the pure enantiomers. This leads us once again to call attention to the various ways of expressing solubility (Section 3.1.1). The ratio of solubilities of the racemate to that of the enantiomers, designated as α, depends upon the manner in which the concentrations are defined. Different ways of formulating this ratio are given in Table 1; it is only in dilute solutions that all of these equations become equivalent.

The equation of Schröder–Van Laar indicates that the solubility of L remains constant and equal to x_A whatever the concentration of D. This requires that the solubility curve for L in the triangular phase diagram D, L, S (Fig. 8) be a straight

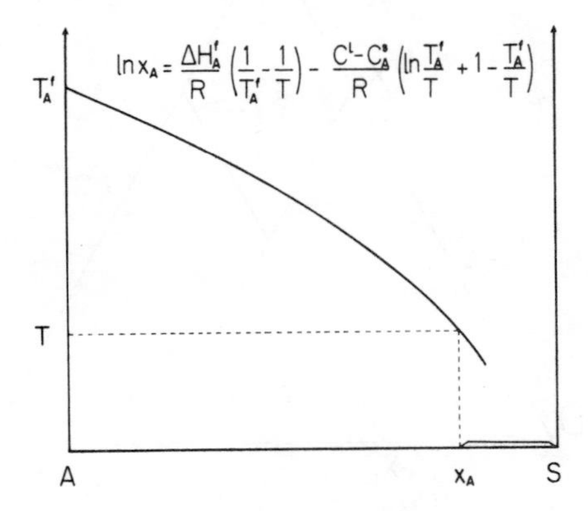

Figure 7 Ideal solubility of compound A in solvent S; the solubility curve of A as a function of temperature is given by the equation of Schröder–Van Laar. At temperature T, the solubility of A is x_A.

Table 1 Some expressions of the ratio α = solubility of racemate/solubility of enantiomers

A. Definitions

M, M_s: Molecular weights of the solute and of the solvent
Saturated solution:
W_A, W_R: Weight of enantiomer or of racemate
W_S: Weight of solvent
d_A, d_R: Density of the solution of enantiomer or of racemate

B. Solubility Expressions

x_A, x_R: Mole fraction of the solute
γ_A, γ_R: Grams solute per 100 grams solution
$\mathscr{M}_A, \mathscr{M}_R$: Moles solute per liter solution

C. Solubility Ratio α

$$\alpha_x = \frac{x_R}{x_A} = \frac{W_R}{W_A} \times \frac{W_A + W_S(M/M_s)}{W_R + W_S(M/M_s)} \qquad \alpha_\gamma = \frac{\gamma_R}{\gamma_A} = \frac{W_R}{W_A} \times \frac{W_A + W_S}{W_R + W_S}$$

$$\alpha_{\mathscr{M}} = \frac{\mathscr{M}_R}{\mathscr{M}_A} = \frac{\gamma_R}{\gamma_A} \times \frac{d_R}{d_A}$$

In dilute solution: $\alpha_x \simeq \alpha_\gamma \simeq \alpha_{\mathscr{M}}$

line parallel to side SD (just as the solubility curve for D must be the straight line parallel to SL corresponding to $x_D = x_L$). If concentrations are expressed other than as mole fractions, straight lines are still obtained but these are no longer parallel to the sides of the triangle.

The arguments just made are valid only if the species present in solution are the same as those present in the crystals, that is, provided that the dissolved molecules

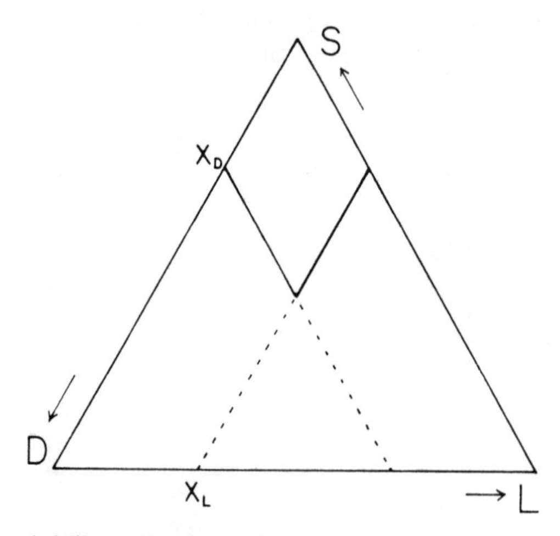

Figure 8 Ideal solubility diagram of a conglomerate of nondissociable enantiomers. If concentrations are expressed as mole fractions, the solubility curves are straight lines parallel to sides SD and SL.

of the enantiomers are not dissociable (or at least not dissociated). This is actually the case for the majority of organic compounds that are not salts.

Let us now turn to the behavior of dissociable molecules AX in which A is a chiral ion (or molecule) and X is an achiral ion (or molecule). The solubility and dissociation equilibria may be written as follows so as to define the respective constants K_s and K_d:

$$AX_{solid} \rightleftharpoons AX_{soln} \quad K_s = [AX] \tag{1}$$

$$AX_{soln} \rightleftharpoons A + X \quad K_d = \frac{[A][X]}{[AX]} \tag{2}$$

Equation (3) is then derived from eqs. (1) and (2):

$$K_s K_d = [A][X] \tag{3}$$

In order to simplify the following arguments, let us equate activities to concentrations (ideal solutions). If S_A is the solubility of the chiral molecule AX, we have

$$S_A = [AX] + [A] = K_s + [A], \qquad \text{and} \qquad [A] = S_A - K_s$$

since $[A] = [X]$,

$$K_s K_d = (S_A - K_s)^2 \tag{4}$$

For the case in which $AX + \bar{A}X$ exists as a conglomerate, equilibria (1) and (2) still hold, and we may suppose that constants K_s and K_d maintain the same values. There remains the need to determine [A] and [X]. If S_R is the solubility of the racemate, that of the enantiomer is $S_R/2$, and thus

$$\frac{S_R}{2} = [AX] + [A] = K_s + [A] \qquad \text{and} \qquad [A] = \frac{S_R}{2} - K_s$$

Since in the racemate $[X] = 2[A]$, eq. (3) gives

$$K_s K_d = 2\left(\frac{S_R}{2} - K_s\right)^2 \tag{5}$$

Combining (4) and (5), we have

$$(S_A - K_s)^2 = 2\left(\frac{S_R}{2} - K_s\right)^2 \tag{6}$$

from which we may obtain the following:

$$S_R = \sqrt{2}(S_A - K_s) + 2K_s \tag{7}$$

This equation would be rigorously valid for ideal solutions by expressing solubilities S_R and S_A as mole fractions x_R and x_A. In practice, it is generally valid only for rather dilute solutions. Under these conditions, we have seen, in Table 1, that the ways in which solubilities are expressed become equivalent. Returning to eq. (7), we see immediately that for zero dissociation, that is, $K_s = S_A$, one finds again the double solubility $S_R = 2S_A$. On the other hand, in the case of total dissociation ($K_s = 0$), we arrive at

$$S_R = \sqrt{2}S_A \qquad (8)$$

This is a relationship which was independently arrived at by Yamanari et al.[4] Equation (7) shows that the solubility of a conglomerate is equal to twice that of the undissociated enantiomer fraction (K_s) plus $\sqrt{2}$ times the dissociated fraction ($S_A - K_s$). More generally, for conglomerates of structure AX_n, one has

$$S_R = \sqrt[n+1]{2}(S_A - K_s) + 2K_s \qquad (9)$$

Note that n in AX_n may be fractional, such that for A_2X, for example, one would write $AX_{1/2}$ ($n = 1/2$). The subscript n must be associated with the *achiral* ion (or molecule).

The preceding arguments may be applied also to those important cases in which acid–base equilibria intervene.[10] Since the principal conclusions for the latter cases qualitatively follow what has been stated above, we need but reiterate the main points.

Carboxylic acids in aqueous solution are only slightly dissociated, hence $S_R \simeq 2S_A$. This is certainly true, *a fortiori*, in organic solvents.

Salts of strong acids and weak bases, or of weak acids and strong bases, are highly dissociated in water, and the solubility of the racemate tends toward $S_R \simeq \sqrt{2}S_A$ (or $\sqrt[n+1]{2}S_A$).

Finally, salts formed from weak acids and bases lead, for the conglomerate case, to a solubility S_R bracketed between $\sqrt{2}S_A$ and $2S_A$ (or $\sqrt[n+1]{2}S_A$ and $2S_A$).

Let us now examine the effects which the equilibria just decribed have upon the *form* of the solubility curves. We limit ourselves to analysis of the case corresponding to eq. (8), that is to say, a conglomerate of enantiomers AX and $\bar{A}X$ which is completely dissociated in solution. The other cases lead to similar conclusions. The solubility/dissociation equilibrium, eq. (10), allows one to define constant K_{sd}:

$$AX_{solid} \rightleftharpoons A_{soln} + X_{soln} \qquad K_{sd} = [A][X] \qquad (10)$$

For the pure enantiomer, whose solubility is S_A, we have $S_A = [A] = [X]$ and $K_{sd} = S_A^2$. Let us consider a saturated solution of enantiomeric purity p whose concentration is $S_p = [A] + [\bar{A}]$. The solubility of enantiomer AX in this saturated solution is given by

$$S_{Ap} = [A] = S_p\frac{1+p}{2} \qquad [X] = S_p \qquad (11)$$

From eq. (10), it follows that

$$[A][X] = S_p^2\frac{1+p}{2} = S_A^2$$

which in turn yields the solubility of the mixture of enantiomeric purity p:

$$S_p = S_A\sqrt{\frac{2}{1+p}} \qquad (12)$$

For the racemate ($p = 0$), eq. (12) reduces to $S_R = \sqrt{2}S_A$ as required. The solubility

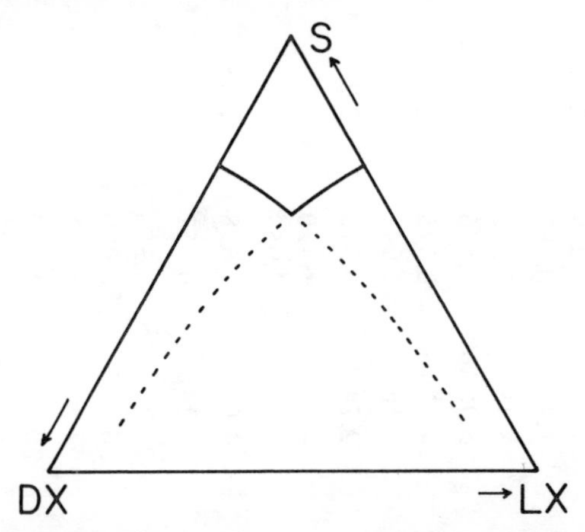

Figure 9 Calculated solubility curves for the case of a conglomerate of DX and LX which are completely dissociated in solution. The solubility of the racemate is $\sqrt{2}$ times that of the pure enantiomers.

curve is defined by the concentration of an enantiomer in its saturated solution, eq. (11). Combining (11) and (12) leads finally to

$$S_{Ap} = S_A \sqrt{\frac{1+p}{2}} \tag{13}$$

As an illustration of the foregoing, we have plotted in Fig. 9 the solubility curve AEB calculated by means of eq. (13) taking $S_A = 0.3$ (mole fraction) for the solubility of the pure enantiomer. The upper part of the curves (solid line) corresponds to the stable solubility equilibrium. Their extension (dotted line) cannot be observed except under conditions in which one of the enantiomers remains supersaturated (metastable equilibrium). It is evident that, contrary to the case of conglomerates of undissociated enantiomers giving ideal solutions (see Fig. 8), the solubility curves *are no longer straight*. Nevertheless, the curvature found is rather slight.

In a general way, these conclusions remain valid even when weaker or less specific interactions are manifested between enantiomers in solution. In the case we have elaborated in detail, the interactions are due to the existence of a common ion. In the general case, the intermolecular association or the acid–base exchanges that involve solvent give rise to equilibria which all lead to the same consequences: the solubility of each enantiomer is affected by the presence of the other and the solubility curves are no longer straight.

The experimental data available on the relative solubilities of enantiomers and of their conglomerates are given in Tables 2, 3, and 4. Table 2 deals with undissociated compounds. We see there that most of the α_x values, including those of carboxylic acids such as N-acetylglutamic acid and N-acetylleucine, are quite close to the theoretical value ($\alpha_x = 2$). Departures from the theoretical value correspond to a slight increase in the solubility of the racemate ($\alpha > 2$). These become important only in some cases of high solubility.

Table 2 Solubility ratio α for conglomerates of undissociated compounds.[a]

Name	Solvent	T(°C)	γ_A	$x_A \times 100$	γ_R	$x_R \times 100$	α_x	Refs.
2-(1-Naphthoxy)propionamide	Acetone	25	1.38	0.376	2.80	0.770	2.05	1
		35	1.87	0.511	3.86	1.072	2.10	
	Ethanol	25	0.60	0.129	1.20	0.259	2.01	
		35	0.90	0.194	1.79	0.388	2.00	
threo-1-*p*-Nitrophenyl-2-amino-1,3-propanediol	Methanol	25	1.63	0.249	3.35	0.520	2.09	1
		35	2.39	0.368	5.05	0.796	2.16	
Anisylidenecamphor	Methanol	25	2.17	0.262	4.65	0.575	2.19	1
		35	3.16	0.385	8.16	1.042	2.71	
N-Acetyl-α-methylbenzylamine	Water	35	1.23	0.137	2.74	0.310	2.26	1
N-Acetylleucine	Acetone	25	1.86	0.631	4.12	1.420	2.25	1
		35	2.66	0.908	5.90	2.059	2.27	
2-*p*-Methoxyphenylpropiophenone	Hexane	25	1.89	0.686	4.60	1.698	2.48	1
N-Acetylglutamic acid	Water	25	4.14	0.410	8.28	0.852	2.08	7
Dilactyldiamide	Water	35.5	12.4	1.568	23.7	3.376	2.15	3
		60	20.1	2.752	36.2	6.000	2.18	
		80	27.5	4.093	47.5	9.238	2.26	
N-Acetylproline monohydrate	Water	20	13.12	1.702	30.265	4.470	2.78	8
		30	19.16	2.645	45.80	8.832	3.34	
N-Chloroacetylproline	Acetone	20	11.51	3.790	23.85	8.662	2.29	8
		30	15.02	5.082	33.835	13.41	2.64	
N-Butyrylproline	Water	20	5.66	0.580	13.79	1.533	2.64	8
		30	7.15	0.744	20.255	2.412	3.24	

[a] Definitions and symbols are given in Table 1.

Table 3 Solubility ratio α for conglomerates of ionic compounds in aqueous solution[b]

Name	T (°C)	γ_A	$x_A \times 100$	γ_R	$x_R \times 100$	α_x[a]		α_γ	Refs.
Amonium N-acetyltryptophanate	25	6.28	0.457	10.25	0.776	1.70		1.63	6
Glutamic acid hydrochloride	25	32.21	4.453	42.39	6.732	1.51		1.32	6
Ammonium glutamate	25	45.86	8.506	55.16	11.896	1.40		1.21	6
Histidine monohydrochloride	45	23.30	2.776	33.00	4.425	1.59		1.42	1
Lysine 3,5-dinitrobenzoate	30	6.96	0.375	13.17	0.757	2.02		1.89	5
	50	13.50	0.779	26.53	1.783	2.29		1.97	
	65	24.00	1.563	43.68	3.753	2.40		1.82	
Alanine p-chlorobenzenesulfonate	15	19.48	1.522	33.42	3.108	2.04		1.72	7[b]
	30	27.11	2.321	46.32	5.226	2.25		1.71	
	45	40.30	4.135	58.19	8.167	1.97		1.44	
DOPA naphthalenesulfonate · $1.5H_2O$	10	1.09	0.044	1.57	0.064	1.45	‡	1.44	7
	30	1.96	0.080	3.19	0.132	1.65	‡	1.63	
	50	5.03	0.212	8.26	0.360	1.70		1.64	
Leucine benzenesulfonate	15	14.53	1.047	28.26	2.392	2.28		1.94	7
	25	18.63	1.405	36.71	3.483	2.48		1.97	
Lysine p-aminobenzenesulfonate	15	25.26	1.869	35.06	2.953	1.58		1.39	7
	25	29.92	2.350	39.80	3.592	1.53		1.33	
	45	38.69	3.434	47.53	4.857	1.41		1.23	

Table 3 Continued.

Name	T (°C)	γ_A	$x_A \times 100$	γ_R	$x_R \times 100$	α_x[a]		α_γ	Refs.
Serine *m*-xylenesulfonate	15	19.03	1.347	31.08	2.553	1.90		1.63	7
	25	28.32	2.244	44.72	4.489	2.00		1.58	
	40	46.47	4.802	63.69	9.249	1.93		1.37	
Tryptophan benzenesulfonate	15	3.38	0.173	5.39	0.282	1.63	‡	1.59	7
	35	5.30	0.277	9.67	0.529	1.91		1.82	
	50	8.17	0.440	17.29	1.028	2.34		2.12	
$[Co(ox)(en)_2]Cl \cdot H_2O$	5	1.00	0.060	1.55	0.094	1.56	‡	1.55	4
$[Co(ox)(en)_2]Br \cdot H_2O$	5	0.475	0.025	0.743	0.039	1.56	‡	1.56	4
$[Co(NO_2)_2(en)_2]Cl$ cis	5	2.55	0.153	3.97	0.242	1.58	‡	1.56	4
$[Co(NO_2)_2(en)_2]Br$ cis	5	0.713	0.037	0.948	0.049	1.33	‡	1.33	4

[a] See Table 1 for definitions. Cases in which $\alpha_\gamma \simeq \alpha_x$, i.e., dilute solutions (see text) are identified by the symbol ‡.
[b] The Japanese authors[7] express concentrations in g% without further explanation. We suppose, by reason of analogy, that this refers to g/100 g *solvent*; values cited in this table have been recalculated on this basis.

Table 4 Solubility ratio α in water for free amino acids that form conglomerates[a]

Name	T (°C)	γ_A	$x_A \times 100$	γ_R	$x_R \times 100$	α_x	α_γ	Refs.
Asparagine	25	2.69	0.376	5.61	0.804	2.14	2.09	6
	30	3.75	0.528	7.33	1.067	2.02	1.95	
Threonine	30	9.09	1.490	14.53	2.507	1.68	1.60	6
Homocysteic acid	30	15.26	2.345	20.00	3.225	1.38	1.31	6

[a] See Table 1 for definitions.

The solubilities of a variety of ionic compounds in aqueous solution are given in Table 3. In the examples of compounds of very low solubility for which $\alpha_x = \alpha_\gamma$, identified by means of the symbol ‡, we find values of α close to that predicted ($\alpha = 1.41$) for completely dissociated substances. The other cases have variable α values, though they are mostly smaller than 2. Note that in concentrated solution of electrolytes departures from ideality are large, and consequently the validity of theoretical solubility relationships becomes highly tenuous.

Finally, the behavior of free amino acids which form conglomerates (Table 4) seems to conform accurately to the statement of Greenstein and Winitz:[11] "The amino acids possess an essentially dual personality, for they carry electric charges and hence behave to some degree as ions, but they also carry a wide variety of chemical groupings and because of this may behave very much like other organic molecules which happen to be electrically uncharged."

REFERENCES AND NOTES 3.2

1 J. Jacques and J. Gabard, *Bull. Soc. Chim. Fr.*, 1972, 342.

2 Y. Shimura and K. Tsutsui, *Bull. Chem. Soc. Jpn.*, 1977, **50**, 145.

3 P. Vieles, *Ann. Chim. (Paris)*, 1935, 181.

4 K. Yamanari, J. Hidaka, and Y. Shimura, *Bull. Chem. Soc. Jpn.*, 1973, **46**, 3724.

5 N. Sato, T. Uzuki, K. Toi, and T. Akashi, *Agr. Biol. Chem.*, 1969, **33**, 1107.

6 T. Watanabe and G. Noyori, *Kogyo Kagaku Zasshi (J. Chem. Soc. Jpn., Ind. Chem. Sect.)*, 1969, **72**, 1083.

7 S. Yamada, M. Yamamoto, and I. Chibata, *J. Org. Chem.*, 1973, **38**, 4408.

8 C. Hongo, M. Shibazaki, S. Yamada, and I. Chibata, *J. Agr. Food Chem.*, 1976, **24**, 903.

9 W. Meyerhoffer, *Ber.*, 1904, **37**, 2604.

10 M. J. Brienne, unpublished results.

11 J. P. Greenstein and M. Winitz, *Chemistry of the Amino Acids*, Wiley, New York, 1961, p. 550.

12 Description of the lower part D_sL_sLD of Fig. 3. There are two cases to consider: (1) If the unsolvated binary mixture D,L forms a racemic compound, then within the triangular zone D_sL_sR there will be present a mixture of solvated solid D_s, L_s and of the racemic compound R. In zone D_sDR, the solid will consist of D_s, D and R; and, similarly, in zone L_sLR it will consist of L_s, L and R. At constant pressure, the three cases correspond to a variance of 1 ($v = C - \Phi + 1$). (2) In the case where unsolvated mixture D,L forms a con-

glomerate as does D_S and L_S, then in area D_SL_SDL of the diagram a mixture of crystals of D_S, L_S, D, and L will be present. Unqualified application of the phase rule to this system of four solid phases at constant pressure gives a variance of zero. This clearly is impossible, inasmuch as the four solids must be able to coexist in a *range* of temperatures. This paradox is explained away upon consideration of the energies of solvation and desolvation of crystals of D_S and L_S by the achiral solvent S. These energies are identical, and consequently here the pairs of enantiomeric solids D_S/L_S and D/L become thermodynamically indiscernible. This system is equivalent to one of *two* components (D/L and S) distributed between two *phases* (D_S/L_S and D/L) which yields a variance of 1. We emphasize that this reasoning is not a simple subterfuge but that it rests upon fundamental considerations based on the phase rule. In particular, we may not apply it to *solubility* equilibria which obtain in other regions of the phase diagram; while it is true that enantiomeric solid phases always have the same free energy, crystallization of enantiomer D from solution takes place exclusively in the crystal lattice D_S (or D) and not indiscriminately in a D or L lattice.

3.3 SOLUTIONS OF RACEMIC COMPOUNDS

3.3.1 Theoretical Phase Diagrams

The solid phases observed in saturated solutions of partially resolved racemic compounds are the crystals of the latter (pure or solvated) in addition to those of the enantiomer present in excess (also either pure or solvated).

When neither the racemic compound nor the enantiomers are solvated with S, the solubility diagram is represented by Fig. 1; A and A' represent the solubilities of the pure enantiomers (which are equal to SA). The solubility of the racemic compound R, represented by point r, is equal to $2Sd$.

Just as we have done for systems in which enantiomers form conglomerates, we can analyze the various regions of the phase diagram by means of the phase rule.* If a small amount of solvent is added to a partially resolved mixture (M_1 or M_2), the system at equilibrium will contain three phases: the solid racemic compound, the solid enantiomer present in excess, and the saturated solution. The variance of the system is equal to 1; hence, at a given temperature T_0 no degrees of freedom remain. The composition of the solution is thus fixed and corresponds to point E (or E'). The two areas which represent the three-phase systems are the triangles DER and L$E'R$. Detailed examination of area DER, for example, reveals two different situations. In the first (Fig. 1a), the enantiomeric purity of the initial mixture M_1 is greater than that of the eutectic (given by point e). We see on the phase diagram that system P_1, formed by addition of a relatively small amount of solvent, will deposit at equilibrium crystals of D + R of overall composition N_1, which corresponds to an increased enantiomeric purity. Conversely, in the case of system P_2 (Fig. 1b), whose enantiomeric purity is less than that of the eutectic, the solid deposited upon crystallization (N_2) will have an enantiomeric purity which is lower than that of the initial mixture M_2. This difference between systems whose enantiomeric purity is smaller or larger than that of the eutectic of the ternary phase

* Our reasoning still assumes that pressure is constant and that the vapor phase may be neglected, hence, $v = C - \Phi + 1$ (see Section 3.2.1).

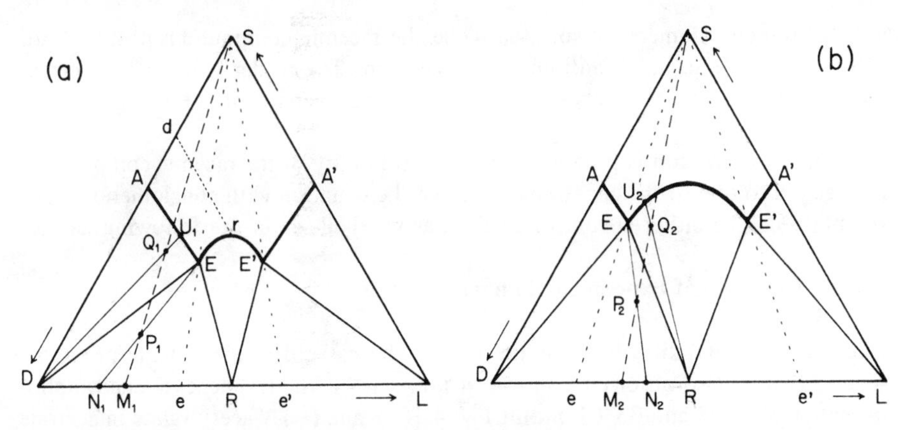

Figure 1 Solubility isotherm for an unsolvated racemic compound.

diagram is a very important feature of enantiomer mixtures forming racemic compounds.

Upon addition of increasing amounts of solvent to system P_1 (Fig. 1*a*), all of the racemic compound R goes into solution, and one passes into region AED in which crystals of the pure enantiomer D are in equilibrium with the saturated solution, whose composition varies along the solubility curve AE (for example, solution U_1 for system Q_1). On the other hand, for system P_2 (Fig. 1*b*), addition of solvent causes the diappearance of crystals of the excess enantiomer D. One reaches area ERE' of the phase diagram in which the pure racemic compound crystals R are in equilibrium with saturated solutions whose composition vary along line EE' (e.g., solution U_2 for system Q_2). Regions AED, $A'E$L, and ERE' of the diagram contain two phases; they are bivariant. At a fixed temperature there remains one degree of freedom, namely, that of the composition of the solution. And, lastly, above the solubility curves, a single phase remains, that of unsaturated solutions.

There are many examples known of systems in which the solvent S forms definite combinations (or solvates) with the racemic compound and/or with the enantiomers. The several conceivable variants of solvation are illustrated in the diagrams of Fig. 2: (a) the racemic compound is solvated while the enantiomers are

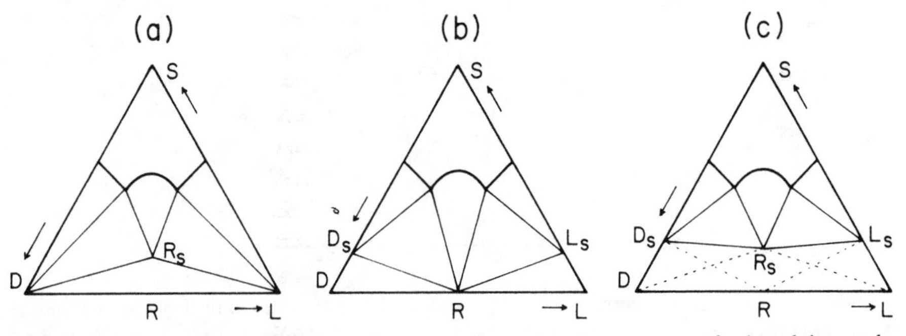

Figure 2 Isothermal solubility diagrams for racemic compounds involving solvation (see text).

not; (b) the enantiomers are solvated while the racemic compound is not; (c) both racemic compound and enantiomers are solvated. The reader will easily ascertain the number and nature of phases present in the various parts of these diagrams (see note 8).

Whether solvation is present or not, the solubility of the racemic compound is not related to that of the enantiomers, unlike the situation with conglomerates. The solubility can be either greater or smaller, as we shall see in the following section.

3.3.2 Examples of Experimental Phase Diagrams

Experimental solubility diagrams illustrating the cases described in the preceding section (without solvation) are shown in Fig. 3 [(+)- and (−)-benzylidenecamphor in methanol at 25 and 35°C] and in Fig. 4 [(+)- and (−)-*N*-acetylvaline in acetone at 25 and 35°C].[1]

In the first case (Fig. 3), the solubility of the racemic compound is substantially greater than that of the enantiomers ($\alpha_x = 2.16$ at 25°C). Note that the eutectic composition found in this diagram can be shown to be equal to that found in the *binary* diagram of the enantiomers (22% of optically active substance and 78% of racemic compound).

In the second example (Fig. 4), the racemic compound is much less soluble than the enantiomers ($\alpha_x = 0.39$ at 25°C). The ratios of solubilities of racemic compounds and corresponding enantiomers can vary within rather wide limits as is evident from the data of Table 1. Nevertheless, one can deduce from the ternary phase diagram that the solubility of a racemic compound in *stable* equilibrium with saturated solution cannot exceed that of the *conglomerate* of the enantiomers. In an ideal solution, the maximum value of α_x is thus equal to 2 for undissociated solutes and $\sqrt{2}$ for dissociated solutes (electrolytes) (see Section 3.2.3).

We are unaware of the existence of experimental ternary phase diagrams for

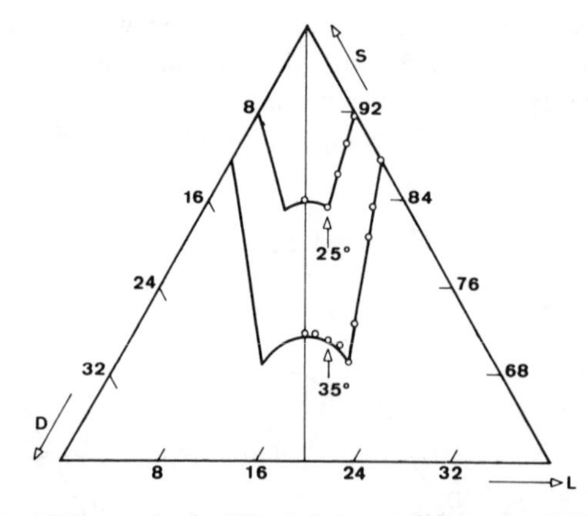

Figure 3 Solubility diagram of (+)- and (−)-benzylidenecamphor in methanol (concentrations in g/100g mixture). J. Jacques and J. Gabard, *Bull. Soc. Chim. Fr.*, 1972, 342. Reproduced by permission of the Société Chimique de France.

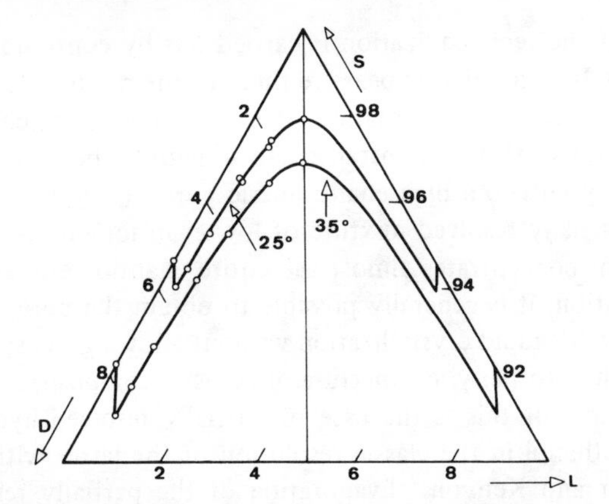

Figure 4 Solubility diagram of (+)- and (−)-*N*-acetylvaline in acetone (concentrations in g/100 g mixture). J. Jacques and J. Gabard, *Bull. Soc. Chim. Fr.*, 1972, 342. Reproduced by permission of the Société Chimique de France.

solvated racemic compounds. Nonetheless, we know that such solvated systems have been identified. The salts of tartaric acid studied by Van't Hoff[3-6] are examples of cases that generally tend to be complicated by the incidence of polymorphism. The existence of solvates is also generally limited to certain temperature ranges. This problem, as well as that of polymorphism of unsolvated racemic compounds, is examined in Section 3.5.

3.3.3 Solubility Rules for Partially Resolved Mixtures

Just as was true for conglomerates, solubility rules for partially resolved racemic compounds may be deduced from the phase diagrams discussed in Section 3.3.1.

Returning to Fig. 1*b*, we see that isothermal recrystallization of a mixture M_2, of enantiomeric purity smaller than that of the eutectic (*e*), gives crystals of *reduced* enantiomeric purity with a relatively small quantity of solvent ($P_2 \rightarrow N_2$), while crystals of pure *racemic compound* are obtained if the quantity of solvent is

Table 1 Relative solubilities of enantiomers and racemic compounds[a]

Compound	Solvent	T (°C)	Solubilities: Enantiomer $x_A \times 100$	Solubilities: Racemate $x_R \times 100$	$\alpha_x = \frac{x_R}{x_A}$	Refs.
Benzylidenecamphor	Methanol	25	1.137	2.459	2.16	1
N-Ethyl tartramide	Methanol	25	1.375	1.685	1.22	1
N-Acetylvaline	Acetone	25	1.358	0.531	0.39	1
α-(2-Naphthyl)propionic acid	Benzene	25	27.28	0.582	0.02	1
$[Co(ox)(en)_2]I$	Water	5	0.011	0.00394	0.34	2

[a] x_A and x_R are the respective mole fractions of the enantiomer and the racemic compound in their corresponding saturated solutions.

larger ($Q_2 \rightarrow R$). If the recrystallization is carried out by continuous cooling of the solution, the crystals which first appear are those of the racemic compound.

In this *a priori* unfavorable situation, the enantiomeric enrichment takes place in the mother liquors. This circumstance can sometimes be taken advantage of if the eutectic corresponds to a high enantiomeric purity (as in Fig. 4). In such cases, treatment of a partially resolved mixture of low enantiomeric purity with a small quantity of solvent concentrates almost the entire enantiomeric excess in the solution. In this situation, it is generally possible to obtain the pure enantiomer from this solution through a rapid crystallization without allowing the system to reach its solubility equilibrium (otherwise a mixture of eutectic composition would result).

A good example of this is the case of partially resolved hydrogen phthalate ester of 1-phenylethanol in the classic resolution of the latter with brucine as described by Downer and Kenyon.[7] Evaporation of the partially resolved ester, mp 82°C, in cold carbon disulfide deposits hard crystals of the racemate, mp 108°C. The pure levorotatory isomer (mp 86°C) is obtained when petroleum ether is added to the warm CS_2 filtrate.

When the partially resolved mixture M_1 (see Fig. 1*a*) is enantiomerically purer than the eutectic, isothermal recrystallization increases the enantiomeric purity when a small quantity of solvent is used ($P_1 \rightarrow N_1$) and, as in the case of conglomerate-forming systems, leads to the pure enantiomer D with a larger quantity of solvent ($Q_1 \rightarrow$ D). When crystallization of the mixture M_1 is carried out by cooling, the crystals first to appear are those of the pure enantiomer.

REFERENCES AND NOTES 3.3

1 J. Jacques and J. Gabard, *Bull. Soc. Chim. Fr.*, 1972, 342.

2 K. Yamanari, J. Hidaka, and Y. Shimura, *Bull. Chem. Soc. Jpn.*, 1973, **46**, 3724.

3 J. H. Van't Hoff, H. Goldschmidt, and W. P. Tomssen, *Z. Phys. Chem.*, 1895, **17**, 49.

4 J. H. Van't Hoff and W. Muller, *Ber.*, 1899, **32**, 857.

5 J. H. Van't Hoff and H. Goldschmidt, *Z. Phys. Chem.*, 1895, **17**, 505.

6 J. H. Van't Hoff and W. Muller, *Ber.*, 1898, **31**, 2206.

7 E. Downer and J. Kenyon, *J. Chem. Soc.*, 1939, 1156.

8 The only point which merits attention in these diagrams of Fig. 2 which differ little from those of Fig. 1 and hence are easily interpretable as to number and nature of phases present in various regions is that of isotherm (*c*) and specifically the composition of the solids present in area $D_sR_sL_sLD$. Two cases obtain: (1) the solvate of racemic compound R_s is *more stable* than that of its enantiomers D_s or L_s in which case both solvated racemic compound and one solvated enantiomer will be found in areas D_sDR and L_sLR along with the corresponding unsolvated enantiomer. Solvated racemic compound and unsolvated enantiomer would be present in area R_sLD; and (2) the solvate of racemic compound R_s is *less stable* than that of the enantiomers. Unsolvated racemic compound exists in the presence of solvates D_s and L_s in area RD_sR_sL while the corresponding unsolvated enantiomer *and* D_s and L_s as well as the unsolvated racemic compound R are present in areas L_sLR and R_sRD. Generally, the part of the phase diagram in which no liquid phase is present is only of academic interest.

3.4 SOLUTIONS OF PSEUDORACEMATES

3.4.1 Theoretical and Experimental Phase Diagrams

When enantiomers cocrystallize throughout the entire range of concentrations, the crystals formed in a saturated solution at equilibrium constitute a single solid phase (mixed crystals). The variance of this diphasic system (at constant pressure) is thus equal to 2. At a given temperature, one degree of freedom remains. It follows that the two variables, the composition of the solution and the composition of the crystals, are interdependent.

The theoretical isothermal phase diagrams corresponding to the three types of solid solutions described in Section 2.4 are illustrated in Fig. 1. The solubilities of the pure enantiomers (equal to SA) are represented by points A and A', while r describes the solubility of pseudoracemate R (equal to $2Sd$). In diagram (*a*), the solubility of the pseudoracemate is equal to that of the enantiomers ($2Sd = SA$) and the solubility curve is the horizontal line AA'. In diagrams (*b*) and (*c*), the pseudoracemate is more and less soluble, respectively, than the enantiomers.

Contrary to the cases described in Sections 3.2 and 3.3, the part of the phase diagram lying below the solubility curves constitutes a single region in which the tie lines do not converge to any apex. Nevertheless, in diagram (*a*) these lines form a sheaf which converges to the outside of the region toward point *S*. Where the pseudoracemate is more soluble than the enantiomers (Fig. 1*b*), tie lines such as *UN* are less inclined (have smaller slopes) than the lines passing through *S*, such as *SPM*. Conversely, when the pseudoracemate is less soluble than the enantiomer (Fig. 1*c*), tie lines *UN* are more inclined (have greater slopes) than lines *SPM*.

Construction of experimental ternary phase diagrams for nonideal solid solutions presents the same practical difficulties as those described in Section 2.4 for binary phase diagrams. Since the composition of the solid phase and that of the liquid phase are interdependent, establishment of a true solubility equilibrium would require that there be a continuous change in the composition of the solid phase as a whole concomitantly with that of the solution during the process of crystallization (or dissolution). As a rule, this is clearly impossible, and the con-

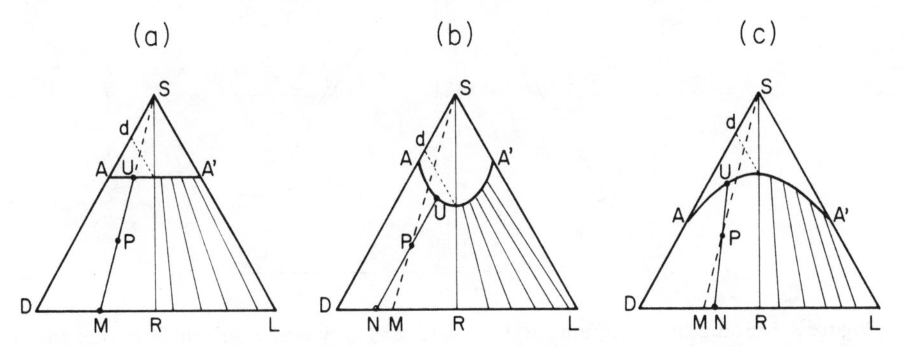

Figure 1 Isothermal solubility diagrams of the three types of pseudoracemates. The appearance of the tie lines is shown in the right half of the triangles.

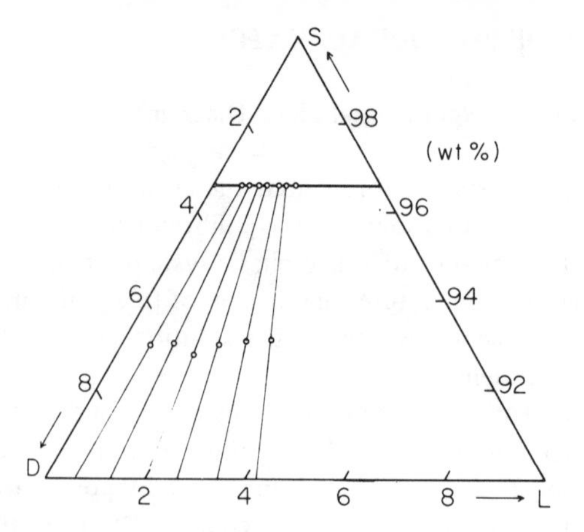

Figure 2 Solubility diagram of mixtures of (+)- and (±)-camphor at 25°C. Solvent: absolute ethanol 40 ml diluted with water to 100 ml.[1]

sequence is that the crystals of the solid solution will not be homogeneous, and experimental tie lines will only have a qualitative significance.

In order to obtain experimental ternary diagrams close to the expected theoretical ones, one needs to employ mixtures from which the quantity of crystals present at equilibrium is as small as possible with respect to the quantity of liquid. In such a way, the change in concentration of the solution during equilibration is negligible and the crystals are thus likely to be homogeneous.

Examples of such experimental solubility diagrams are rare indeed. Moreover, in addition to the difficulties just envisioned, the occurrence of polymorphism, which is particularly common in this type of racemate, sometimes complicates the behavior of solid solutions by leading to unexpected results, as we shall see.

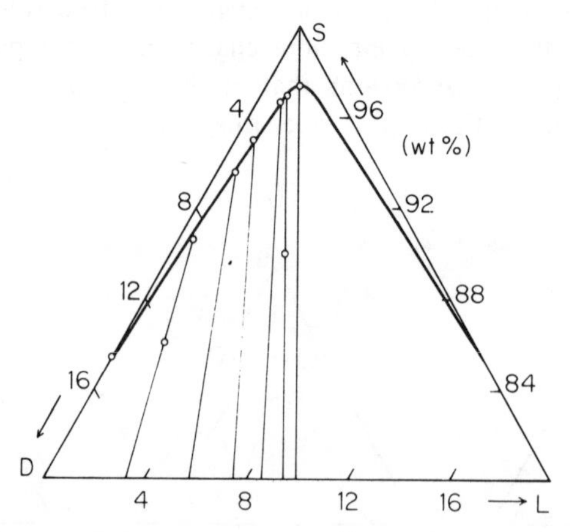

Figure 3 Solubility diagram of (+)- and (−)-carvoxime mixtures in hexane at 25°C. J. Jacques and J. Gabard, *Bull. Soc. Chim. Fr.*, 1972, 342. Reproduced by permission of the Société Chimique de France.

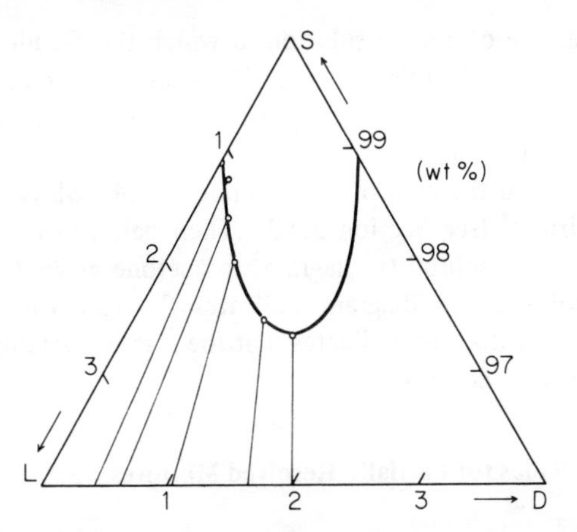

Figure 4 Solubility diagram of enantiomers of 2,2,5,5-tetramethyl-1-pyrrolidinoxy-3-carboxylic acid in chloroform at 25°C. B. Chion, J. Lajzerowicz, D. Bordeaux, A. Collet, and J. Jacques, *J. Phys. Chem.*, 1978, **82**, 2682. Reproduced by permission of the publisher. Copyright 1978, American Chemical Society.

The solubility isotherm at 25°C of mixtures of natural (+) and racemic camphor in aqueous ethanol[1] (Fig. 2) illustrates the case of ideal solid solutions. The tie lines were drawn by the method of Hill and Ricci (Section 3.1.6).

The solubility diagram of the carvoxime system in hexane at 25°C (Fig. 3) illustrates the behavior of solid solutions with maximum melting point for the pseudoracemate.[2] The tie lines were drawn by the method of wet residues. The very large difference between the solubility of the pseudoracemate (2.74 g/100 g solution at 25°C) and that of the enantiomers (7.4 g/100 g) is noteworthy.

The ternary solubility isotherm of mixtures of enantiomers of 2,2,5,5-tetramethyl-1-pyrrolidinoxy-3-carboxylic acid in chloroform at 25°C is shown in

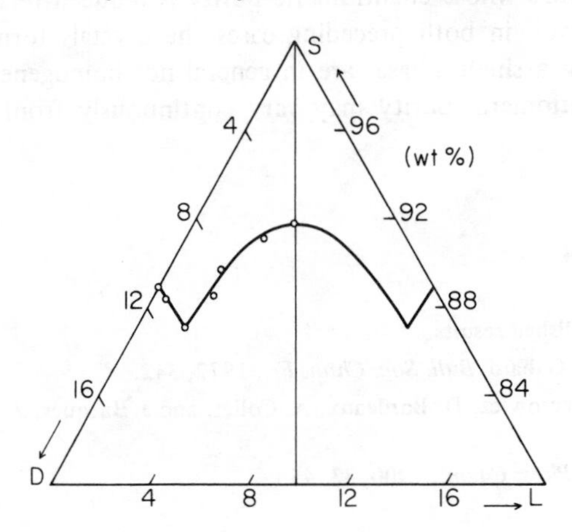

Figure 5 Solibility diagram of (+)- and (−)-camphoroxime mixtures in acetone at 25°C. J. Jacques and J. Gabard, *Bull. Soc. Chim. Fr.*, 1972, 342. Reproduced by permission of the Société Chimique de France.

Fig. 4. It is an example of a solid solution in which the pseudoracemate exhibits a minimum melting point.[3] In this case, the binary phase diagram could not be determined precisely because of the decomposition of the compound during melting ($T_A^f \sim 204°C$; $T_R^f \sim 200°C$).

Lastly, we come to the case of the camphoroximes whose polymorphism was established by Adriani[4] (see Section 2.4.3). Their behavior is characterized unambiguously also by their solubility diagram[2] in acetone at 25°C (Fig. 5). While the binary melting point phase diagram indicates the formation of an ideal solid solution, the solubility diagram indicates that the stable crystalline form at ambient temperature is a racemic compound.

3.4.2 Solubility Rules for Partially Resolved Mixtures

We limit ourselves to a qualitative description of the behavior of partially resolved solid solutions of enantiomers upon recrystallization.

When the solubility curve is a horizontal line (Fig. 1*a*), isothermal recrystallization, or crystallization by slow cooling or evaporation, leads to no alteration of the enantiomeric purity of a solid mixture initially of composition *M*. In effect, system *P* obtained upon addition of solvent to mixture *M* is found on line *MS* and on tie line *MU* at the same time, that is to say, tie lines are superposed on enantiomer mixture–solvent composition lines. As a consequence, at equilibrium, homogeneous solid *M* will be present in mother liquors *U* of the same enantiomeric purity.

When the solubility curve possesses a minimum for the racemate (Fig. 1*b*), corresponding to maximum solubility, recrystallization leads to an increase in enantiomeric purity relative to the composition of the initial solid *M*. Conversely, the case where the solubility curve possesses a maximum (Fig. 1*c*) yields upon recrystallization crystals whose enantiomeric purity is reduced relative to that of the initial solid. However, in both preceding cases the crystals formed, while constituting theoretically a single phase, are in general not homogeneous; that is, for a given crystal enantiomeric purity may vary continuously from its center to the surface.

REFERENCES 3.4

1 A. Collet, unpublished results.

2 J. Jacques and J. Gabard, *Bull. Soc. Chim. Fr.*, 1972, 342.

3 B. Chion, J. Lajzerowicz, D. Bordeaux, A. Collet, and J. Jacques, *J. Phys. Chem.*, 1978, **82**, 2682.

4 J. H. Adriani, *Z. Phys. Chem.*, 1900, **33**, 453.

3.5 POLYMORPHISM IN TERNARY SYSTEMS

In Section 2.5 we have seen that some binary systems of enantiomers exhibit polymorphism that mediates the transformation of various racemate forms into one another. The experimental study of this polymorphism through thermal analysis of binary mixtures, for example, turns out to be anything but simple. The results are difficult to interpret, particularly because of the slowness with which equilibria involving solids are established. The difficulty of preparing in a reproducible way a specific crystal form of a substance exhibiting polymorphism also contributes to these difficulties.

On the other hand, such difficulties largely disappear when studying ternary mixtures of enantiomers and solvent rather than binary ones. This is because crystallization from solution more rapidly yields thermodynamically stable crystalline species which also rapidly equilibrate. Such transformations between crystals as a function of temperature may be evidenced and characterized by the construction of several isotherms. Complicating such studies is the occurrence of a new type of polymorphism which is peculiar to ternary systems. This polymorphism arises when solvent intervenes to solvate racemate or enantiomers crystals. Inasmuch as solvate stability is temperature dependent, examination of several isotherms here also allows one to establish the behavior of these systems.

If we limit ourselves simply to the determination of solubility curves without concern for possible solvation of crystalline species, then the two types of polymorphism just described may be treated in the same way as described in Section 3.5.1.

3.5.1 Description of the Polymorphism of Ternary Systems

In this section we specifically consider cases of polymorphism in which conglomerates interconvert with racemic compounds. These are among the most frequently observed instances of polymorphism. From a practical point of view they are also among the most important. We shall focus on *isothermal solubility curves*, which define the nature of the racemate quite independently of crystal solvation, and on *solubility curves* of the racemate *as a function of temperature*, which in turn reveal transition temperatures between polymorphic forms.

When the solubility of a polymorphic racemate is measured as a function of temperature and the results plotted, a curve such as that shown in Fig. 1 is obtained. The transition, which takes place at T_0, is revealed by a break in the solubility curve. Below T_0, crystals of form C_1 constitute the solid phase in equilibrium with the saturated solution. Above T_0, the crystals which are stable are those of form C_2. This type of diagram does not reveal the nature of the C_1 or C_2 crystals. In the case illustrated, the possibilities are as follows: (a) it is the conglomerate (solvated or otherwise) which is stable at low temperature and which is converted to a racemic compound above T_0, or the inverse may be the case; (b) transition at T_0 is nothing more than a change in the crystal form, or in the degree of solvation of the crystals, without the occurrence of a change in the nature of the racemate. The second of these cases is elaborated in Section 3.5.2.

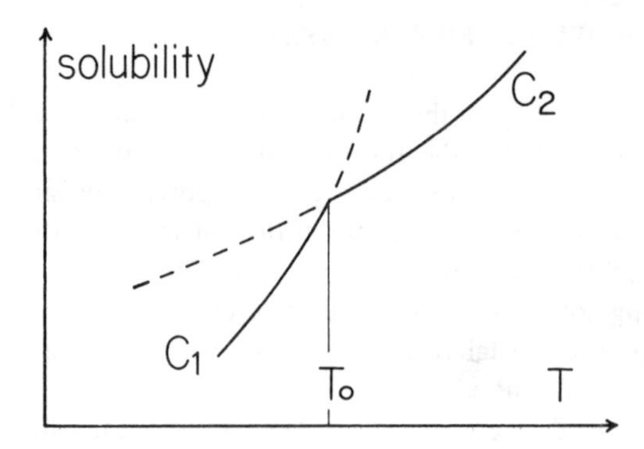

Figure 1 Solubility curve as a function of temperature. Evidence for polymorphism.

To distinguish between the two alternatives of possibility (a), it is necessary to examine ternary isotherms such as those illustrated in Fig. 2. These must be established at temperatures lower than, eventually equal to, and higher than the transition temperature T_0. The isotherms (a), at left, illustrate the case in which the

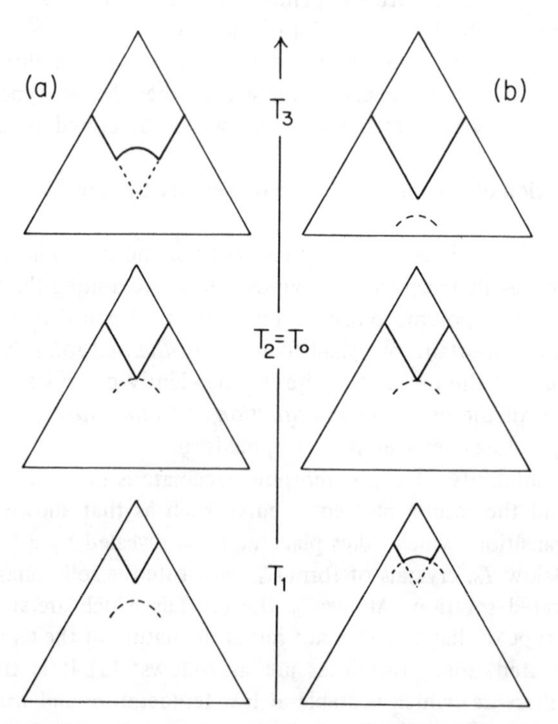

Figure 2 Conglomerate–racemic compound transformations as a function of temperature. Characterization of the system through determination of ternary isotherms: (a) conglomerate stable at low temperature; (b) racemic compound stable at low temperature.

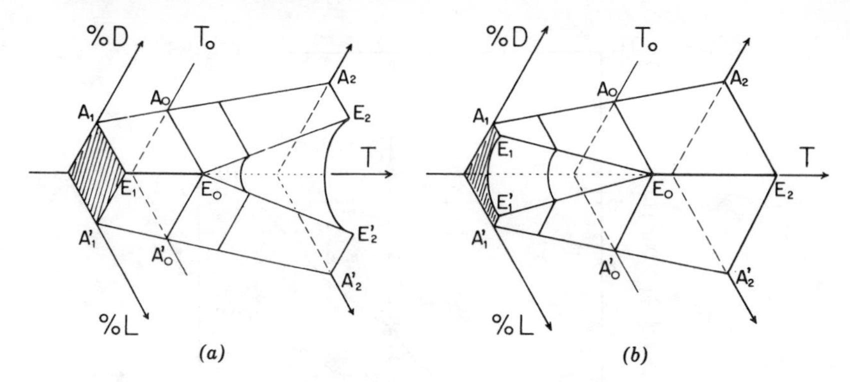

Figure 3.

conglomerate is stable at low temperature, while case (b), at right, is that in which the racemic compound is the stable low-temperature form. Systems in stable thermodynamic equilibrium give rise to the solubility curves drawn as solid lines. The partial curves drawn with dotted lines represent metastable solubilities* which are rarely observable under experimental conditions normally employed in the construction of solubility diagrams.

Each of the two series of isotherms of Fig. 2 may be summarized in a single diagram which corresponds to the perspective in a plane of a three-dimensional model whose three axes represent, respectively, temperature and the quantities of each enantiomer in solution. This type of diagram may be conveniently drawn on the same kind of triangular graph paper used in the construction of triangular phase diagrams. The two axes representing the enantiomer concentrations are related to one another through an angle of 120°.

Diagram (*a*) of Fig. 3 represents the case in which the conglomerate is stable at low temperature; $A_1E_1A'_1$ is the solubility curve for the conglomerate (stable) at T_1 and $A_2E_2E'_2A'_2$ is that of the racemic compound (stable) at T_2. At T_0, where the racemate and the conglomerate are at equilibrium, there is the limiting solubility curve $A_0E_0A'_0$. In the same manner, diagram (*b*) corresponds to the case in which the racemic compound is stable at low temperature.

3.5.2 Polymorphism and Solvation of Crystals

Numerous examples of polymorphism accompanying changes in the degree of solvation of crystals are known. In some cases, changes in the degree of solvation do not cause any change in the nature of the racemate. Thus, the solvated conglomerate $[Co(ox)(en)_2]Cl \cdot H_2O$ is transformed into an anhydrous conglomerate above 36°C.[1] The solubility curves of the racemate and of one of the enantiomers as a function of temperature are shown in Fig. 4. Inasmuch as the racemate is a conglomerate, the latter and the pure enantiomer show transitions at the same temperature.

* The solubility of metastable crystalline forms is always greater than that of the corresponding stable forms.

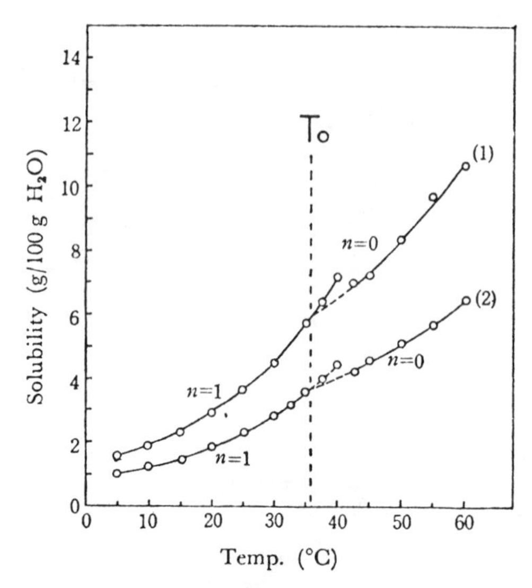

Figure 4 Polymorphism of $[Co(ox)(en)_2]Cl$ in presence of water. Adapted from K. Yamanari, J. Hidaka, and Y. Shimura, *Bull. Chem. Soc. Jpn.*, 1973, **46**, 3724 by permission of the Chemical Society of Japan.

Most often, however, passage of a solvated form to a completely desolvated one changes the nature of the racemate. For example, rubidium tartrate exists as a racemic compound $Rb_2C_4H_4O_6 \cdot 2H_2O$ below 40.4°C and is transformed into a conglomerate of anhydrous enantiomers above this temperature.[2] As other examples, we may cite the case of ammonium malate,[3,4] a racemic compound which is monohydrated below 74.6°C and an anhydrous conglomerate above this temperature, and that of dilactyldiamide, a racemic compound monohydrate which cleaves to an anhydrous conglomerate above 35.5°C.[5,6]

Cases in which the two crystalline racemic and enantiomeric species are differently solvated are quite numerous. However, few of these cases have been thoroughly studied. An example of such a study is that of histidine hydrochloride.[7,8] Racemic histidine monohydrochloride crystallizes in water at 20°C as a dihydrate while the pure enantiomers are monohydrates. According to Duschinsky,[7] the racemic compound which is stable at low temperature changes to a conglomerate between 40 and 55°C. The ternary isotherms of the histidine hydrochlorides were determined in water at 25, 35, and 45°C by Jacques and Gabard[8] and are shown in Fig. 5. While the 25° and 35°C isotherms confirm the existence of a racemic compound, the data measured at 45°C reveal the formation of a conglomerate. It is clear that the transition temperature above which the racemic compound decomposes is quite close to 45°C. The existence of the conglomerate is consistent with resolution by entrainment to which this compound has been subjected (see Section 4.2.).

In a similar way, potassium tartrate, which forms a racemic compound dihydrate, converts to a conglomerate hemihydrate ($0.5H_2O$) above 71.8°C.[9] Zinc di(2-pyrro-

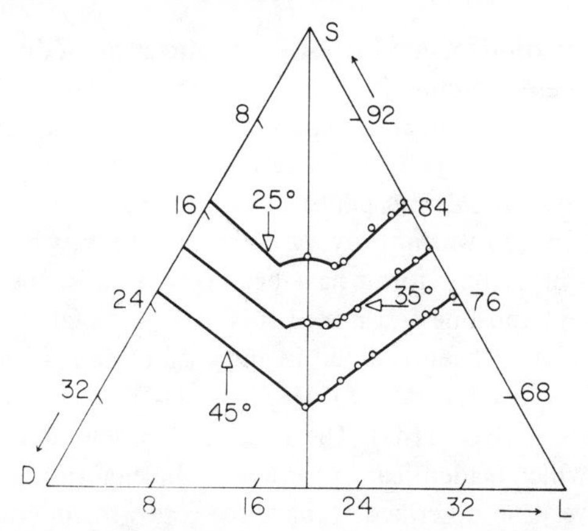

Figure 5 Solubility isotherms of (+)- and (−)-histidine hydrochlorides in water. J. Jacques and J. Gabard, *Bull. Soc. Chim. Fr.*, 1972, 342. Reproduced by permission of the Société Chimique de France.

lidine-5-carboxylate), an important intermediate in the synthesis of proline, is a racemic compound octahydrate which is transformed into a conglomerate dihydrate upon heating.[10] The transition temperature is unknown.

The double salts of tartaric acid, in which the dianion is associated with two different metallic cations, were the subject of numerous studies at the turn of the century.[12] These still have some historic value in part due to their relation to sodium ammonium tartrate, the subject of the first resolution by Louis Pasteur.[11] This double salt exists as a conglomerate tetrahydrate at low temperature; it is converted above 28°C to a racemic double salt monohydrate which is known under the name *Scacchi's salt*. If heating is continued, above 35°C one observes a disproportionation of the racemic double salt to anhydrous ammonium tartrate and anhydrous sodium tartrate, both racemic compounds. The following summarizes these transformations:

$$(NaNH_4C_4H_4O_6 \cdot 4H_2O) \rightleftharpoons 28°C \rightleftharpoons (NaNH_4C_4H_4O_6 \cdot H_2O) + 3H_2O$$

Conglomerate "Pasteur's salt" — Racemic compound "Scacchi's salt"

$$2(NaNH_4C_4H_4O_6 \cdot H_2O) \rightleftharpoons 35°C \rightleftharpoons Na_2C_4H_4O_6 + (NH_4)_2C_4H_4O_6 + 2H_2O$$

Racemic compounds

A case analogous to the one just mentioned is that of sodium potassium tartrate,[13] which also has two transition temperatures. After a first transformation from conglomerate to racemic compound at −6°C, the latter changes at 41°C to a mixture of anhydrous Na tartrate and K tartrate dihydrate both of which are racemic compounds:

$$(NaKC_4H_4O_6 \cdot 4H_2O) \rightleftharpoons -6°C \rightleftharpoons (NaKC_4H_4O_6 \cdot 3H_2O) + H_2O$$

Conglomerate "Seignette's salt" — Racemic compound "Wyrouboff's salt"

$$(NaKC_4H_4O_6 \cdot 3H_2O) \rightleftharpoons 41°C \rightleftharpoons Na_2C_4H_4O_6 + (K_2C_4H_4O_6 \cdot 2H_2O) + H_2O$$

Racemic compounds

The disproportionation of double salts introduces an additional complication in the behavior of these systems. It certainly must have contributed to the belief that transitions between conglomerates and racemic compounds are more complicated than they actually are. In fact, one rarely encounters situations that are nearly intractable, such as that of calcium pantothenate, whose pharmacologic properties are of sufficient interest to warrant a very complete study, which was carried out by Inagaki.[15,16] Four crystalline forms have been observed for the enantiomer; three are unsolvated (α, β, γ) and one is a mixed solvate – $4CH_3OH + H_2O$. No less than *nine* crystalline forms have been isolated in the case of the racemate: five unsolvated (A, B, C, D, and E), two hydrates ($3/4H_2O$ and H_2O), one solvate (CH_3OH), and one mixed solvate ($4CH_3OH + H_2O$). The latter is a conglomerate, as is true of the unsolvated form C (which is identical to form α of the enantiomer).

The examples we have described all have the following in common: transitions observed while *raising* the temperatures always are accompanied by a *reduction* in the degree of solvation of the crystals. This has been generalized by Van't Hoff into the empirical rule according to which the *least solvated compound is the most stable at high temperature*. This rule is justified by the principle of Le Chatelier. In

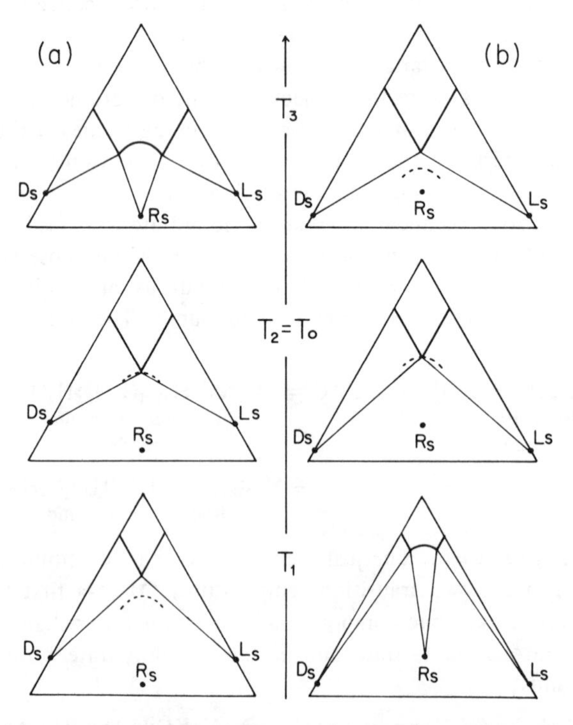

Figure 6 Conglomerate–racemic compound transformations associated with solvation. D_s and L_s describe the enantiomer solvates and R_s that of the racemic compound: (a) The conglomerate which is more highly solvated than the racemic compound is stable at low temperature. (b) The racemic compound which is more highly solvated disproportionates into a conglomerate on heating.

effect, as Findlay indicates,[14] the enthalpy of solvation/desolvation of salts is generally much more important than that which is associated with other crystalline modifications. It is thus this enthalpy which determines the *sign* of the overall heat effect in this type of polymorphism.

If solvation is exothermic, as is generally the case, at least for hydration, desolvation (an endothermic process) is favored as the temperature is raised, from which Van't Hoff's rule follows. In other words, when the enantiomers are more highly solvated than the racemic compound, the conglomerate is stable at low temperature. This is true, for example, for the double salt sodium ammonium tartrate. Conversely, a racemic compound which is more highly solvated than the enantiomers will form a conglomerate *above* a transition temperature as in the case of ammonium malate. The isotherms of Fig. 6 summarize these two situations in a simple way.

REFERENCES 3.5

1 K. Yamanari, J. Hidaka, and Y. Shimura, *Bull. Chem. Soc. Jpn.*, 1973, **46**, 3724.

2 J. H. Van't Hoff and W. Muller, *Ber.*, 1898, **31**, 2206.

3 F. B. Kenrick, *Ber.*, 1897, **30**, 1749.

4 J. H. Van't Hoff and H. M. Dawson, *Ber.*, 1898, **31**, 528.

5 P. Vieles, *Ann. Chim. (Paris)*, 1935, **3**, 147.

6 P. Vieles, *C. R. Acad. Sci.*, 1934, **198**, 2102.

7 R. Duschinsky, *Chem. Ind. (London)*, 1934, 10. *Festschrift Emil Barrell. 1936*, F. Reinhardt Verlag, Basel, p. 375.

8 J. Jacques and J. Gabard, *Bull. Soc. Chim. Fr.*, 1972, 342.

9 J. H. Van't Hoff and W. Muller, *Ber.*, 1899, **32**, 857.

10 Japanese Patent, (1968), 68 27,859 (to Ajinomoto Co.); *Chem. Abstr.*, 1969, **70**, 57639.

11 J. H. Van't Hoff, H. Goldschmidt, and W. P. Tomssen, *Z. Phys. Chem.*, 1895, **17**, 49.

12 J. H. Van't Hoff, *Vorlesungen über Bildung und Spaltung von Doppelsalzen*, W. Engelmann, Leipzig, 1897.

13 J. H. Van't Hoff and H. Goldschmidt, *Z. Phys. Chem.*, 1895, **17**, 505.

14 A. Findlay, *The Phase Rule*, 9th ed., revised by A. N. Campbell and N. O. Smith, Dover, New York, 1951, p. 353.

15 M. Inagaki, H. Tukamoto, and O. Akazawa, *Chem. Pharm. Bull.*, 1976, **24**, 3097.

16 M. Inagaki, *Chem. Pharm. Bull.*, 1977, **25**, 1001.

3.6 ENANTIOMERIC PURITY DETERMINATION FROM SOLUBILITY MEASUREMENTS

Analysis of the curve describing the solubility of a mixture as a function of the quantity of solvent present constitutes an interesting method for the determination of enantiomeric purity, just as is true for the fusion curve. This is a method of general utility[1–4] which has the specific advantage of being applicable to compounds that are thermally unstable, infusible, or even exhibit polymorphism sufficiently complex as to render impossible the analysis of melting point curves. Just as is true for the thermal method, this procedure is not usable with mixtures of substances that cocrystallize. For purposes of discussion here, we show how it may be applied to conglomerates and to racemic compounds.

The use of this analytical procedure is related to the method of determining solubility diagrams. No special equipment is used, and the amount of substance required depends only on the technique employed in the determination of concentrations. Less than ca. 100 mg of substance should suffice for the analysis if the refractometric method described above (Section 3.1.6) is used. Moreover, it is generally possible to recover almost all of the sample employed in the analysis through evaporation of the solvent.

3.6.1 Conglomerates

Let us suppose that we have a partially resolved conglomerate M of unknown enantiomeric purity p (D > L). Let us add increasing amounts of this mixture to known quantities of solvent and determine the compositions of the solutions after attainment of equilibrium at constant temperature. The diagrams in Fig. 1 represent this process in two different ways: (a) in the usual triangular isotherm where the overall composition of successive systems is given by line $S \rightarrow M$ and (b) in the rectangular diagram in which C, the quantity of substance in solution, is plotted as a function of Q, the total quantity of mixture M (see Section 3.1.5). The unit in which compositions are expressed is, in principle, immaterial. However, for reasons of convenience, they generally are expressed as g substance/100 g solvent, which simplifies the calculations.

Let us first examine the case of the pure D enantiomer. When crystals of D are progressively added to solvent, these begin to dissolve to give unsaturated solutions located on segments SA of both diagrams of Fig. 1. Beginning with point A, the solution becomes saturated and its concentration remains constant whatever the quantity of D added subsequently. One obtains the horizontal step AA' (Fig. 1b),

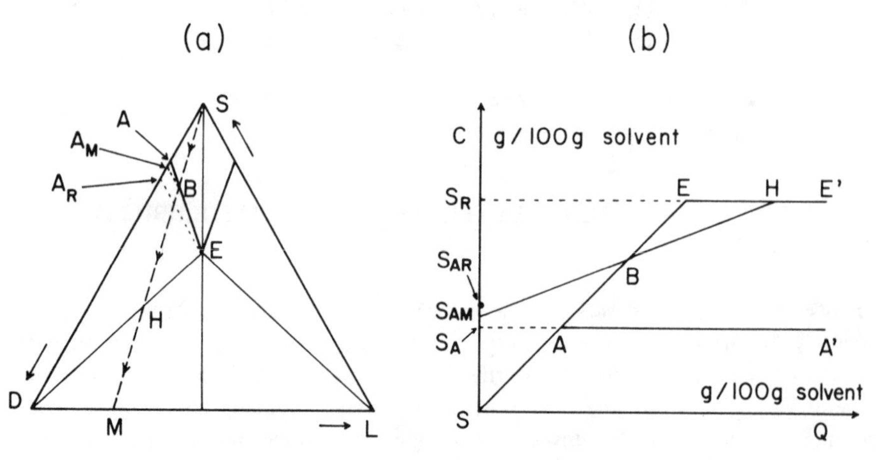

Figure 1.

corresponding to solubility S_A of the pure enantiomer. Similarly, for the racemate, the line SE gives the solubility of unsaturated solutions; this is followed by the horizontal step EE' which indicates the solubility S_R of the conglomerate. For a partially resolved mixture M, the successive solutions beginning with S and ending in M are described by line SM (Fig. 1*a*). Up to point B, the solution remains unsaturated (segments SB in Figs. 1*a* and 1*b*). Beyond B, the excess enantiomer (D) exceeds saturation and one observes the line BH of reduced slope corresponding to the dissolution of L alone (Fig. 1*b*). Extrapolation of BH to $Q = 0$ yields solubility S_{AM} of enantiomer D in mixture M. At H, enantiomer L also reaches saturation, and beyond this one penetrates the three-phase region (D*E*L in Fig. 1*a*) in which the composition of the solution remains constant (and racemic), corresponding to horizontal step EE' of Fig. 1*b*.

The slope of line BH permits the determination of the enantiomeric purity. Effectively, the total quantity of the more abundant enantiomer (D) in the system, q_{D}, is equal to the fraction present in solution (S_{AM}) plus the part of the solid consisting of pure D and equal to $C - Q$:

$$q_{\mathrm{D}} = S_{AM} + C - Q \tag{1}$$

If p is the enantiomeric purity of mixture M, then it also follows that

$$q_{\mathrm{D}} = \frac{1+p}{2} Q \tag{2}$$

A simple algebraic argument allows one to show that[5]

$$S_{AM} = \frac{\alpha}{2} \frac{1+p}{1+(\alpha-1)p} S_A \tag{3}$$

in which $\alpha = S_R/S_A$ (recall that for low solubilities, α is independent of the way in which solubility is expressed; see Section 3.2.3). Combining the preceding expressions gives the equation for the straight line BH:

$$C = \frac{\alpha}{2} \frac{1+p}{1+(\alpha-1)p} S_A + \frac{1-p}{2} Q \tag{4}$$

whose slope s is dependent only on the enantiomeric purity of the mixture M:

$$s = \frac{1-p}{2} \quad \text{and} \quad p = 1 - 2s \tag{5}$$

Three cases must be considered according to the value of α; these cases are schematically described in Fig. 2. When $\alpha = 2$, eq. (4) becomes

$$C = S_A + \frac{1-p}{2} Q \tag{6}$$

which shows that, whatever p may be, all BH lines converge to $S_{AM} = S_A$ (Fig. 2*a*). Equation (6) is none other than that of Mader,[1] which implies strict additivity of the solubilities of the substance under study and of its impurity (in this context, the less abundant enantiomer).

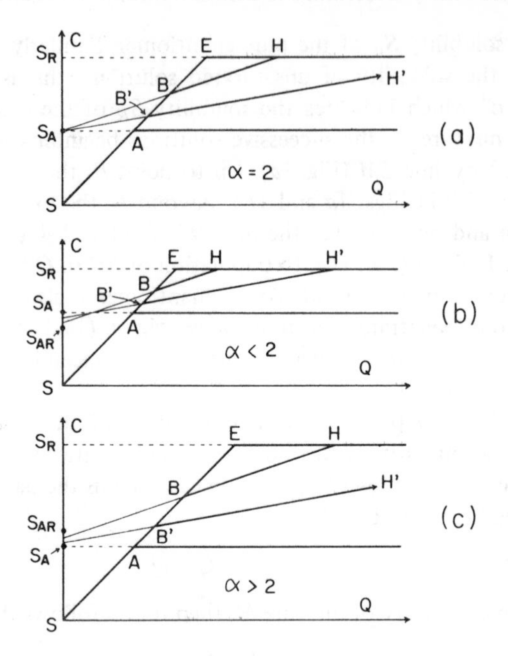

Figure 2 Determination of enantiomeric purity from solubility measurements for the case of a conglomerate and according to the value of α. BH is the line observed for $p = 0.33$, and $B'H'$ is that for $p = 0.67$.

If $\alpha < 2$ (Fig. 2*b*) or $\alpha > 2$ (Fig. 2*c*), the BH lines do not converge but cross axis SC at points S_{AM} which are dependent upon p, eq. (3), but which fall between S_A and $S_R/2$.

We know of no application of this process to the determination of enantiomeric purity of a conglomerate. This is perhaps not surprising in view of the fact that there is little need of it for this type of racemate from which pure enantiomers are so easily obtained

3.6.2 Racemic Compounds

Just as we have done for conglomerates, let us compare triangular (Fig. 3*a*) and rectangular (Fig. 3*b*) isothermal solubility diagrams of racemic compounds. First of all, let us look at those of pure compounds.

Progressive addition of pure enantiomer D to solvent may be traced in Fig. 3*a* by the displacement of the system on line $S \rightarrow D$. In Fig. 3*b*, one follows line SA and horizontal step AA' (saturated solution). For the pure racemate, line SRR' in Fig. 3*a* describes the changes occurring when racemic compound is dissolved; in Fig. 3*b*, this corresponds to line SR followed by RR'. The behavior of mixture E' having a composition corresponding to the eutectic is given by line SEE' (Fig. 3*a*) or by SE followed by EE' (Fig. 3*b*). We see that, as for the fusion process, the solubility curve of a eutectic is not different from that of a pure substance.

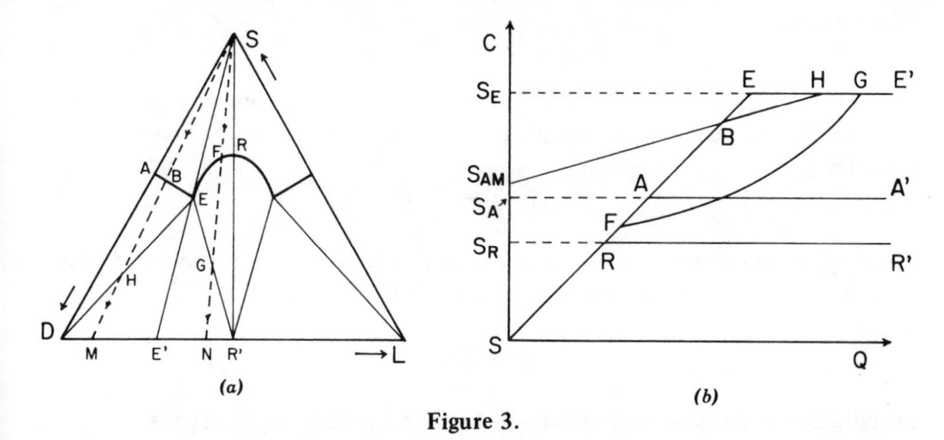

Figure 3.

A mixture M of composition intermediate between D and E' has a behavior similar to that exhibited by a comparable mixture in the case of a conglomerate (Fig. 1). Line *SBHM* of Fig. 3*a* corresponds in Fig. 3*b* to a succession of segments *SB* and *BH* and then to horizontal line HE'. As was true above, the equation of line *BH* is

$$C = S_{AM} + \frac{1-p}{2} Q \tag{7}$$

in which S_{AM} is the solubility of the enantiomer present in excess in mixture M and p is the enantiomeric purity of the mixture. Slope $s = (1-p)/2$ of line *BH* yields the enantiomeric purity.

Let us now examine the behavior of mixture N falling between E' and R'. Beginning with point F, the composition C of the solution gradually changes along curve *FE* of the triangular diagram which corresponds to curve *FG* of Fig. 3*b*. This is followed by step GE', which indicates attainment of saturation with respect to racemic compound and to enantiomer (eutectic). The information necessary for the determination of enantiomeric purity is contained in curve *FG*. This curve represents systems saturated with respect to racemic compound (which constitutes the sole solid phase present) and unsaturated with respect to the enantiomer in excess. The theory of the analysis of enantiomeric purity in this case is also relatively simple.[1] It presumes that the solution behaves ideally, namely, that in practice one is dealing with dilute solutions.

In a saturated solution of pure racemic compound, the solubility equilibrium is given by

$$(\mathrm{DL})_{\mathrm{crystal}} \rightleftharpoons (\mathrm{D}+\mathrm{L})_{\mathrm{solution}}$$

from which

$$K_s = [\mathrm{D}][\mathrm{L}] \qquad \text{with} \qquad [\mathrm{D}] = [\mathrm{L}] = \frac{S_R}{2}$$

from which $K_s = S_R^2/4$ follows. Given a solution derived from mixture N (Fig. 3*a*), saturated in racemic compound but not in D, the concentration is given by

$$C = [\mathrm{D}] + [\mathrm{L}] = [\mathrm{D}] + \frac{K_s}{[\mathrm{D}]} \qquad \text{or} \qquad [\mathrm{D}]^2 - [\mathrm{D}]C + K_s = 0$$

that is to say,

$$2[\mathrm{D}] = C + \sqrt{C^2 - 4K_s} \tag{8}$$

If p is the enantiomeric purity of the mixture N and Q is the quantity of this mixture per unit of quantity of solvent, then the total amount of D present in the system is

$$q_\mathrm{D} = \frac{1+p}{2} Q \tag{9}$$

Since q_D is also equal to the quantity of D in solution, [D], in addition to that present in solid racemic compound, $(Q - C)/2$, it follows that

$$\frac{1+p}{2} Q = [\mathrm{D}] + \frac{Q-C}{2} \tag{10}$$

Introduction of the value of [D] extracted from eq. (8) into eq. (10) gives

$$pQ = \sqrt{C^2 - 4K_s} \qquad \text{or} \qquad C^2 = p^2Q^2 + 4K_s$$

and finally

$$C^2 = p^2Q^2 + S_R^2 \tag{11}$$

Plotting C^2 as a function of Q^2 (Fig. 4), a straight line is obtained between F and G of slope $s = p^2$, and segments SF and GE' remain straight lines as in Fig. 3*b*. Note in particular that the enantiomeric purity of the eutectic p_E is obtained from the slope of line EZ:

$$s_E = \frac{S_E^2 - S_R^2}{S_E^2} \qquad \text{and} \qquad p_E = \sqrt{s_E}$$

We are aware of but one example of the use of this type of determination for racemic compounds (unpublished work by F. A. Bacher, cited by Mader[1]). Compared to other methods for the determination of enantiomeric purity (by dsc or by nmr spectroscopy, for example), the one just described suffers from a weakness, namely, relative slowness, since the establishment of solubility equilibria may sometimes take several days. It is thus logical to consider the procedure as a useful alternative for those cases in which other methods are, for one reason or another, impracticable.

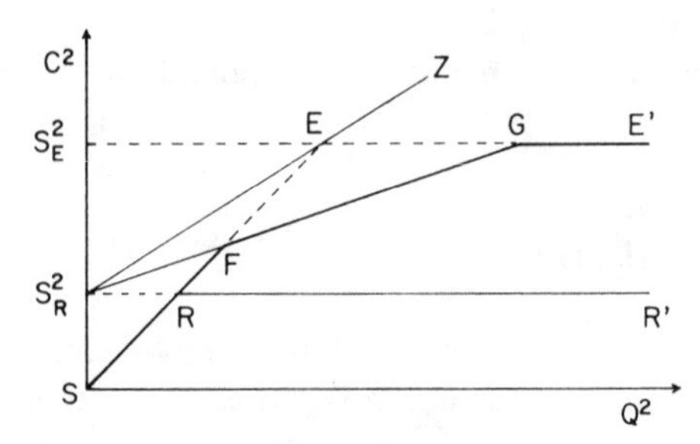

Figure 4 Graph for the determination of the enantiomeric purity of racemic compounds.

REFERENCES AND NOTES 3.6

1 W. J. Mader, in *Organic Analysis,* J. Mitchell, Jr., I. M. Kolthoff, E. S. Proskauer, and A. Weissberger, Eds., Vol. 2, Interscience, New York, 1954, p. 253.

2 R. M. Herriott, *Fed. Proc.*, 1948, 7, 479, and references cited.

3 H. A. Frediani, *Ann. Chim.* (*Rome*), 1952, **42**, 692.

4 D. Giron and C. Goldbronn, *Analusis*, 1979, 7, 109.

5 M. J. Brienne, unpublished results. It is assumed that the solubility of partially resolved mixtures varies linearly with p between S_A (for $p = 1$) and $S_R = \alpha S_A$ (for $p = 0$). This is tantamount to considering the solubility curve AE (Fig. 1*a*) to be a straight line. We have already seen (Section 3.2.3) that this is a good approximation.

REFERENCES AND NOTES

1. [illegible]
2. [illegible]
3. [illegible]
4. [illegible]
5. [illegible]
6. [illegible]

PART 2

Resolution of Enantiomer Mixtures

In the first part of this book we have described some important physical properties of racemates, enantiomers, and mixtures of the two. We shall now see how it is possible to separate the two enantiomers from mixtures in which they are present in equal parts. In particular, we shall focus upon separations that are based on crystallization techniques.

A *resolution* is a separation whose point of departure is a *racemate* and at whose conclusion at least one of the enantiomers present in the initial mixture is recovered. In our analysis, we exclude those asymmetric syntheses that convert an achiral compound into a partially resolved mixture of enantiomers and also those that transfer the chirality of a reactant to a given achiral or racemic substrate to form a *new* compound possessing a given enantiomeric excess. The consequence of this rather conventional limitation may well be that the best optical activation route for a given compound may not be found in this book. However, we wish to emphasize that the crystallization processes described here are broadly applicable to the preparation of *pure enantiomers* from partially resolved mixtures derived from asymmetric syntheses or, indeed, from any other process.

We first examine those resolution methods which require no auxiliary optically active agent but are based upon direct crystallization of enantiomer mixtures (Chapter 4). Subsequently, we describe processes which involve the formation and separation of diastereomers (Chapter 5) and the special separations involving asymmetric transformations of diastereomers and of enantiomers (Chapter 6). Finally, in the last chapter we are more specifically concerned with experimental aspects, or the art, of resolutions (Chapter 7).

CHAPTER 4

Resolution by Direct Crystallization*

The possibility of separating two enantiomers by direct crystallization of their mixture implies that the racemate is a conglomerate. If this is the case, then essentially two resolution methods are available.

The first of these takes advantage of the spontaneous resolution which occurs when a conglomerate crystallizes. This crystallization may be followed by the mechanical separation of the crystals of the two enantiomers which form simultaneously in the mother liquor which itself remains racemic. Various techniques have been devised to facilitate this separation, and these are described in Section 4.1.

The second, called resolution by entrainment, depends on differences in the rates of crystallization of the enantiomers in a solution supersaturated with respect to the racemate. We shall see that, under appropriate conditions, it is generally possible to favor the crystallization of a *single* enantiomer. Here, the system of crystals and solution is not allowed to come to equilibrium, and time plays an important role. The underlying theory of this process and illustrations of it are given in Sections 4.2 (entrainment in a supersaturated solution) and 4.3 (entrainment in a supercooled melt).

Lastly, the possibility of resolution by preferential crystallization of one enantiomer in an optically active solvent, which has been the subject of a fair number of experiments since the end of the 19th century, is discussed in Section 4.4

4.1 SEPARATION BASED UPON THE SIMULTANEOUS CRYSTALLIZATION OF THE TWO ENANTIOMERS

4.1.1 Manual Sorting of the Conglomerate. Triage

Allusion has already been made to the memorable experiments of Louis Pasteur who, during the month of May 1848 and under the watchful and incredulous eyes

*Reprinted in part from A. Collet, M. J. Brienne, and J. Jacques, *Chem. Rev.*, 1980, 80, 215 by permission of the editor. Copyright 1980, American Chemical Society.

of Biot, separated the dextrorotatory and levorotatory crystals constituting the racemic double salt sodium ammonium tartrate.[1–3]

While these historical experiments retain considerable value as examples of simplicity and economy both with respect to concept and resources, today they do not serve as useful models of practical manipulations of general applicability. This Pasteurian separation is extremely laborious. It permits one to collect only crystals that are well formed and exhibit well-defined morphological characteristics (hemihedrism) which distinguish "left" from "right" crystals (see Section 1.2.4), a situation that does not always obtain even if relatively large crystals are available. Nevertheless, the manual sorting (triage) of conglomerates of small and poorly formed crystals is feasible if one takes advantage of all the properties, which may be more or less evident, to differentiate the enantiomers.

We cite the experiments carried out by Jungfleisch on the resolution of the same sodium ammonium tartrate wherein he took advantage of the insolubility of calcium racemate (i.e., racemic calcium tartrate) to recognize, without dissolving them, whether two crystals of tartrate are or are not of the same sign:[4]

> It suffices to detach by means of a sharp needle a small portion of the crystal to be examined, and to deposit it along with a drop of water on a glass slide which itself rests on black paper. Once the crystal is dissolved, half of the solution is removed to another part of the slide by means of a stirring rod, and a drop of aqueous dextrorotatory calcium tartrate . . . is added to it. A drop of reactant . . . , prepared from levorotatory calcium tartrate, is added to the other half of the original solution. The precipitate of calcium racemate appears immediately in that drop for which the reactant added possesses a rotatory power of opposite sense to that of the crystal examined. Since no reaction would be expected in the other drop, the test would thus be subject to control. Moreover, this test can be carried out rapidly for many crystals on the same glass slide.

When one is dealing with substances having accessible melting points (which was not the case with the mineral tartrate salts studied by Pasteur), then an alternative analysis is to remove some fragments from an isolated crystal, by means of a razor blade, for example; these may be tested by mixing them with a small sample derived from other crystals to see whether melting point lowering takes place.[14]

The sign of rotation can also be determined polarimetrically by dissolving a crystal in a given solvent without weighing it or measuring the volume, prior to assembling the various solutions having the same sign.[13] The sensitivity of modern polarimeters has significantly increased the utility of this technique.

Other stratagems for distinguishing between crystals of opposite signs may be envisaged, such as that which consists of exposing a conglomerate to the vapor of an optically active reagent so as to form a visible coating which differs for (+) and (−) crystals of substrate[5] (see Section 1.2.4).

Even if manual sorting of crystals is only rarely of preparative value, this possibility is nonetheless of considerable interest. After all, the initial resolution of sodium ammonium tartrate was a crucial experiment in the subsequent development of stereochemistry as a whole. Sometimes, the demonstration that a substance is resolvable may directly provide stereochemical information about its structure. Thus, the observation that spontaneous resolution takes place allowed an immediate distinction to be made between the *meso*- and *dl*-forms of hydroveratroin[6] (1) and *o*-hexaphenylene[7] (2), to cite only two examples.

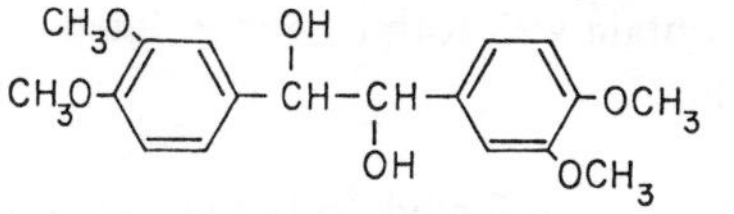

Most often, manual sorting is utilized to collect the first crystals of enantiomer required to apply the technique of resolution by entrainment, which we describe below (Section 4.2). This is particularly useful when resolution by formation and separation of diastereomers (Chapter 5) is impossible, unsuitable, or simply susceptible to complications as, for example, the case of enolizable ketones such as 3.[8,14]

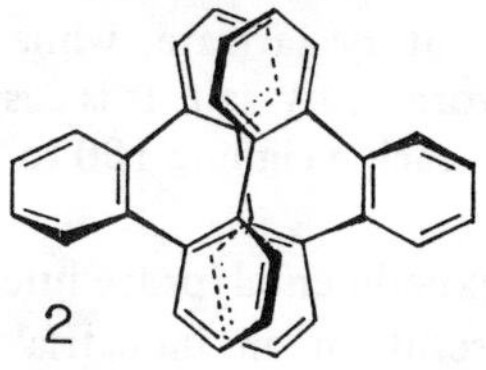

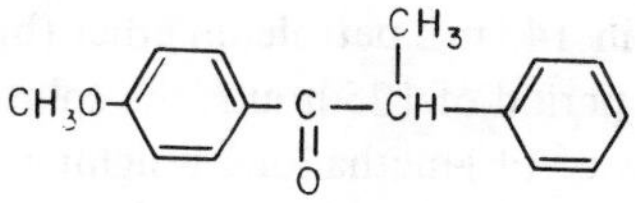

4.1.2 Simultaneous and Separate Crystallization of Enantiomers

A more attractive variant of the process which we have just described consists in the *localization* of crystallization of individual enantiomers on suitably disposed seeds within a racemic supersaturated solution. This process was conceived by Jungfleisch,[4] again in connection with the resolution of sodium ammonium tartrate. A racemic solution is prepared such that its supersaturation is ca. 150 to 160 $g \cdot L^{-1}$ at the temperature of crystallization:

> One operates in crystallizing dishes containing one to two liters of liquid whose ground edges permit an air-tight enclosure to be maintained by covering the dishes with flat glass plates. The saturated solution is placed in the vessel while still warm. Upon condensation, the water vapor emitted by the warm solution wets the edges of the dish as well as the glass cover by capillary action thus yielding a hydraulic closure which allows the solution to cool completely while remaining supersaturated . . . . The solution having attained room temperature, one carefully wets the hands and places a small fragment of dextrorotatory sodium ammonium tartrate between the fingers; one washes the crystal by exposing it momentarily to the stream of a washbottle and allows it to fall in the right hand side of the crystallizing dish. One does the same with a crystal of levorotatory salt which is allowed to fall in the left

side and then immediately replace the glass cover . . . . The introduction of crystalline dust must be carefully avoided which would otherwise rapidly lead to a mixed crystallization. The wetted crystals do not yield this phenomenon: they momentarily dilute the liquid layer which surrounds them, enlarge themselves slowly and attain their maximal size only after two to three days by which time the solution has ceased to be supersaturated. They remain perfectly isolated . . . and the dextrorotatory crystal has been enlarged only by dextrorotatory tartrate, while the levorotatory crystal is formed exclusively with levorotatory salt. It is easy to obtain well formed and isolated crystals in this way each weighing 180 to 200 g.

This experimental procedure has been perfected to obtain separations which may be useful on an industrial scale. Zaugg[9] has described an apparatus which allows the simultaneous crystallization of the enantiomers of methadone: 50 g racemic substance is dissolved in 145 mL petroleum ether (bp 63–68°C). Through slow evaporation at 40°C over a period of 125 hours, the solution loses about one quarter of its volume. Two crystals of (+)-methadone weighing 13.0 g and two crystals of (−)-methadone weighing 13.1 g develop from seeds deposited in the solution at the onset.

Another variant of this procedure consists in allowing the supersaturated solution of the racemate to circulate over the suitably arranged (+) and (−) seeds (fluidized bed system). The apparatus employed for the resolution of lysine 3,5-dinitrobenzoate is constructed from two columns each separable into halves to permit loading of seeds and collection of crystals.[10] Columns of dimensions 8 x 2 cm are fitted with fritted glass plates at each end. They are loaded each with 1 g of seeds, one enantiomer to each column. The supersaturated solution of racemate is then allowed to circulate through the two columns which are arranged in parallel. After a given time, the crystals that have developed are removed from each compartment. Table 1 gives an idea of the results obtained.

Table 1 Resolution of (±)-lysine 3,5-dinitrobenzoate in a fluidized bed system[a,b]

Run No.	Racemate Concentration (g/100 g H_2O)	T (°C)	Time (minutes)	Yield (g)		Optical Purity (%)	
				(−)	(+)	(−)	(+)
1	23.0 ± 0.2	35	45	2.2	2.4	78.5	72.2
2	23.0 ± 0.2	35	85	2.8	2.8	71.2	68.3
3	23.0 ± 0.2	35	160	3.8	3.4	58.1	59.5
4	36.0 ± 0.2	55	125	2.8	2.8	95.1	95.1
5	36.0 ± 0.2	55	200	3.2	3.2	92.2	93.1
6	36.0 ± 0.2	55	300	4.4	4.2	89.7	92.5

[a] N. Sato, T. Uzuki, K. Toi, and T. Akashi, *Agric. Biol. Chem.*, 1969, **33**, 1107. Reproduced by permission of Dr. T. Akashi.
[b] Flow rate 250 mL · min^{-1}; seed crystals 1.0 g, 60 to 115 mesh.

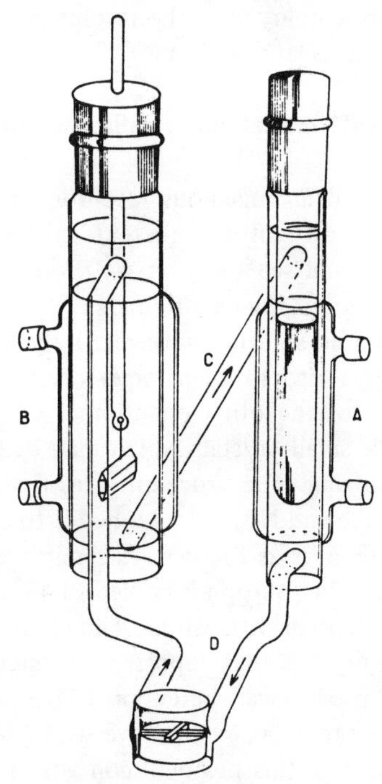

Figure 1 Apparatus for the resolution of conglomerates. J. Brugidou, H. Christol, and R. Sales, *Bull. Soc. Chim. Fr.*, 1974, 2033. Reproduced by permission of the Société Chimique de France.

Along the same lines, Brugidou et al.[11] have described an apparatus which allows one to obtain monocrystals of one of the enantiomers starting from a conglomerate. This device (Fig. 1) is made up of two jacketed tubes A and B which are maintained at different temperatures by circulation of appropriate thermostatted fluids through the jackets. A minipump is interpolated in tube D to provide for the slow circulation of the solution from the warm and toward the cold enclosure. The cycle is completed by means of tube C, the return line. A seed of one of the enantiomers suspended from a wire is introduced in cold tube B. The crystal of the corresponding enantiomer grows from the seed with the solution originating in the warm tube providing the "nourishment." The crux of the process is, of course, that the warm solution becomes supersaturated upon arrival in the cold tube.

This apparatus was used in the resolution of (±)-hydrobenzoin. Ethyl acetate was employed as solvent and the temperature of the warm tube was maintained at 22 to 24°C while that of the cold tube was kept at 14 to 15°C. Monocrystals weighing between 1.0 and 2.9 g were obtained having an optical purity of 98 to 100%. Of course, the racemate crystals contained in the thimble (in A) progressively increase in optical purity with respect to the other enantiomer as the crystallization proceeds in the cold tube.

A device combining the key elements of the Sato device and the Brugidou device has been successfully employed in the resolution of 3-fluoro-D,L-alanine-2-*d*-benzenesulfonate on a scale as large as 13 kg.[16]

4.1.3 Simultaneous and Differentiated Crystallization of Enantiomers

In 1960, Dowling[17] described an ingenious resolution procedure through seeding which avoids some of the difficulties inherent in resolutions by entrainment and which depend, as we shall see (Section 4.2), on the maintenance of supersaturation with respect to one of the enantiomers. The process consists in the seeding of a racemic supersaturated solution with relatively large seeds of one enantiomer whose growth will give rise to even larger crystals. At the same time, the spontaneous crystallization of the other enantiomer or its crystallization induced by small seeds will produce small crystals which may be separated from the larger enantiomeric crystals by sifting. The procedure was first applied to glutamic acid salts. Subsequently, Watanabe and Noyori[12] applied it to acetylglutamic acid.

The experimental protocol is as follows: Racemic acetylglutamic acid (30 g) is dissolved in 150 g water and the solution is cooled to 43°C; (−)-acetylglutamic acid crystals (10 g) of size greater than 30 mesh are added, and stirring is carried out at 43°C for 30 minutes. The crystals that deposit are filtered, washed, and dried. The solid is sifted through a 30-mesh sieve whereupon 13.6 g of practically pure product $[\alpha]_D^{17} - 16.1°$ ($c = 2$, water) remains in the sieve while 2.6 g of acid $[\alpha]_D^{17} + 10.8°$ passes through. A refinement of this protocol consists in the seeding with the (−) enantiomer, as above, but simultaneously with 1 g of (+) enantiomer of crystal size less than 200 mesh. Sifting yields 13.7 g of acid $[\alpha]_D^{17} - 15.0°$ and 4.8 g of acid $[\alpha]_D^{17} + 13.6°$. As is evident from this balance sheet, one is dealing with a resolution in which equilibrium between the solution and the two enantiomers is practically undisturbed (taking into account the weight of seeds, 3.7 g of one and 3.8 g of the other enantiomer are actually obtained).

A similar process has been applied successfully by Brienne and Jacques to the separation of a mixture of two *diastereomers* forming a eutectic. Pure and differently sized seeds of each diastereomer were obtained from their mixture by hand sorting followed by crystallization and sifting.[15]

REFERENCES 4.1

1 L. Pasteur, *C. R. Acad. Sci.*, 1848, **26**, 535.
2 L. Pasteur, *Ann. Chim. Phys.*, 3 ème Série, 1850, **28**, 56.
3 G. B. Kauffman and R. D. Myers, *J. Chem. Educ.*, 1975, **52**, 777.
4 M. E. Jungfleisch, *J. Pharm. Chim.*, 5 ème Série, 1882, **5**, 346.
5 C.-T. Lin, D. Y. Curtin, and I. C. Paul, *J. Am. Chem. Soc.*, 1947, **96**, 6199.
6 J. Grimshaw and J. S. Ramsey, *J. Chem. Soc.*, (C), 1966, 653.
7 G. Wittig and K. D. Rümpler, *Liebigs Ann. Chem.*, 1971, **751**, 1.
8 A. Collet, M. J. Brienne, and J. Jacques, *Bull. Soc. Chim. Fr.*, 1972, 336.
9 H. E. Zaugg, *J. Am. Chem. Soc.*, 1955, **77**, 2910.

10 N. Sato, T. Uzuki, K. Toi, and T. Akashi, *Agric. Biol. Chem.*, 1969, **33**, 1107.

11 J. Brugidou, H. Christol, and R. Sales, *Bull. Soc. Chim. Fr.*, 1974, 2033.

12 T. Watanabe and G. Noyori, *Kogyo Kagaku Zasshi*, 1969, **72**, 1083.

13 M. Delépine, R. Alquier, and F. Lange, *Bull. Soc. Chim. Fr.*, 1934, 1250.

14 Mme Bruzau, *Ann. Chim.*, 11ème Série, 1934, **1**, 257 (see also p. 319).

15 M. J. Brienne and J. Jacques, *Bull. Soc. Chim. Fr.*, 1973, 190.

16 U.-H. Dolling, A. W. Douglas, E. J. J. Grabowski, E. F. Schoenewaldt, P. Sohar, and M. Sletzinger, *J. Org. Chem.*, 1978, **43**, 1634.

17 B. B. Dowling, U.S. Patent 2,898,358 (1959) (to International Minerals & Chemical Corp.). *C. A.*, 1960, **54**, 17284g.

4.2 RESOLUTION BY ENTRAINMENT

Except for the direct isolation of natural products such as sucrose, the largest amounts of optically active substances are produced industrially mainly in two ways: by fermentation processes, which are limited to the synthesis of several amino acids of natural configuration; and by the technique to which Amiard gave the name *resolution by entrainment*[5] (dédoublement par entraînment). The latter is often called resolution by preferential crystallization. To cite but one example of the scale in which the process has been employed, 13,000 tons of L-glutamic acid were produced annually in the period 1963 to 1973 through resolution by entrainment of the synthetic acid derived from acrylonitrile.*

We do not wish to imply that this technique is only of interest to industrial-scale resolutions. Resolution by entrainment on a laboratory scale sometimes constitutes the method of choice when several grams or tens of grams of two enantiomers of an optically active substance are required. Nevertheless, this method is little used. The often cited but now dated review of Secor[16] concerns itself with general principles. On the other hand, many recent articles and patents have appeared which apply resolution by entrainment to specific cases. In this section, we concern ourselves with both of these aspects.

4.2.1 History and First Examples

The first observation showing the way to resolution by entrainment is due to Gernez, who was a student of Pasteur. The discovery was announced in all of 12 lines in a letter[1] addressed to Pasteur in 1866:

> I have observed that a supersaturated solution of levorotatory double salt sodium ammonium tartrate does not crystallize in the presence of a fragment of this salt which is hemihedric in the dextrorotatory sense; and vice versa, the supersaturated solution of the dextrorotatory salt yields no crystals when seeded with levorotatory salt.

*The resolution of glutamic acid by entrainment has been discontinued. Nevertheless, this process has been said to still be economically competitive with the fermentation process which is now employed in Japan (see note 22).

> This fact led me to study the inactive solution of the double salt sodium ammonium racemate. I prepared a supersaturated solution of this salt from the racemic acid. . . . When seeded by a particle of dextrorotatory salt, it yielded only dextrorotatory crystals. A portion of the same liquid in contact of a levorotatory crystal produced a deposit of levorotatory salt. Here then is a simple means for separating at will one or the other of the two salts which constitute the double salt sodium ammonium racemate.

In a paper dated 1882, Jungfleisch, while confirming the observations of Gernez, also cited the disadvantages of the method while recognizing the role of supersaturation:[2] if the solution is not strongly supersaturated, only a low yield is obtained in each operation. Contrariwise, it is difficult to prevent the crystallization of the nonseeded salt from a strongly supersaturated solution.

The method of Gernez was forgotten for a long period of time, no doubt due to the fact that so few spontaneous resolutions were known. It was not until 1914 that Werner rediscovered the same phenomenon though in a somewhat different form.[3] Werner had observed a large difference in solubility of the optically active and racemic (oxalato)bis(ethylenediamine)cobalt bromides, $[Co(ox)(en)_2]Br$. He tried to recover the less soluble optically active salt by precipitation with alcohol from an aqueous solution of this complex partially enriched with the *dextrorotatory* enantiomer. The attempt succeeded; but, much more remarkably, examination of the residual mother liquors showed that they had changed sign and that they now contained an excess of *levorotatory salt*. A further addition of alcohol furnished a crop of *levorotatory* crystals along with new and now dextrorotatory mother liquors, and so forth. This observation was subsequently confirmed with similar success on the dinitrobis(ethylenediamine)cobalt chloride, whose optically active form is also less soluble than the racemic form.

Just as was true of the work of Gernez a half-century earlier, the results of Werner were also forgotten for quite some time. It was only some 20 years later, through a demonstration of the efficiency of the resolution by entrainment of histidine monohydrochloride by Duschinsky,[4] that the method really began to interest chemists, in particular those in industry.

4.2.2 Description of the Process of Resolution by Entrainment

The use of resolution by entrainment such as it is practiced nowadays may be exemplified by the case of hydrobenzoin. Racemic hydrobenzoin (11 g) is dissolved along with 0.37 g (−)-hydrobenzoin in 85 g of 95% ethanol and the solution is cooled to 15°C. Seeds of (−)-hydrobenzoin (10 mg) are added and the stirred solution is allowed to crystallize for 20 minutes. The weight of (−)-hydrobenzoin recovered after filtration (0.87 g) is roughly double that of the (−) enantiomer introduced in excess at the beginning of the experiment. Racemic hydrobenzoin is then added to the remaining solution in an amount equal to that of the (−) crystals collected. The solution is heated to complete dissolution of the solid and is then cooled to 15° and crystallized as above, after seeding with 10 mg (+) enantiomer, to yield a weight of (+)-hydrobenzoin nearly equal to that of the (−) isomer earlier

Table 1 Resolution of (±)-hydrobenzoin by entrainment[a]

Run No.	Hydrobenzoin added (g)		Yield of Resolved Hydrobenzoin (g)	
	Racemic	(−)	(−)[b]	(+)[b]
1	11.0	0.37	0.87	
2	0.9			0.9
3	0.9		0.8	
4	0.8			0.75
5	0.7		0.7	
6	0.7			0.75
7	0.75		0.8	
⋮	⋮	⋮	⋮	⋮
Total (15 runs)	23.5[c]	0.37	6.5	5.7

[a] Experimental conditions are given in the text. Adapted from A. Collet, M. J. Brienne, and J. Jacques, *Chem. Rev.*, 1980, 80, 215 by permission of the editor. Copyright 1980, American Chemical Society.
[b] Enantiomers having ca. 97% optical purity.
[c] 11 g racemic hydrobenzoin recovered.

collected. The same cycle of operations, namely, loading with racemic hydrobenzoin and collection of (+) and (−) crystals is carried out 15 times, yielding 6.5 g of (−) and 5.7 g of (+) enantiomer having ca. 97% optical purity. Table 1 gives a summary of the process.

Even though individual cases differ somewhat from one another and thus require specific procedures, most resolution by entrainment described in the literature, including patents, require this sequence of alternate crystallization of the two enantiomers. The initial system (or its "mirror image") is reconstituted between each cycle through addition of a quantity of racemate equal, or approximately equal, to that of the enantiomer just isolated (for other examples of this process, see refs. 5 and 6.).

4.2.3 Interpretation Based on Solubility Diagrams

A resolution by entrainment can take place only if the enantiomers can crystallize separately. In other words, the solubility diagram of the racemate must be that of a conglomerate. We return in Section 4.2.4 to this essential requirement which has given rise to much confusion in the literature.

Consider for a moment the triangular diagram of Fig. 1 and observe first of all what takes place when a mixture of composition *P* is crystallized. This mixture is supersaturated with respect to the two enantiomers, but the extent of supersaturation of L is greater than that of D. We know that once the solubility equilibrium is attained, the mother liquor will have composition *E* (racemic) and the solid will have composition Q. What matters now is to understand *how* this equilibrium

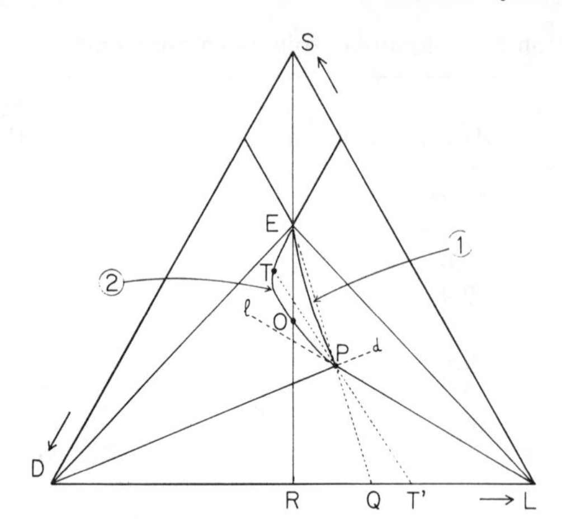

Figure 1 A. Collet, M. J. Brienne, and J. Jacques, *Chem. Rev.*, 1980, **80**, 215. Reproduced by permission of the editor. Copyright 1980, American Chemical Society.

will be attained. If enantiomer L crystallized *by itself*, then the composition of the mother liquor would be displaced along line *Pl*. By the same token, if D crystallized by itself, then the composition of the mother liquor would be displaced along *Pd*. With D and L crystallizing simultaneously, the composition of the mother liquor is displaced along a resultant curve which finally terminates at point *E*. Since the appearance of this curve depends upon the ratio of the crystallization rates of the two enantiomers, when these rates are comparable, a curve of type 1 would obtain. If, on the other hand, L crystallizes more rapidly than D at first, one might find a curve of type 2. In the latter case, the rotation of the mother liquor would change sign during the crystallization. Initially it would be (−) at point *P*; it would vanish at point *O* and be (+) at point *T*, for example. From this fact, the solid which would deposit just then (and which would have composition *T'*) necessarily would contain more of enantiomer L than there was initially in excess (*P*).

It must be clear that the best resolution conditions will be those for which curve *POT* will remain close to line *Pl* for the longest possible time corresponding to a large difference in crystallization rates between the enantiomers. In the extreme, this is equivalent to the precipitation of a single enantiomer, namely L. We shall see later how these conditions may be optimized (Section 4.2.5). To reinforce these ideas, consider Fig. 2, which shows an experimental curve obtained from a supersaturated solution of *N*-acetylleucine in acetone. Even though crystallization was carried out under routine conditions, the change of sign of the mother liquor was quite evident. An example of an optimized crystallization curve, corresponding to the resolution by entrainment of (±)-hydrobenzoin, is shown in Fig. 3.

We are now ready to interpret a resolution by entrainment by means of a ternary diagram such as that described above. The initial solution *M* (Fig. 4) contains the racemate and a slight excess of enantiomer L. It is prepared at an elevated temperature and then cooled to T_0, a temperature for which the solubility curves

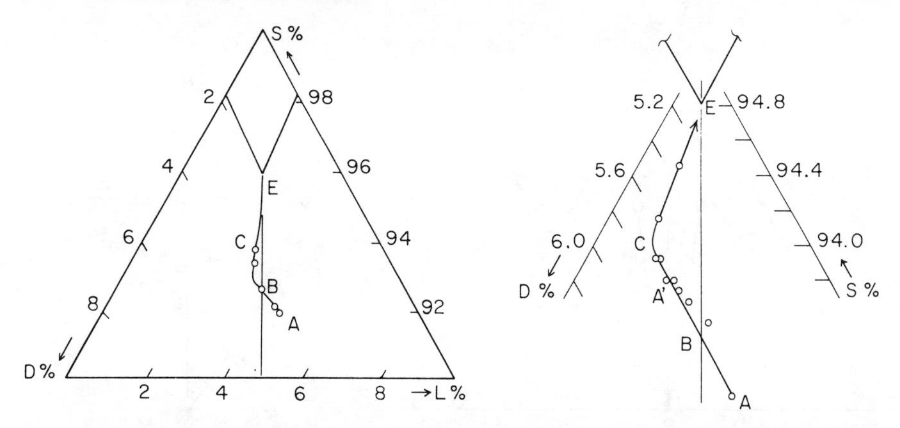

Figure 2 Crystallization curve of *N*-acetylleucine in acetone at 25°C. Initial optical purity 11% (*A*). The solution was seeded with a very small amount of pure crystalline L. Point *B* (change of sign of the solution) is attained at the end of ca. 20 minutes and point *C* at the end of 40 minutes. Only the upper part of the ternary diagram is shown (see scales at the sides of the triangle). A. Collet, M. J. Brienne, and J. Jacques, *Chem. Rev.*, 1980, **80**, 215. Reproduced by permission of the editor. Copyright 1980, American Chemical Society.

Figure 3 Crystallization curve optimized for the resolution by entrainment of hydrobenzoin in ethanol at 15°C. Initial conditions (A): concentration 11.7 g/100 g solution, optical purity 2.56%; seeding by 10 mg of finely ground (−) crystals; stirring speed 215 rpm. The change of sign (*B*) takes place after 25 minutes, with the maximum (*C*) at the end of 180 minutes. The point symmetrical to *A*(*A'*) is attained in 85 minutes (enlargement of the part of the ternary diagram located in the vicinity of the eutectic point *E*). A. Collet, M. J. Brienne, and J. Jacques, *Chem. Rev.*, 1980, **80**, 215. Reproduced by permission of the editor. Copyright 1980, American Chemical Society.

are *AEA'*. The solution is then supersaturated with respect to *both* enantiomers, but the extent of supersaturation is greater for L. Upon seeding, L is induced to crystallize alone, and the point representing the mother liquor composition is displaced along the line *LMN* and toward *N*. For convenience, the duration of the crystallization is adjusted so as to yield a quantity of crystals of L which is *double* that taken initially in excess. The composition of the mother liquor is then given by *N*, and its rotation attains a value effectively equal and of opposite sign to the starting value.* After removing the crystals of L, an equivalent weight of racemate is added to the now dextrorotatory mother liquor so as to yield system *P* which is symmetric with *M*. This mixture is heated to dissolve the solid, cooled to T_0, and seeded with enantiomer D which then crystallizes. The composition of the mother

*Strictly speaking, the absolute value of the rotation of the solution at point *N* must actually be slightly larger than at *M* as a result of the change in concentration of the solution due to the deposition of crystals of one enantiomer. In practice, this correction is negligible. Once the best experimental conditions for the resolution have been established, the measurement of the rotation remains the simplest way to monitor the resolution.

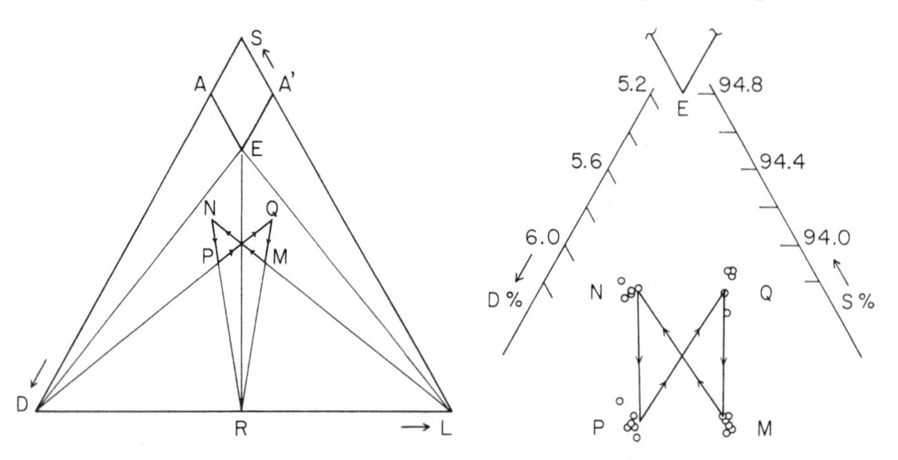

Figure 4 Resolution by entrainment. Cyclic alternate crystallizations of the two enantiomers. A. Collet, M. J. Brienne, and J. Jacques, *Chem. Rev.*, 1980, **80**, 215. Reproduced by permission of the editor. Copyright 1980, American Chemical Society.

Figure 5 Successive cycles of resolution of hydrobenzoin by entrainment (same scale and conditions as in Fig. 3). A. Collet, M. J. Brienne, and J. Jacques, *Chem. Rev.*, 1980, **80**, 215. Reproduced by permission of the editor. Copyright 1980, American Chemical Society.

liquor varies from P to Q after which it is collected. Racemate is added once again to return to M, and the cycle may then be repeated.

In practice, points $M_i, N_i \ldots$ corresponding to successive cycles need not be rigorously superposable. Fig. 5 gives a bird's-eye view of the dispersion of compositions observed in the case of hydrobenzoin under conditions approximating those shown in Fig. 3. Provided the mode of seeding and all the conditions of crystallization are standardized, the duration of the process $M \rightarrow N$ or $P \rightarrow Q$ may be considered as a constant that may be established during preliminary experiments.

4.2.4 Racemates Resolvable by Entrainment

Let us now return to the requirement stated earlier regarding the nature of the racemate and how this affects its ability to be resolved by entrainment.

Werner[3] stipulated as a necessary condition the simple fact that the enantiomer be less soluble than the racemate, without specifying the nature of the latter. We are aware that this conclusion was based upon the results obtained by him in the resolution of several complexes which later were recognized to be *conglomerates*. Other authors[7,8] have generalized the statement of Werner to mean that such resolution may also be applied to racemic compounds, provided that these are more soluble than either pure enantiomer. However, Duschinsky[4] earlier had actually been the first to correctly identify the requirement, which follows from the properties of the phase diagram as a whole and not only from the individual solubilities of *pure* racemate and enantiomer, namely, a necessary condition for resolution by entrainment to be successful is that the racemic compound, if it exists at all, not crystallize during the operation. This may correspond to several possibilities:

1 The conglomerate is the stable crystalline form, and no racemic compound exists. The above necessary condition is then always fulfilled.

2 Even when the conglomerate is the stable form, the existence of a metastable racemic compound is likely to make the entrainment difficult if not impracticable. For example, though the thermodynamically stable form of acid (±)-**1** is a conglomerate (See Section 2.5.4), its crystallization in a solvent always yields a metastable racemic compound whose solubility is around *seven times greater than that of the enantiomers*. All attempts at resolving this compound by entrainment have been in vain, however.

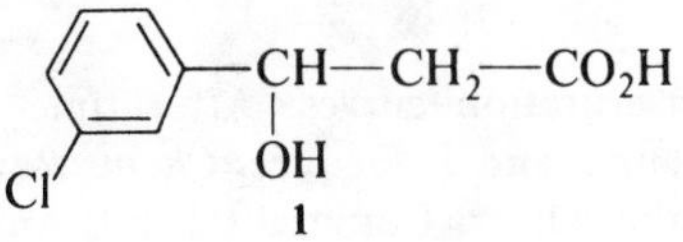

$$\text{3-Cl-C}_6\text{H}_4\text{-CH(OH)-CH}_2\text{-CO}_2\text{H}$$

1

3 A substance may exist as a racemic compound or as a conglomerate according to the temperature. In this case, the resolution is generally feasible in the region of stability of the conglomerate. The best known example of this possibility is that of sodium ammonium tartrate, which may be resolved by entrainment below 27°C. Sometimes, such resolution is possible also in the region of stability of the racemic compound; the conglomerate is then metastable, a situation which may introduce difficulties. The case of histidine hydrochloride[4] illustrates this possibility.

4 The stable crystalline form is a racemic compound under all conditions. As a matter of fact, this is the most common case; here, entrainment is virtually unrealizable even if the racemic compound is more soluble than either enantiomer.

4.2.5 The Search for Conditions Favoring Entrainment. The Method of Amiard

The use of a resolution by entrainment procedure requires the mastery of the factors that influence the *rate of crystallization*. Some of these factors are in turn largely dependent on the properties of the ternary diagram itself which, in particular, limit the useful concentration ranges. Others, more experimental in character, such as the rate of stirring or mode of seeding, lend themselves to the possibility of relatively greater variation.

The first detailed analysis of the mechanism of this type of resolution, carried out by Duschinsky,[4] deals with the case of histidine monohydrochloride which, as we know, is complicated by a polymorphism associated with solvation. More recently, in connection with their work on the resolution of threonine and *threo-p*-nitrophenyl-2-amino-1,3-propanediol,* Amiard and his collaborators[5, 9, 10, 11] applied

*The (−) enantiomer is an intermediate in the synthesis of the antibiotic chloramphenicol. The biologically inactive (+) isomer is also frequently used as resolving agent.

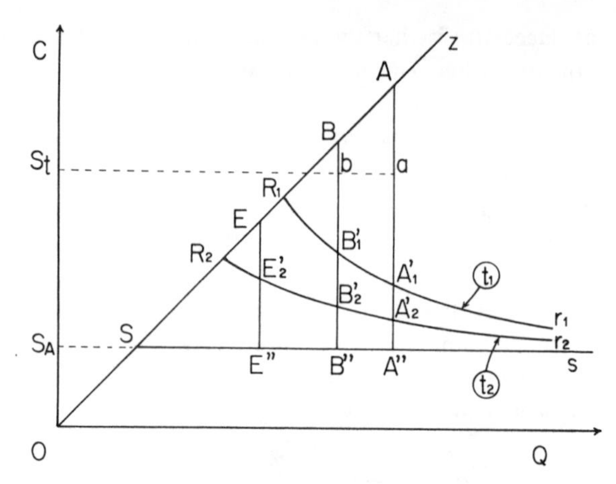

Figure 6 Residual supersaturation curves. After time t_1: R_1r_1; after time t_2: R_2r_2. A. Collet, M. J. Brienne, and J. Jacques, *Chem. Rev.*, 1980, **80**, 215. Reproduced by permission of the editor. Copyright 1980, American Chemical Society.

themselves to a systematic search for conditions permitting the maintenance of supersaturation of the nonseeded enantiomer during the course of the crystallization of the other. Their contribution rests on two ideas, that of *residual supersaturation* (sursaturation rémanente), and that of the *entrainment* of crystallization, from which the name of the method was coined.

Consider the isothermal solubility diagram of a pure enantiomer shown in Fig. 6. The quantity of *dissolved* substance C is shown as a function of the *total* quantity Q present in a given volume of solvent (this type of representation is described in Section 3.1.5). The straight line Ss corresponds to the solubility S_A of the compound at the temperature of the experiment. The unsaturated solutions are described by segment OS, while the supersaturated solutions prior to crystallization are given by line Sz. During the course of the crystallization of a solution represented initially by A, the point representing the state of the system is displaced vertically until the solubility equilibrium is finally attained at A''. In reality, this type of equilibrium occasionally takes a very long time to establish. Thus, in our example (Fig. 6), the solution of initial composition A may be represented at time t_1 and after seeding by A'_1, at time t_2 by A'_2, and so on. These points correspond to the *residual supersaturations*, σ_{t_1}, σ_{t_2}... of solution A at the times specified. If the values of the residual supersaturations are determined as a function of the quantity of compound initially dissolved (Q), under specified conditions for crystallization, one obtains curves such as R_1r_1 (for a time t_1) or R_2r_2 (for t_2). The shape of these curves shows that the residual supersaturation varies, after a while, in a sense inverse to that of the extent of the initial supersaturation ($\sigma_{t_1} = A'_1A''$ for A, B'_1B'' for B). This phenomenon reveals the second idea, that of the *entrainment* of crystallization: after a given time, the larger the initial concentration of the system, the closer the system is to solubility equilibrium.

We may now make an important inference from our analysis of Fig. 6. At time

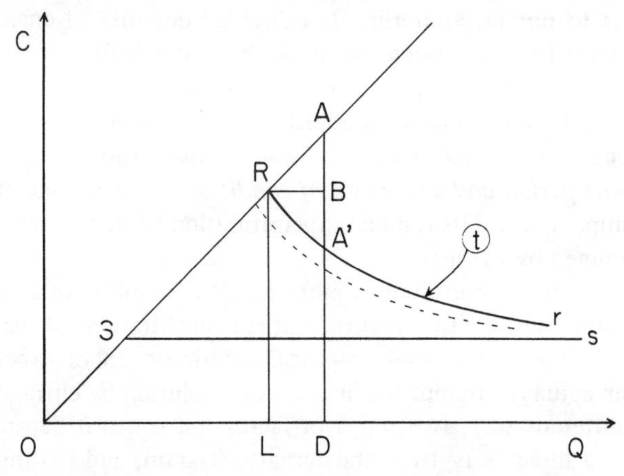

Figure 7 Alternative view of a resolution by entrainment showing the use of a residual supersaturation curve. A. Collet, M. J. Brienne, and J. Jacques, *Chem. Rev.*, 1980, **80**, 215. Reproduced by permission of the editor. Copyright 1980, American Chemical Society.

t_2, a supersaturated solution E will have attained composition E_2'. However, the fact that E is found below the residual supersaturation curve R_1r_1 also means that after a time t_1, this solution will have shown no evidence of crystallization in the presence of seeds. This last criterion clearly distinguishes ***residual*** supersaturation from the *metastable* supersaturation of Ostwald and Miers (see p. 235). The latter defines a concentration limit below which crystalliation of a supersaturated solution never takes place spontaneously but is triggered by the introduction of seeds, while residual supersaturation can only be defined with respect to ***a given time*** of crystallization. It is important that these two states of supersaturation not be confused with one another.

The notion of residual supersaturation allows one to specify, in a rather simple way, some conditions under which resolution by entrainment may be achieved. In the first instance, the residual supersaturation curve of an enantiomer is determined for a given set of conditions in *racemic* supersaturated solution. A curve (*Rr*, Fig. 7) is thus obtained whose coordinates may be slightly different from those for the pure enantiomer. The separation between the two curves reflects the influence of the presence of one enantiomer on the behavior of the other. Based on this curve, a supersaturated solution of racemate is prepared such that the concentration of L and D each is equal to OL (Fig. 7). The point which represents enantiomer L prior to crystallization is thus R. An additional quantity of enantiomer D (equal to LD) is then dissolved. Thus, the total quantiy of D is OD, while the solution is represented by A. If crystallization is now induced under conditions in which curve *Rr* was obtained, it follows from the preceding considerations that compound L remains residually supersaturated, while D is deposited until the point A' on curve *Rr* is attained. The quantity of D thus precipitated, which is given by AA', is necessarily larger than that originally invested (corresponding to LD $= AB$). With the aid of the diagram, one must choose the initial excess LD in such a way that

$AB = BA'$ so as to obtain, after time t, a doubled quantity of enantiomer. After these crystals have been separated, reconstitution of conditions inverse to those which obtained previously (R representing the concentration of enantiomer D and A that of enantiomer L) requires only the addition of a weight of racemate equal to AA'.

Summarizing, the method consists of the crystallization of a supersaturated solution *in a time period and a given set of conditions* in which one of the enantiomers is at the upper limit of its residual supersaturation while the other is present in an excess determined by the first.

This analysis actually corresponds more to the description of a process than to a theory. That it allows one to *adjust* the initial concentrations of the two enantiomers optimally is due to the prior empirical determination of experimental conditions. One can actually attempt the mastery of resolution by entrainment by considering, in a different way, two series of factors on which it depends: (1) those which derive in a direct way from the ternary diagram, and (2) those which are more concerned with kinetic factors in the crystallization. Let us not, however, lose sight of the fact that in reality these two groups of factors are never completely independent of one another.

4.2.6 Derivation of Favorable Conditions for Resolutions by Entrainment from the Ternary Diagram

Ternary diagrams allow us to *locate* the region of concentration in which resolution by entrainment is possible. They also provide information on the *course* of the resolution by showing the effect of the crystallization of one enantiomer on the supersaturation of the other.

The solubility diagram of a conglomerate we need to examine here is shown in Fig. 8. Let us consider the isotherm at T_0 under circumstances such that only one of

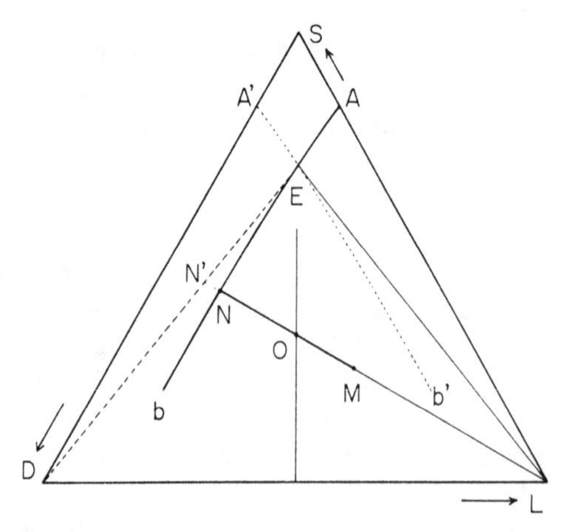

Figure 8 Ternary isotherm corresponding to the solubility equilibrium of enantiomer L in solvent D + *S*. A. Collet, M. J. Brienne, and J. Jacques, *Chem. Rev.*, 1980, **80**, 215. Reproduced by permission of the editor. Copyright 1980, American Chemical Society.

the enantiomers, say L, crystallizes. The parts of the diagram that correspond to the solubility equilibrium of enantiomer D (the solibility curve $A'E$ and the tie line ED) no longer are involved in the behavior of the system. (In other terms, this diagram describes the solubility of L in the binary solvent S,D.) Thus, the crystallization of L will continue up to its *solubility curve* which is no longer limited to segment AE but extends all the way to b. Point E, which here does not signify a eutectic, nonetheless has a physical meaning: it delimits the curves of *stable* (AE) and of *metastable* solubility equilibrium (Eb). Only the latter equilibrium interests us inasmuch as entrainment is not possible in the region AEL.

Let us now return to Fig. 8. During the course of the crystallization of enantiomer L beginning with a system of composition M, the point which describes the solution moves along line LM beyond M up to the solubility curve at N (and not at N' on line ED as is sometimes indicated erroneously[8]). Solubility curve Eb is therefore a limit in resolution by entrainment. One sees immediately that the form of the solubility diagram and, in particular, the value of the solubility ratio $\alpha = S_R/S_A$ directly influences the extent of the usable region of the diagram, as we can see from Fig. 9. Contrary to what we might intuitively have imagined, it is the case for which the racemate is *much more soluble* than the enantiomers ($\alpha > 2$) that corresponds to the *least favorable* situation for entrainment. On the other hand, a ratio $\alpha < 2$ substantially enlarges the usable part of the diagram.

There is another important piece of information which may be extracted from these diagrams and which has a bearing on the alteration of the degree of supersaturation of the nonseeded enantiomer during the course of the crystallization of the other. Our reasoning is summarized in Fig. 10. During the crystallization of L,

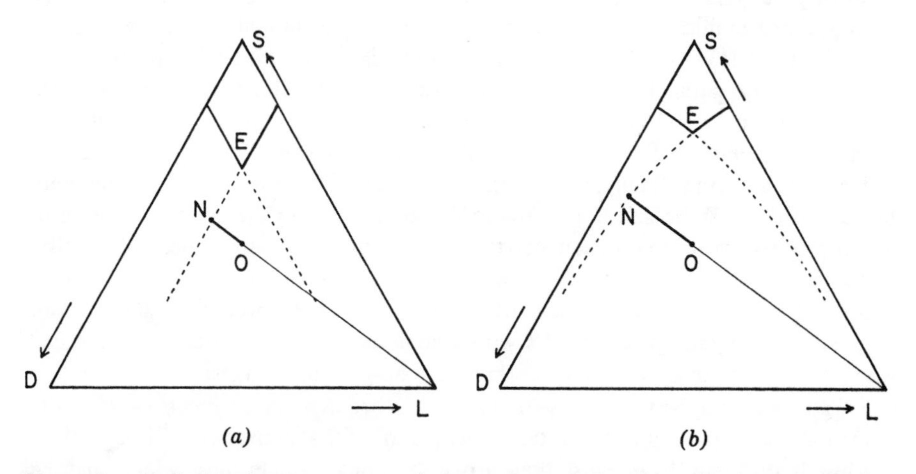

Figure 9 Variation in the extent of the region of the phase diagram usable for resolution by entrainment as a function of the solubility ratio $\alpha = S_R/S_A$. The solubilities of the pure enantiomers and the coordinates of point O are identical in both diagrams: (a) $\alpha = 2$; (b) $\alpha = \sqrt{2}$. A. Collet, M. J. Brienne, and J. Jacques, *Chem. Rev.*, 1980, **80**, 215. Reproduced by permission of the editor. Copyright 1980, American Chemical Society.

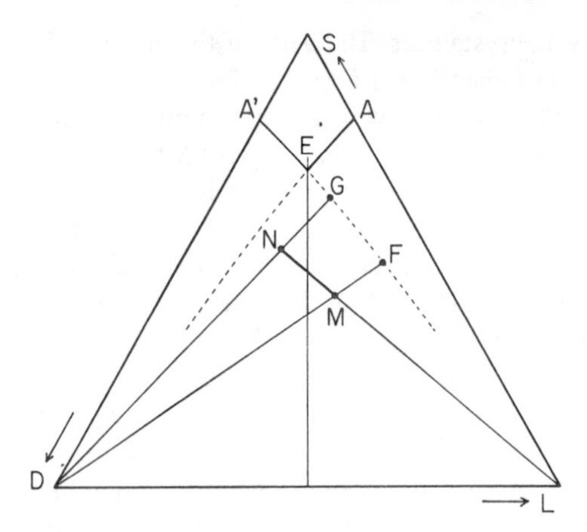

Figure 10 Extent of supersaturation of the nonprecipitating enantiomer during resolution by entrainment. A. Collet, M. J. Brienne, and J. Jacques, *Chem. Rev.*, 1980, **80**, 215. Reproduced by permission of the editor. Copyright 1980, American Chemical Society.

the composition of the solution may, for example, vary from *M* to *N*. The degree of supersaturation of D, which must remain in solution, corresponds to the distance between the point which describes the solution and the extension of solubility curve *A'E*. Thus, at *M*, the supersaturation in enantiomer D is represented by *MF*, and at *N*, by *NG*.

The alteration of this degree of supersaturation during the process of entrainment may be qualitatively probed for dilute solutions as a function of the solubility ratio α. Since in dilute solutions the region of interest is pushed back toward vertex *S* of the triangle, line L*MN* on one hand and lines D*MF* and D*NG* on the other become nearly parallel to sides L*S* and D*S*, respectively. The conclusions are sketched out in Fig. 11. For $\alpha > 2$, the degree of supersaturation of the undesired enantiomer D *increases* during the crystallization of the other ($MF < NG$). If $\alpha = 2$, it does not vary. And, finally, it *decreases* when $\alpha > 2$ ($MF > NG$). In more concrete terms, with $\alpha > 2$ the solution tends to destabilize during the course of the resolution process, with the consequence that the risk of spontaneous nucleation of the undesired enantiomer is increased. On the other hand, when $\alpha < 2$, the solution becomes increasingly stable during the development of the process. It should then be possible to obtain purer crystals with a large extent of resolution since it is also when $\alpha < 2$ that the area in which entrainment may occur is largest.

Watanabe and Noyori[15] have studied this problem in connection with the industrial-scale resolution of glutamic acid and of its derivatives. Though their analysis is different from ours, they arrive at similar conclusions, which may be illustrated by several examples. For the case of glutamic acid hydrochloride ($\alpha \sim 1.4$) or ammonium glutamate ($\alpha \sim 1.5$), it is shown that one easily crystallizes virtually all of the supersaturated fraction of the seeded enantiomer in an optically pure state. On the other hand, with free glutamic acid ($\alpha \sim 2.35$) or with *N*-acetylglutamic acid ($\alpha \sim 2.18$), one observes that the other enantiomer crystallizes before

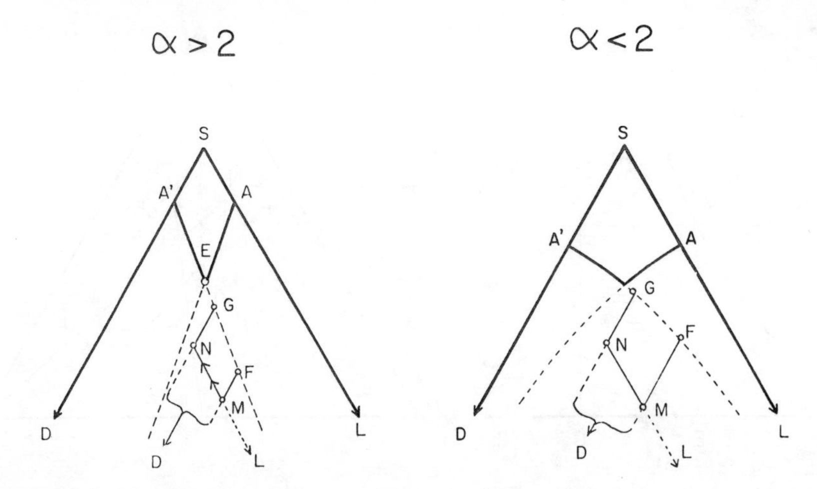

Figure 11 A. Collet, M. J. Brienne, and J. Jacques, *Chem. Rev.*, 1980, **80**, 215. Reproduced by permission of the editor. Copyright 1980, American Chemical Society.

maximal resolution has been attained. As a matter of fact, it is precisely in order to overcome this problem that these authors recommend, in this case, the simultaneous seeding of the solution with seeds of the two enantiomers having different sizes (Section 4.1.3).

Finally, we may deduce from the foregoing analysis that resolution by entrainment must generally be easier to carry out with salts than with undissociable organic compounds. It is for salts that there is the greater possibility of finding a ratio α that is substantially smaller than 2 (see Section 3.2.3). This has led to systematic searches for *dissociable derivatives* which may exist as conglomerates by combining substrates to be resolved with a variety of achiral reagents.[6, 17]

4.2.7 Control of Crystallization Rates

There are two opposing requirements associated with attempts to control the rate of resolution by entrainment. It is necessary to reconcile the relatively rapid growth of the crystals of the desired enantiomer with the lowest possible rate of crystallization of the other.

Let us first examine the second of these requirements. The *spontaneous* crystallization of a supersaturated solution comprises two steps: (1) the appearance of crystalline seeds (nucleation), and (2) development of crystals from these seeds. Ostwald and Miers have defined two ranges of supersaturated states.[23] One is a region of "labile supersaturation," I (Fig. 12*a*), found at high concentration in which spontaneous nucleation is possible. The other is a region of "metastable supersaturation," II, found between the former region and the solubility curve in which the spontaneous generation of seeds would not seem possible. In reality,

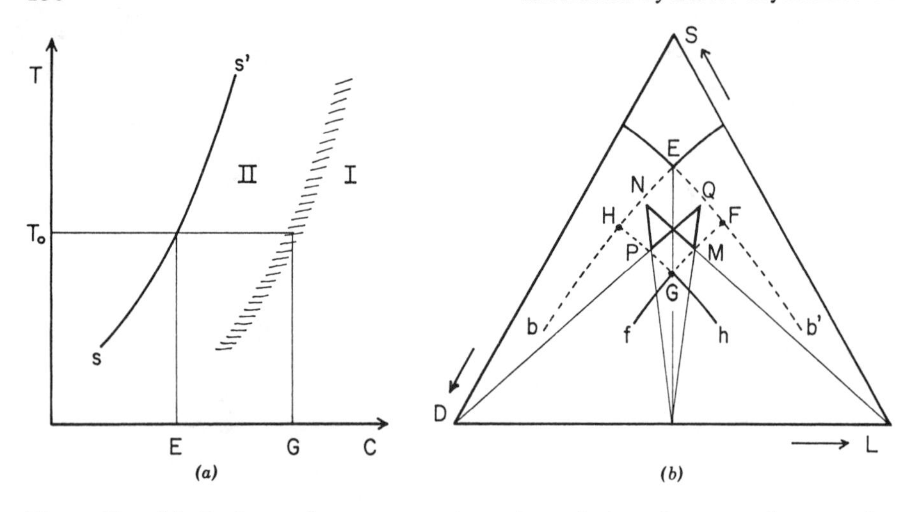

Figure 12 (a) Regions of supersaturation of a solution, for example, racemic: ss', solubility curve as a function of temperature; E, solubility of the racemate at T_0; G, maximum concentration of a solution which does not spontaneously crystallize at T_0 within a stated time interval (see text). (b) Determination of the region of the phase diagram which is useful for resolution by entrainment. A. Collet, M. J. Brienne, and J. Jacques, *Chem. Rev.*, 1980, **80**, 215. Reproduced by permission of the editor. Copyright 1980, American Chemical Society.

there is no sharp demarcation line between these two regions;[14] nevertheless, this analysis is useful in practice and justified by theory (see below).

While the appearance of crystalline seeds depends upon numerous factors[21] which are not all easily controllable,* it is nevertheless possible to determine empirically, if not the limit of metastable supersaturation, at least a concentration limit beyond which crystallization cannot be prevented. Moreover, in the context of the method of Amiard (Section 4.2.5), it is logical to take *time* into account. In practice, we shall determine the maximum concentration G of a racemic solution (at a temperature T_0) in which no spontaneous crystallization will have taken place during a given period of time, for example 1 hour, and under specified conditions of stirring. The region of the phase diagram (Fig. 12*b*) in which resolution by entrainment is favorable is approximately given by drawing lines parallel to the solubility curves such as *FGf* and *HGh*. Within these limits, we may draw a crystallization cycle *MNPQ* from which the conditions of resolution may be further refined if required.

The practical problem remaining is related to the fact that the limit of labile supersaturation of the undesired enantiomer is not really a well-defined line. This is particularly true in the presence of growing crystals of the seeded enantiomer which might well catalyze the nucleation of the other. While theories seeking to describe

*Nucleation may be induced by stirring, by impact, by high pressure, by means of electric and magnetic fields, by sound waves and ultrasonic irradiation, by X- and γ-rays, by soluble impurities, or by dust.

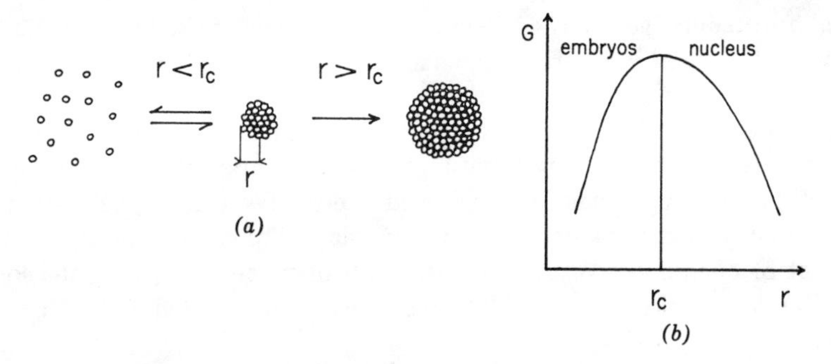

Figure 13 The process of nucleation. A. Collet, M. J. Brienne, and J. Jacques, *Chem. Rev.*, 1980, **80**, 215. Reproduced by permission of the editor. Copyright 1980, American Chemical Society.

the rather complex mechanism of nucleation are as yet imperfectly supported by experiment in the case of solutions, some of their implications may nevertheless be useful. (At present, these theories quite satisfactorily explain nucleation in supersaturated vapors.[18]) Consequently, it may be worthwhile to present them briefly here.[18]

These theories propose that solute molecules first undergo collision to form very small aggregates of several molecules called "embryos" through a reversible process sketched in Fig. 13*a*. The change in free energy accompanying the formation of these aggregates increases with their size, passes through a maximum at a critical size r_c, and then diminishes (Fig. 13*b*). This means that aggregates smaller than r_c have a greater tendency to fall apart than to grow, while those exceeding the critical size may develop and are called "nuclei." The mean size r of the aggregates in equilibrium with the solution increases with the free energy of the solution itself, that is, with its supersaturation. There is a critical concentration for which equilibrium corresponds to aggregates of size r_c; this is how theory interprets the "metastable supersaturation" limit. The physical reality of this critical size r_c has been experimentally brought to light by Ostwald whose results indicate that the minimum size of aggregates able to grow corresponds to a sphere of several microns in diameter containing 10^{13} molecules. Theoretical calculations, which are generally considered more realistic, give a considerably smaller size, that is, of several dozen Å and several hundred molecules. All of this boils down to the proposition that the minimum size for efficient nucleation is for the most part smaller than the limits of visual observation. Consequently, it will not suffice to obtain a clear supersaturated solution to be quite certain that one has eliminated seeds of the enantiomer whose crystallization one wants to avoid.

In order to obtain reproducible results, it is essential to rid the system of aggregates of size greater than r_c. This requirement may consist of prolonged heating of the solution at a temperature at which it is no longer supersaturated, that is, a temperature which is distinctly higher than that at which crystallization is to take place. Depending upon the substrate and the solvent used, the duration of this preheating may range from a few minutes to tens of minutes. This simple "sterilization" procedure generally guarantees reproducible crystallization rates.[19,20] Undissolved seeds may also be eliminated along with foreign particles by filtration, ultrafiltration, or centrifugation of the solution employed.[21]

In principle, we know some of the factors tending to slow down the crystallization of the enantiomer which must remain in solution if our resolution is to succeed. Let us now examine what we need to do to favor the crystallization of the other. From the experimental point of view, the rate of growth of a crystal under a given set of conditions depends on the extent of supersaturation of the solution, $(S_t - S_A)$, and of the area $\mathscr{A}$ of the crystals exposed to the growth. An equation such as

$$-\frac{dS_t}{dt} = k\mathscr{A}(S_t - S_A)^n$$

is in most cases compatible with the kinetics observed.[12] In this equation, S_t represents the concentration of the dissolved enantiomer at time t and S_A represents its equilibrium solubility at the temperature considered; n is the order of the reaction; and k is a constant that incorporates, *inter alia*, the stirring rate. The equation makes it clear that the extent of supersaturation of the desired enantiomer should be as high as possible at the onset. In order to optimally carry out a resolution by entrainment it is necessary to begin with a substance which already is partially resolved. In practice, the usual ranges of initial optical purity and of supersaturation are both closely linked to the dimensions of the usable part of the phase diagram (points M or P, Fig. 12b).

The area $\mathscr{A}$ of the growing crystals depends upon their shape, their number N, and their total weight W. Since the weight of a spherical crystal of density d is $w = 4/3\pi r^3 d$, their number $N = W/w$, and their total area $\mathscr{A} = 4\pi r^2 N$, it follows that

$$\mathscr{A} = \frac{3W}{rd}$$

In a very general way, the total area of a group of crystals of any shape may be estimated by an expression such as

$$\mathscr{A} = \frac{fW}{ld}$$

in which f is a "geometric" constant, for example, $f = 3$ for a sphere, and l is an average linear dimension for the crystals. From this, one can deduce the requirement that, in order to attain a large initial area $\mathscr{A}_0$ for a given overall weight, the seeds should have a particle size as small as possible. In order to have reproducible kinetics over several successive cycles of crystallization, it is important to carefully systematize the conditions of seeding. This requires the accumulation of a sample of seeds for each enantiomer which is enantiomerically pure and uniformly ground and sifted.

The industrial-scale resolution of glutamic acid hydrochloride has involved the development of a special seeding technique. Formation of a very large number of optically pure microcrystalline seeds is induced by ultrasonic irradiation (10 to 100 kHz). The resulting crystallization occurs very rapidly, with the entire resolution requiring less than 15 minutes.[13]

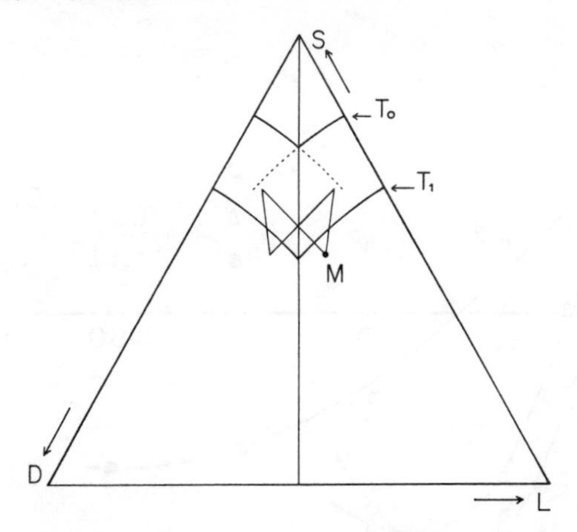

Figure 14 A. Collet, M. J. Brienne, and J. Jacques, *Chem. Rev.*, 1980, **80**, 215. Reproduced by permission of the editor. Copyright 1980, American Chemical Society.

A more classical procedure consists of the following: The solution is heated to T_1 which is a temperature slightly higher than that at which resolution is to take place. At this temperature the solution is supersaturated only with respect to the enantiomer present in excess. In this situation, seeding or spontaneous crystallization generates crystals which are necessarily optically pure and which will serve as seeds when the solution is cooled at temperature T_0 at which the resolution is to take place (see Fig. 14). The resolution of threonine illustrates this technique.[5]

Finally, let us consider the rate of stirring of the solution in which the crystallization is taking place. This is also an important variable. Schematically the growth of a crystal in solution may be broken down into a number of steps: (a) diffusion of solute molecules to the crystal–solvent interface; (b) adsorption of the molecules to the surface of the crystal, and (c) incorporation of these molecules into the growing crystal layer. At the same time, the inverse of this series of steps also takes place (dissolution). In an unstirred solution, it is often the diffusion rate, step (a), which is the limiting factor in the crystallization rate since steps (b) and (c) are rapid. The effect of stirring would then be to accelerate diffusion and consequently the growth of the crystals.

When the stirring rate is gradually increased, an increase in crystal growth is observed at first, which is usually followed by a horizontal step. The latter arises when the diffusion rate exceeds the rate of the following step, (b) or (c), which then determines the rate. The range of stirring speeds useful in controlling the crystallization rate is thus limited. The resolution of hydrobenzoin illustrates this in a particularly clear way. Figure 15 shows the change in rotation of identical solutions in the course of a resolution for different rates of stirring. The crystallization rate of the (+) enantiomer rapidly increases as the stirring rate is increased from 30 to 60 and then 115 rpm. Higher stirring rates no longer influence the crystallization; the curve corresponding to a sitrring rate of 215 rpm is identical to that of 115 rpm.

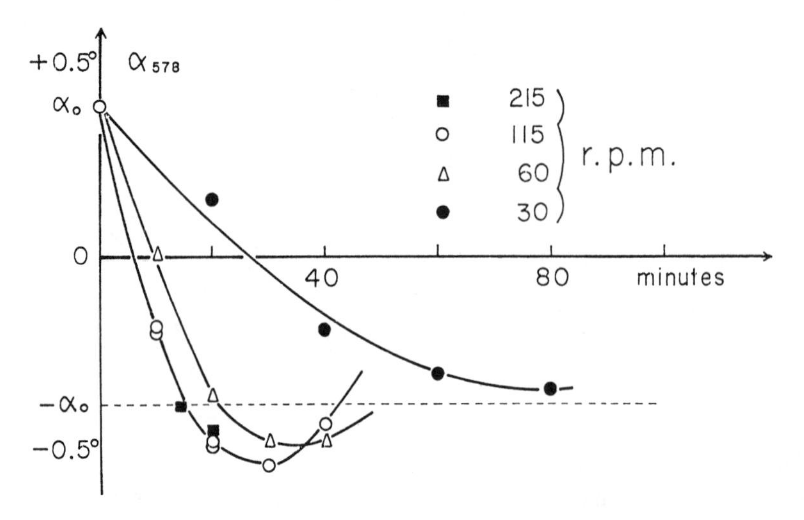

Figure 15 Resolution of hydrobenzoin by entrainment as a function of stirring rate. Initial conditions: 11.8 g/100 g solution; optical purity 3.75%. A. Collet, M. J. Brienne, and J. Jacques, *Chem. Rev.*, 1980, **80**, 215. Reproduced by permission of the editor. Copyright 1980, American Chemical Society.

Also note that the stirring rate has but a small effect on the maximum rotation between 60 and 215 rpm. Table 2 gives the times required for the rotation of the solution to drop to zero ($\alpha = 0$ at t_1) and those required to reach a value equal but opposite in sign to the starting value ($\alpha = -\alpha_0$ at t_2) which determine the optimal duration of the resolution.

An additional remark is that too great a stirring rate increases the risk of spontaneous nucleation of the other enantiomer by generating vibrations and shocks within the solution. It is pointless, therefore, if not detrimental, to increase the stirring rate beyond that which determines the maximum rate of crystallization (115 rpm in the example described).

Table 2 Duration of the resolution of hydrobenzoin as a function of stirring rate[a]

	Stirring Rate (rpm)			
	30	60	115	215
t_1 ($\alpha = 0$)	28 min	10 min	7 min	7 min
t_2 ($\alpha = -\alpha_0$)	–	21	15	15

[a] A. Collet, M. J. Brienne, and J. Jacques, *Chem. Rev.*, 1980, **80**, 215. Reproduced by permission of the editor. Copyright 1980, American Chemical Society.

REFERENCES AND NOTES 4.2

1 D. Gernez, *C. R. Acad. Sci.*, 1866, **63**, 843.

2 M. E. Jungfleisch, *J. Pharm. Chim.*, 5 ème Série, 1882, **5**, 346.

3 A. Werner, *Ber.*, 1914, **47**, 2171.

4 R. Duschinsky, *Festschrift Emil Barell. 1936*, F. Reinhardt Verlag, Basel, p. 375. *Chem. Ind. (London)*, 1934, 10.

5 G. Amiard, *Bull. Soc. Chim. Fr.*, 1956, 447.

6 S. Yamada, M. Yamamoto, and I. Chibata, *J. Org. Chem.*, 1975, **40**, 3360.

7 E. L. Eliel, *Stereochemistry of Carbon Compounds*, McGraw–Hill, New York, 1962, p. 48.

8 M. Inagaki, *Chem. Pharm. Bull.*, 1977, **25**, 2497.

9 L. Velluz, G. Amiard, and R. Joly, *Bull. Soc. Chim. Fr.*, 1953, 342.

10 L. Velluz and G. Amiard, *Bull. Soc. Chim. Fr.*, 1953, 903.

11 G. Amiard, *Experientia*, 1959, **15**, 38.

12 G. H. Nancollas and N. Purdie, *Q. Rev. Chem. Soc.*, 1964, **18**, 1.

13 French Patent 1,389,840 (1965) (to the Noguchi Research Foundation). *Chem. Abstr.*, 1965, **63**, 5740f.

14 H. H. Ting and W. L. McCabe, *Ind. Eng. Chem.*, 1934, **26**, 1201.

15 T. Watanabe and G. Noyori, *Kogyo Kagaku Zasshi*, 1969, **72**, 1083.

16 R. M. Secor, *Chem. Rev.*, 1963, **63**, 297.

17 S. Yamada, M. Yamamoto, and I. Chibata, *J. Org. Chem.*, 1973, **38**, 4408.

18 A. Van Hook, *Crystallization*, Reinhold, New York, 1961, p. 92.

19 R. Gopal, *J. Ind. Chem. Soc.*, 1947, **24**, 279.

20 A. C. Chattersi and A. N. Bose, *J. Ind. Chem. Soc.*, 1949, **28**, 94.

21 R. S. Tipson, "Crystallization and Recrystallization," in *Technique of Organic Chemistry*, 2nd ed., Vol. III, Part 1, A. Weissberger, Ed., Wiley, New York, 1956, pp. 395–562.

22 According to the *Chemical Economics Handbook*, Stanford Research Institute, August 1979, Ajinomoto Co. is the only company that has used the process commercially in the production of L-glutamic acid. See also A. Yamamoto in M. Grayson, Ed., *Kirk–Othmer Encyclopedia of Chemical Technology*, 3rd ed., Vol. 2, Wiley, New York, 1978, p. 338; and Y. Izumi, I. Chibata, and T. Itoh, *Angew. Chem.*, 1978, **90**, 187; *Angew. Chem. Int. Ed.*, 1978, **17**, 176.

There are currently (1979) at least two commercially important compounds which are resolved by the direct crystallization of enantiomers; these are L-α-methyldopa (*Chem. Eng.*, 1965, 247) and *l*-menthol, the latter via the benzoate ester (*Chem. Eng.*, 1978, 62).

23 W. Ostwald, *Lehrbuch der Allgemeinen Chemie*, 2nd ed., Vol. 2, Engelmann, Leipzig. 1903.

4.3 RESOLUTION BY ENTRAINMENT IN A SUPERCOOLED MELT

Just as is true for resolutions by entrainment from solution, so is entrainment in the molten state, that is, for a supercooled conglomerate, an entirely valid alternative for the resolution of enantiomers. This type of resolution has been carried out in only a few instances.

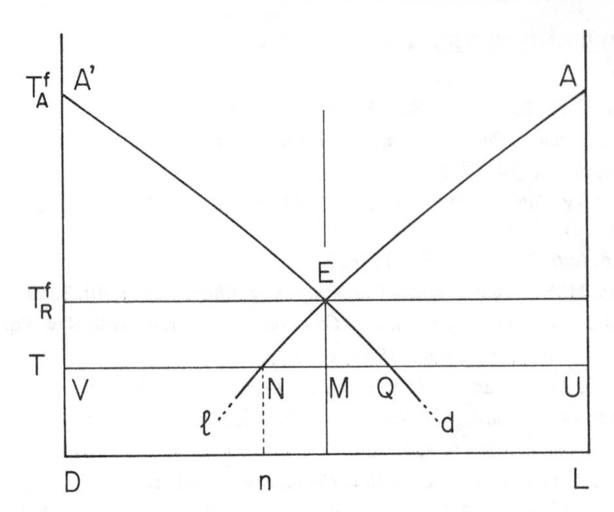

Figure 1 Resolution by entrainment in a supercooled melt. A. Collet, M. J. Brienne, and J. Jacques, *Chem. Rev.*, 1980, **80**, 215. Reproduced by permission of the editor. Copyright 1980, American Chemical Society.

4.3.1 Theory

The theory of this process may easily be deduced from the binary diagram in Fig. 1. The liquidus curves AE and $A'E$, which correspond to stable equilibria, retain their meaning beyond point E. Point N, on AEl, indicates the composition of the liquid at equilibrium with crystals of L at temperature T. Of course, N describes a metastable equilibrium which may subsist only as long as crystals of D are absent. In practice, this condition is almost always met if the extent of supercooling ($T_R^f - T$) is not too large. In a conglomerate which has been supercooled to temperature T (point M), seeding with enantiomer L, for example, will induce the crystallization of the latter. The point describing the composition of the liquid is displaced from M toward the equilibrium curve at N. For 1 mole of racemate, one obtains $X_s^1 = MN/NU$ moles of pure crystals of L and $X_l^1 = MU/MN$ moles of liquid enriched with respect to D (of composition n). If the X_l^1 moles of supercooled liquid N are seeded with D after separation of the crystals of L, D crystallizes right up to the point where the liquid attains the composition Q on the liquidus curve $A'Ed$. The amount of the crystals equals $X_s^2 = X_l^1(NQ/VQ)$, and that of the liquid (now enriched in L) equals $X_l^2 = X_l^1(NV/VQ)$. Seeding of this X_l^2 mole of liquid Q by enantiomer L would lead to crystals of L in liquid N, and so on. Alternate crystallization of (+) and (−) enantiomers would lead, in principle, to the exhaustion of all of the initial racemate.

The yield of each crystallization depends upon the ratio $\rho = NQ/VQ$ which increases with the extent of supercooling ($T_R^f - T$). This is a problem similar to that observed in the case of resolution in solution: as the degree of supercooling is increased, that is, as the temperature is lowered, the risk of spontaneous nucleation by the nonseeded enantiomer is increased. What is required is to find optimal con-

Table 1 Theoretical maximum resolution yield as a function of the degree of supercooling,[a] $\Delta T = T_R^f - T$

ΔT (°C)	ρ	Crystals of Enantiomers Isolated (mol) X_s^1	X_s^2	X_s^3	X_s^4	X_s^5	X_s^6	Residue X_l^6
		L	D	L	D	L	D	
1	0.031	0.016	0.031	0.030	0.029	0.029	0.028	0.837
2	0.073	0.037	0.070	0.065	0.060	0.056	0.052	0.660
3	0.113	0.057	0.107	0.095	0.084	0.074	0.061	0.518
4	0.152	0.076	0.140	0.119	0.101	0.086	0.073	0.405
5	0.219	0.110	0.195	0.152	0.119	0.093	0.073	0.259

[a] Liquidus curve of the binary diagram calculated with the equation of Schröder–Van Laar with $T_A^f = 60°C$, $\Delta H_A^f = 4\ \text{kcal} \cdot \text{mol}^{-1}$; the calculated T_R^f is 26°C (see Section 2.2.3). A. Collet, M. J. Brienne, and J. Jacques, *Chem. Rev.*, 1980, **80**, 215. Reproduced by permission of the editor. Copyright 1980, American Chemical Society.

ditions of temperature and of stirring which allow the maximum harvest of pure crystal in the briefest possible time and without disturbing the metastable equilibrium upon which the entire process depends.

Table 1 gives the theoretical maximum yield X_s^i of crystalline L and D obtainable from 1 mole of racemate subjected to six successive crystallizations, and this as a function of the degree of supercooling. The calculations were carried out for a specific case; however, the import of the results is general. Beginning with the racemate, the first crystallization (say of L) would give $X_s^1 = \rho/2$ (where $\rho = NQ/VQ$ in Fig. 1) and $X_l^1 = 1 - X_s^1$. Subsequently, $X_s^2 = \rho X_l^1$, and $X_l^2 = X_l^1 - X_s^2$, and so on. The last column of the table gives the amount of liquid remaining after six crystallization (X_l^6). Examination of the data shows that the yield increases very rapidly as a function of the degree of supercooling. Thus, 1° below T_R^f, the total amounts of L and D formed during six crystallizations are, respectively, 0.075 and 0.088 mole and there remains 0.837 mole of liquid. At a temperature 3° below T_R^f, these amounts become 0.226 and 0.252 mole, respectively, of L and D, and 0.518 mole of residual liquid. Otherwise stated, with 3° of supercooling *half* of the racemate taken could be used up in only six steps.

4.3.2 Application of the Procedure

The first examples of the application of resolution by entrainment in a supercooled melt are apparently those described by Jensen[1] in 1970. The publication is a patent dealing with resolution of β-lactam derivatives **1**, **2**, and **3**. These compounds form conglomerates having low melting points and which can be easily maintained in a supercooled state.

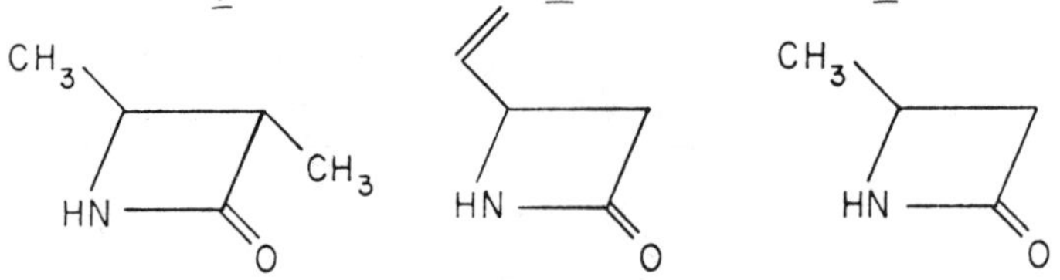

	1	2	3
T_A^f	60.6°	26.8°	26.7°
T_R^f	23.7°	-1°	-12.4°

The apparatus employed allows the resolution to be carried out semiautomatically. The supercooled conglomerate is placed in equal parts in the vessels 1 and 2 (Fig. 2) which are fitted with stirrers and thermostatted at a temperature slightly lower than T_R^f. At first, seeds of enantiomers L are placed in vessel 1 and seeds of D

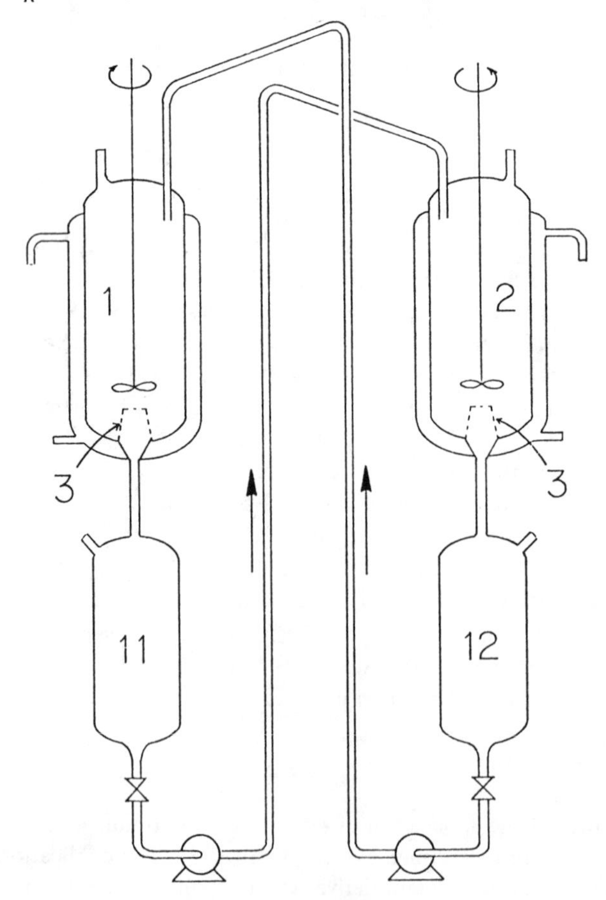

Figure 2 Apparatus for resolution by entrainment in a supercooled melt. H. H. Jensen, *Ger. Offen.* 1,807,495 (1970). Reproduced by permission of Dr. H. Jensen.

in vessel 2. After a certain time during which crystallization takes place, the liquid phases in each compartment are separated by filtration through filters 3 from the crystals formed and collected in vessels 11 and 12. The liquid in 11, containing a slight excess of enantiomer D, is transferred to vessel 2 in which the first formed crystals of D remain, while the liquid in 12 is simultaneously transferred to vessel 1 where it comes in contact with crystals of L. After another period of time to allow further crystallization, the same operations (filtration and transfer of the liquids) are carried out again, and so on. Enantiomer L will eventually be collected from compartment 1 and D from compartment 2.

In the resolution of β-lactam **1**, 250 g racemate is divided between vessels 1 and 2 and supercooled to 20°C (corresponding to 3.7° below T_R^f). After seven stages of crystallization, each lasting 30 minutes, about 100 g of the two enantiomers having optical purities around 85% is collected. There remains 150 g of nearly racemic substance. The yield of enantiomers is estimated to be $80\,g \cdot L^{-1}$ of racemate per hour.

Fleischer et al. have described[2] the resolution of supercooled menthyl benzoate by entrainment (T_A^f 54.5°C, T_R^f 24.5°C). The technique they employed differs from the preceding one by being more closely related to that of resolution by entrainment in solution.

REFERENCES 4.3

1 H. Jensen, German Offen. 1,807,495 (1970) (to Farbwerke Hoechst. A.-G.). *Chem. Abstr.*, 1970, **73**, 77222p.

2 J. Fleischer, K. Bauer, and R. Hopp, German Offen. 2,109,456 (1972) (to Haarmann and Reimer GmbH). *Chem. Abstr.*, 1972, **77**, 152393h.

4.4 CRYSTALLIZATION IN OPTICALLY ACTIVE SOLVENTS

In this section we have summarized data concerning the crystallization of enantiomeric substances in optically active solvent and in achiral solvents containing variable amounts of optically active co-solutes.

4.4.1 Solubility of Pure Enantiomers in Optically Active Solvents

In a letter dated 1893, Van't Hoff wrote to Meyerhoffer that it should be possible to find a difference in the solubility of enantiomers in an optically active solvent.[1]* He envisaged the use of this property as a new resolution method. Recently, Amaya[2] has justified this solubility difference on the basis of theory. However, no order of magnitude was given nor was experimental support provided.

*In a footnote in his pamphlet *Gleichgewichte der Stereomeren*, (1906, p. 54), Meyerhoffer gives interesting historical details regarding this correspondence: "At that time, Professor Van't Hoff communicated to me that LeBel might already have carried out experiments in this direction, and that previously Pasteur also conceived this possibility. He advised me to work with the sodium ammonium racemate in a solution of sugar water. I undertook experiments but I did not pursue them."

The older and more recent experiments designed to reveal a difference in solubility between enantiomers in an optically active solvent can clearly be divided into two groups:

1 All the measurements concerning *stable* enantiomers dissolved in an inert optically active solvent were negative in outcome. Jones[3] found identical soluilities for (+)- and (—)-camphor and (+)- and (—)-camphoroximes in (+)-pinene and (—)-amyl bromide. Goldschmidt and Cooper[4] reached the same conclusion with (+)- and (—)-carvoximes in (+)-limonene. Nor do the more recent experiments of Ebert and Kortum[5] detect a significant difference between the solubilities of (+)- and (—)- sodium potassium tartrates in an aqueous solution of mannitol.

2 In contrast with these data, a fair number of experiments carried out with ionic organometallic complexes in hydroxylic optically active solvents (or in achiral solvents containing optically active ions) have shown more or less large differences in solubility for *d* and *l* species. Bosnich and Watts[6] have reported that the solubilities of (+)- and (—)-*cis*-$[Co(en)_2Cl_2]ClO_4$ in (—)-2,3-butanediol at 30°C are 2.6 and $1.25 \times 10^{-3}\,mol \cdot L^{-1}$, respectively. Solubility differences of the same order of magnitude were found for (+)- and (—)-$[Ru(phen)_3](ClO_4)_2$ and for (+)- and (—)-$[Ru(bipy)_3](ClO_4)_2$ in (—)-2-methyl-1-butanol,[7] and a surprisingly high solubility ratio (+)/(—) = 4.2 was reported for enantiomeric $[Co(en)_3]Cl_3$ in (+)-diethyl tartrate at 25°C.[8] These differences are ascribed to the formation of strong diastereomeric associations between the solvent and the (+) and (—) ions. It is noteworthy that enantiomeric complexes exhibiting such solubility differences are also those which may undergo "enantiomerization" in the presence of optically active solvent. According to the authors just cited,[6–8] it is assumed that no enantiomerization has taken place during the course of the solubility measurements, however. In our opinion, we cannot exclude the possibility that the reported differences do not really reflect solubility equilibria. Thus we do not completely rule out the operation of kinetic phenomena (dissolution or crystallization rates) in these experiments.

4.4.2 Resolution Experiments

In contrast to the results cited in (1) above, there are a number of examples reported in the literature where recrystallizations of racemates in optically active solvents lead to induction of partial resolution. The oldest systematic experiments of this type are probably those of Kipping and Pope (1899).[9] They observed that, in the crystallization of sodium ammonium tartrate in dextrose solution, the first precipitate contains an excess of (+)-salt, even though the solubilities of the two salts are identical within the precision of the measurements.[5] Similarly, Ostromisslenskii reported in 1908 that the same (+)-sodium ammonium tartrate preferentially crystallized from an aqueous solution of the racemate in the presence of (—)-sodium ammonium malate as chiral co-solute.[10] More recently, Lüttringhaus and Berrer obtained diols **1** and **2** in optically active form through simple crystallization in

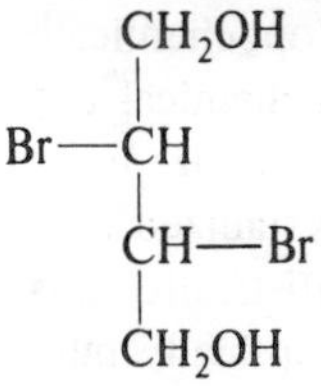

CH_2OH
Br—CH
CH—Br
CH_2OH

1

N (pyridyl)
CH—OH
HO—CH
N (pyridyl)

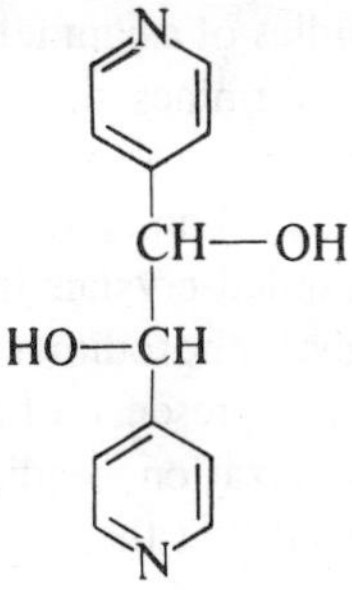

2

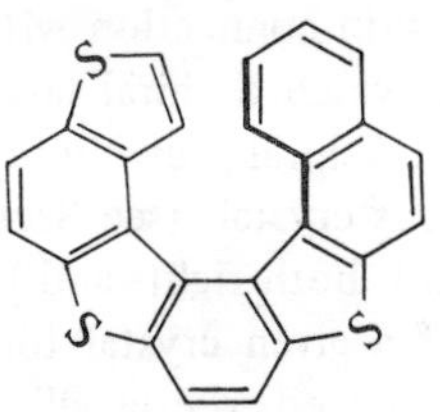

3

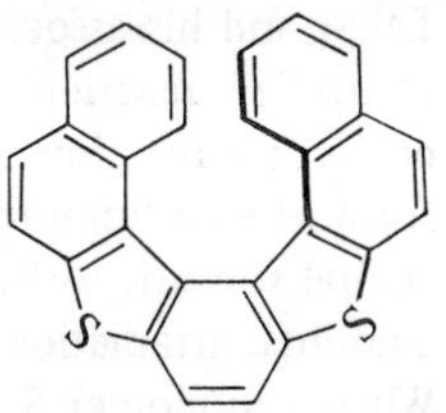

4

isopropyl *d*-tartrate,[11] and Wynberg et al. resolved the heptaheterohelicene **3** by crystallization in (—)-α-pinene.[12]

It is noteworthy that these examples of successful resolutions of racemates are conglomerates. Indeed, Wynberg has indicated that the heptaheterohelicene **4**, which exists as a racemic compound, is not resolved by crystallization in (—)-α-pinene.[12]

4.4.3 Origin of the Preferential Crystallization Phenomenon

Although we are unaware of the existence of experimental ternary diagrams (D,L,S*) describing systems of the above type, the fact that the solubilities of *pure* enantiomers in optically active solvents S* are very close, if not equal, indicates that the ternary diagrams would be nearly symmetric. That is, the eutectic of the conglomerate system would be very close to the 1–1 (racemic) composition in the ternary diagram. A very small dissymmetry cannot account, under equilibrium conditions, for the relatively large amounts of resolved substances obtained in the examples described above. It is quite likely that the results observed reflect the operation of a *kinetic* phenomenon, and are related to resolution by preferential crystallization which is discussed in Section 4.2.

The chirality of the solvent or of a cosolute can influence the rate of crystallization of the substrate in several ways, depending, *inter alia*, on the nature, the stereospecificity, and the strength of the solvent-substrate interactions. One possible mechanism involves intervention of the solvent at the *nucleation* step (see Section 4.2.7), leading to "chiral seeding."[10] There is, at present, no experimental support for this hypothesis.

The chiral solvent or cosolute can also play a role in modifying the rate of growth of enantiomeric crystals. *Adsorption* of a chiral cosolute on the growing crystals has been invoked by Barton and Kirby[13] as a possible cause of the preferred crystallization of one enantiomer.* This hypothesis has been adopted recently by

*These authors isolated the alkaloid (+)-narwedine during the crystallization of the racemate in the presence of (—)-galanthamine as the chiral cosolute. This experiment is complicated by the simultaneous racemization of narwedine in the solution leading to a second order asymmetric transformation.

Lahav and his associates, in connection with studies of asymmetric topochemically controlled reactions in which a chiral lattice determines the stereochemical outcome of a reaction.[14, 15] Achiral *p*-divinylbenzenes **5** are known to crystallize in the form of enantiomorphous crystals (see Section 1.2.3) and, when crystallized in an achiral solvent, to furnish both right- and left-handed crystals in statistically equal amounts. Irradiation of a given crystal form gives photodimers **6**, optically pure. When a monomer **5** is allowed to crystallize in the presence of any of the resolved photodimers **6**, one observes asymmetric crystallization, leading to 30 to 100% excess of crystals of **5** of one handedness, provided that the inducing additive itself pack in a crystal form having the same three-dimensional network as the monomer crystal.

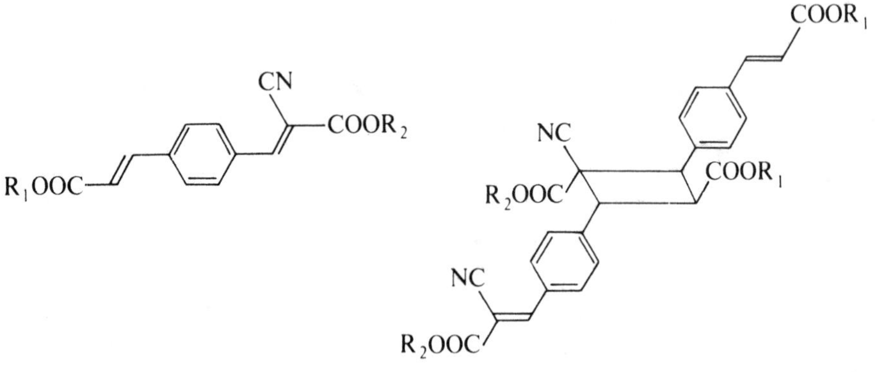

CN COOR$_2$ R$_1$OOC

COOR$_1$ NC R$_2$OOC COOR$_1$ NC R$_2$OOC

a. R$_1$ = 3-pentyl R$_2$ = Me
b. R$_1$ = 3-pentyl R$_2$ = Et
c. R$_1$ = 3-pentyl R$_2$ = *n*-Pr
d. R$_1$ = *sec*-butyl R$_2$ = Et
e. R$_1$ = *sec*-butyl R$_2$ = *n*-Pr

A significant finding was that the enantiomorphous crystals precipitating in excess in the presence of a (+)-cyclobutane additive always led, upon irradiation, to a (−)-cyclobutane dimer and vice versa. These observations led Lahav et al. to conclude[15] that the preferential crystallization is due to a stereoselective adsorption of the additive on the surface of one of the two growing enantiomorphous phases. In fact, the (+)-cyclobutane additive preferentially adsorbs on that monomer crystal which, on irradiation, will give a (+)-cyclobutane dimer. This selective adsorption effectively inhibits, or at least decreases, the crystal growth of that enantiomorphous crystal form. As a consequence, crystals of the other enantiomorphous form grow preferentially and dominate the precipitate.

A series of experiments has been carried out by the Lahav group to test the mechanism, which for the time being requires the use of selected or "tailor-made" additives. For example, racemic threonine has been crystallized in the presence of 10% (S)-glutamic acid. Rapid crystallization leads to deposition of (2R,3S)-threonine [D-threonine, which is configurationally related to (R)-glutamic acid] with an enantiomeric excess as high as 94%. The second crop is enriched in L-threonine. Crystallization with (R)-glutamic acid leads to a deposition of L-threonine. As expected, dissolved (S)-glutamic acid inhibits the *dissolution* of L-threonine from crystalline racemic threonine and the remaining crystals are enriched in L-threonine.

This explanation is consistent with the earlier findings of Ostromisslenskii regarding the crystallization of (R,R)-sodium ammonium tartrate from the racemate in the presence of (S)-sodium ammonium malate,[10] and those of Purvis, who reported the preferential crystallization of (R)-glutamic acid in the presence of (S)-aspartic acid,[16] as well as those of Green and Heller to the effect that crystallization of achiral 4,4′-dimethylchalcone in the presence of its chiral bromination product (Section 1.2.3) led to preferential crystallization of those chalcone crystals that would give the bromine additive of opposite configuration.[17]

It is noteworthy that the effect of *soluble* additives is inverse to that of crystalline substances acting as seeds in a resolution by entrainment (Section 4.2). In the latter case, seeding always leads to the growth of that enantiomorphous form which is identical or isomorphous with that of the seed crystals.

A corollary of the above is that stereoselective adsorption occurring during crystal growth leads to morphological changes at one or more of the crystal faces of the affected enantiomer which, in many cases, are quite easily visible while the other enantiomer is almost unaffected. Such changes may be the basis of modified Pasteurian resolutions of conglomerates by triage (Section 4.1.1).[15]

4.4.4 Resolution with Inclusion of an Optically Active Solvent

A somewhat different resolution involving an optically active solvent has been described by Mislow.[18] The racemic tetraarylethane 7 which forms crystalline inclusion compounds with benzene or with cyclohexylamine, among other solvents, also forms a 1–1 complex by crystallization in (+)-α-pinene. After removal of the solvent, the complex furnishes partially resolved 7 with $[\alpha]_{356}^{27} = -4.7°$ ($CHCl_3$). The unknown optical purity is believed to be low, and it does not increase upon successive recrystallizations.

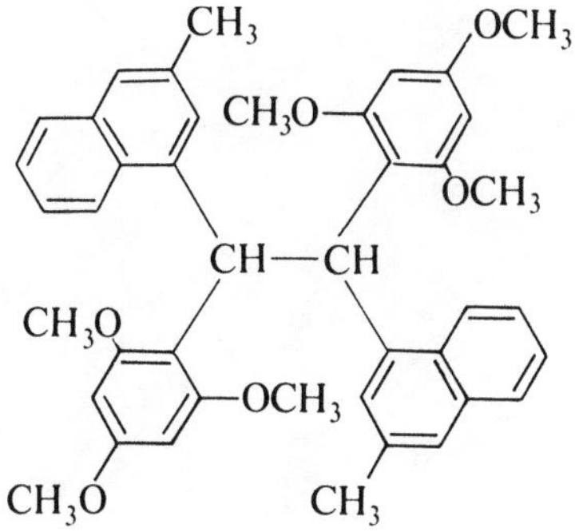

7

It is known that certain racemates may be resolved by inclusion in an enantiomorphous crystal (Section 5.1.8). According to Mislow, the above results correspond to the inverse of this type or resolution since racemic **7** constitutes the host molecules which are resolved by inclusion of optically active guest molecules in their lattice.

REFERENCES 4.4

1 J. H. Van't Hoff, *Die Lagerung der Atome im Raume*, Vieweg, Braunschweig, 2nd ed., 1894, p. 30; 3rd ed., 1908, p. 8.

2 K. Amaya, *Bull. Chem. Soc. Jpn.*, 1961, **34**, 1803.

3 H. O. Jones, *Proc. Cambridge Philos. Soc.*, 1907, **14**, 27; quoted by L. Ebert and G. Kortüm (ref. 5); see also E. Schröer, *Ber.*, 1932, **65**, 966.

4 H. Goldschmidt and M. C. Cooper, *Z. Phys. Chem.*, 1898, **26**, 711.

5 L. Ebert and G. Kortüm, *Ber.*, 1931, **64**, 342.

6 B. Bosnich and D. W. Watts, *J. Am. Chem. Soc.*, 1968, **90**, 6228.

7 K. Mizumachi, *J. Coord. Chem.*, 1973, **3**, 191.

8 M. Yamamoto and Y. Yamamoto, *Inorg. Nuclear Chem. Lett.*, 1975, **11**, 833.

9 F. S. Kipping and W. J. Pope, *Proc. Chem. Soc. London*, 1897–1906, 113. *J. Chem. Soc.*, 1898, **73**, 606.

10 L. Ostromisslenskii, *Ber.*, 1908, **41**, 3035.

11 A. Lüttringhaus and D. Berrer, *Tetrahedron Lett.*, 1959, 10.

12 M. B. Groen, H. Schadenberg, and H. Wynberg, *J. Org. Chem.*, 1971, **36**, 2797.

13 D. H. R. Barton and G. W. Kirby, *J. Chem. Soc.*, 1962, 806.

14 (a) L. Addadi and M. Lahav, *J. Am. Chem. Soc.*, 1978, **100**, 2831. (b) *ibid.*, 1979, **101**, 2152. (c) *Pure Appl. Chem.*, 1979, **51**, 1269.

15 J. van Mil, E. Gati, L. Addadi, and M. Lahav, submitted for publication (1981). We are greatly indebted to Professor M. Lahav for permitting us to cite this work prior to publication.

16 J. L. Purvis, U.S. Patent 2,790,001 (1957); *Chem. Abstr.*, 1957, **51**, 13911a.

17 B. S. Green and L. Heller, *Science*, 1974, **185**, 525.

18 K. S. Hayes, W. D. Hounshell, P. Finocchiaro, and K. Mislow, *J. Am. Chem. Soc.*, 1977, **99**, 4152.

CHAPTER 5

Formation and Separation of Diastereomers

In the preceding chapter we have principally examined methods for the resolution of enantiomers that do not require the intervention of chiral agents. We shall now examine those processes which depend on the formation of *diastereomeric compounds* derived from the enantiomers to be separated. Unlike enantiomers, diastereomer pairs may have significantly different physical properties which may be the basis of their separation from one another. We consider, in particular, *crystalline* diastereomeric compounds. We examine two broad categories in succession: *dissociable compounds*, or *complexes*, and *covalent compounds*. This classification is convenient even if somewhat arbitrary.

The number of resolutions mediated by diastereomers described in the literature is quite large, and we have not felt it necessary to cite all examples known to us. The cases cited are representative and cover the principal resolving agents and functional groups.

While covalent diastereomers are increasingly separated by chromatography, the separation of other types of diastereomeric substance depends entirely on crystallization techniques that are based upon differences in solubility. Thus, in this chapter we apply several of general concepts developed earlier and, in particular, those involving the use of phase diagrams.

Before taking up these matters, let us briefly examine the nontrivial matter of diastereomer specification and of the way in which the different mixtures derived from diastereomers may be distinguished.

The bimolecular combination of two chiral substances A and B may lead to four diastereomers (Scheme 1). We have adopted the terminology in which the letter p is used to designate the diastereomers resulting from reaction of the two constituents having like *sign* of rotation and the letter n to designate the diastereomers formed from constituents of unlike sign. This convention stems from a suggestion made by I. Ugi (*Z. Naturforsch.*, 1965, **20b**, 405) for covalent compounds possessing but two chiral centers and which is based on the nomenclature of Cahn, Ingold, and Prelog: $RR = SS = p$ and $RS = SR = n$. In our convention, which is applicable to all types of dissociable as well as to covalent diastereomers, no account is taken of the absolute configurations of the chiral centers, however.

The p and n designations take into account only the signs of the rotations of species A and B.* In the case of diastereomeric salts, for example, reference may be made to Tables 1 and 2 in Section 5.1, which give the signs of the rotatory power of the principal alkaloids and the naturally occurring acids used in resolutions.

SCHEME 1

①	$dAdB = p_+$ (or p_-)	③	$dAlB = n_+$ (or n_-)
②	$lAlB = p_-$ (or p_+)	④	$lAdB = n_-$ (or n_+)

Given the above, the four diastereomers $[AB]$ are found to consist of two enantiomeric p compounds, p_+ and p_- and, by the same token, two enantiomeric n compounds, n_+ and n_-. It should be evident that the sign of rotation of a given diastereomer p or n is not necessarily related directly to those of its constituents A and B. In Scheme 1, the diastereomers $dAdB$ and $dAlB$ have been arbitrarily designated as p_+ and n_+; they could just as well (experimentally) have been found to be p_- and n_-, or even p_- and n_+, for example.

In a resolution, which brings into play a racemic substrate and a resolving agent which is by definition a single enantiomer, the formation of diastereomers leads to a mixture of only *two* compounds: p and n. It is important to observe, as Scheme 2 makes clear, that the mixtures derived from racemic A and optically active B (case ①) are not identical to those derived from the inverse operation, namely, racemic B and optically active A (case ②). In one case, the diastereomers p and n have the same sign, while in the other they have unlike signs.

SCHEME 2

① dlA $\xrightarrow{dB}$ $dAdB + lAdB$ (e.g., p_+, n_-)

dlA $\xrightarrow{lB}$ $dAlB + lAlB$ (n_+, p_-)

② dlB $\xrightarrow{dA}$ $dBdA + lBdA$ (p_+, n_+)

dlB $\xrightarrow{lA}$ $dBlA + lBlA$ (n_-, p_-)

The consequence of this lack of symmetry between the two cases is taken up in Section 5.1.16. In the sections which follow, we generally deal with mixtures of diastereomeric pairs (p,n) without need of further specification of their sign of rotation.

* In those cases – fortunately relatively rare – in which the sign inverts upon a change in solvent, it is necessary to stipulate the solvent used (preferably that solvent in which the salts are best formed).

5.1 DISSOCIABLE COMPOUNDS AND COMPLEXES

The most widely used resolution method remains the formation and separation of crystalline diastereomeric *salts* between racemic substrates and optically active resolving agents. Other usable dissociable crystalline combinations do exist, nonetheless; these are Lewis acid–base complexes, inclusion compounds, and quasiracemates. These diverse addition compounds have in common their ease of usage. They are obtained generally by simple mixing of the constituents in an appropriate solvent. Regeneration of the constituents is most often immediate and the resolving agent is almost always recovered in a form that allows its reuse.

The resolution method consisting of the formation of a salt between a racemic acid and an optically active base was discovered by Pasteur[1–3] in 1853[1]:

> I have shown that the absolute identity of the physical and chemical properties of nonsuperposable right and left bodies ceased to exist when these substances were put in the presence of [optically] active bodies. Thus, the right and left tartrates of the same [optically] active organic base are entirely distinct in their crystalline forms, in their solubility, etc. . . ; it was thus to be hoped that one could take advantage of this difference to isolate the two tartaric acids which comprise the racemate: after much fruitless research attempted on various bases, this is the service done by the two bases quinicine and cinchonicine. When, for example, one prepares the racemate of cinchonicine [i.e., in modern terms the cinchotoxine salt of racemic tartaric acid], then for a given concentration of the solution it is always the case that the first crystallization consists for the most part of left tartrate of cinchonicine; the right tartrate remains in the mother liquor. A similar result is obtained with quinicine; however, in this case it is the right tartrate which deposits at first. Thus, when a binary composition analogous to that of racemic acid be suspected, its resolution should be attempted by placing it in the presence of an [optically] active product which, as a consequence of the necessary dissimilarity of the properties of the combinations which it will be possible to make from the components of the complex group, will allow the separation of the latter.

The process may be summarized by Scheme 1, which corresponds to the treatment of a racemic acid *dlAH* with an optically active base to form salts *p* and *n* or to its inverse (racemic base and active acid).

SCHEME 1

$$dlAH \xrightarrow{dB} \underbrace{\{dA^-,dBH^+\}}_{p\ \text{salt}} + \underbrace{\{lA^-,dBH^+\}}_{n\ \text{salt}}$$

When they are prepared separately from previously resolved components, salts *n* and *p* have different crystalline forms and frequently also different degrees of solvation. The possibility of separating such diastereomeric salts when they are allowed to crystallize from a mixture of a racemate and an optically active resolving agent presupposes the occurrence of a number of conditions all present together which we examine in the following sections: salts *p* and *n*, or at least one of these, must be crystalline; their solubilities must differ; they must not cocrystallize (form solid solutions); and they must not form double salts (addition compounds [*p*, *n*]).

While any optically active acid or base is in principle usable, its use as a resolving agent is limited by its availability and cost.

The naturally occurring alkaloids, which were virtually exclusively utilized for about 100 years, are still much utilized in the resolution of acids. They are gradually being displaced by synthetic bases and by derivatives of natural products. The totally synthetic bases are, most often, primary amines and consequently are stronger bases than the common alkaloidal resolving agents, which are all tertiary amines and this may in some instances facilitate salt formation. On the other hand, synthetic bases all have the disadvantage that they themselves need to be resolved. This resolution, however, furnishes both enantiomers, which are thus available for use as resolving agents, quite unlike what obtains with alkaloids.*

The use of diastereomeric salts in resolutions of acids or bases in preference to that of covalent diastereomers is traditional. It stems from the simplicity with which diastereomeric salts are formed and from the ease of their cleavage to resolution substrates.

The frequency of use of the several basic resolving agents is quite unequal. For some 230 cases of resolutions of acids described in the literature between 1960 and 1970,[4] about one third were carried out with brucine and quinine. During the same interval, tartaric acid and its derivatives accounted for about half of all resolving agents used in the resolution of bases.

The cost of a resolving agent is also of some interest, although it is clearly not an independent variable since one does not necessarily have the option of choice. Brucine, cinchonidine, cinchonine, strychnine, dehydroabietylamine, (+)- and (−)-ephedrine, (−)-2-amino-1-butanol, (+)- and (−)-α-methylbenzylamine, (+)-amphetamine, and (+)-deoxyephedrine are the least expensive bases available commercially; each cost less than $100 per mole in 1979.

Comparable least expensive resolving acids commercially available are: (+)-camphor-10-sulfonic acid, (+)-camphoric acid, (−)-dibenzoyltartaric acid, diacetoneketogulonic acid, (+)- and (−)-mandelic acid, (−)-malic acid, and (+)- and (−)-tartaric acid. Each of these cost less than $90 per mole in 1979.

Tables 1 and 2 list the principal bases and acids used as resolving agents through salt formation. Virtually all are commercially available. The leading firms furnishing organic compounds for laboratory use also supply most of the resolving agents on a relatively small scale and at relatively high prices. There are also suppliers more or less specialized in certain types of compounds (e.g., alkaloids, camphor derivatives, or synthetic amines for pharmaceutical use) who may furnish kilogram quantities of some resolving agents at lower prices.

The practical aspects of the use of resolving agents, which are only briefly given in the following sections, are taken up in Chapter 7. The purification of resolving agents and the cleavage of diastereomeric salts are discussed in Section 7.4.

*Quinine and quinidine, on one hand, and cinchonidine and cinchonine, on the other, may be considered as pairs of quasi-enantiomers. Their use in the resolution of acids so as to lead to both substrate enantiomers is discussed in Section 5.3.2.

Table 1 Principal bases used as resolving agents via salt formation

Formula No.	Name	Mol. Formula	Mol. Weight	$[\alpha]_D$ (deg)
1	Brucine	$C_{23}H_{26}N_2O_4$	394.4	− 127 ($CHCl_3$)
2	Strychnine	$C_{21}H_{22}N_2O_2$	354.4	− 139 ($CHCl_3$)
3	Quinine	$C_{20}H_{24}N_2O_2$	324.4	− 117 ($CHCl_3$)
4	Quinidine	$C_{20}H_{24}N_2O_2$	324.4	+ 230 ($CHCl_3$)
5	Cinchonidine	$C_{19}H_{22}N_2O$	294.4	− 109 (EtOH)
6	Cinchonine	$C_{19}H_{22}N_2O$	294.4	+ 229 (EtOH)
9	Yohimbine	$C_{21}H_{26}N_2O_3$	354.4	+ 108 (py)
10	Morphine	$C_{17}H_{19}NO_3$	285.3	− 132 (MeOH)
11	Dehydroabietylamine	$C_{20}H_{31}N$	285.5	+ 46 (MeOH)
13	Ephedrine (−), (+)	$C_{10}H_{15}NO$	165.2	± 6.3 (EtOH)
14	Deoxyephedrine (+), (−)	$C_{10}H_{15}N$	149.2	± 17.9 (H_2O, hydrochloride)
15	Amphetamine (+), (−)	$C_9H_{13}N$	135.2	± 38 (C_6H_6)
16	*threo*-2-Amino-1-(*p*-nitrophenyl)-1,3-propanediol (+), (−)	$C_9H_{12}N_2O_4$	212.2	± 22.6 (MeOH)
17	*threo*-2-(*N,N*-Dimethylamino)-1-(*p*-nitrophenyl)-1,3-propanediol (+), (−)	$C_{11}H_{16}N_2O_4$	240.3	± 26 (EtOH)
18	*threo*-2-Amino-1-phenyl-1,3-propanediol (+), (−)	$C_9H_{13}NO_2$	167.2	± 18 (H_2O)
20	α-Methylbenzylamine (+), (−)	$C_8H_{11}N$	121.2	± 40 (neat)
21	α-(1-Naphthyl)ethylamine (+), (−)	$C_{12}H_{13}N$	171.2	± 82 (neat)
22	α-(2-Naphthyl)ethylamine (+), (−)	$C_{12}H_{13}N$	171.2	± 19 (neat)

Table 2 Principal acids used as resolving agents via salt formation

Formula No.	Name	Mol. Formula	Mol. Weight	$[\alpha]_D$ (deg)
23	Tartaric acid (+), (–)	$C_4H_6O_6$	150.1	± 12 (H_2O)
24	*O,O′*-Dibenzoyltartaric acid (+), (–)	$C_{18}H_{14}O_8$	358.3	± 118 (EtOH)
25	*O,O′*-Di-*p*-toluoyltartaric acid (+), (–)	$C_{20}H_{18}O_8$	386.4	± 140 (EtOH)
26	2-Nitrotartranilic acid (+), (–)	$C_{10}H_{10}N_2O_7$	270.2	± 90 (H_2O)
27	Mandelic acid (+), (–)	$C_8H_8O_3$	152.2	± 157 (H_2O)
28	Malic acid (+), (–)	$C_4H_6O_5$	134.1	± 2.3 (H_2O)
29	2-Phenoxypropionic acid (+)	$C_9H_{10}O_3$	166.2	+ 40 (EtOH)
30	Hydratropic acid (+), (–)	$C_9H_{10}O_2$	150.2	± 79 (EtOH)
31	*N*-Acetylleucine (–), (+)	$C_8H_{15}NO_3$	173.2	± 24 (MeOH)
32	*N*-(α-Methylbenzyl)succinamic acid (+), (–)	$C_{12}H_{15}NO_3$	221.3	± 112 (EtOH)
33	*N*-(α-Methylbenzyl)phthalamic acid (+), (–)	$C_{16}H_{15}NO_3$	269.3	± 48 (EtOH)
34	Camphor-10-sulfonic acid (+)	$C_{10}H_{16}O_4S$	232.3	+ 103 (H_2O)
35	3-Bromocamphor-9-sulfonic acid (+), (–)	$C_{10}H_{15}O_4SBr$	311.2	± 86 (H_2O)
36	Camphor-3-sulfonic acid (+)	$C_{10}H_{16}O_4S$	232.3	+ 103 (H_2O, 546 nm)
37	Quinic acid (+), (–)	$C_7H_{12}O_6$	192.2	± 43 (H_2O)
38	Di-*O*-isopropylidene-2-oxo-L-gulonic acid (–)	$C_{12}H_{18}O_7$	274.3	– 20.4 (MeOH)
39	Lasalocid (–)	$C_{34}H_{54}O_8$	590.8	– 39 ($CHCl_3$)
40	1,1′-Binaphthyl-2,2′-phosphoric acid (+), (–)	$C_{20}H_{13}O_4P$	348.3	± 580 (MeOH)
41	Cholestenonesulfonic acid	$C_{27}H_{44}O_4S$	464.7	–

5.1.1 Resolution of Acids

Brucine, **1**, remains one of the most widely used alkaloids for the resolution of acids in spite of its toxicity. The commercial product is generally hydrated (di- or tetrahydrate) and must be dried (at ca. 120°C) prior to use.

Strychnine, **2**, though much less used, resembles brucine in usage. Among the four alkaloids isolated from cinchona bark, **3**, **4**, **5**, and **6**, the ones most commonly used in the resolution of acids are quinine, **3**, and cinchonidine, **5**.

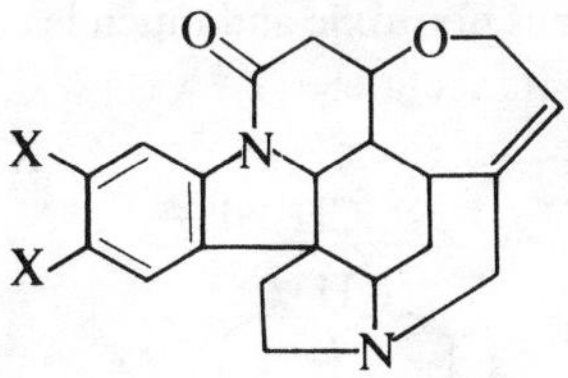

1 X = OCH_3 Brucine
2 X = H Strychnine

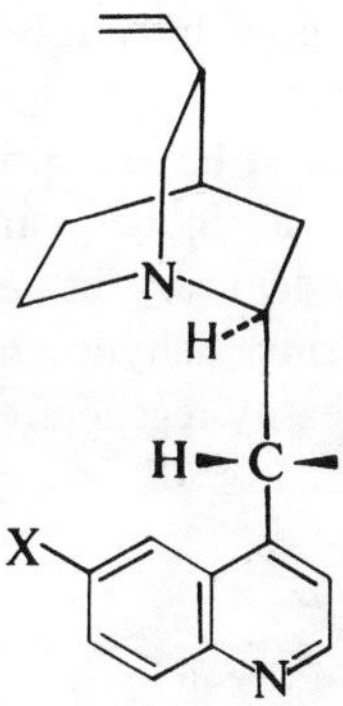

3 Quinine X = OCH_3
5 Cinchonidine X = H

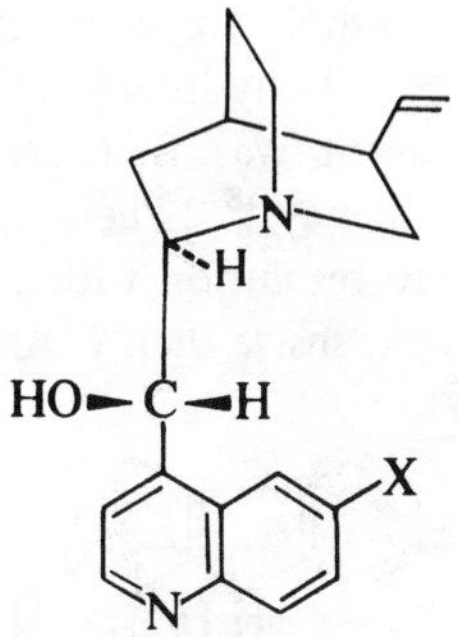

4 Quinidine
6 Cinchonine

It was demonstrated by Pasteur[5] that the heating of any salt of either quinine or of quinidine, **4**, yields an isomeric alkaloid named quinicine, **7**. In the same manner, Pasteur obtained cinchonicine, **8**, from either cinchonine, **6**, or cinchonidine, **5**. These two new bases were the ones with which Pasteur demonstrated the first examples of resolutions through formation of diastereomeric salts.[1] Quinicine and cinchonicine have identical absolute configurations and differ only in the presence of a methoxy group in the former. Quinicine (or quinotoxine) is the key intermediate in the total syntheses of quinine and of quinidine described by Woodward[6,7] and Uskovic[8] and their collaborators. The alkaloids **7** and **8** have, to our knowledge, not been used as resolving agents in modern times.

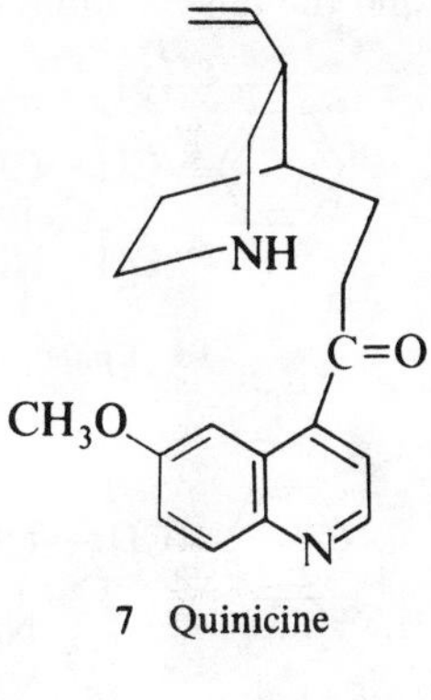

7 Quinicine

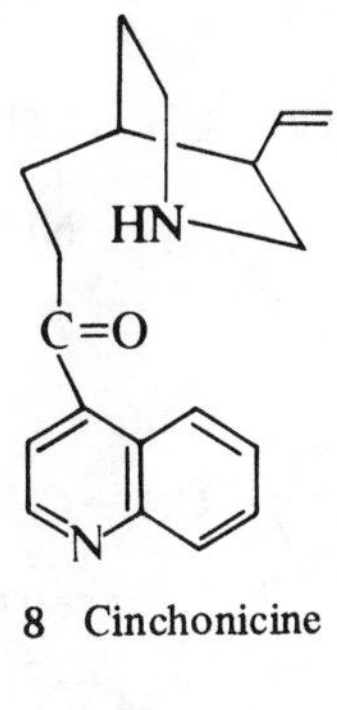

8 Cinchonicine

Yohimbine, **9**, was employed by Fourneau and Sandulesco[9] for the resolution of 2-phenoxypropionic acid. This is a resolution in which all the usual alkaloids are relatively ineffective as resolving agents. Since yohimbine hydrochloride is rather insoluble in water, the diastereomeric salts are decomposed with NaOH and the alkaloid is then extracted.

Morphine, **10**, is relatively little used as a resolving agent because of its toxicity, high cost, and inaccessibility.[10] Mandelic[11] and atrolactic acids,[12] both resolved with morphine, can now be resolved with ephedrine[13–15] and with α-methylbenzylamine,[16] respectively.

Dehydroabietylamine, **11**, is a terpene primary amine whose use as a resolving agent was first described by B. Sjöberg and S. Sjöberg[17] and Gottstein and Cheney.[18] 2-Phenoxypropionic acid may be resolved with **11**, and this is preferable to resolution with yohimbine since dehydroabietylamine is nontoxic and much less expensive than **9**. Amine **11** is easily regenerated through its acetate.

9 Yohimbine **10** Morphine **11** Dehydroabietylamine

For amines of natural origin (or their derivatives), only one enantiomer is accessible. Amino acid derivatives are exceptions in that both enantiomers are usually available. For example, leucine methyl ester, **12**, has been used in the resolution of carboxylic acids (see below). Synthetic amines are evidently available in both forms.

Ephedrine, **13**, and its synthetic derivatives deoxyephedrine, **14**, and amphetamine, **15**, find wide use as resolving agents, as does **16**, an intermediate in the synthesis of the antibiotic chloramphenicol, and the related amines **17**, **18**, and **19**.

$(CH_3)_2CH{-}CH_2{-}CH(NH_2){-}CO_2CH_3$

12

$C_6H_5{-}CH(OH){-}CH(NHCH_3){-}CH_3$

13 Ephedrine

$C_6H_5{-}CH_2{-}CH(NHCH_3){-}CH_3$

14

$C_6H_5{-}CH_2{-}CH(NH_2){-}CH_3$

15

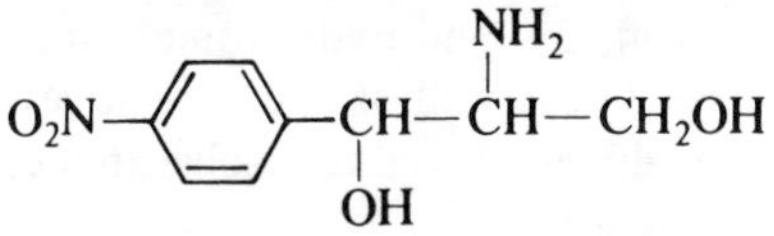

16

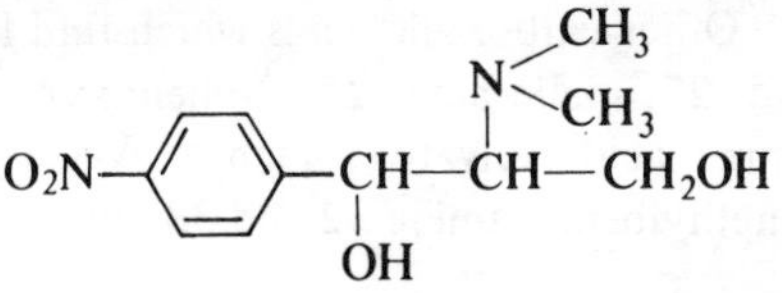

17

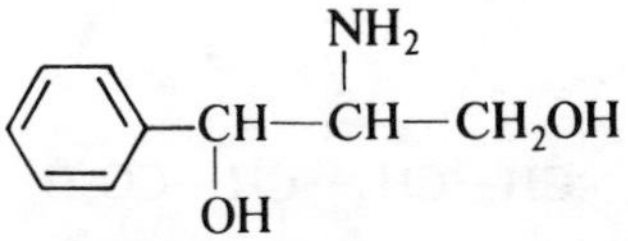

18

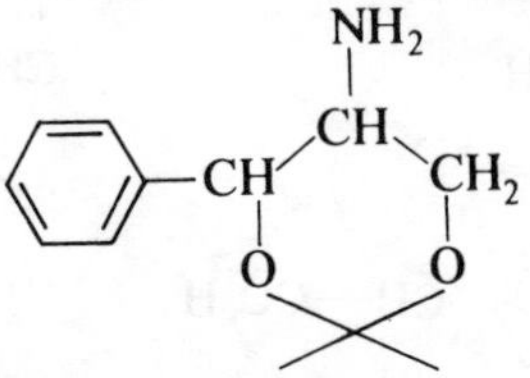

19

α-Methylbenzylamine, **20**, is the most widely used synthetic resolving agent. After use, it is easily regenerated and purified via its sulfate.[20] The analogous naphthylethylamines **21** and **22** also find use as salt-forming resolving agents; however, they are considerably less accessible and more expensive than **20**. Another useful basic resolving agent is tyrosine hydrazide.

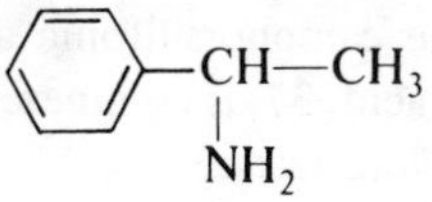

20

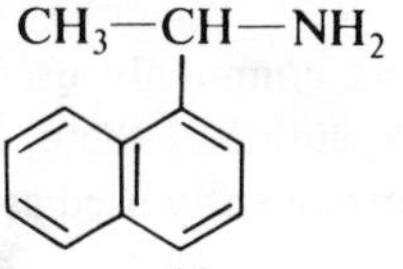

21

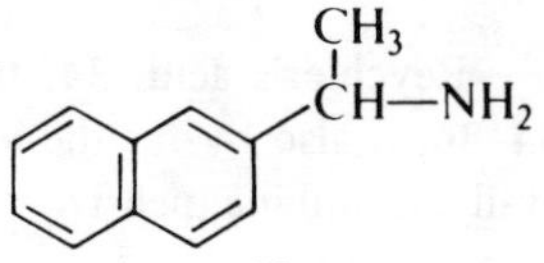

22

5.1.2 Resolution of Bases

The choice of acidic resolving agents forming crystalline salts with racemic amines is more limited than that of basic resolving agents.

Tartaric acid, **23**, in particular, and its diaroyl derivatives **24** and **25**, are among the most widely used chiral acids. Both can form either neutral salts or acid salts according to the stoichiometry and conditions employed; this complication is taken up in Section 7.2.4. The *o*-nitrotartranilic acid **26** is also employed in some instances.

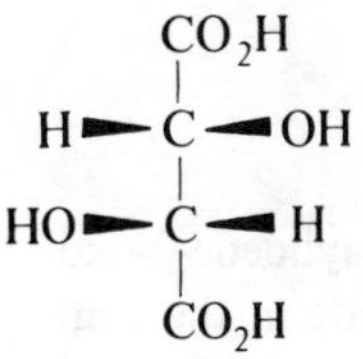

23 *d*-Tartaric acid

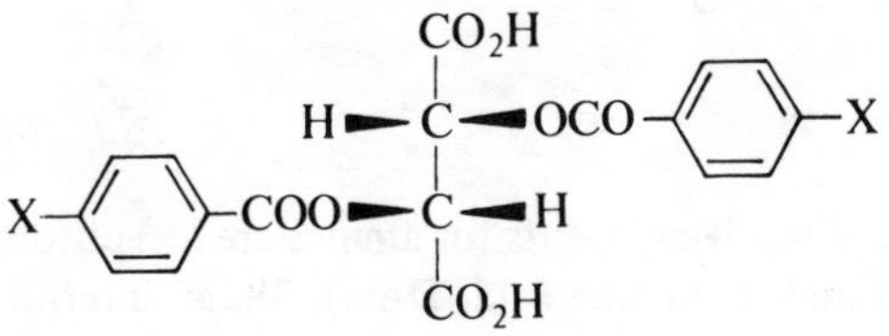

24 X = H
25 X = CH_3

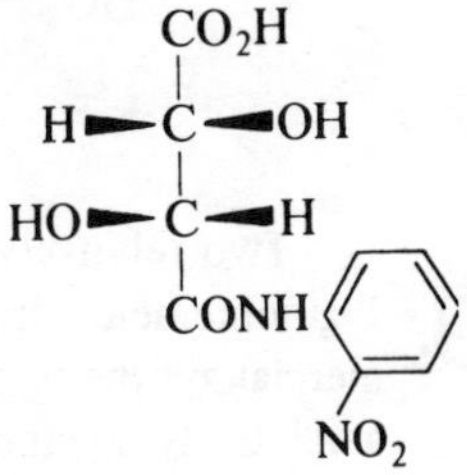

26

Other carboxylic acids which find frequent use as resolving agents are mandelic acid, **27**, malic acid, **28**, 2-phenoxypropionic acid, **29**, and hydratropic acid, **30**, Amino acid derivatives such as *N*-acetylleucine, **31**, as well as the derivatives of α-methylbenzylamine **32** and **33** are also occasionally employed to resolve amines.

C_6H_5–CH(OH)–CO_2H

27

HO–CH(CO_2H)–CH_2–CO_2H

28

C_6H_5–O–CH(CH_3)–CO_2H

29

C_6H_5–CH(CH_3)–CO_2H

30

$(CH_3)_2$CH–CH_2–CH($NHCOCH_3$)–CO_2H

31

C_6H_5–CH(CH_3)–NH–C(=O)–CH_2–CH_2–CO_2H

32

C_6H_5–CH(CH_3)–NH–C(=O)–C_6H_4–CO_2H

33

Reychler's acid, **34**, the most commonly used of the camphorsulfonic acids **34**–**36**, is also by far the least expensive of these. Quinic acid, **37**, is commercially available and inexpensive yet is infrequently used as a resolving agent.

O, CH_2SO_3H

34

O, Br, CH_2SO_3H

35

O, HO_3S

36

HO, CO_2H, HO, OH, OH

37

Two relatively new resolving agents for amines are (−)-diisopropylidene-2-oxo-L-gulonic acid (diacetoneketogulonic acid, DAG), **38**, an intermediate in the commercial synthesis of ascorbic acid, and lasalocid, a polyether antibiotic. DAG, **38**, is a relatively inexpensive, water-insoluble acid which readily forms crystalline salts with amines such as **20**, **21**, and **22** and easily resolves them.

Lasalocid, **39**, first described in 1977, also water insoluble, readily resolves **20** and **21** as well as other amines.[19] The resolving agent is recovered from its diastereomeric salts by extraction of the amine hydrochloride from CH_2Cl_2 or EtOAc solutions with dilute HCl and evaporation of the organic layer.

The binaphthylphosphoric acid, **40**, is a relatively new resolving agent easily accessible from binaphthol. The two enantiomers are obtained by resolution of racemic **40** with cinchonine, **6**, and with cinchonidine, **5**. This strong acid has been found to be particularly effective in salt-forming resolutions of feebly basic amines.[21]

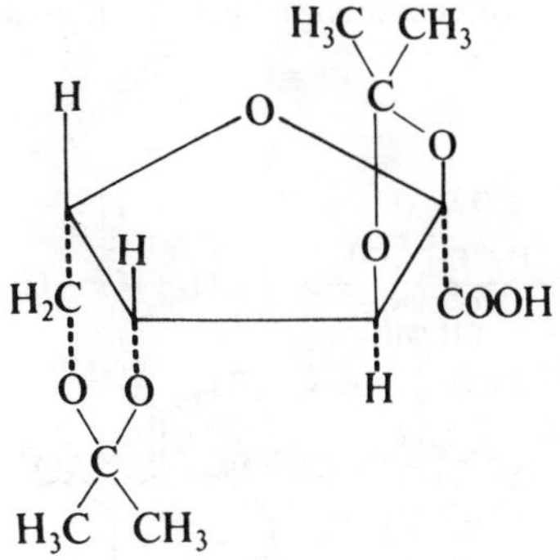

38

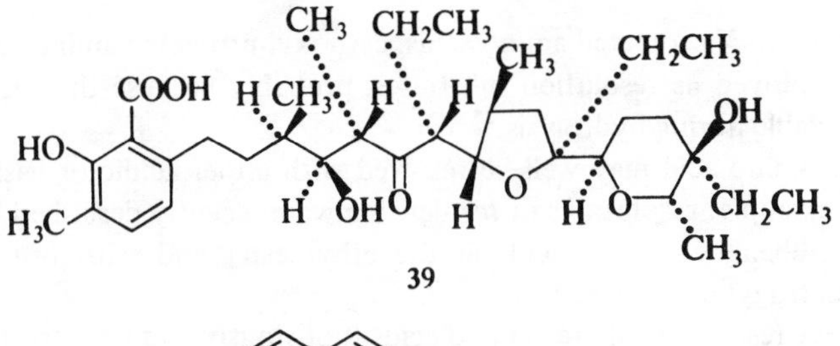

39

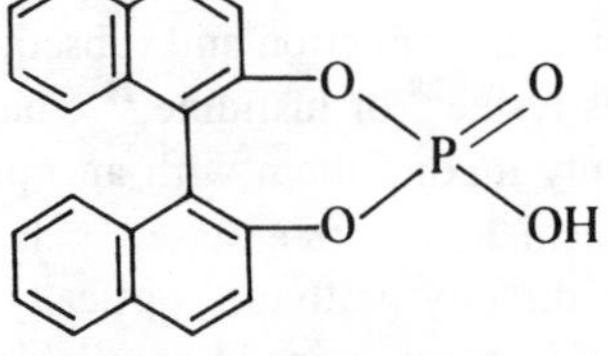

40

5.1.3 Resolution of Amino Acids

Classical resolutions of amino acids and their derivatives mediated by formation of salts continues to be widely practiced along with other optical activation routes. The processes employed are of two types: (1) The bifunctional amino acid is transformed into an ordinary acid by protection of the amino group, or into an ordinary base by protection of the carboxyl group. The derivatized substrate is then resolved like other acids or bases. (2) Underivatized amino acids are combined *directly* with an acidic or basic resolving agent.

The resolution of amino-protected amino acids was originally described by Emil Fisher in 1899.[23] *N*-Benzoyl derivatives of racemic alanine, glutamic acid, and aspartic acid were resolved with brucine or strychnine. Resolved benzoylamino acids were converted into free amino acids by refluxing with strong HCl. Fisher subsequently employed *N*-formyl groups, which are more easily hydrolyzed by refluxing in dilute HCl or HBr. A number of amino acids have been resolved via *N*-benzoyl, *N*-acetyl, *N*-formyl, *N*-benzyloxycarbonyl, *N*-benzyl (cleaved by catalytic hydrogenation), or *N*-tosyl derivatives (the latter cleaved with sodium in liquid ammonia).[24, 25] A recent procedure [26] is the simultaneous derivatization and diastereomer salt formation of enamine-protected amino acids (Scheme 2).

SCHEME 2

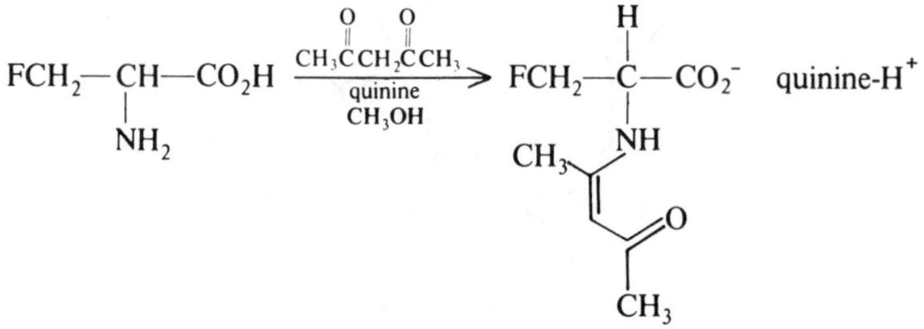

$$FCH_2{-}CH(NH_2){-}CO_2H \xrightarrow[\text{quinine},\ CH_3OH]{CH_3C(O)CH_2C(O)CH_3} FCH_2{-}CH[NH{-}C(CH_3){=}CH{-}C(O)CH_3]{-}CO_2^- \ \text{quinine-}H^+$$

In contrast to *N*-protected amino acids, carboxyl-protected amino acids (esters) are rarely employed as resolution substrates, probably because they are relatively easily racemizable during hydrolysis.[27]

A given amino acid may well be resolved with either acidic or basic resolving agents. For example, resolutions of *tert*-leucine were recently described in the same article with dibenzoyltartartic acid on the ethyl ester and with brucine on the *N*-*p*-toluenesulfonyl derivative.[22]

The direct resolution of free amino acids is obviously a more attractive process than that which involves prior introduction and subsequent cleavage of a protecting group. Such amino acids as lysine[28] or histidine,[29] which are basic in the free state, have been resolved by simply mixing them with an optically active acid. Similarly, those which are already acid in the free state, such as glutamic[30a] and aspartic acids,[31] may be resolved directly with an optically active base. (Incidentally, optically active glutamic acid, which is readily available at a low price may itself be used as an acidic resolving agent.[30b])

Unprotected *neutral* amino acids also are often directly resolvable with strong acidic resolving agents such as the camphorsulfonic acids[32] **34** and **35**, cholestenonesulfonic acid, **41**,[33] or the new terpenesulfonic acid **42**, prepared from (+)-*cis*-limonene oxide.[34] The binaphthylphosphoric acid, **40**, has been employed with success for the resolution of *o*-tyrosine.[35] The relatively strong tartaric acid has been occasionally used with success.[36, 37] These acids do in fact form salts that are analogous to amino acid hydrochlorides, the achiral chloride ion being replaced by the optically active anion of the acidic resolving agent.

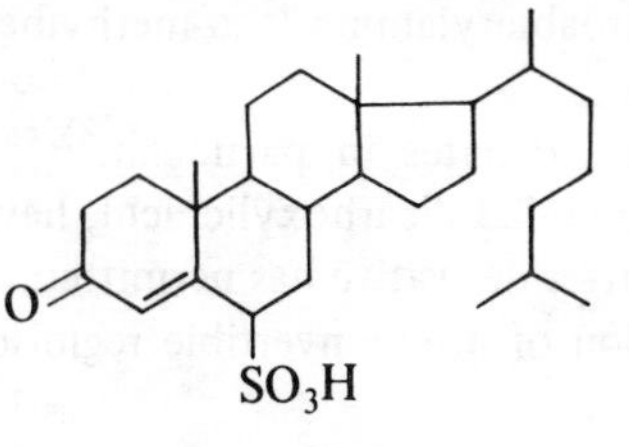

41

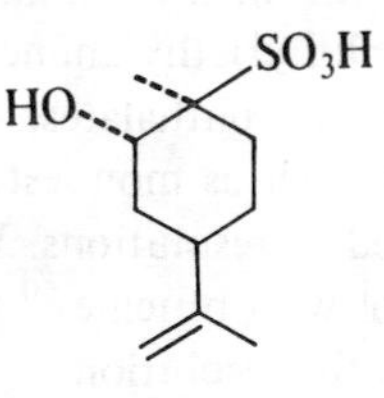

42

5.1.4 Resolution of Alcohols. Transformation of Alcohols Into Salt-Forming Derivatives

Given the relative abundance of optically active acids and bases able to form crystalline salts, the desirability of converting neutral functionalities such as alcohols or carbonyl compounds into bases or acids resolvable via diastereomeric salts is self-evident. A variety of methods based upon this principle have been developed and remain useful

Pickard and Littlebury were the first to describe, in 1907, the resolution of chiral alcohols through salt formation between their hydrogen phthalates and optically active bases. Borneol and isoborneol were resolved as hydrogen phthalates with menthylamine or cinchonine.[41] In fact, this possibility had been clearly anticipated in 1904 by Erlenmeyer and Arnold[38] who explicitly suggested the resolution of alcohols via phthalates and oxalates and their alkaloid salts. Experiments along these lines were evidently undertaken by them but do not appear to have been published. This proposal was lost sight of and is not cited either in the classic review by Ingersoll[39] or in the recent one by Klyashchitskii and Shvets.[40]

The preparation of phthalates of primary and secondary alcohols is quite straightforward,* and the reversal of this derivatization is generally carried out by saponification (this cleavage may be attended by difficulties; see Section 7.4). The efficiency of the method is attested to by the very large number of successful resolutions of hydrogen phthalates described in the literature.[4, 39]

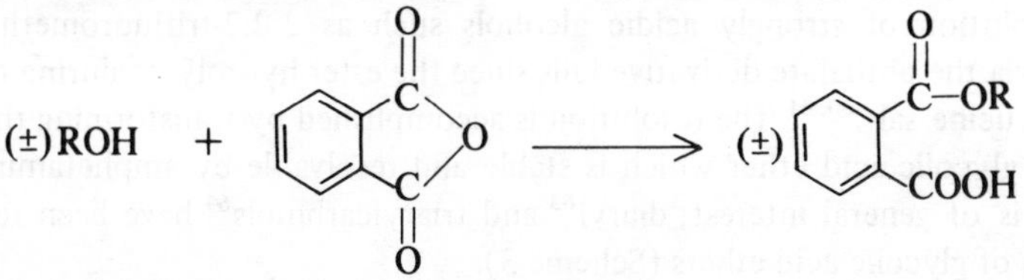

(±)ROH + phthalic anhydride ⟶ (±) $C_6H_4(COOR)(COOH)$

Hydrogen phthalates have been resolved with the classical alkaloids, brucine being the preeminent choice. However, there is no impediment to the use of primary amine resolving agents, potential complication of ester/amide interchange being practically nonexistent under typical conditions for salt formation. Thus, successful resolutions of alcohols via hydrogen phthalates have been reported, for

*Some phthalates are difficult to crystallize. They may be purified and later easily crystallized efficiently via their piperazine salts.

example, with menthylamine,[41] dehydroabietylamine,[42] α-methylbenzylamine,[43] and α-(2-naphthyl)ethylamine.[44]

Substituted phthalates, the 3-nitrophthalates in particular,[45,46] and related compounds such as monoesters of pyridine-2,3-dicarboxylic acid, have been occasionally used in resolutions. While the latter derivative has permitted the resolution of 2-butanol with brucine,[50] the formation of interconvertible regioisomeric esters complicates the resolution.

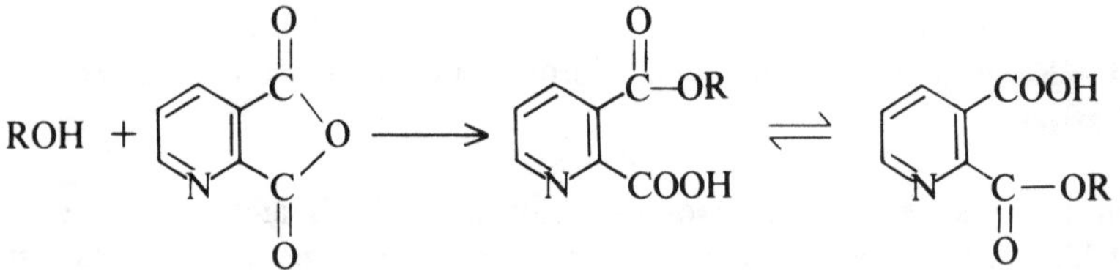

ROH + (pyridine-2,3-dicarboxylic anhydride) ⟶ (pyridine with C(=O)–OR at 3-position and COOH at 2-position) ⇌ (pyridine with COOH at 3-position and C(=O)–OR at 2-position)

Other acid derivatives of alcohols have occasionally been used. A recent application of hydrogen succinate esters[39] has been the resolution of phenyl(trichloromethyl)carbinol. Quinine was the resolving agent.[47] Oxalate derivatives have been recommended for the resolution of alcohols which are unable to tolerate acidic or alkaline hydrolysis during the recovery process.[48] While the use of covalent hydrogen sulfates is uncommon due to the limited stability of such compounds, a number of resolutions mediated by sulfates have been reported.[39] Pantolactone has been resolved via sulfate formation on its α-hydroxyl group[49] (see also below).

The resolution of tertiary alcohols via their hydrogen phthalates has been applied to trialkyl-,[51] aryldialkyl-,[52,53] and diarylalkyl carbinols.[54] The resolution of a triaryl carbinol by this method has been described by Thaker.[55] However, this resolution could not be reproduced by other workers.[56]

Earlier problems in the synthesis of the sometimes unstable hydrogen phthalates of tertiary alcohols have been overcome, for example, by reaction of phthalic anhydride with sodium salts of alcohols,[57] by use of bases such as triethylamine in place of pyridine,[58] and by catalysis by the hypernucleophile 4-(*N,N*-dimethylamino)pyridine.[59]

Resolution of strongly acidic alcohols such as 2,2,2-trifluoromethylphenylethanol via the phthalate derivative fails since the ester hydrolyzes during formation of the brucine salt.[60,61] The resolution is accomplished by transforming the alcohol into the glycolic acid ether which is stable and resolvable by amphetamine.[62] This method is of general interest; diaryl[63] and triarylcarbinols[64] have been resolved in the form of glycolic acid ethers (Scheme 3).

SCHEME 3

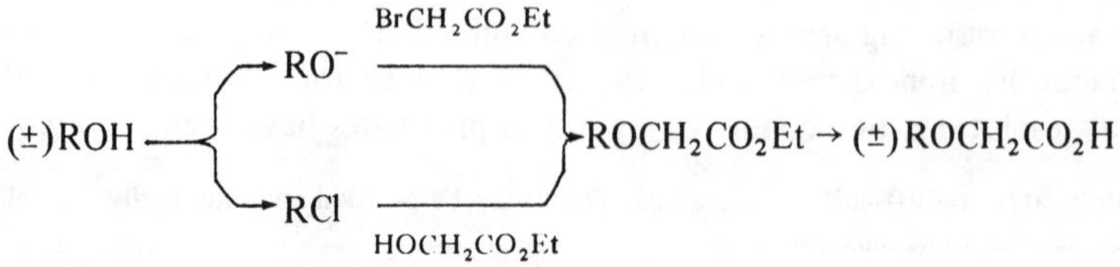

$(\pm)ROH \rightarrow RO^-$ $\xrightarrow{BrCH_2CO_2Et}$ $ROCH_2CO_2Et$

$(\pm)ROH \rightarrow RCl$ $\xrightarrow{HOCH_2CO_2Et}$ $ROCH_2CO_2Et$

$ROCH_2CO_2Et \rightarrow (\pm)ROCH_2CO_2H$

The glycolic acid derivatives are prepared either by reaction of the alcoholate with ethyl bromoacetate or by conversion of the alcohol into the alkyl chloride followed by reaction with ethyl glycolate. Cleavage of glycolic acid ethers to the alcohols following resolution may be carried out (a) via the α-bromo derivative of the glycolic acid ether,[62] (b) by Hoffmann rearrangement of its amide,[64] or (c) by transformation of the acid into a primary alkyl chloride which may be decomposed by a metal[65] (Scheme 4).

SCHEME 4

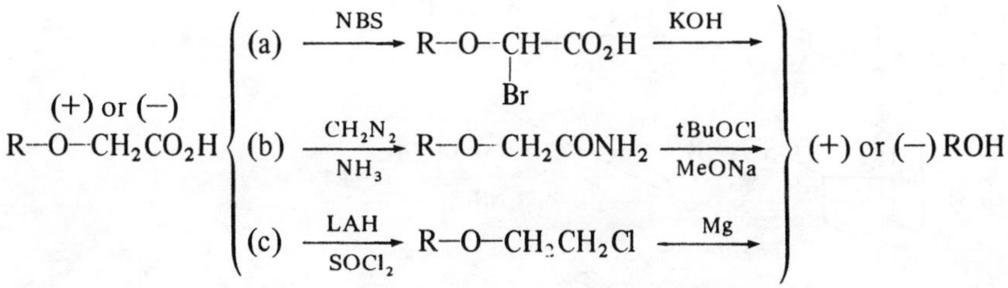

$$(+)\text{ or }(-)\ R\text{—O—}CH_2CO_2H \begin{cases} (a) \xrightarrow{NBS} R\text{—O—}\underset{\displaystyle Br}{\underset{|}{CH}}\text{—}CO_2H \xrightarrow{KOH} \\ (b) \xrightarrow[NH_3]{CH_2N_2} R\text{—O—}CH_2CONH_2 \xrightarrow[MeONa]{tBuOCl} \\ (c) \xrightarrow[SOCl_2]{LAH} R\text{—O—}CH_2CH_2Cl \xrightarrow{Mg} \end{cases} (+)\text{ or }(-)\ ROH$$

The resolution of alcohols via thioglycolic acid derivatives appears to be less useful than the glycolic acid route.[66] On the other hand, thioglycolic acids would seem to be acceptable intermediates in the resolution of thiols.[67]

Another process applicable to acidic alcohols consists in their transformation into ethers of hydracrylic acid. Triaryl carbinols[65] and 1,1,1-trifluoro-2-propanol[68] have been resolved by this method. The hydracrylic acid derivatives can be prepared from the alcohol and methyl 3-hydroxypropionate, from propiolactone, or from acrylonitrile (Scheme 5). Cleavage of the derivatives consists of the retro-Michael reaction carried out on the acid or its methyl ester. This is not a general procedure; it takes place only with derivatives of acidic alcohols. For example, *β-sec*-butoxypropionic acid is not cleaved to 2-butanol under conditions which yield 1,1,1-trifluoro-2-propanol from its derivative.[68]

SCHEME 5

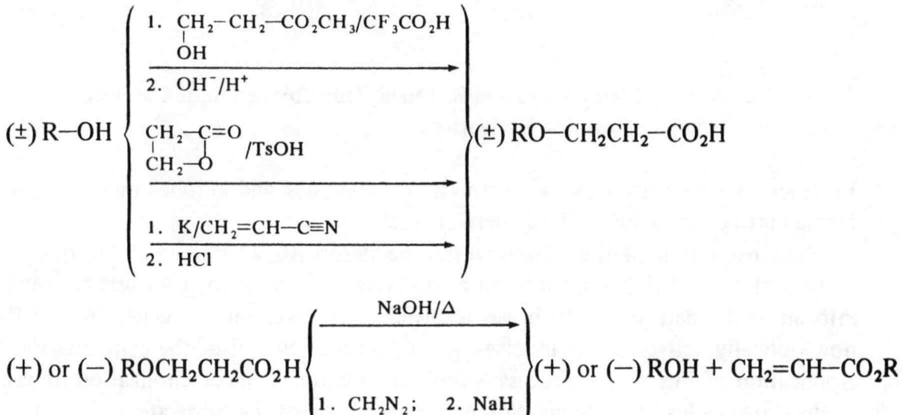

$$(\pm)\ R\text{—OH} \begin{cases} \xrightarrow[2.\ OH^-/H^+]{1.\ \underset{\displaystyle OH}{\underset{|}{CH_2}}\text{—}CH_2\text{—}CO_2CH_3/CF_3CO_2H} \\ \xrightarrow{\begin{matrix}CH_2\text{—}C{=}O \\ CH_2\text{—}O\end{matrix}\ /TsOH} \\ \xrightarrow[2.\ HCl]{1.\ K/CH_2{=}CH\text{—}C{\equiv}N} \end{cases} (\pm)\ RO\text{—}CH_2CH_2\text{—}CO_2H$$

$$(+)\text{ or }(-)\ ROCH_2CH_2CO_2H \begin{cases} \xrightarrow{NaOH/\Delta} \\ \xrightarrow{1.\ CH_2N_2;\quad 2.\ NaH} \end{cases} (+)\text{ or }(-)\ ROH + CH_2{=}CH\text{—}CO_2R$$

Some steroidal alcohols have been resolved as hemisulfate salts. This method, which would seem to be a general one, requires the conversions of the alcohol to a triethylammonium hemisulfate derivative. Exchange of the cation for a chiral ammonium ion yields diastereomeric salts which may be separated by crystallization.[69]

(±) CH_3O — OH $\xrightarrow{Et_3N \cdot SO_3}$ OSO_3^-, Et_3NH^+

$CH_3-\overset{*}{C}H-CH_2OH$, NH_3^+ (= $\overset{*}{R}^+$) → (p,n) OSO_3^-; $\overset{*}{R}^+$

The resolution of glycols has been effected via the acid derivative obtained with *p*-boronobenzoic acid. The derivative is resolved by crystallization of the quinine salt and the optically active glycol recovered after hydrolysis in aqueous dioxane.[70] This method, which is due to Agosta, has been applied to the resolution of 2,4-pentanediol[71] and is evidently a general one.

OH, OH, NO_2 $\xrightarrow{HOOC-C_6H_4-B(OH)_2}$ O–B–C_6H_4–COOH, O, NO_2

5.1.5 Resolution of Aldehydes and Ketones. Transformation of Carbonyl Compounds Into Salt-Forming Derivatives

In order to be resolved by salt formation, aldehydes and ketones must be transformed into either acidic or basic derivatives.

The resolution of the ketoprogesterone derivative **43** illustrates the transformation of a methyl ketone into an acid derivative. The ketone **43** was converted into an oxalyl derivative which was resolved with strychnine. Regeneration of the now optically active ketone involves base-catalyzed reversal of the condensation.[72] Application of this simple process is limited to ketones whose enolization in basic media does not lead to racemization or epimerization of the substrate.

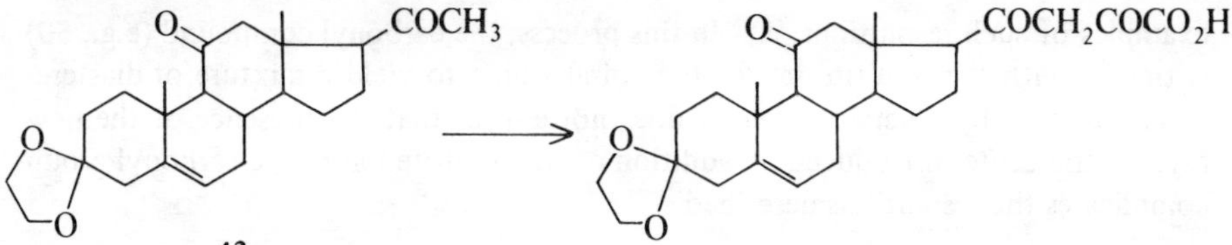

O COCH$_3$ → O COCH$_2$COCO$_2$H

43

Conversion of carbonyl compounds to *N*-substituted imino derivatives constitutes a more widely applied method for their resolution. The racemic aldehyde or ketone is linked to an auxilary acidic reactant as follows:

$$>C=O \xrightarrow{H_2N-(X)-AH} >C=N-(X)-AH$$

Apart from 4-sulfophenylhydrazine, **44**, a reagent proposed for this purpose by Neuberg and Federer in 1905,[73] the first such reagent prepared specifically for ketone resolutions was 4-(4-carboxyphenyl)semicarbazide, **45**, which was effective in the resolution of 3-methylcyclohexanone.[74] Nerdel and co-workers[75] have proposed the hydrazinobenzoic acids **46** and **47** as well as oxalic acid monohydrazide, **48**, as adjuvants for resolution purposes.

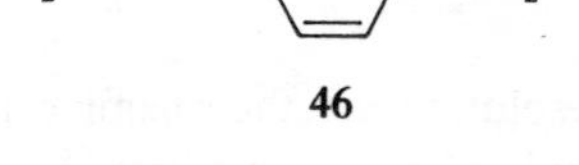

$H_2N-NH-C_6H_4-SO_3H$

44

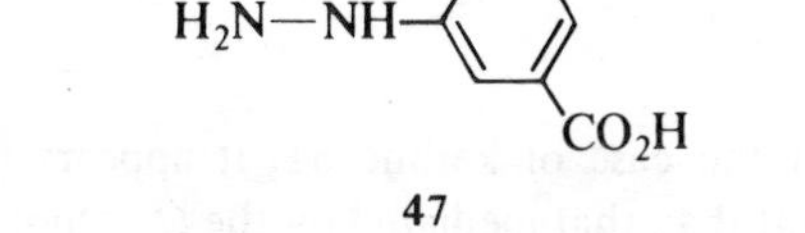

$H_2N-NH-C(=O)-NH-C_6H_4-CO_2H$

45

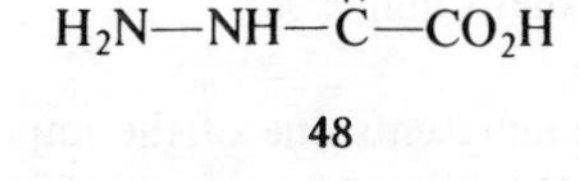

$H_2N-NH-C_6H_4-CO_2H$

46

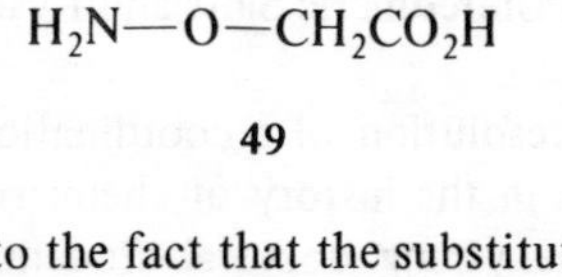

$H_2N-NH-C_6H_4-CO_2H$ (meta)

47

$H_2N-NH-C(=O)-CO_2H$

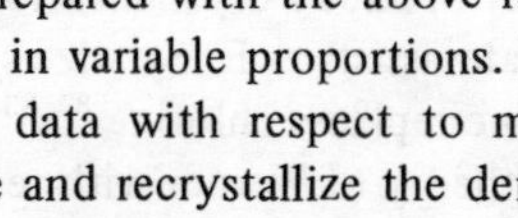

48

$H_2N-O-CH_2CO_2H$

49

The work of Nerdel et al. calls attention to the fact that the substituted hydrazones prepared with the above reagents may be present as mixtures of syn and anti isomers in variable proportions. It is therefore necessary, in order to obtain reproducible data with respect to melting points and optical rotations, to carefully separate and recrystallize the derivatives under well defined conditions. Touboul et al. have employed *O*-carboxymethyl oximes, prepared with aminooxyacetic acid, **49**, in the same way for the resolution of polycyclic α,β-ethylenic ketones.[76] In the cases described, it appears that the *O*-carboxymethyl oximes prepared are isomerically homogeneous.

The application of amine bisulfite addition compounds of aldehydes and ketones to resolutions is a relatively simple procedure. Nonetheless, there are few examples of such resolutions.[77, 78] In this process, the carbonyl compound (e.g., **50**) is treated with the bisulfite salt of the chiral amine to yield a mixture of diastereomers separable by crystallization. It does not appear that the presence of the new asymmetric center introduced by addition of the bisulfite ion to the carbonyl group complicates the resolutions described.

(±) **50** + C_6H_5–$\overset{*}{C}H(NH_2)$–CH_3/H_2SO_3 ⟶ (p,n) [bisulfite adduct: $\overset{*}{C}(OH)(SO_3^-)$] C_6H_5–$\overset{*}{C}H(NH_3^+)$–CH

The transformation of a ketone into an enamine followed by reaction with a strong and optically active acid leads to a mixture of diastereomeric iminium salts which may be separated by crystallization. Hydrolysis of such salts yields the separated ketone enantiomers under very mild conditions.[79] Ketones that do not readily form enamines may be directly transformed into iminium perchlorates[80] whose inorganic anion may be subsequently exchanged by that of an optically active acid.[79]

51 ⟶ [enamine, N] ⟶ [iminium, N^+] [camphor-10-sulfonate, $CH_2SO_3^-$]

In the case of ketone **51**, it appears that resolution via the enamine is less efficient than that mediated by the *O*-carboxymethyl oxime intermediate.[76]

5.1.6 Diastereomeric Salts and Resolution of Werner Complexes

The first resolution of a coordination compound represents one of the important milestones in the history of chemistry. In June 1911, after 15 years of effort and failure, the young doctoral student V. L. King, working with Alfred Werner in Zurich, succeeded in resolving *cis*-$[Co(en)_2(NH_3)Cl]^{2+}$, **52**, by crystallization of the 3-bromocamphor-9-sulfonates,[81] and Werner's private assistant E. Scholze similarly resolved the analogous complex ion in which bromine replaces chlorine.[82, 83]

These first examples brilliantly confirmed the theory of Werner with respect to octahedral Co(III) while at the same time demonstrating that tetrahedral structures did not exert a monopoly over chirality. Shortly thereafter, Werner demonstrated the chirality of "hexol" **54a**, a binuclear complex which was totally inorganic, that is, devoid of carbon. These first examples were followed by a large number of resolutions of complexes of this type. We examine here only those mediated by diastereomeric *salt* formation in which optically active counterions intervene.

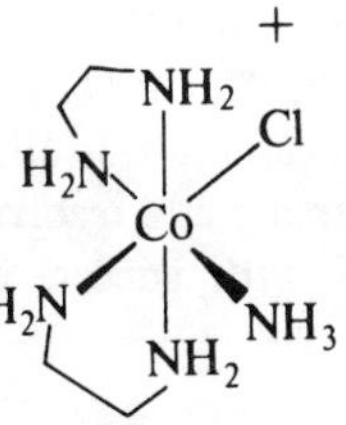

52 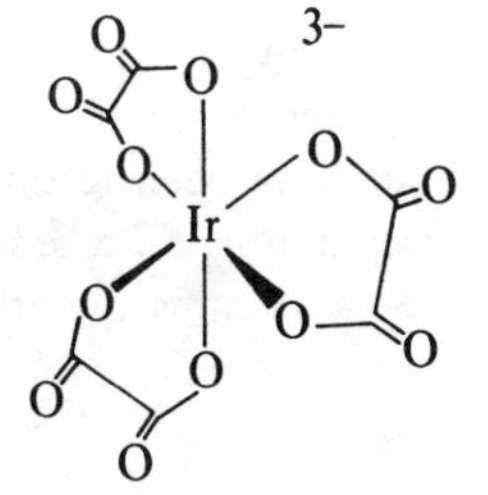 53 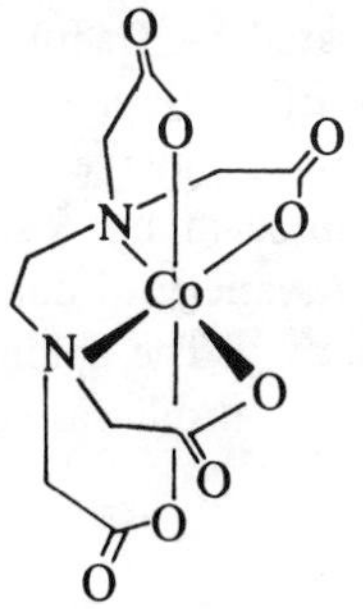 54

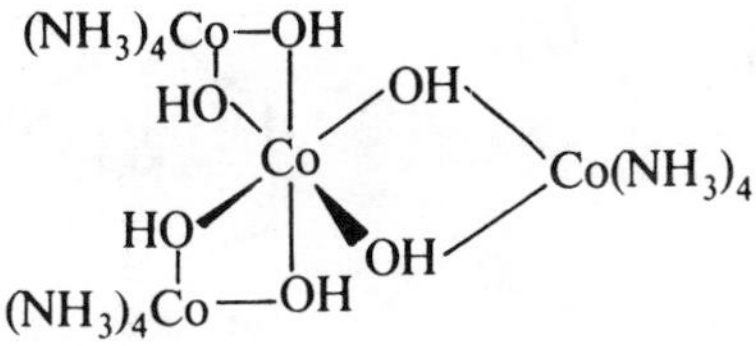

54a

The resolution of *anionic* complexes may be carried out by means of *cations* which are conjugate acids of the optically active bases described in Section 5.1.1. The resolution of the trioxalato iridium anion **53** described by Delépine is a good illustration of this type of process.[84] The potassium salt of **53** is treated with strychnine sulfate in water. After separation by crystallization, the *p* and *n* diastereomers are decomposed by potassium carbonate or barium oxide to yield the resolved anions in the form of their potassium or barium salts. The process is shown schematically in the following equations:

$$(\pm)[\mathrm{Ir(ox)_3}]^{3-}, 3\mathrm{K}^+ + 3(\mathrm{St\text{-}H}^+, \mathrm{SO_4H}^-) \rightarrow p \text{ and } n\ [\mathrm{Ir(ox)_3}]^{3-}, 3\mathrm{St\text{-}H}^+ + 3\mathrm{KHSO_4} \quad (1)$$

$$p \text{ or } n \text{ salt} + 3\mathrm{KOH} \rightarrow (+) \text{ or } (-)\ [\mathrm{Ir(ox)_3}]^{3-}, 3\mathrm{K}^+ \quad (2)$$

The ethylenediaminetetraacetatocobaltate(III) anion, **54**, is resolvable in the same manner with strychnine sulfate.[85] A much more efficient process, developed by Gillard et al., uses histidine salts.[86] The barium salt, $\mathrm{Ba[Co(EDTA)]_2}$ is first transformed into the *acid* H[Co(EDTA)] by passage through a cation exchanger. The resulting solution is treated with histidine and the diastereomeric histidine salts separated by crystallization. Conversion of the latter into the enantiomeric potassium salts is carried out by ion exchange through an appropriate cationic resin.

In a similar manner, resolution of *cationic* complexes may be carried out by combining them with conjugate bases of available optically active acids. Only a few such acids appear to give satisfactory results. 3-Bromocamphor-9-sulfonic acid is often used. In the resolution of cation **52** by Werner, it was the silver salt of this acid which was combined with the cation in the form of its chloride.[82] The precipitated silver chloride was filtered and the diastereomeric salts remaining were separated by fractional crystallization. The resolved ions were regenerated in the form of dithionates $[(\mathbf{52})^+, S_2O_6^-]$. Contemporary resolutions of **52** and other complexes in most cases use the less costly and commercially available ammonium 3-bromocamphor-9-sulfonates in place of the silver salt.[87, 88]

A variety of complex cations have been resolved with camphor-3-sulfonic acid anion.[89, 90] The anion of 3-nitrocamphor, **55**, used by Werner[91] as a resolving agent for the rhodium complex **56**, was subsequently used with success in other cases.[92, 93]

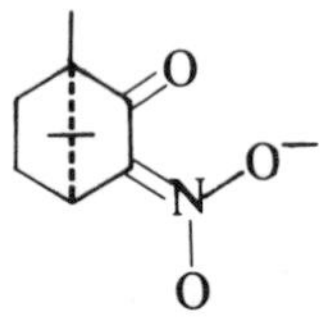

55

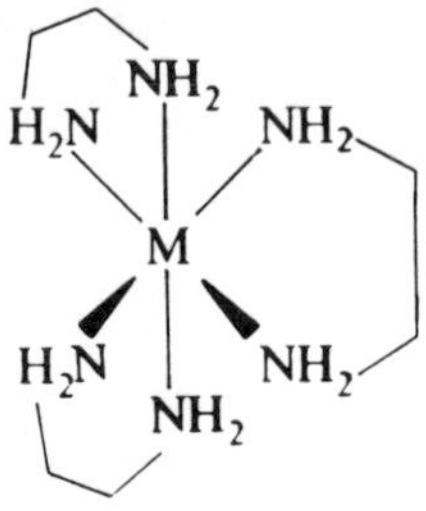

56 M = Rh(III)

57 M = Co(III)

Tartartic acid merits special mention as a resolving agent for cationic coordination compounds. It may be used in the form of (+)-tartrate anion itself (from barium or silver tartrate), as in the resolution of rhodium complex **56** and of its cobalt analogue **57**[91, 94, 95, 97] or in the form of a complex ion of arsenic $(AsOC_4H_4O_6)^-$ or of antimony $(SbOC_4H_4O_6)^-$.[98–100] The latter (commonly called antimonyl tartrate), in particular, is widely used in the resolution of octahedral, for example, *cis*-$[Pt(en)_2Cl_2]^{2+}$, and of binuclear, for example, $[L_2Cr(OH)_2CrL_2]^{2+}$, cationic complexes of structure **58**.[101, 102]

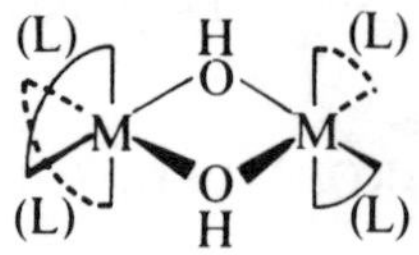

58 L = 1,10-Phenanthroline or 2,2′-bipyridyl

The resolution of complex ions often takes on an unusual character when the optically active counterion used as resolving agent is itself a complex ion that has previously been resolved. This situation is somewhat reminiscent of the resolution of amino acid derivatives with the aid of resolving agents that are themselves derivatives of amino acids. Thus, beginning with an easily accessible optically pure ion, it is possible to realize successive resolutions in which the resolved substrate becomes resolving agent for a second resolution and so on. Note that it is not necessary that all complexes in such a chain of resolutions derive from the same transition element. For example, the cation *cis*-$[Co(en)(NO_2)_2]^+$, **59**, which is resolved as its (+)-antimonyl tartrate,[103] may in turn be used as resolving agent in the resolution of the anion $[Co(en)(ox)_2]^-$, **60**. According to Worrell, **60** is an excellent resolving agent for numerous cationic complexes of cobalt(III).[104]. Cation **59** also allows anion **54** to be resolved in a more expeditious manner than that mentioned above.[105]

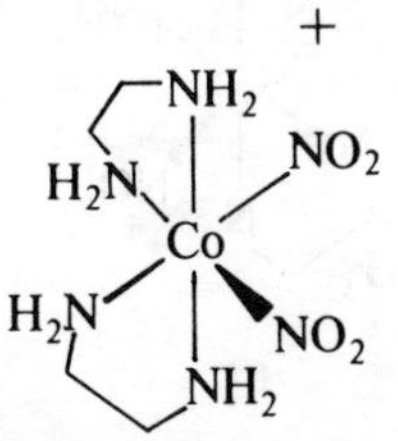

59 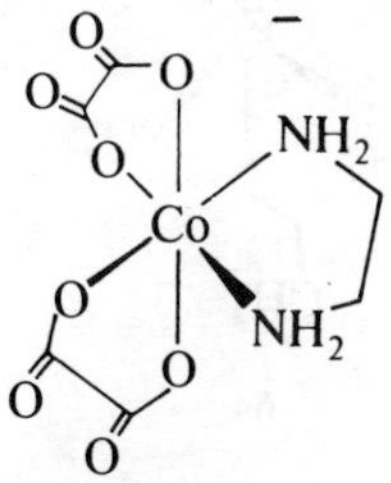 **60** 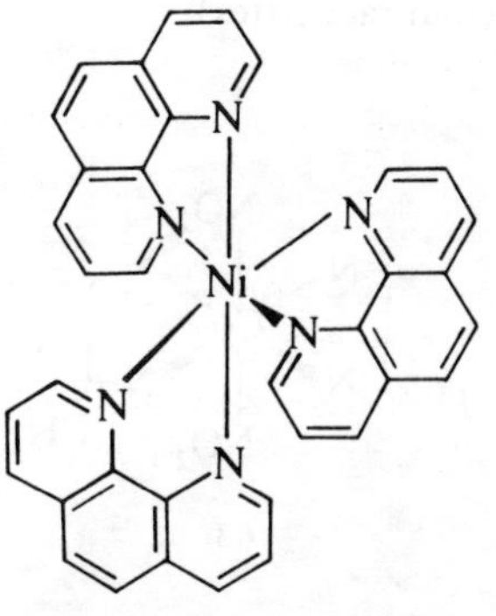 **61**

In a similar way, cation [tris(1,10-phenanthroline) Ni(III)]$^{2+}$, **61**, which is easily resolved with antimonyl tartrate ion, may be used in the resolution of anion **53**.[100] It is evident that many more examples of such resolutions could be found.

The ready availability of optically active complex ions leads one to envisage their use as resolving agents outside the realm of inorganic chemistry. The resolution of *rac*-2,3-dimethylsuccinic acid as its silver salt, described by Werner and Basyrin,[106] illustrates such an application; the resolving agent was optically active ion **57** in the form of its tribromide $[Co(en)_3]Br_3$. In the same spirit, one may cite the resolution of *N*-benzoylvaline, of phthaloylleucine, and of other *N*-protected amino acids through salt formation of their conjugate bases with optically active cation **59**.[107]

An analogous type of resolution, which takes a somewhat less direct approach, requires that the resolution substrate be first converted into a complex ion which is then associated with an optically active counterion. Thus, propylenediaminetetraacetic acid (PDTA), **62**, is transformed into complex (±)-$[Co(PDTA)]^-$ (see **54**) which is resolved through salt formation with optically active cation **59**. Optically pure (+)- and (−)-**62** are finally obtained after decomposition of (+)- and (−)-$[Co(PDTA)]^-$.[108]

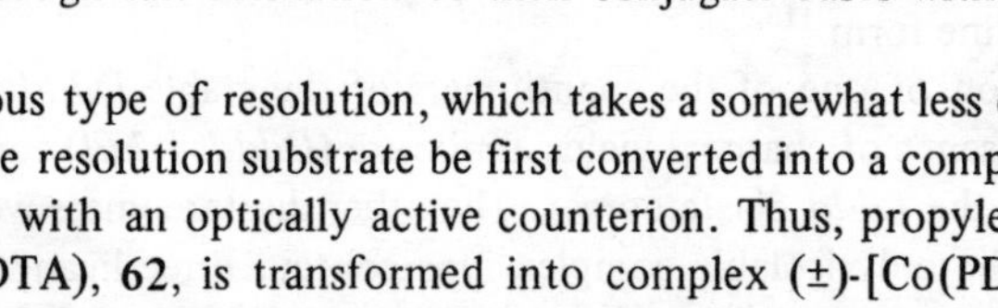

$$(HO_2C{-}CH_2)_2N{-}CH(CH_3){-}CH_2{-}N(CH_2CO_2H)_2$$

62

Apart from the last example given, the resolutions described in this section involve the separation of complex salts in which chiral ions are built up from achiral ligands (en, EDTA, ox, etc.). When the ligands themselves are chiral, a new type of diastereoisomerism appears, to which Werner first called attention in 1918.[109]

Ion $[Co(en)(pn)(NO_2)_2]^-$, in which the propylenediamine (pn) ligand may be either dextro- or levorotatory, is a particularly well studied example of this type of isomerism. In the trans "croceo" series, **63**, there are but two enantiomers; but in the cis ("flavo") series, **64** and **65**, eight diastereomers are theoretically possible (four racemates).

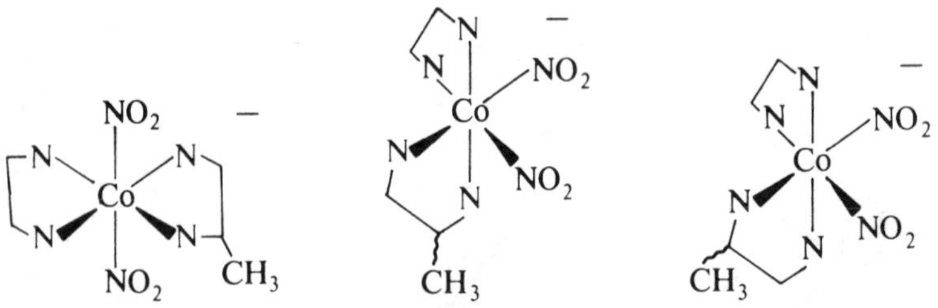

63 **64** **65**

Without going into details for this particular case, it is expected and, indeed, known that the stabilities of these different ions are unequal and that the formation of the ions from the individual ligands (chiral and achiral) therefore takes place stereoselectively leading to a small number of preferred diastereomers. This may be illustrated with the ion $[Co(pn)_3]^{3+}$ as example. Using the conventional and arbitrary terminology of Jaeger,[110] the symbols *D* and *L* represent the signs of rotation of the complex ions at the D line of sodium and *d* and *l* the signs of rotation of the propylenediamine ligand enantiomers. The eight possible diastereomers (four racemates) are then designated *Dddd*, *Dddl*, . . . *Llll*. Two of these forms, *Dddd* and *Llll*, which comprise a racemate, appear to be more stable than the others and were for a long time the only ones that had been isolated.[111, 112, 114] Subsequently, a second and less stable isomer formulated as *Dlll* (and its enantiomer *Lddd*) was isolated in crystalline form.[113]

Equilibration of one of the enantiomers of the stable *Dddd* form in water at 20°C in the presence of charcoal yields a mixture (*Dddd*, *Lddd*) containing approximately 15% of the *Lddd* diastereomer. The other diastereomers were not detected. The fact that the only isolable complex ions contain three ligands (pn) having the same configuration is turned to advantage by use of $[Co(pn)_3]^{3+}$ in the resolution of propylenediamine.[112,113] Corey and Bailar have successfully accounted for the observed stereoselectivity on the basis of conformational analysis.[115]

Many other chiral cations and anions whose chirality is associated with atoms other than carbon have been resolved by methods analogous to those described in this section. Among these are ammonium, phosphonium, and arsonium compounds. Leading references to the extensive literature describing such resolutions are given in the reviews of Boyle and Wilen.[4, 116]

The practice of resolution of complex salts possesses certain characteristics which distinguishes it from that familiar to organic chemists. It is marked, among others, by the frequent recourse to precipitation in place of conventional crystallization; this is effected by addition of nonpolar solvents in which the diastereomers have low solubility. Moreover, use of ion exchange resins in the preparation and decomposition of such diastereomers is common. Also, the infusibility of such salts makes it difficult to follow the progress of resolutions by means other than optical rotation (however, see Section 3.6). For the rest, resolutions of inorganic and organic compounds have much in common; a knowledge of phase diagrams, notably solubility diagrams, is equally useful here.

5.1.7 Lewis Acid–Base Complexes

The formation of Lewis acid–base complexes between an optically active reactant and a racemic substrate constitutes another approach to formation of dissociable crystalline diastereomers. α-(2,4,5,7-Tetranitrofluorenylideneaminooxy)propionic acid, **66** (TAPA), was specially conceived for such use by Newman, Lutz, and Lednicer.[117,118] It is prepared[119] via α-isopropylideneaminooxypropionic acid, **67**, which is resolved with ephedrine; **67** is converted to (+)- or (−)-TAPA by trans-oximation of 2,4,5,7-tetranitrofluorenone.

(+) **67**

(−) **66**

TAPA forms crystalline π- (or charge transfer [CT]) complexes with numerous aromatic compounds which have figured in the resolution of aromatic ethers and carboxylic esters[117,118] as well as phosphorus esters.[120] However, its most spectacular uses surely are in the resolutions of chiral hydrocarbons devoid of functional groups,[124,125] penta-,[122] hexa-,[117] and 2-bromohexahelicene,[123] among others, as well as certain meta-[126] and paracyclophanes.[121] Nitrogen-containing heterohelicenes are also resolved with TAPA apparently through complexation.[127] In all these cases chirality is a consequence of molecular overcrowding or restricted rotation.

TAPA has also been used in the resolution by complexation of aromatic amines **68** and **69** whose chirality resides in the side chain. Amines **68** and **69** are too feebly basic to yield crystalline salts with the usual acidic resolving agents. In these examples, the formation of π-complexes is attested to by the bright orange color of the crystals formed.[128] 1- and 2-Methylcholanthrene, aromatic hydrocarbons whose chirality is due to asymmetric carbon atoms, have also been resolved with TAPA.[125]

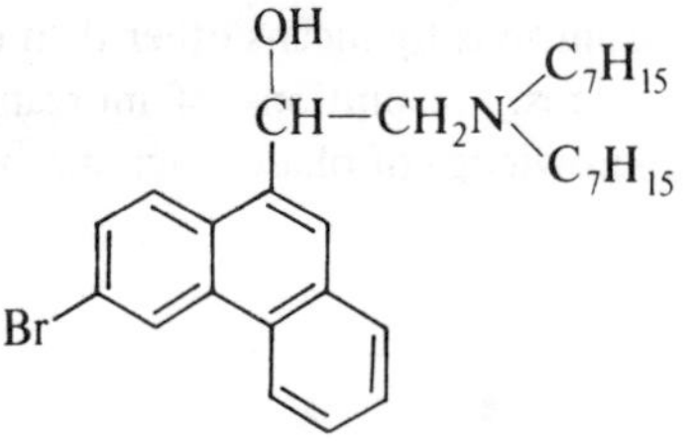

68

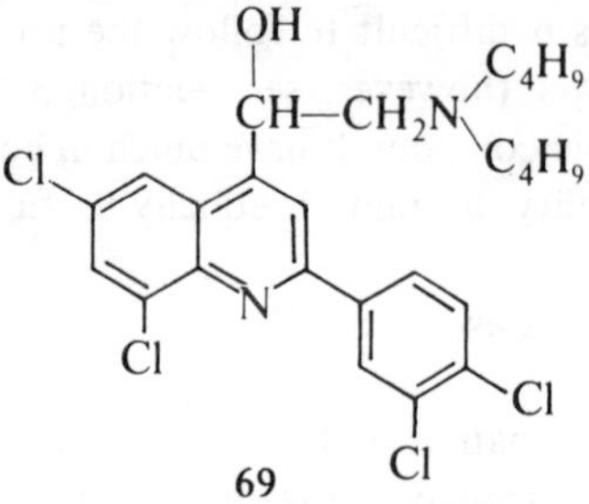

69

Many π-complexes are only moderately stable, and resolution through formation of such complexes is complicated by decomposition of the adducts. Occasionally, only one diastereomeric complex precipitates while the other, less stable one remaining in solution dissociates. Conversely, occasionally it is the dissociated substrate which precipitates while the stabler diastereomeric complex stays in solution. The latter case corresponds to the resolution of [6]-helicene with TAPA.[117, 129]

The equilibration between the charge transfer complexes and their components is the basis of chromatographic cleavage of these complexes to regenerate the electron donor substrate. An elaboration of this process is the direct resolution of chiral electron-rich substances by complexation chromatography numerous examples of which have involved TAPA-impregnated silica gel or alumina columns. An early report of the partial resolution of hydrocarbons and aromatic ethers on such columns was given by Klemm et al.[130, 131] One of the *o*-hexaphenylene atropisomers was partially resolved by chromatography on TAPA-impregnated silica gel.[132] Significantly, a synthetic optically active polyester covalently bound to an aminooxytetranitrofluorenone moiety failed to resolve either *sec*-butyl α-naphthyl ether or hexahelicene.[133]

However, resolution of a whole range of helicenes has been achieved by high-pressure liquid chromatography on TAPA-coated silica gel.[271] Gil-Av and his associates also have achieved resolutions on TAPA covalently bound to silica gel.[272] Other chiral stationary phases of this type have been suggested,[273, 274] and some have already been successfully applied to resolutions.[275]

Only one other π-complex-forming resolving agent has been examined, namely, 2-naphthylcamphylamine. This Lewis base has been employed in the resolution of *N-sec*-butylpicramide, **70**, by complexation.[117] The latter, itself a potential π-acid resolving agent, does not seem to have been used for this purpose.

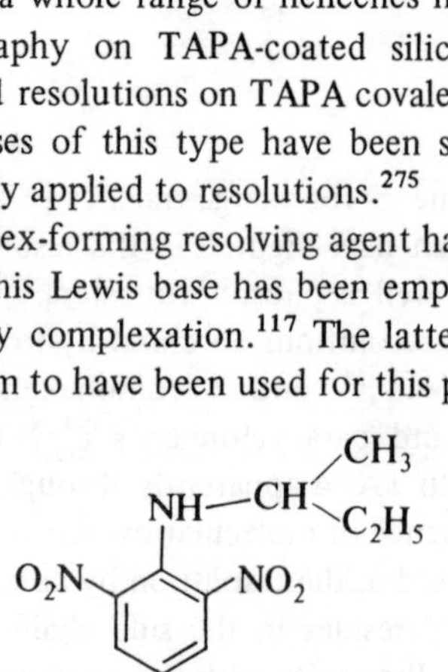

70

The properties of chiral charge transfer complexes have been little studied. One of the few studies is that of charge transfer bands in the circular dichroism spectra of the diastereomeric complexes of TAPA and hexahelicene.[129]

5.1.8 Crystalline Inclusion Compounds

Certain molecules crystallize in such a way as to leave spaces within the lattice that permit other molecules of appropriate shapes and sizes to be included.[134,135] The cavities wherein included molecules lodge may be lamellate (graphites), tubular (urea), or in the form of cages (clathrates).

In some cases, the cavities which exist in the interior of the crystal lattice of the host substance are *chiral*.[136] This makes possible the resolution of racemates by more or less preferential inclusion of one of the enantiomers. The selectivity of the process depends upon the relative stability of the crystalline diastereomeric complexes formed. A tight fit between guest and host molecules would seem to be a necessary condition for enantioselectivity. Thus, a chiral inclusion compound in which guest molecules have considerable freedom to move has little chance of discriminating between the guest enantiomers. A second condition for selectivity is evidently tied to the topological compatibility, that is, the lock-and-key relationship of the chirality of the cavity and that of the guest molecule. Additionally, the selectivity is related to the flexibility of the guest in adapting itself to the configuration of the vacant space.

Relatively few substances form optically active inclusion compounds. Those which have been applied to resolutions and have the most interesting properties and utility are urea, the cyclodextrins, tri-*o*-thymotide, and some of the bile acids.

(a) Urea

Urea is an achiral molecule which normally crystallizes with tetragonal symmetry. However, in the presence of compounds capable of inclusion, enantiomorphous crystalline structures are formed containing helical tubular cavities which are right-handed or left-handed in different crystals[136, 148, 149, 150] (Fig. 1).

The study of urea inclusion compounds in connection with resolutions has been extensively pursued by Schlenk, Jr.[137–145] since 1952 and has given rise to a substantial number of patents (assigned to BASF).[146] The substances that can be resolved by this process are linear aliphatic molecules (unbranched hydrocarbons, acids, and esters) whose chirality is due to a substituent of very small size (CH_3, Cl, Br, NH_2, SH, OH, epoxide, etc.).*

In the presence of a racemate *dl*, two pairs of diastereomeric inclusion compounds, D*d*, L*l* and D*l*, L*d*, can be formed from the two enantiomorphous urea lattices D and L:

D*d*	L*l*
D*l*	L*d*

Mirror

*The mean diameter of the canals of urea is 5.5 Å, while the cross section of a paraffin chain stretched into an elongated zigzag is 4.8 Å (see ref. 136).

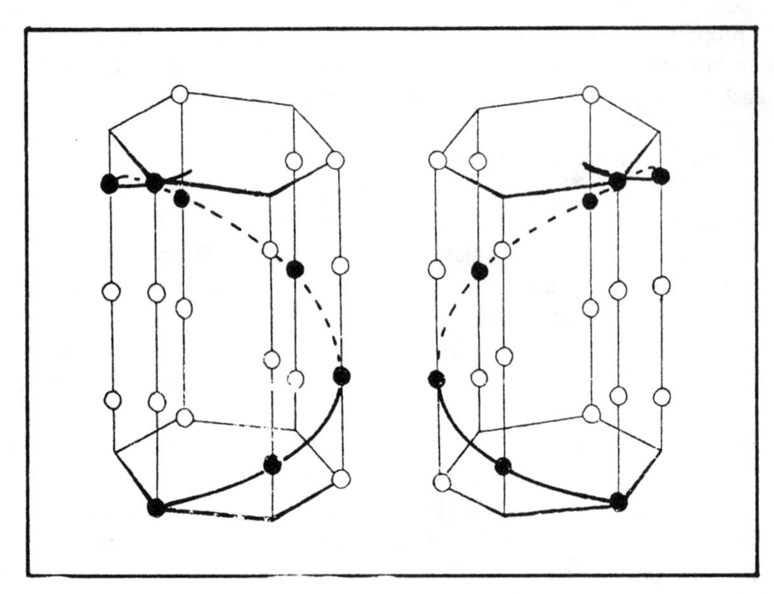

Figure 1 Mirror image unit cells in the lattice of inclusion compounds of urea. The black circles represent those molecules belonging to the unit cell proper; the $C(NH_2)_2$ moieties of these molecules are oriented to the outside of the canal. C. Asselineau and J. Asselineau, "Use of Some Inclusion Compounds in Organic Chemistry," *Ann. Chim.* (Paris), 1964, **9**, 461. Reproduced by permission of Masson, S.A., Paris.

In practice, one of the pairs is stabler than the other and is preferentially formed with a selectivity which depends on the compound included, for example D*d* and L*l* stabler than D*l* and L*d*. For resolution to occur, it is necessary to have only *one* of the D or L lattices crystallize, by an appropriate technique, in the presence of the racemic substrate. As an example, lattice D preferentially includes enantiomer *d* (D*d* more stable than D*l*); its crystallization will yield product enriched in *d*, while the mother liquors are enriched with *l*.

Generally, the complexes are formed in methanol. Schlenk has shown[141] that crystallization of a urea lattice having exclusively one configuration can take place spontaneously or can be artificially provoked in various ways: by use of a racemate slightly enriched with respect to one enantiomer or seeding with the inclusion compound of any pure enantiomer or even with enantiomorphous crystals of an achiral substance, for example, glycine[147]; by contact with asymmetric surfaces of solid natural substances (hemp, flax, casein, etc.); or in the presence of an optically active cosolute.

Since the optical enrichment attained in one operation is not generally large, numerous successive crystallizations are required to attain optical purity.[138] This process is therefore only useful in those cases in which other resolution methods are impracticable, for example, in the case of chloroalkanes[137] or long-chain epoxy-esters.[147]

(b) *Cyclodextrins*

The cyclodextrins[151–153] are cyclic glucose oligomers arising from the bacterial degradation of starch and contain anywhere from 6 to 10 molecules of D-glucopyranose per cycle (Fig. 2).

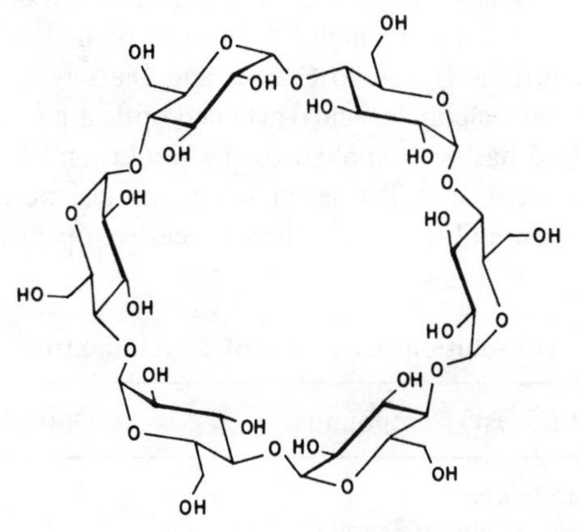

Figure 2 α-Cyclodextrin.

α-, β-, and γ-Cyclodextrins (hexamer, heptamer, or octamer; cycloamyloses) form crystalline inclusion compounds with many substances. Those formed with α-cyclodextrin have cage structures when the included molecules are small and neutral, and have channel structures when the guest molecules are ionic and long

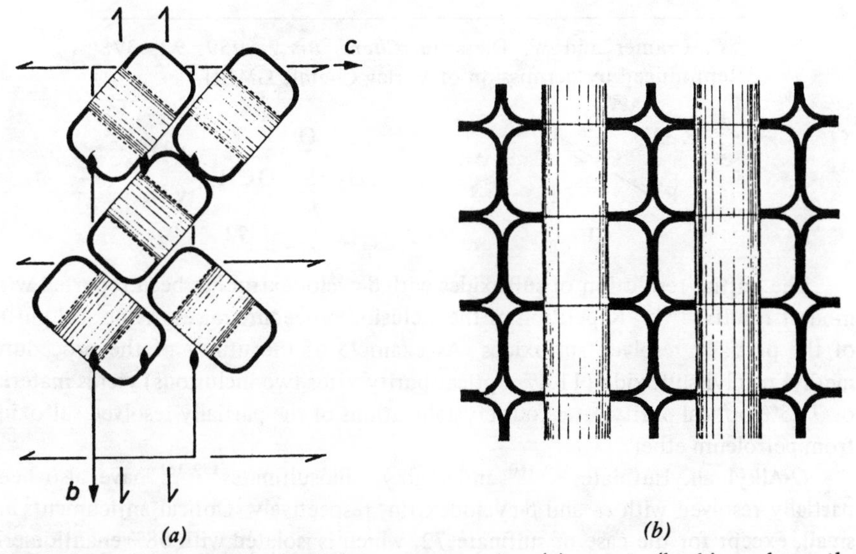

Figure 3 α-Cyclodextrin inclusion compounds: (a) cages (looking along the crystallographic *a*-axis; α-cyclodextrin molecules are seen from the side); (b) canals (side view). R. K. McMullan, W. Saenger, J. Fayos, and D. Mootz, *Carbohydr. Res.*, 1973, **31**, 37. Reproduced by permission of Elsevier Scientific Publishing Company and the authors.

(Fig. 3).[152] Note that in the first type of structure, the cyclodextrin molecule itself constitutes the chiral cage. (The α-cyclodextrin molecule takes the shape roughly of a section of a cone with a height of 9 Å and a maximum outer diameter of 13.5 Å. The interior cavity is also conical with a diameter ranging from 5 to 8 Å.)

The crystal structures of inclusion compounds of the other cyclodextrins do not seem to have been studied much.[154] The use of inclusion compounds of β-cyclodextrin in resolutions is due to Cramer and Dietsche.[155, 156] The compounds resolved by them (esters, alcohols, acids) attained optical purities of 12% maximum (Table 3). The method has been applied to the resolution of phosphinates **71** with either α- or β-cyclodextrins.[157] The latter led to optical purities has high as 66% after one crystallization and 84% after three successive operations (**71**, R_1 = iso-Pr; R_2 = Me).

Table 3 Resolutions by means of β-cyclodextrin[a]

Resolved (Guest) Compounds	Optical Purity (%)
Ethyl mandelate	3.3
Ethyl α-chlorophenylacetate	3.2
Ethyl α-bromophenylacetate	5.8
Menthol	4.9
Ethyl atrolactate	8.4
Atrolactic acid	5.6
2,3-Dibromo-3-phenylpropionic acid	11.3
2,3-Dibromosuccinic acid	8.2

[a] F. Cramer and W. Dietsche, *Chem. Ber.*, 1959, **92**, 378. Reproduced by permission of Verlag Chemie GMBH.

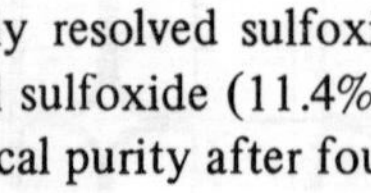

$R_1O(R_2)P(=O)H$

71

$CH_3-S(\rightarrow O)-OCH(CH_3)_2$

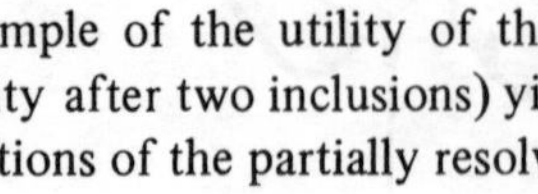

72

The partial resolution of sulfoxides with β-cyclodextrin has been reported with modest results.[158, 159] Repetition of the inclusion procedure raises the optical purity of the partially resolved sulfoxides. As example of the utility of the procedure, methyl *p*-tolyl sulfoxide (11.4% optical purity after two inclusions) yields material of 71.5% optical purity after four crystallizations of the partially resolved sulfoxide from petroleum ether.

O-Alkyl alkylsufinates[159, 160] and *S*-alkyl thiosulfinates[159, 161] have also been partially resolved with α- and β-cyclodextrin, respectively. Optical enrichments are small, except for the case of sulfinate **72**, which is isolated with 68% enantiomeric excess after a single crystallization with β-cyclodextrin.[159]

Finally, Knabe and Agarwal[162] have described an attempt at resolving $CF_3CHBrCl$ (the anesthetic Halothane®) by inclusion in α-cyclodextrin. Enantiomeric enrichment was observed but its magnitude was less than 1%.

(c) *Trio-o-thymotide*

Tri-*o*-thymotide (TOT), **73**, is a cyclic trilactone obtained by dehydration of thymotic acid.[163] In solution, TOT exists as an equilibrium of several chiral conformers the most stable of which possesses C_3 (propeller) symmetry.[164, 165] TOT crystallizes with spontaneous resolution while forming an inclusion compound with the solvent.[166] Since the interconversion of the two C_3 conformations is relatively rapid in solution[164] ($\Delta G^{\ddagger} = 20.9$ kcal · mol^{-1} at 0°C), it is possible for a single enantiomer to crystallize, under appropriate conditions, while the solution remains racemic (see Chapter 6). TOT can also crystallize in a *nonsolvated* form as a racemic compound.[167]

COOH OH → P_2O_5 or $POCl_3$, 15-40% → **73**

The inclusion compounds of TOT, which form with very many substances, belong to two categories according to the size of the guest molecules.[167] Cage structures (clathrates) are formed with small molecules (shorter than 9.5 Å) with a 2–1 stoichiometry (TOT–guest molecule). Longer molecules insert in channels with variable stoichiometry as a function of the size of the guest molecule. Both types of structure are spontaneously resolved. Lawton and Powell[167, 168] succeeded in partially resolving 2-bromobutane and 2-chlorooctane by inclusion in TOT.

Resolution trials with bromochlorofluoromethane have been claimed to give "evidence of success," but the quantity of haloform isolated from the complex was insufficient to measure the optical rotation.[169, 170]

In the clathrate form, the cavities of TOT have a C_2 symmetry. Optical enrichments as high as 47% may be obtained with guest molecules having the same symmetry as that of the cavities (e.g., with *trans*-2,3-epoxybutane).[171] This result is a good illustration of the lock-and-key relationship involved in the chiral discrimination process.

(d) *Choleic acids*

Desoxycholic acid, **74**, one of the bile acids, forms inclusion compounds called *choleic acids* with many acids, esters, alcohols, ethers, phenols, and hydrocarbons.[172] Sobotka and Goldberg[173] have suggested the use of inclusion compounds of desoxycholic acids to resolve compounds devoid of functional groups. Inclusion compounds of *dl*-dipentene and of *dl*-camphor actually contain a slight excess of the levorotatory forms; however, choleic acids do not seem to have been applied to other resolutions of this type.

Note that desoxycholic acid has been successfully used as an *acidic* resolving agent in the resolution of *erythro*-2,3-bis(*p*-methoxyphenyl)butyl- and pentylamines[174] and in the resolution of the extrememly water-soluble *trans*-3-amino-4-hydroxycyclopentene.[175]

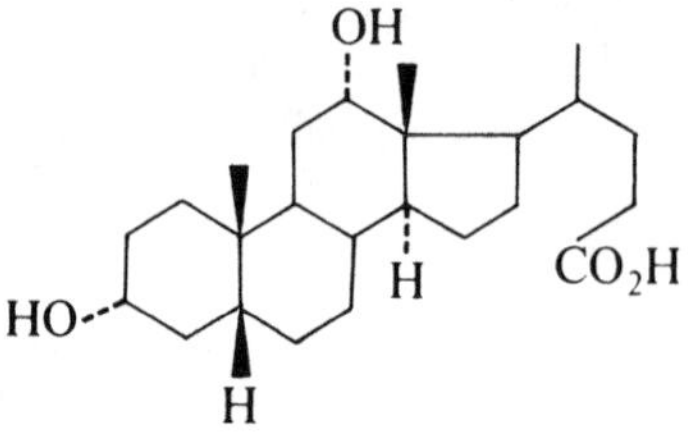

OH HO H H CO_2H

74

(e) *Dianin's compound and its derivatives*

Dianin's compound, **75**, known to form inclusion compounds with a very large variety of organic[176] and inorganic[177] compounds, was suggested by one of us as a potential resolving agent.[4] Crystallographic analysis of its complexes with ethanol and chloroform[178] has shown that the included molecules are in the interior of *cages* of ternary symmetry formed by six molecules of *racemic* **75** (Fig. 4). The same behavior has been observed in the case of analogues of Dianin's compound[179] such as **76**.

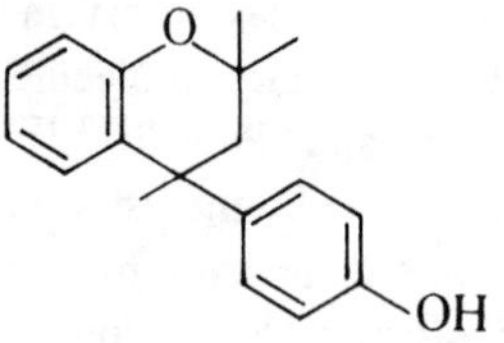

O OH S OH

75 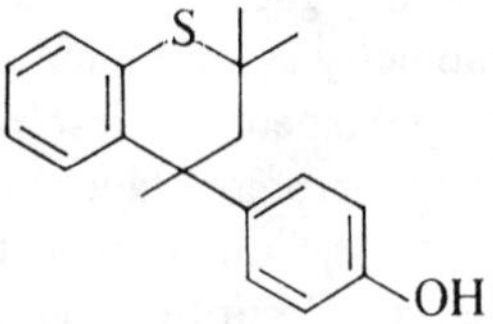 76

The centrosymmetrical structure of these clathrates leads one to understand why *resolved* Dianin's compound no longer forms inclusion compounds and therefore cannot serve as a resolving agent.[180] In order to overcome this stumbling block, an attempt has been made to construct cages of the same type whose chirality is a consequence of a quasi-racemic structure (see Section 5.1.9). Diastereomers **77** and **78**, 2-nor analogues of Dianin's compound, retain the ability to form inclusion compounds in the racemic form.[181, 182] An equimolecular mixture of *resolved* (+)-**77** and (+)-**78** with inverse configurations at C-4 gives a quasi-racemic clathrate in the presence of solvent.[183] Similarly, another quasi-racemic clathrate is formed between resolved Dianin's compound (−)-**75** and (+)-**78**. On the basis of X-ray analysis, these clathrates possess the same organization as that of Dianin's compound except

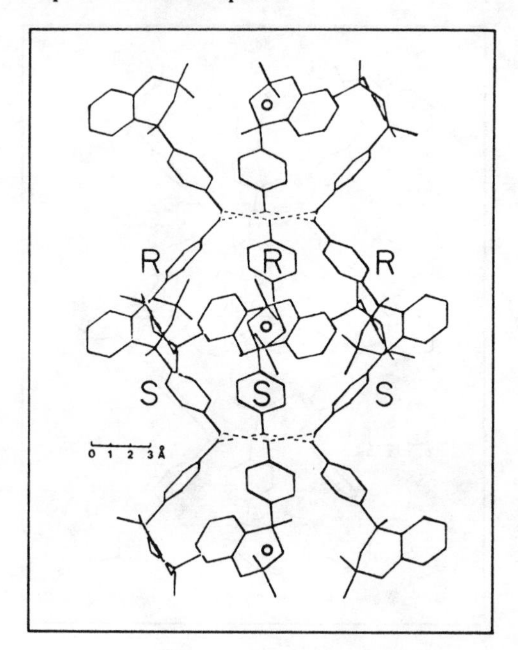

Figure 4 Adapted from D. D. MacNicol, J. J. McKendrick, and D. R. Wilson, *Chem. Soc. Rev.*, 1978, 7, 65 by permission of the Royal Society of Chemistry and the authors.

that the space group becomes chiral[184]: R3 instead of $R\bar{3}$. These chiral cages are built up of three molecules of one of the compounds linked to three molecules of the other by a pseudocenter of symmetry, as shown in Fig. 5.

Nonetheless, the chirality of the cavities of these quasi-racemic clathrates is apparently insufficient to selectively recognize one of the enantiomers of an included compound: inclusion of racemic ethyl α-bromopropionate in the clathrate [(−)-75, (+)-78] yields an ester of optical purity lower than 0.5%.[183] This negative result may be explained either by the quasi-centrosymmetric form of the chiral cavity and/or by the fact that the asymmetric guest molecule is disordered in the interior of a cavity of high (C-3) symmetry.[185]

(+) 2R, 4S **77** (+) 2R, 4R **78** S(−) **75**

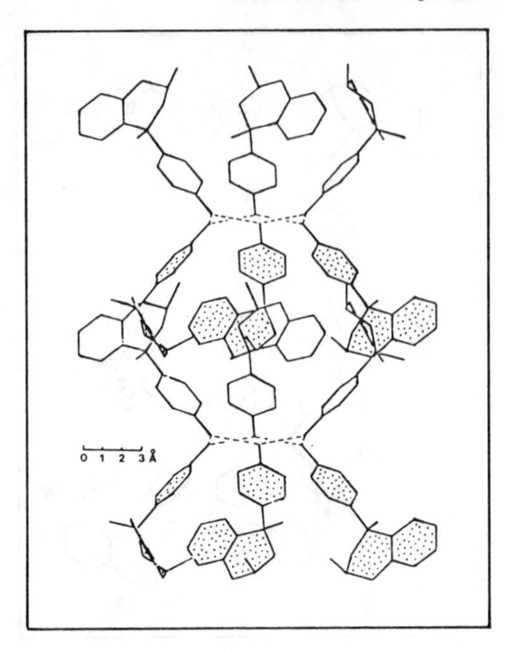

Figure 5 Adapted from D. D. MacNicol, J. J. McKendrick, and D. R. Wilson, *Chem. Soc. Rev.*, 1978, 7, 65 by permission of the Royal Society of Chemistry and the authors.

5.1.9 Quasi-Racemates

We have previously described quasi-racemates (see Section 2.3.5). These have been applied in particular to the correlation of absolute configurations. However, these crystalline addition compounds may also be employed to carry out resolutions. Delépine[186] is perhaps the first to have suggested, in the course of his studies of Werner complexes of rhodium, iridium, and chromium, that formation of "active racemates" might be used as a new resolution method. From his experimental details (these are rather complicated and more qualitative than quantitative), one may retain the fact that reaction of 2 moles of a racemic complex (*A*) with 1 mole of a resolved sample of another complex (*B*) leads to the precipitation of a relatively insoluble quasi-racemate (*AB*) which leaves in solution the uncombined enantiomer (*A*) more or less pure.

Arsenijevic et al. are the only ones who have provided precise experimental details for the formation of such compounds in connection with successful resolutions. Their experiments deal with, for example, the preparation of (+)-malic acid which is made possible by its combination with (+)-tartaric acid[187] or with the resolution of (±)-tartaric acid by (−)-asparagine, or, inversely, of (+)-asparagine by an optically active tartaric acid.[188]

These quasi-racemates are, in general, relatively unstable. Consequently, they do not suffer recrystallizations well (as Pasteur had already observed for the ammonium malates and tartrates). One may, in turn, take advantage of this instability to regenerate the constituents of the quasi-racemate by successive recrystallizations. In the case of the tartaric acid/malic acid quasi-racemate, the complex is more easily decomposed by removal of the tartaric acid in the form of its potassium acid salt.

5.1.10 Physical Properties of Diastereomeric Salts

(a) Crystal structures of diastereomeric salts

Few crystal structures of salts employed in the resolution of enantiomers have been determined by X-ray crystallography. Essentially for technical reasons, that is, the need to have well-formed crystals, the determination of the crystal structures of *both* diastereomeric salts formed in a resolution is unfortunately quite rare.

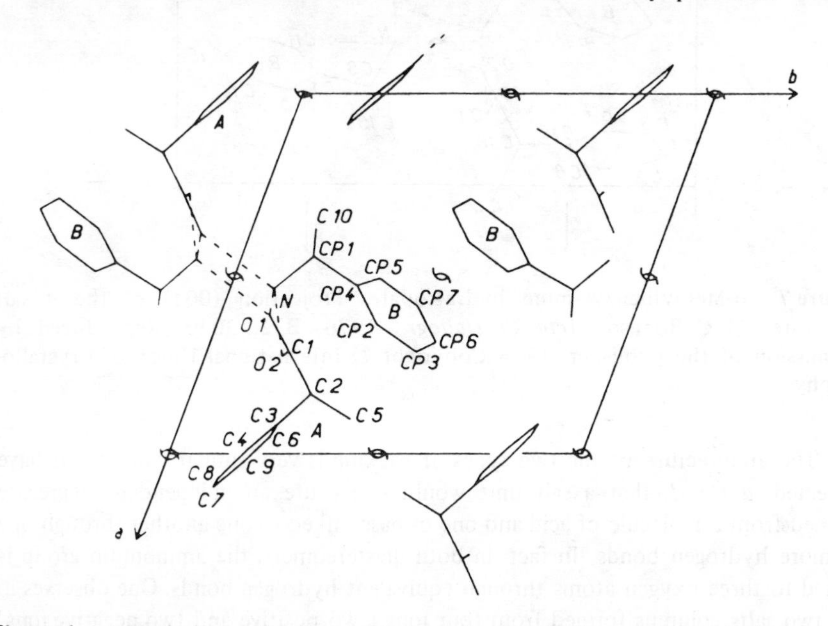

Figure 6 α-Methylbenzylamine hydratropate. Projection (001) of the *p* salt structure. M. C. Brianso, *Acta Crystallogr.*, 1976, **B32**, 3040. Reproduced by permission of the publisher. 1976 Copyright © International Union of Crystallography.

Brianso has described both the *p* and the *n* salts of hydratropic acid with α-methylbenzylamine (Figs. 6 and 7).[189] Comparison of the three-dimensional structures of the two salts leads first of all to the observation that the ions of acid (*A*) and of base (*B*) possess identical conformations in the *n* and *p* salts. In both salts the benzene rings eclipse the hydrogen atoms bonded to the asymmetric carbon atoms. The most obvious difference between the two salts appears in the relative orientation of the aromatic rings of the acid and of the base which are parallel in the *n* salt and perpendicular in the *p* salt.

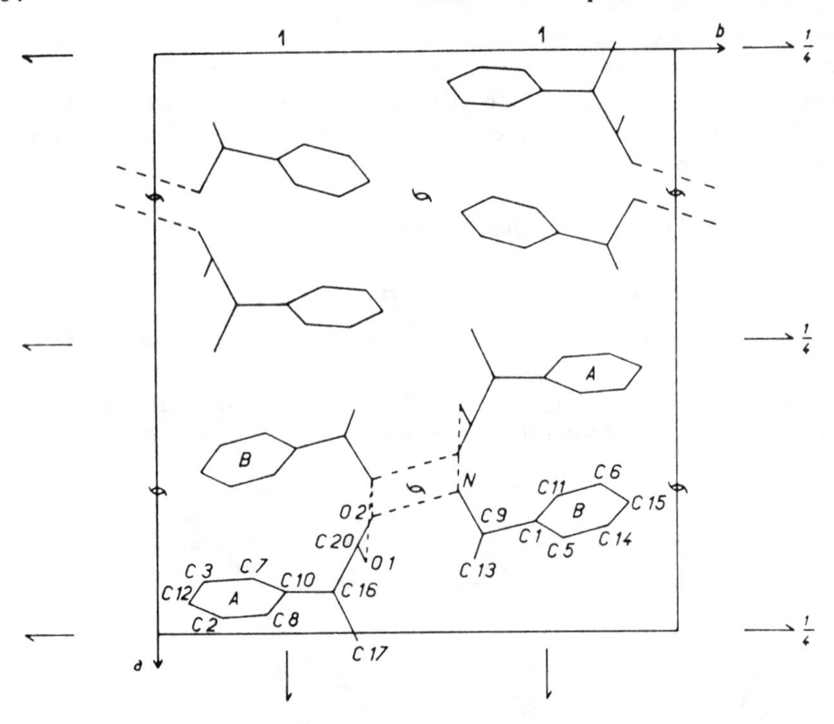

Figure 7 α-Methylbenzylamine hydratropate. Projection (001) of the n salt structure. M. C. Brianso, *Acta Crystallogr.*, 1976, **B32**, 3040. Reproduced by permission of the publisher. 1976 Copyright © International Union of Crystallography.

The architecture of the two types of packing is very similar. One might have expected, *a priori*, that a salt unit would constitute an independent aggregate formed from a molecule of acid and one of base linked to one another through one or more hydrogen bonds. In fact, in both diastereomers, the ammonium group is linked to three oxygen atoms through equivalent hydrogen bonds. One observes in the two salts columns formed from four ions (two positive and two negative ions) whose axis is the helical 2_1 axis which is parallel to the common parameter of the two structures (n salt: $c = 5.800$ Å; and p salt: $c = 6.558$ Å). These columns are juxtaposed through lattice translations which are perpendicular to their axes in the case of the p salt (group $P2_1$) and through the interplay of a binary axis which is also perpendicular to the axis of the n salt (group $P2_12_12_1$). Two types of intermolecular interactions are thus distinguished – strong hydrogen bonds which form the molecular column and weak bonds of van der Waals type which stabilize the general cohesion of the packing of the different columns.

Finally, examination of the superposition of the structures of p and n salts projected in a plane perpendicular to the columnar axis (Fig. 8) simultaneously shows the difference in the positioning of the aromatic rings which prevents the cocrystallization of the two salts and thus makes their separation possible by crystallization.

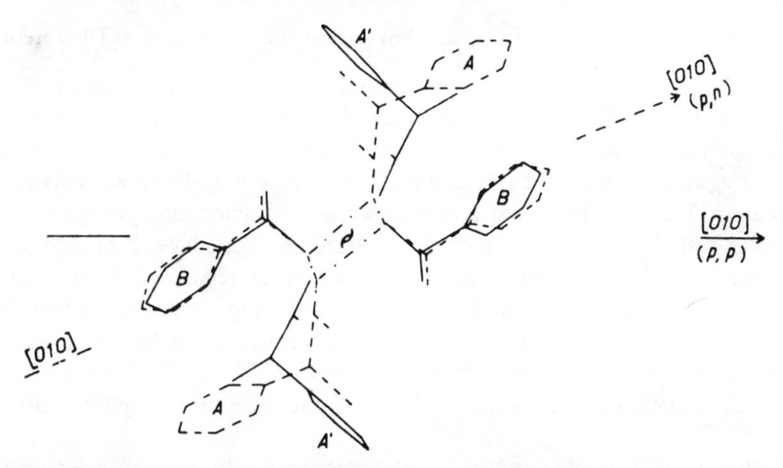

Figure 8 α-Methylbenzylamine hydratropate. Superposition of the (001) projections of the *p* and *n* salt structures in the vicinity of the helical binary axis parallel to the common *c* axis: (——) *p* salt; (----) *n* salt. M. C. Brianso, *Acta Crystallogr.*, 1976, **B32**, 3040. Reproduced by permission of the publisher. 1976 Copyright © International Union of Crystallography.

In the case of two other salts, α-methylbenzylamine mandelate[190] and α-methylbenzylamine α-phenylbutyrate[191] (both *n* salts), one finds an analogous system of columns built up of chains of hydrogen bonds wrapped around a helical 2_1 axis.

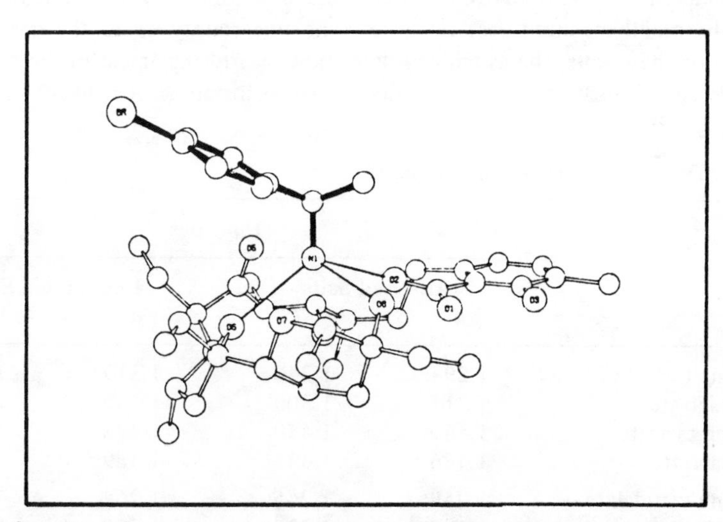

Figure 9 Crystal structure of the α-methyl-(*p*-bromobenzyl)amine (solid bonds) salt of lasalocid (open bonds). View is perpendicular to the amine C–N bond. Reproduced from ref. 19 by permission. J. W. Westley, R. H. Evans, and J. F. Blount, *J. Am. Chem. Soc.*, 1977, **99**, 6057. Reproduced by permission of the publisher and authors. Copyright 1977, American Chemical Society.

A different sort of X-ray crystallographic study attempts to rationalize the preferential formation of one salt between the acid lasalocid (see p. 261) and α-methylbenzylamine derivatives which is responsible for the utility of lasalocid as a resolving agent. The *n* salt of (−)-lasalocid with α-methyl-(*p*-bromobenzyl)amine is monomeric. The three hydrogen bonds between the ammonium hydrogens and the O–2, O–6, and O–8 oxygen atoms of the lasalocid conjugate base, the conformation about the C–N bond, and the steric fit of the aryl group relative to the shape of the large lasalocid ion determine the enantioselectivity[19] (Fig. 9). Along the same lines, Yoneda et al. have carried out a number of crystal structure determinations of the less-soluble diastereomeric salts formed between several cobalt(III) complex cations and the *d*-tartrate, *d*-hydrogen tartrate, and α-bromocamphor-π-sulfonate anions.[264–270]

X-Ray crystallographic studies of diastereomeric salts may go a long way in rationalizing the hitherto mysterious preferred pairings of substrates with some but not other resolving agents as well as explaining the enantioselectivity of the resolving agent. Such studies are in their infancy as of this writing.

(b) Densities of diastereomeric salts

Given their different modes of packing, it would not be surprising to find that two diastereomeric salts *p* and *n* have different densities in the crystal state. We are indebted to Bergman[192] for a series of careful measurements of such densities. These are given in Table 4. The densities of these tartrates and mandelates of various alkaloids, crystallized from water, were determined by flotation (e.g., in xylene-dibromoethane mixtures). Despite some uncertainty as to the degree of hydration of these salts, the assembled data show clearly that most of the *p* and *n* pairs have significantly different densities. In some instances, it is likely that the

Table 4 Densities of diastereomeric salts[a]

	Densities			
	Air-dried Salts		Dedydrated Salts	
Salts	*d*	*l*	*d*	*l*
Cinchonine tartrate	1.294	1.340	1.319	1.355
Quinine tartrate	1.311	1.300	1.332	1.303
Strychnine tartrate	1.449	1.450	1.433	1.422
Brucine tartrate	1.486	1.435	1.489	1.459
Cinchonine mandelate	1.259	1.245	1.264	1.249
Quinine mandelate	1.195	1.320	1.358	1.341
Strychnine mandelate	1.357	1.346	1.355	1.356
Brucine mandelate	1.345	1.342	1.350	1.345

[a] *d* or *l* refers to the rotation of the acid. S. W. Bergman, *Ark. Kemi, Min. Geol.*, 1927, **9**, No. 42. Reproduced by permission of the Royal Swedish Academy of Sciences and the author.

difference is large enough to allow a separation by centrifugation or even by decantation. We are not aware of examples of such separations, except for one observation by Graf and Boedekker[193] who pointed out that *p* and *n* ethyl β-phenylalaninate hydrogen tartrates could be partially resolved by decantation.

(c) *Melting points and enthalpies of fusion of diastereomeric salts*

The study of diastereomeric salt mixtures through the measurement of melting points and enthalpies of fusion is uncomplicated, provided that the diastereomer salts are sufficiently stable to melt without decomposition. In fact, in our experience and contrary to our original preconceptions, diastereoisomeric salts of nearly all types melt cleanly and at relatively low (<200°C) temperatures. A frequently encountered characteristic which does sometimes complicate the study of diastereomer salt mixtures is their solvation.

It is uncommon to find melting points of both diastereomeric salts that mediate resolutions given in the experimental sections of research articles. We have assembled, in Table 5, the enthalpies of fusion of a few diastereomeric salts measured by microcalorimetry and, in Table 6, some data concerning the melting points of diastereomeric salts culled from the literature. These data are too sparse as yet to permit useful analysis and interpretation (see also Section 5.1.11).

Table 5 Enthalpies of fusion of diastereomeric salts[a]

		ΔH^f (kcal·mol^{-1})	
Acid	Amine	*p* salt	*n* salt
Hydratropic acid	α-Methylbenzylamine	11.9	10.1
α-Phenylbutyric acid	α-Methylbenzylamine	8.2	12.4
α-Phenylvaleric acid	α-Methylbenzylamine	6.45	13.2
Mandelic acid	α-Methylbenzylamine	6.6	11.3
Hydratropic acid	Amphetamine	10.7	8.2

[a] Taken in part from ref. 194 and from unpublished results obtained at the Collège de France.

(d) *Solubilities of diastereomeric salts*

This last of the physical properties to be considered is by no means the least important inasmuch as it is the key that governs their separation. As we have already seen in several instances, the solubility is directly related to the enthalpies of fusion and the melting points (except when association occurs with solvent molecules either in the solid state or in solution).

The greater the difference in solubility of the diastereomers, the easier the resolution. However, quantitative data illustrating this generalization are hard to come

Table 6 Comparison of melting points of various diastereomeric salts[a]

Acid	Amine	mp (°C)		Refs.
		p salt	*n* salt	
Mandelic acid	Amphetamine	166	162	195
Mandelic acid	Adrenaline	139	oil	195
Mandelic acid	α-Methylbenzylamine	109	178	195
Mandelic acid	Ephedrine	170	91	196
Mandelic acid	Propadrine	172	165	196
Mandelic acid	ψ-Propadrine	164	170	196
Mandelic acid	β-Phenylpropylamine	127	119	196
Mandelic acid	Strychnine	184	115	197
Mandelic acid	Quinidine	110	100	197
Mandelic acid	Brucine	97	135	197
Mandelic acid	Quinine	202d	180d	197
Tartaric acid	Pipecoline[b]	111	126	197
Tartaric acid	Amphetamine	170	145	195
Tartaric acid	Adrenaline	150	144	195
α-(4-Methylphenoxy)propionic acid	Amphetamine	187	156	195
α-Naphthoxypropionic acid	Amphetamine	105	oil	195
α-(4-Chlorophenoxy)propionic acid	Amphetamine	179	148	195
α-Phenoxybutyric acid	Amphetamine	153	148	195
α-Phenoxypropionic acid	Amphetamine	151	147	195
α-(2-Methoxyphenoxy)propionic acid	α-Methylbenzylamine	121	oil	195
Hydratropic acid	α-Methylbenzylamine	165	146	194
α-Phenylbutyric acid	α-Methylbenzylamine	134	165	194
α-Phenylvaleric acid	α-Methylbenzylamine	129	166	194
p-Methoxymandelic acid	Cinchonine	160	174	197
Valeric acid	Brucine[b]	88	63	197
Mandelic acid	α-Methyl-*p*-methylbenzylamine	140	146	198
6,6′-Dinitrodiphenic acid	α-Methylbenzylamine	219	205	199

[a] The melting points indicated in the table are rounded off.
[b] In the case where solvation of a given salt has been recognized, the table lists only the melting point of the anhydrous salt.

by. Those of Ingersoll and co-workers[198, 199, 222a] and Leclercq and Jacques[194] are among the few known (Table 7).

The difference in solubility between two salts may be quite substantial. For example, the salts of 2,2′-dihydroxybinaphthyl-3,3′-dicarboxylic acid with leucine methyl ester studied by Kuhn and Vogler[222b] have the following solubilities in methanol at 28.5°C: *p* salt (mp 217.5°C) 0.44 g/100 g; *n* salt (mp 176°C) 101 g/100 g solvent.

The polymorphism of salts, whether or not associated with changes in degree of solvation, may also lead to important changes in relative solubility which, however, remain unpredictable. Such a case was described by Zambito, Peretz, and Howe[222c] for the salts of brucine and *p*-nitrobenzoylthreonine. It is remarkable in that the two polymorphic forms of one diastereomer are of different colors.

Table 7 Comparison of solubilities of various pairs of diastereomeric salts

Acid	Amine	Solubilities[a]		Solvent	T (°C)	Ref.
		p salt	n salt			
		(a)				
Hydratropic acid	α-Methylbenzylamine	4.8	6.4	Ethanol	10	194
		7.8	11.3	Ethanol	30	
Phenylethylacetic acid	α-Methylbenzylamine	22	3.6	Ethanol	10	194
		30	5.6	Ethanol	30	
Phenylpropylacetic acid	α-Methylbenzylamine	10.3	3.4	Ethanol	10	194
		17.4	6.0	Ethanol	30	
Mandelic acid	α-methylbenzylamine	23	4.0	Ethanol	10	194
		(b)				
Mandelic acid	α-Methylbenzylamine	18	4.91	Water	30	198
Mandelic acid	α-Tolylethylamine	7.12	5.18	Water	25	198
Camphoric acid	α-Tolylethylamine	2.49	1.05	Water	25	222a
6,6′-Dinitrodiphenic acid	α-Methylbenzylamine	0.61	13.0	Acetone	30	199
		3.95	4.74	95% Ethanol	25	
		(c)				
Mandelic acid	Ephedrine	6.0	77.48	Water	25	196
		6.27	80.61	Water	37	
Mandelic acid	Propadrine	6.08	7.70	Water	25	196
		7.39	10.69	Water	37	
Mandelic acid	ψ-Propadrine	11.95	6.22	Water	25	196
		18.74	8.43	Water	37	
Mandelic acid	Benzedrine	5.98	3.93	Water	25	196
		7.25	4.63	Water	37	
Mandelic acid	Isobenzedrine	14.59	51.15	Water	25	196
		25.30	63.46	Water	37	

[a] In (a), solubilities are in g/100 g solution; in (b), in g/100 g solvent; in (c), solubility units are not given in ref. 196; we may suppose that the data are expressed in g/100 g solvent.

5.1.11 Binary Melting Point Diagrams of Diastereomeric Salts

Binary phase diagrams which describe a system consisting of two diastereomeric salts are constructed and may be described just as are those earlier seen in connection with enantiomer mixtures. In the most general case, such salts form eutectics, and it is only exceptionally that they form addition compounds (double salts equivalent to racemic compounds). However, the occurrence of solid solutions (resembling pseudoracemates) is certainly less rare in systems of diastereomeric salts than it is for enantiomer mixtures.

The only point differentiating these diagrams of p and n salts from those formed by enantiomers is that they do not exhibit the symmetry shown by the latter. The composition of a eutectic formed by p and n salts is equimolar only by accident.

Several phase diagrams of pairs of diastereomeric salts have been determined by Leclercq and Jacques.[194] In the case described in Fig. 10 (salt formed by α-methylbenzylamine and hydratropic acid), one observes that the liquidus curves calculated by means of the Schröder–Van Laar equation (Section 2.2.3) from the enthalpies of fusion and melting points of the pure salts match the experimental points very satisfactorily. Figure 11 concerns the salts of α-phenylvaleric acid with α-methylbenzylamine which cocrystallize partially, that is, they form solid solutions over part of the composition range. The Tammann diagram curve (lower part of the figure) reveals the region of miscibility of the solid diastereomers.

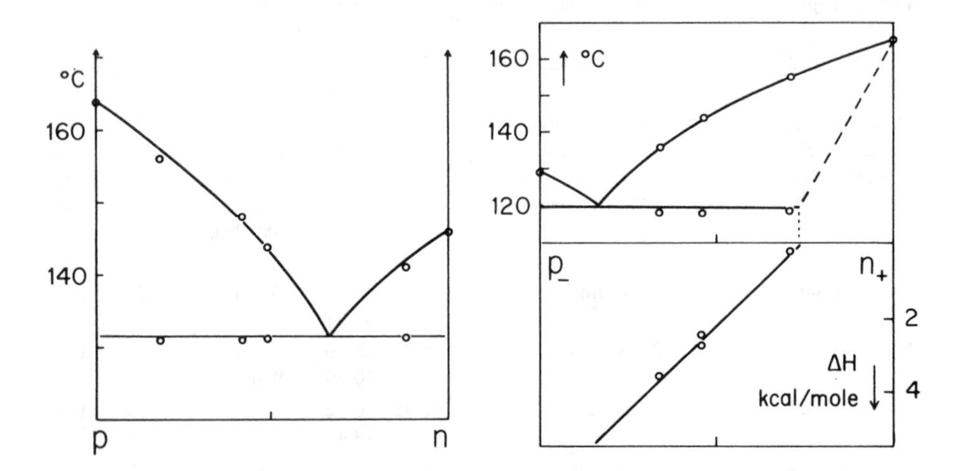

Figure 10 Phase diagram for p and n salts derived from α-methylbenzylamine and hydratropic acid.[276]

Figure 11 Phase diagram for p and n salts derived from α-methylbenzylamine and α-phenylvaleric acid.[194] Lower part: Tammann diagram for the same salts. M. Leclercq and J. Jacques, *Bull. Soc. Chim. Fr.*, 1975, 2052. Reproduced by permission of the Société Chimique de France.

To the extent to which they are uncomplicated by polymorphism, these binary diagrams yield two useful kinds of information bearing on the progress of resolutions: (1) they give the composition of the eutectic which one finds again as a solubility maximum at virtually the same composition in the solubility diagram; and (2) from the melting points (fusion interval) it is possible to estimate the purity of the salts during the course of separation. This technique may, of course, be applied only to those salts which melt without decomposition and are unsolvated.

5.1.12 Solubility Diagrams of Diastereomer Salt Mixtures

Few diastereomeric salt pairs have been studied from the standpoint of solubility difference or other property with a view to applying the measured difference rationally to the improvement of a resolution. However, particularly in the case of industrial resolutions where optimization is economically justifiable, such investigation may lead to the desired improvement as shown in Section 7.3.

Very few experimental solubility diagrams of diastereomeric salt mixtures have been determined.

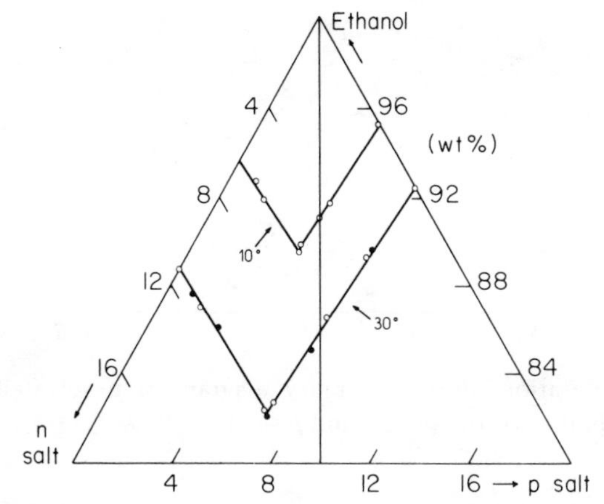

Figure 12 Solubility diagram of α-methylbenzylamine hydratropates p and n. Quantities of the three components are given in grams. The 30° isotherm: white circles = salts of (+)-hydratropic acid and (+) and (−) bases; black circles = salts of (−)-hydratropic acid and (−) and (+) bases. M. Leclercq and J. Jacques, *Bull. Soc. Chim. Fr.*, 1975, 2052. Reproduced by permission of the Société Chimique de France.

(a) Unsolvated salts

The diagram of Fig. 12, which describes the system of α-methylbenzylamine/hydratropic acid (p and n salts) in ethanol, illustrates the simplest case one may encounter – that of unsolvated salts which do not form solid solutions. The composition of the eutectic as determined from the diagram ($x = 0.41$ at 10°C; 0.38 at 30°C) is relatively close to that derived from the melting point diagram ($x = 0.33$)[194] The interpretation of the ternary diagram is given in Fig. 13. From an equimolar mixture of p and n (labeled M) in the presence of solvent at temperature T_0, the following situations obtain after the attainment of equilibrium:

1 A concentrated solution (given by A) deposits a solid mixture (n,p) of composition A_s in mother liquors of composition E and whose enantiomeric composition is given by E_s.

2 A more dilute solution (C) deposits crystals of pure p salt. The maximum yield of pure p salt is obtained when the overall composition of the mixture is given by B.

3 Above the isotherm PEN, the solution is unsaturated. Crystallization carried out under equilibrium conditions thus always allows one to obtain the less soluble salt pure. The techniques required to isolate the more soluble salt remaining in the eutectic are described in Section 5.1.15.

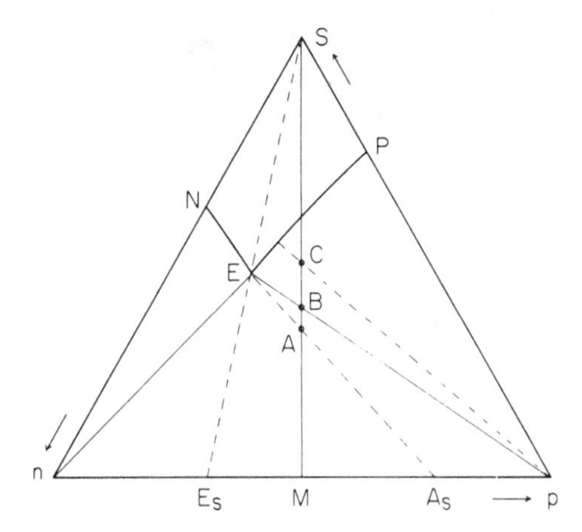

Figure 13 Interpretation of the solubility diagram of unsolvated salts. *N* and *P* represent the solubilities of the pure *n* and *p* salts.

(b) Solvated salts

Solvation of salts (generally by water or by polar solvents) is a relatively common phenomenon. Some rare examples of experimental ternary diagrams corresponding to this situation have been determined by Shimura and Tsutsui.[200] One of these cases, in which both salts are solvated, is represented by Fig. 14. The interpretation of such a diagram is quite similar to that given in part (a) of this section, immediately above, with the exception that the crystals which deposit here are the

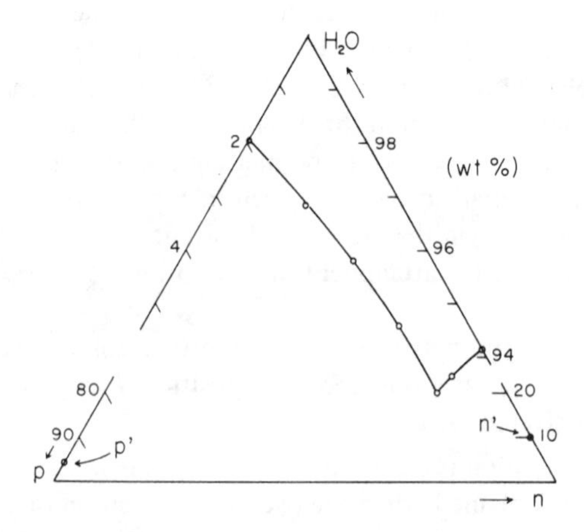

Figure 14 Solubility isotherm of the ternary system H_2O–Δ-[Co(ox)(en)$_2$](*d*-$C_4H_5O_6$)(*p*)–Λ-[Co(ox)(en)$_2$](*d*-$C_4H_5O_6$)(*n*) at 25°C. The solid phase *p*′ is 2.5 hydrate of *p*, and *n*′ is the monohydrate of *n*. Adapted from Y. Shimura and K. Tsutsui, *Bull. Chem. Soc. Jpn.*, 1977, **50**, 145 by permission of the Chemical Society of Japan.

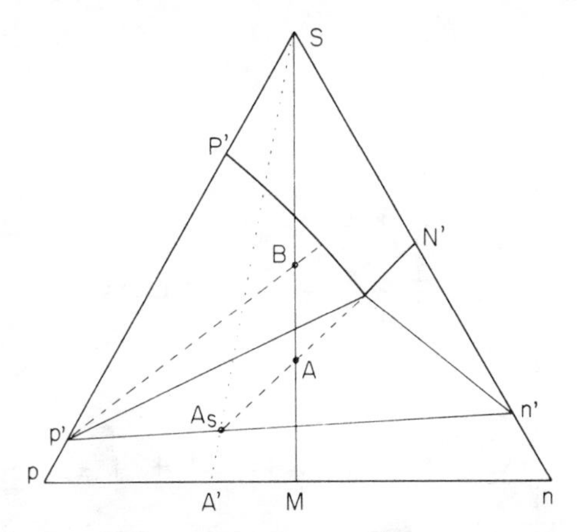

Figure 15 Solubility diagram of a system consisting of salts p and n in solvated form in the presence of solvent S; p' and n' represent the solvates; P' and N' are the solubilities of the pure solvates.

hydrated p' and n' salts and not the anhydrous ones (Fig. 15). From a concentrated solution (A) at temperature T_0, one obtains a mixture of solid p' and n' represented by A_s whose enantiomeric composition is given by A'. A more dilute solution (B) of the same diastereomer mixture deposits crystals of pure solvate p'.

The interaction between diastereomeric constituents in solution is neither simple nor easy to deal with. Nevertheless, it is hardly likely that the nature of the solvent would exercise a very large selective influence on the entities *in solution*. Rather, the solvent might well play a decisive role on the structure of *crystalline* products, particularly when the two salts present in a mixture are solvated to different degrees.

Unfortunately, it is but rarely that analyses of salts isolated in a resolution are published; hence the solvation of most diastereomeric salts is unknown. We must content ourselves with the observation that occasionally the relative solubilities of p and n salts are reversed as a consequence of a change in crystallization solvent and particularly in the presence of water. Several examples of the influence of solvent on the relative solubility of diastereomeric salts are given in Table 8.

The interpretation of such phenomena in terms of solubility diagrams is simple. In the absence of experimental data, a typical case may be represented by the hypothetical isothermal diagrams (a) and (b) of Fig. 16. These describe what may take place when a pair of salts p and n crystallize in two different solvents.

The degree of solvation of diastereomeric salts may change as a function of temperature. While to our knowledge such cases have not been explicitly reported, they would be described by the same diagrams of Fig. 16 with the solvent remaining fixed and (a) and (b) representing the solubilities of the system below and above a given transition temperature.

Table 8 Influence of solvent on the relative solubilities of diastereomeric salts. Solvation of salts

Acid	Base	Enantiomer Obtained from the Less-Soluble Salt		Degree of Solvation of the Less-Soluble Salt	Refs.
		Solvent	Sign		
2-Phenyladipic acid	Strychnine	Aqueous ethanol	+		201
		Methanol	−		
2-(*n*-Propylthio)succinic acid	Strychnine	Aqueous ethanol	+	Neutral salt, $3H_2O$	202
		Water	−	Neutral salt, $4H_2O$	
trans-1,2-Dicarboxy-Δ^4-bicyclo[2.2.2]hexene	Brucine	Aqueous methanol	+	Neutral salt, $12H_2O$	203
		Methanol	−	Neutral salt, $3CH_3OH$	
2,2-Dimethyl-3-phenylpentanoic acid	Quinine	Ethanol	+	Nonsolvated salt	204
		Aqueous ethanol	−		
N-(8-Nitro-1-naphthyl)-*N*-(phenylsulfonyl)glycine	Brucine	Acetone	−	$1H_2O$	205
		Methanol	+	$3H_2O$	
N-Formyl-*m*-tyrosine	Brucine	95% Ethanol	−	$1H_2O$	206
		Water	+	$4H_2O$	
1-*endo*-Norbornanecarboxylic acid	Cinchonidine	Ethanol	−	$1C_2H_5OH$	207
		95% Ethanol	+	$1/2C_2H_5OH$	
α-(2-Naphthyl)propionic acid	Cinchonidine	Methanol	−		208
		Aqueous ethanol	+		

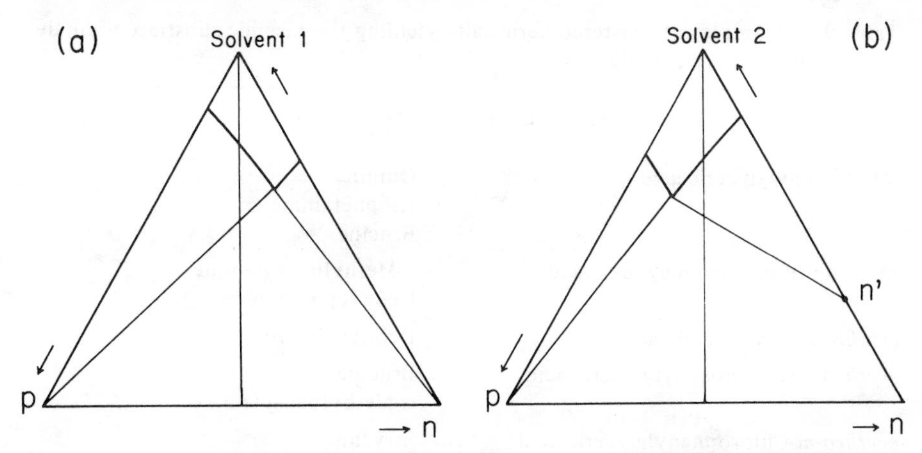

Figure 16 Inversion of relative solubility of diastereomeric salts with change in solvent: (a) In solvent 1, the two salts (p and n) are unsolvated and n is more soluble than p. (b) In solvent 2, salt n forms solvate n' which is less soluble than p.

5.1.13 Double Salts

Just as a conglomerate may be transformed into a racemic compound above (or below) a given transition temperature, an ordinary mixture of two salts may change into a *double salt*.

It is not exceptional to find that, during the preliminary trials of a resolution involving diastereomeric salts, a well-crystallized and little-soluble mixture is obtained which nonetheless regenerates the racemic substrate upon work-up. These observations may be ascribed to the formation of double salts.

In theory, several types of definite combinations between two diastereomeric salts, p and n, may exist. These double salts may be described by the formula $[p_x, n_y]$, in which x and y are whole numbers.

(a) 1:1 Double salts

We first examine the more common case in which a *racemic* substance yields a stable and relatively insoluble combination corresponding to formula $[p, n]$ (i.e., $x = y = 1$) with an optically active reagent. Diastereomeric salts from which the racemic substrate is regenerated belong to this category. Some examples are given in Table 9. Careful culling of the literature would produce many more examples of this phenomenon.

The formation of double salts of this type gave rise at the turn of the century to a relatively extensive study by Ladenburg and his associates.[209–211] This author designated the phenomenon by the term *partial racemy*, a term earlier employed by E. Fischer to describe pairs of isomeric sugar possessing several carbon atoms of opposite configuration.

In an interesting summary,[209] Ladenburg recalls the observation which drew his attention to this unusual type of salts. In the course of resolution trials on racemic β-pipecoline with tartaric acid, the tartrate salt of low solubility formed at steam-

Table 9 Examples of diastereomeric salts yielding the racemic substrate upon decomposition. Putative double salts

A. Racemic Acid/Optically Active Base		Refs.[a]
threo-Phenylglyceric acid	Quinine Amphetamine Brucine	*
threo-o-Chlorophenylglyceric acid	α-Methylbenzylamine Deoxyephedrine	*
erythro-Phenylglyceric acid	Deoxyephedrine	*
erythro-o-Chlorophenylglyceric acid	Brucine α-Methylbenzylamine	*
erythro-m-Chlorophenylglyceric acid	Strychnine Funtumine	*
o-Chlorophenylhydracrylic acid	Amphetamine	*
(3,4,5-Trichlorophenoxy)propionic acid	Brucine	214
2-Methylsuccinic acid	Quinine	209
Tartaric acid	Brucine Strychnine	209
2,2-Dimethyl-3-phenylbutyric acid	Cinchonidine	*
trans-9,10-Dicarboxyethanoanthracene	Brucine	*
B. Racemic Base/Optically Active Acid		
β-Pipecoline	Tartaric acid	209
Hydroquinaldine	Tartaric acid	
Colchinol methyl ether	Tartaric acid Malic acid	213

[a] Those marked by an asterisk are observations made at the Collège de France in connection with unsuccessful resolution trials. There is no direct proof for the formation of double salts.

bath temperature was found, upon decomposition, to regenerate racemic β-pipecoline. On the other hand, at room temperature, the resolution proceeds normally.*

The salt formed during the resolution of tartaric acid with strychnine is one of the better studied examples of double salts. The formation of a racemic neutral tartrate corresponds to the following equilibrium (St = strychnine; Tr = tartaric acid):

$$[St_2,d\text{-}Tr,\ 7H_2O] + [St_2,l\text{-}Tr,5H_2O] + 2.5H_2O \rightleftharpoons [St_4,d\text{-}Tr,l\text{-}Tr,13H_2O] \quad (1)$$

*The conditions under which this resolution takes place, and in particular the importance of the temperature, seem to have given some difficulties in its interpretation on the part of some authors.[212]

Here, it is the double salt which is stable at room temperature. Cleavage of this salt into simple salts takes place above 29.5 to 30°C. Conversely, for the already cited case of the acid tartrates of β-pipecoline, the transformation that provokes the *formation* of the double salt takes place when the temperature is raised (transition at 39°C) (Pi = β-pipecoline):

$$[d\text{-Pi},l\text{-Pi},d\text{-Tr}_2] + 1.5H_2O \rightleftharpoons [d\text{-Pi},d\text{-Tr},0.5H_2O] + [l\text{-Pi},d\text{-Tr},H_2O] \qquad (2)$$

In these examples, the formation or decomposition of the double salt is accompanied by a change in the degree of solvation. This is an entirely analogous situation to that found in the case of racemate–conglomerate transformations which has been examined in Section 3.5.2. In particular, the empirical rule of Van't Hoff,[215] namely, "it is the less solvated salt which is stable at high temperature" remains valid here, as may be readily verified from the examples cited above.

As in the case of enantiomers, the double salt–simple diastereomer salt transformations may also be illustrated by solubility diagrams. An example of such an experimental diagram,[200] where the p and n salts as well as the double salt are solvated, is shown in Fig. 17.

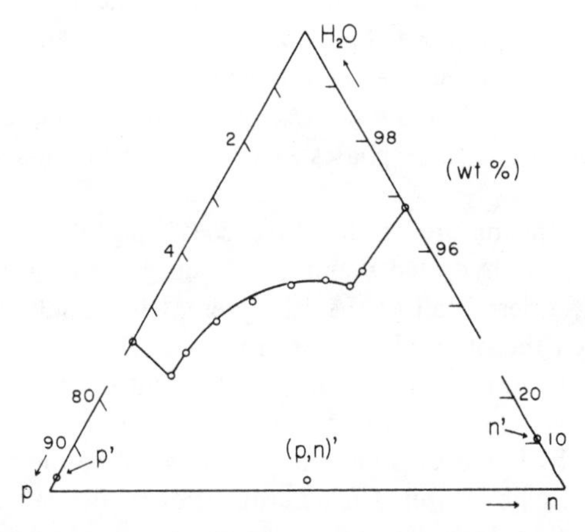

Figure 17 Solubility isotherms showing the formation of a double salt. Ternary system H_2O–Δ-[Co(ox)(en)$_2$](d-$C_{10}H_{14}OBrSO_3$)(p)–Λ-[Co(ox)(en)$_2$](d-$C_{10}H_{14}OBrSO_3$)(n) at 25°C. The solid phase p' is the tetrahydrate of p, n' is the monohydrate of n, and $(p, n)'$ is the double salt (p, n, $2H_2O$). Adapted from Y. Shimura and K. Tsutsui, *Bull. Chem. Soc. Jpn.*, 1977, **50**, 145 by permission of the Chemical Society of Japan.

The three hypothetical isothermal diagrams of Fig. 18 represent a system such as that which must correspond to the β-pipecoline tartrate case (precise experimental data are unfortunately lacking). Figure 18 shows the diastereomeric p and n salts to be unequally solvated by solvent W (water). The compositions of these solvates are designated by symbols A and B, respectively, on the diagrams and their solubilities, by C and D. Above a transition temperature T_2, the two solvated salts form an anhydrous double salt $[p,n]$ represented by R.

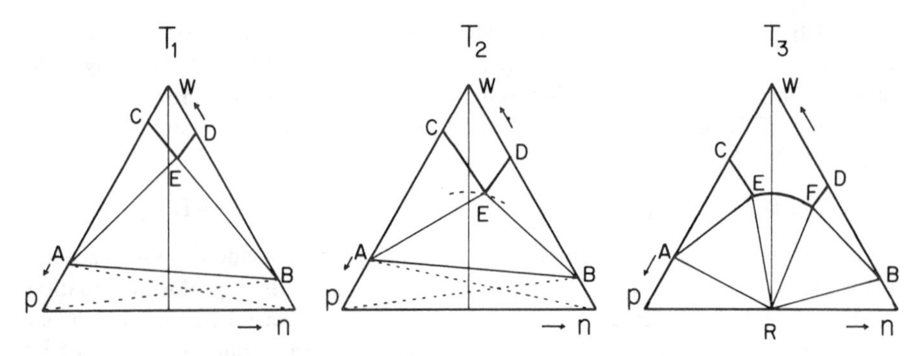

Figure 18 Transformations between double and simple diastereomeric salts. Case where the double salt is less solvated than the simple salts and is stable at high temperature (β-pipecoline tartrate).

Let us analyze the composition of the phases at the lower temperature, T_1. In area *DWCE* there is but a single liquid phase; in *CAE* there are two phases, the hydrated *p* salt (*A*) and the mother liquors of variable composition; in *DBE* there are also two phases, the hydrated *n* salt (*B*) and the mother liquors of variable composition; in *ABE* one finds three phases: hydrated *p* and *n* salts and mother liquors of composition *E*.

In area *ApnB*, the nature of the three coexisting phases depends upon the relative stabilities of the hydrated *p* and *n* salts since the quantity of solvent available is insufficient to permit all of the salts present to be solvated. If the *p* salt is the *stabler* one, then the other phases present in area *Abn* are the hydrated *p* and *n* phases and anhydrous *n*; in area *Apn*, hydrated *p* and the two anhydrous salts coexist. If, on the other hand, the *p* salt is the *less* stable one, then *ApB* is the region containing the two hydrated salts and anhydrous *p*, while the two anhydrous salts and hydrated *n* are found in *Bpn*. The determination of the data necessary to establish those parts of the phase diagram below line *AB* would be very difficult; it is unlikely that such an analysis would have a practical application.

Above transition temperature T_2, the isothermal diagram at temperature T_3 may be analyzed in a similar manner. The phases present in each area in Fig. 18 (T_3) are as follows: *EFR*, anhydrous double salt and mother liquors of variable composition; and *EAR* or *FRB*, both containing mixtures of the anhydrous double salt and of a hydrated salt (*A* or *B*, respectively) in the presence of mother liquors *E* or *F*, respectively. Areas *ApR* and *BRn* contain only solid phases: double salt *R* with hydrated and anhydrous *p* salt on one hand and hydrated and anhydrous *n* salt on the other.

The three isotherms of Fig. 19 illustrate the case in which the double salt is more highly solvated than the individual diastereomeric salts and dissociates into the latter upon heating. These isotherms may be subjected to an analysis which is virtually identical to that just performed.

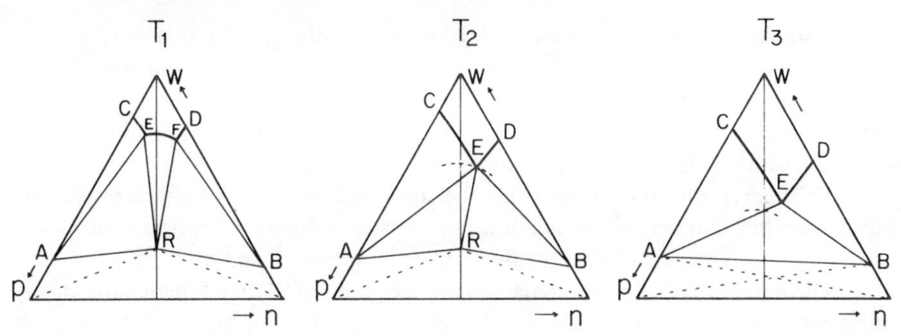

Figure 19 Transformations between double and simple diastereomeric salts. Case where the double salt is more highly solvated than the simple salts and dissociates upon heating (strychnine tartrate).

(b) 1:x Double salts

When a double salt (p,n) is formed between a racemate and an optically active reactant, the lack of success in resolution is evident right from the start. This is not the case when double salts are formed which have the general formula $[p_x, n_y]$, with $x \neq y$. A partial resolution is then possible. However, the optical purity attainable cannot exceed that which may be deduced from the stoichiometry of the salts formed (for $x = 2, y = 1$: 33%; for $x = 3, y = 1$: 50%; for $x = 3, y = 2$: 20%; etc.).

Such limited success may, in spite of the better prognosis for a successful resolution at the onset, actually lead one to lose more time than during a real failure. Fortunately, such double salts appear to be rare indeed. While, in principle, there is no impediment to their existence that we are aware of, experimental data are virtually absent.

One of the rare sets of data dealing with such salts has been provided incidentally by Matell.[216] In the course of the resolution of α-(2-naphthoxy)valeric acid in 30% ethanol, brucine yields a salt from which one may recover an acid $[\alpha]_D = +25°$, a value which is unchanged upon recrystallization of the salt. Since the specific rotation of the pure acid, obtained by resolution with amphetamine, is 73.6°, Matell has suggested that the brucine salt is formed in the proportion $x = 2$, $y = 1$. In fact, repetition of this resolution under the conditions given by Matell instead yields a completely unexceptional result. A relatively insoluble brucine salt is obtained whose analysis indicates that it is hydrated (with $5.5H_2O$); three recrystallizations of this salt yields a product from which optically pure acid may be recovered. This difference in results may possibly be due to a difference in operating temperature (20°C in the latter case). Thus, while the formation of double salts of the $[p_x, n_y]$, $x \neq y$ type is possible, the demonstration of their existence is questionable.

5.1.14 Cocrystallization of Diastereomeric Salts

The formation of solid solutions between diastereomeric salts is anything but rare and is at the root of difficulties encountered in many resolutions. The isomorphism of diastereomeric salts has nevertheless been the object of but a very limited number of studies or even of explicit comment. Those of Pope and Reid,[217] followed by those of Delépine and Larèze,[218] are qualitative and descriptive in character. Moreover, they offer no general remedies to the problems encountered in resolutions with regard to this phenomenon.

The cocrystallization of salts may be suspected, in the context of a resolution, whenever the number of recrystallization of salts required to bring the resolution to a conclusion becomes "abnormally high" (e.g., greater than four).

The determination of several solubility diagrams of diastereomeric salts clearly illustrates the preceding statements. The ternary isotherms of pairs of salts (in water or ethanol) formed between α-methylbenzylamine and α-methylphenylacetic acid, α-ethylphenylacetic acid, α-propylphenylacetic acid, and mandelic acid are among the rare cases of diastereomer salt systems studied in a quantitative manner.[194]

Cocrystallization of p and n salts was shown to take place in three of the four examples cited through application of the method of rests (Section 3.1.6). Figures 20 and 21 represent two characteristic cases; in the first, the appearance of a solid solution between p_- and n_+ is undoubtedly limited to one part of the diagram. In the second, the miscibility of the two salts p_+ and n_+ is complete irrespective of composition.

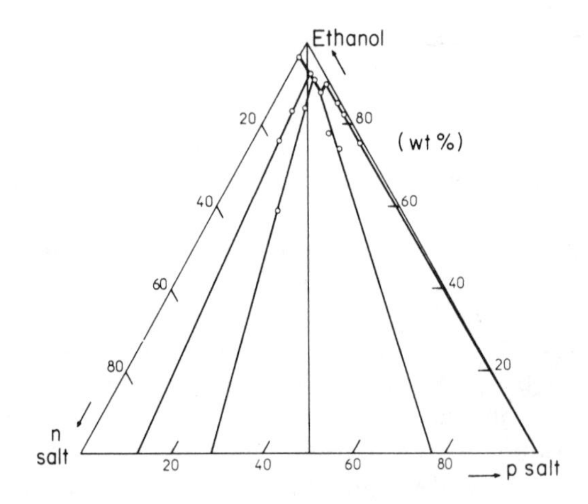

Figure 20 Solubility diagram of the α-methylbenzylamine α-propylphenylacetate p_- and n_+ salts in ethanol at 10°C. Evidence for solid solutions by application of the method of rests. M. Leclercq and J. Jacques, *Bull. Soc. Chim. Fr.*, 1975, 2052. Reproduced by permission of the Société Chimique de France.

In contrast, in the case of the α-methylbenzylamine hydratropates, analysis of the rests shows that these contain only one or the other of the pure salts as solid phase (see Fig. 12, Section 5.1.12).

Upon reflection, the apparently easy formation of solid solutions between diastereomeric salts is not so surprising. The degree of isomorphism between two molecules being defined by the coefficient $\epsilon = 1 - V_{no}/V_o$ (see Section 2.4.9), one

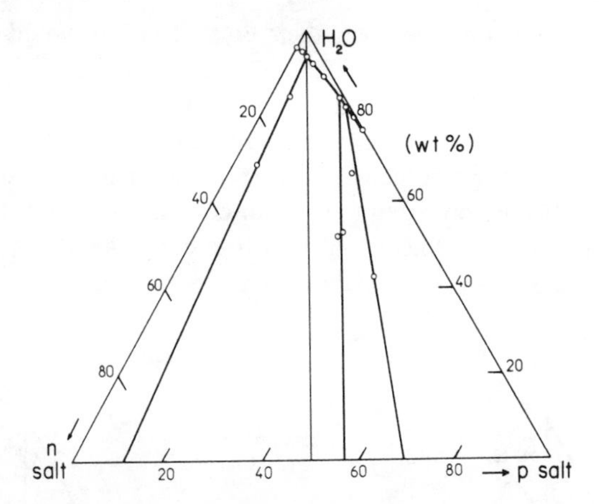

Figure 21 Solubility diagram of α-methylbenzylamine mandelate p_+ and n_+ salts in water at 10°C. Evidence for solid solutions by application of the method of rests. M. Leclercq and J. Jacques, *Bull. Soc. Chim. Fr.*, 1975, 2052. Reproduced by permission of the Société Chimique de France.

can see that for the two salts *p* and *n* formed from a pair of enantiomers the consequence of the presence of the *same* optically active base (or acid) is an increase in the overlapping volume V_o and consequently in ϵ. Since diastereomeric salts perforce contain identical moieties (a common counterion) which make them more identical than are enantiomers, it is understandable that, comparatively, diastereomers much more frequently form solid solutions than do the enantiomers from which they are derived.

This difference immediately suggests one possible way of circumventing the difficulty arising from the formation of mixed crystals between diastereomeric salts. Once a given separation has been achieved in a resolution, it consists of continuing the optical enrichment through recrystallization not of the diastereomeric salts but of the partially resolved enantiomeric substrates recovered from the diastereomeric salts. If a sufficient enantiomeric purity has been attained, a single crystallization can, in principle, permit the isolation of the predominant enantiomer (see Chapter 3).

5.1.15 Isolation of the More Soluble Diastereomeric Salt. The Method of Ingersoll

When one enantiomer is obtained from its less soluble diastereomeric salt, the isolation of the other enantiomer is sometimes difficult. The mother liquors, in effect, contain a mixture of the two diastereomeric salts in proportion close to that of the eutectic of the solubility diagram (see Section 5.1.12). The inherent difficulty in dealing with such a mixture can, however, be circumvented in several ways:

1 If the solubility diagram and the composition of the partially resolved mixture

of the *enantiomers* are favorable, their crystallization would permit the isolation of the desired enantiomer in optically pure form. This operation is schematically shown in Fig. 22. In case of failure, it may be possible to find an enantiomer system that is more favorable to this strategy by transforming the system into one having a eutectic closer to the racemate composition. The mixture to be purified is converted to a salt or covalent derivative by reaction with *achiral* reagents; ideally, this would lead to a conglomerate system, or at least to a racemic compound of low stability.

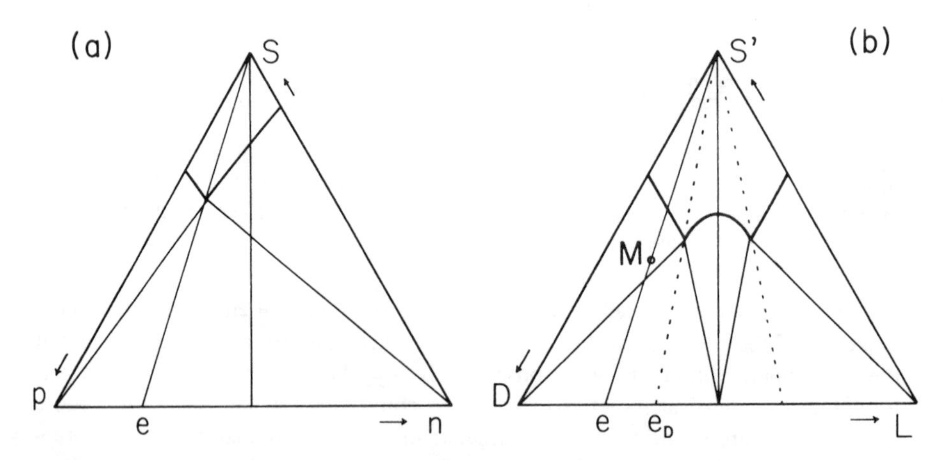

Figure 22 Obtainment of the enantiomer derived from the more soluble diastereomer: (a) After isolation of the less soluble salt (n), there remains a mixture (p, n) of composition close to e. (b) After decomposition of this mixture back to the resolution substrate (racemic compound system assumed), one obtains a mixture of enantiomers of the same composition e whose recrystallization (e.g., system M) leads directly to the pure enantiomer D.

2 The purification of the more soluble salt is sometimes achievable by changing the solvent. In particular, one may hope to invert the relative solubilities of the two salts by playing upon solvation (see Section 5.1.12).

3 An obvious approach is the recombination of the partially resolved enantiomer mixture (or even the racemate) with the enantiomer of the resolving agent initially employed (when it is available), or with another resolving agent which might yield a salt of low solubility with the desired enantiomer. This is in fact the most common procedure used. Generation of a solubility-rotation matrix during resolution trials such as that illustrated in Section 7.2.1 would point to the correct reagent.

4 One can attempt to play upon the *rates of crystallization* of the diastereomeric salts in supersaturated solutions employing seeding or entrainment techniques as required (see Section 7.3).

5 Finally, one can resort to the method of Ingersoll,[219, 220] which takes advantage of the relative solubilities of the several diastereomeric salts that may be formed. It consists of treatment of the partially resolved mixture of enantiomers isolated from the mother liquors enriched in the more soluble salt with the *racemic* form of the resolving agent originally used or with another racemic reagent. The process is illustrated by the resolution of phenylglycine (*B*) with 10-camphorsulfonic acid (*A*).[219] When the racemic amino acid *dlB* is combined with the resolving acid *dA*, one first obtains the less soluble salt *lBdA* as a precipitate. It is isolated and purified. After decomposition of the mother liquors is carried out, a partially resolved mixture of the amino acid (*dB* + *dlB*) corresponding to the more soluble diastereomeric salt is isolated. The mixture has an optical purity of ca. 63%.

This mixture is treated with an equivalent of *racemic* 10-camphorsulfonic acid, *dlA*, which leads to the precipitation of salt *dBlA* (the enantiomer of the salt obtained in the first operation). The salt *dBlA* thus isolated simultaneously yields the enantiomer of the amino acid obtained above and the resolved acid *lA* (from *dlA*). This is, as Ingersoll first called it, a method permitting the mutual resolution of an acid and of a base.

An alternative possibility identified by Ingersoll[220] is that in which it is no longer the enantiomer of the initially separated salt that precipitates but rather a *double salt* between the racemic reagent and the enantiomer present in excess in the mother liquors. Thus, when *threo*-2-hydroxy-1,2-diphenylethylamine *dlB* is resolved with 10-camphorsulfonic acid *dA*, one first obtains salt *dBdA*. The base isolated from the mother liquors (*lB* + *dlB*) is subsequently combined with the racemic acid *dlA*; upon crystallization, the double salt *lBdlA* is obtained from which base *lB* may be isolated. In this case there is no mutual resolution of the acid *dlA*.

These two (and other) favorable examples of this process led Ingersoll and his co-workers[198, 199, 221] to consider the general question: is it *always* possible to isolate and purify the enantiomer derived from the more soluble diastereomer by forming salts with racemic (hence *a priori* easily accessible) reagents?

In order to answer this question, Ingersoll first made up a balance sheet of all the salts that may be formed form the ions present in solution and compared the various orders of solubilities the different salts may exhibit. Let us consider the resolution of a racemic base *dlB* by an acid *lA*; after separation of the less soluble salt, say *lBlA*, the mother liquors are treated so as to free the substrate from the residual dissolved salt. The recovered base is a partially resolved mixture enriched in *dB* which may be represented by (*dB* + *dlB*). This mixture is treated by the racemic form *dlA* of the resolving acid originally used. From the species *dB*, *1B*, *dA*, and *lA* present, Ingersoll envisages the possible formation of the following nine salts*:

*We have used the symbols of Ingersoll in our discussion. In order to accurately reflect the required stoichiometry, the double salts ④, ⑤, ⑧, and ⑨ actually should be formulated as [*dBdA*, *dBlA*], and so on.

dBdA	*dBlA*		*dBdlA*	*lBdlA*
①	②		④	⑤
		dlBdlA		
		③		
lBlA	*lBdA*		*dlBlA*	*dlBdA*
⑥	⑦		⑧	⑨

Note that salts ① and ⑥ on the one hand and ② and ⑦ on the other are, respectively, mirror images; the same is true for double salts ④ and ⑤ and ⑧ and ⑨. Finally, salt ③ is a racemate. Given the postulated, indeed likely (if the original salt precipitate is enriched in *lB*), excess of *dB* in the mixture isolated from the mother liquor, the crystallization of some of these nine salts is, *a priori*, excluded; for example, salts ⑥ and ⑦. According to Ingersoll, only two modes of combination between the acid *dlA* and the partially resolved mixture would remain possible. The first, eq. (1), corresponds to formation of simple salts, and the second, eq. (2), corresponds to that of a double salt with acid *dlA*:

$$(dB + dlB) + dlA \rightarrow dlBdlA + dBdA + dBla \quad (1)$$

$$(dB + dlB) + dlA \rightarrow dlBdlA + dBdlA \quad (2)$$

In each of these situation, the order of relative solubilities of the salts formed must correspond to one of the following combinations (eight in all), from which the conditions of their separation may be deduced:

For eq. (1):

1 $dBdA < dBlA < dlBdlA$
2 $dBlA < dBdA < dlBdlA$
3 $dBdA < dlBdlA < dBlA$
4 $dBlA < dlBdlA < dBdA$
5 $dlBdlA < dBdA < dBlA$
6 $dlBdlA < dBlA < dBdA$

For eq. (2):

7 $dBdlA < dlBdlA$
8 $dlBdlA < dBdlA$

In our example, we know (from the first part of the resolution) that *lBlA* is less soluble than *dBlA*. This alone allows one to eliminate possibilities **2**, **4**, and **6**. If there is no double salt formation (combinations **7** and **8**), isolation of salt *dBdA* rests only on the requirement that condition **5** may not be realized. If, on the other hand, double salts are formed, only condition **7** allows one to isolate *dB* (in the form of *dBdlA*). We see that there is no assurance that the process will succeed.

Moreover, when a racemic reagent *dlA'* is used which does not correspond to the resolving agent (*dA*) employed in the initial resolution, the application of the eight conditions becomes completely unpredictable.

Ingersoll and his associates attempted to find several specific cases which would allow them to illustrate the several possibilities. Combination **1** is observed in the case of the resolution of *dl*-α-(*p*-tolyl)ethylamine by α-bromocamphor-π-sulfonic acid.[198] The following is the order of solubilities of the salts in water at 25°C*:

$$lBlA \cdot H_2O\ (2.10) < lBdA \cdot H_2O\ (2.96) < dlBdlA\ (3.38)$$

Sequence **3** may be observed in the behavior of salts of α-methylbenzylamine and 6,6′-dinitrodiphenic acid[199] for which the solubilities in acetone at 25°C are as follows:

$$dBdA\ (0.61) < dlBdlA\ (1.29) < dBlA\ (13)$$

No doubt as a consequence of solvation phenomena, the same compounds yield salts whose solubilities in 95% ethanol correspond to sequence **5**:

$$dBdlA\ (1.78) < dBdA\ (3.95) < dBlA\ (4.74)$$

Lastly, in the case of the α-methylbenzylamine mandelates, one observes the following order in water at 30°C:

$$dBlA\ (4.91) < dlBdA\ (5.81) < dlBdlA\ (12.29) < dBdA\ (18)$$

which corresponds to sequence **4** when the partially resolved base is treated with racemic acid, or to sequence **7** when partially resolved acid is combined with racemic base.

We must not evade the fact that these arguments, which depend upon a comparison of solubilities of *pure salts* is, in the final analysis, incorrect; it is not only the *order* of solubilities which determines the sequence of precipitations, but also the respective proportions of the several salts in a given mixture. A more detailed analysis such as one carried out with the aid of phase diagrams would be particularly complex as a consequence of the large number of components involved.

Ingersoll presents his method in a more general form than in the summary given here. He considers, in particular, that the second reagent employed may be a partially resolved one instead of a racemate. This possibility considerably increases the number of combinations with respect to the order of relative solubilities of the different salts (20 combinations). In trying to allow for all possible situations, his account leaves the discouraging impression that the proposed method is much more complicated than it really is. It is perhaps for this reason that, to our knowledge, it has never been exploited. This is probably unjustified and unfortunate since, *a priori*, cases must exist where it may be useful. Note also that this exhaustive analysis of solubility relationships between the various diastereomeric salts demonstrates incidentally that a complete resolution may be carried out even with an impure reagent (i.e., an optically impure resolving agent; see Section 7.3.5).

*Solubilities (in parentheses) are given in units of g/100 g solvent.

5.1.16 The Marckwald Principle and Reciprocal Resolutions

(a) The Marckwald principle

The relatively simple and now obvious idea that the two enantiomers of a resolving agent give access in turn to both enantiomers of a resolution substrate is known by the name "Marckwald principle." After having proposed a convenient procedure for obtaining levorotatory (and unnatural) tartaric acid,[223] Marckwald used this acid in the resolution of α-pipecoline. He thus obtained the (−) base while Ladenburg had obtained the (+) base through use of the naturally occurring (+)-tartaric acid.

The Marckwald principle, illustrated in Scheme 1, may be applied if both enantiomers of a resolving agent are accessible. When this is the case, it is convenient and economical to treat the mother liquors of a first separation with the resolving agent which is the enantiomer of that first employed. This is nearly always possible with synthetic resolving agents.

SCHEME 1[a]

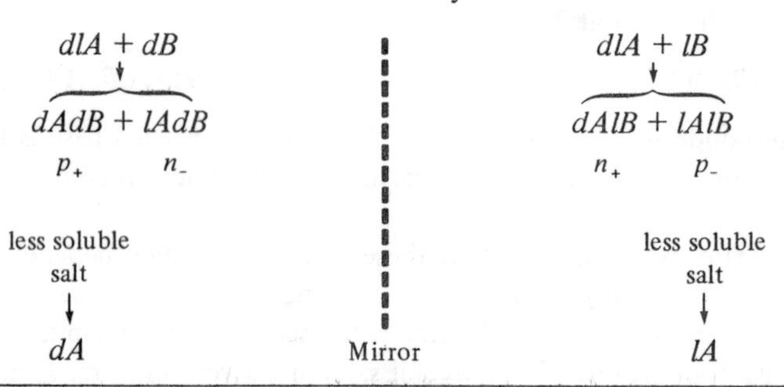

[a] Signs of rotation + and − for salts *dAdB* (p) and *lAdB* (n) have been arbitrarily chosen (see p. 252).

Given the quasi-enantiomeric relationships existing among the quinine alkaloids (Section 5.3), one may envisage the use of these natural products in a manner similar to that outlined for the Marckwald principle. We shall see, however, that the results thus obtained are not always in accord with expectation.

(b) Reciprocal resolutions

When a racemic acid can be resolved by an optically active amine, it is frequently the case that this same amine, as a racemate, may itself be resolvable by the optically active acid. The solubility relationships between the different diastereomeric salts and enantiomers involved in such *reciprocal resolutions* are outlined in Scheme 2. It is immediately apparent that, if one situation leads to salts p and n of unlike signs, the second will yield salts p and n of like sign. Since the relative solubilities of the salts do not depend upon their chirality, one may suppose that the ternary systems (p_+, n_-, S) and (p_+, n_+, S) will be identical (or very nearly so).

Figure 12 (Section 5.1.12) illustrates the legitimacy of this statement. We would thus expect that the reciprocal operations would take place in analogous ways.

SCHEME 2

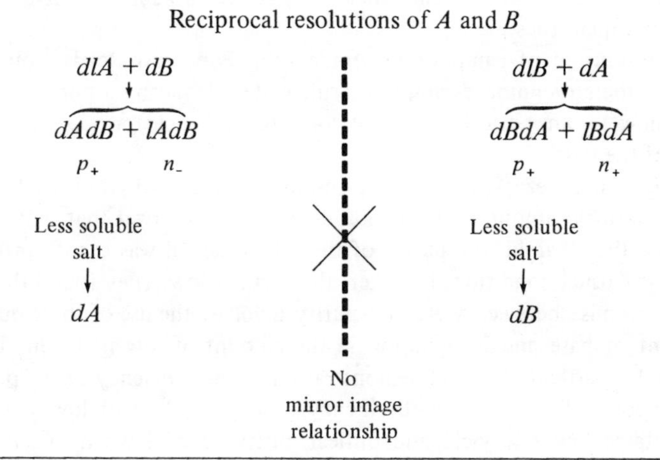

Several authors have earlier been rather insistent[4, 224] in pointing out that reciprocal systems are not mirror images of one another (Scheme 2) and that consequently the success of one of the operations does not absolutely exclude the failure of the other. In effect, even if the salt system p_+,n_- forms a eutectic (corresponding to a successful resolution), nothing prevents system p_+,n_+ from forming a double salt or a solid solution. Either of the latter situations would, of course, make the separation of salts p_+ and n_+ impossible. While this risk does exist,[225] we have already seen that it is rather small. This is equivalent to saying that, in general, one may embark upon reciprocal résolutions with optimism.

5.1.17 Resolution with Nonstoichiometric Quantities of Reagents

The resolution of tartaric acid through crystallization of its water-insoluble salt with cinchonine was described by Bremer in 1880.[226] This resolution was restudied by Marckwald[223] who, in 1896, improved the process while using only half of the cinchonine necessary for the formation of the acid tartrate from 1 mole of the racemic acid. (This resolution was subsequently restudied in detail by Read and Reid.[227]) To our knowledge, this is the first resolution described which is implicitly based not on the separation of two diastereomeric salts formed in equal quantities but rather on that of one of the diastereomers and one of the enantiomers.

This modification of the second Pasteurian resolution method[228] is called "method of half-quantities"* and resembles the procedure of Pope and Peachey

* The note in which this expression is first used describes the resolution of triarylphospholines in the form of α-hydroxymethylphosphonium camphorsulfonates. Rather than illustrating the cited process, it is more likely an example of a kinetic resolution.

which we examine later (following section) but which involves other salts in addition to diastereomeric salts. Neither of these variant procedures has been much used nor have the associated physicochemical problems been studied other than superficially until very recently. The following examples will nevertheless give the reader some feel for the utility and the sometimes remarkable effectiveness of the method of half-quantities.

In the resolution of camphor-π-sulfonic acid, Pope and Read[229] obtained an 84% yield of the strychnine *d*-camphor-π-sulfonate salt having a purity of ca. 97% (corresponding to an optical purity of 94% for the acid) while using half an equivalent of the base.

Delépine and Larèze[230] studied the resolution of α-phenylbutyric acid in the presence of variable amounts of cinchonidine. They observed that, with one-half equivalent of the alkaloid, the purity of the salt obtained was significantly higher than under the usual condition; however, the yield is low. They found that a satisfactory compromise between yield and purity involved the use of three quarters of an equivalent of base and a reduction in the amount of solvent taken. The same acid provided a particularly good demonstration of the efficiency of the process.[231] Its salt with optically active α-methylbenzylamine, which is of low solubility in ether, is obtained in 95% yield and optical purity of 85% when one half of an equivalent of base is used. With one equivalent of optically active base, essentially the same weight of salt is obtained but the acid derived from this salt exhibits an optical purity of only 15%. This technique was generalized by the same authors but with less success for hydratropic, mandelic, and *p*-methoxymandelic acids and for the hydrogen phthalate of methylphenylcarbinol.

dl-*N*-Acetyl- and *N*-chloroacetyl-2-(4-hydroxyphenyl)glycine* have been resolved with half-quantities of dehydroabietylamine.[232, 233] A more complex example of the use of nonstoichiometric amounts of reagents concerns the resolution of ethyl *dl*-2-(4-hydroxyphenyl)glycinate by dibenzoyltartaric acid.[234] Mixtures of the optically active dicarboxylic acid and of the glycinate in the proportions 1–1 (acid salt) or 1–2 (neutral salt) effectively do not yield crystalline products but use of an intermediate quantity, for example, 0.73–1 leads to the isolation of one of the diastereomeric acid salts in good yield.

Another variant for the separation of diastereomeric salts exists in which, unlike the preceding cases where the optically active reagent is *deficient*, the resolving agent is *in excess* with respect to the usual stoichiometry. Thus, the resolution of the spirohydantoin **79** described by Pope[235] requires the use of 2 moles of brucine per mole of racemic **79**. The 2–1 mixture of brucine and of (±)-**79** dissolves in alcohol upon heating; upon cooling, a nearly theoretical quantity of the salt [(−)-**79**.B] is obtained. After separation of the crystals, the mother liquor furnishes crystals of the second salt [(+)-**79**.2B] in ca. 75% yield and optically pure. Note that this is a resolution in which the brucine employed corresponds to a slight excess relative to the stoichiometry of formation of the two salts (1.5 mole brucine

* (4-Hydroxyphenyl)glycine is an unnatural amino acid employed in the synthesis of various derivatives of penicillin and of cephalosporin. It may also be resolved by entrainment (see Section 2.2.5).

is required per mole (±)-**79**). This is not the case in the resolution of amine **68** (p. 274) in which a considerable excess of tartaric acid (20–1 or more) is required to obtain a crystalline salt.[236] Moreover, the same amine may be more simply resolved by complexation with TAPA (see Section 5.1.7).

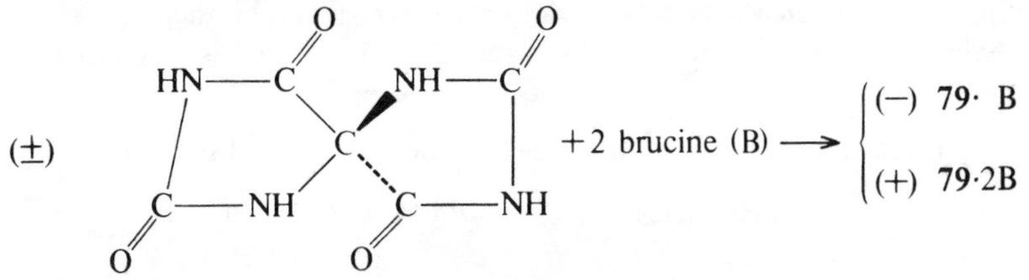

(±) **79** + 2 brucine (B) → (−) **79**· B ; (+) **79**·2B

79

In the preceding examples, whether one or the other of the reactants is in excess, it is always one of the diastereomeric *salts* which is the less soluble product and which crystallizes. Much more exceptionally, in several cases described by Armstrong,[237] a deficiency of resolving agent provokes the crystallization of one of the free *enantiomers* of the substrate. Thus, when racemic *S*-carbomethylhomocysteine is treated with one half equivalent of brucine, the salts formed remain in solution while the amino acid enriched in D enantiomer crystallizes (optical purity 24%). The same phenomenon is also observed with racemic phenylalanine and (+)-10-camphorsulfonic acid but with a lower enrichment in the L form of the phenylalanine which crystallized.

Due to the unpredictable character and low enrichments obtained, this last variant appears to be of limited utility.

5.1.18 The Method of Pope and Peachey

In the method of half-quantities, the racemic base, for example, is incompletely neutralized by the optically active acid (or the inverse). The reaction products are a salt and unreacted substrate. In the process described by Pope and Peachey in 1899,[238] on the other hand, while the resolving agent is used in half-quantity, the excess of racemic base (or acid) is always neutralized by addition of the necessary quantity of an achiral acid (or base). This introduces into the reaction medium other salts than those upon which classical resolutions are based.

Pope and Peachey rationalized their method* by considerations whose apparent simplicity completely mask the reality of the numerous equilibria involved. We shall return to these considerations in Section 5.1.19. Their explanation is as follows:

> The solubilities of the salts (*dBdA* and *lBdA*) of a dextrorotatory acid (*dA*) with a dextro- and a laevo-base (*dB* and *lB*) would hardly be expected to

* Pope and Peachey designated their process the "equilibrium method," to call attention to exact neutralization of the acids and bases present in the system, in contrast to the method of half-quantities (for which, incidentally, Pope implicitly also claims credit; see ref. 229).

> differ considerably, because the solubility is partly a function of the chemical nature of the salts. If, however, the salt, *lBdA*, is the less soluble and only sufficient of the active acid, *dA*, necessary to the formation of this salt is added, the balance of acid required to dissolve the base being made up by adding the requisite amount of an optically inactive acid, such as hydrochloric acid, which forms comparatively soluble salts with the base, it would be expected that on crystallization the greater part of the laevo-base would separate as the sparingly soluble salt, *lBdA*, whilst the mother liquors would retain the dextro-base of which the hydrochloride, *dBHCl*, is very soluble.

The reaction scheme which describes this behavior would then be as follows:

$$dBlB + dA + \mathrm{HCl} \rightarrow lBdA\downarrow + \underset{\text{(In solution)}}{dBH^{+}, \mathrm{Cl}^{-}} \qquad (1)$$

Similarly, the resolution of an acid by an optically active base *dB* in the presence of an achiral base such as KOH would be described by the following equation:

$$dAlA + dB + \mathrm{KOH} \rightarrow lAdB\downarrow + \underset{\text{(In solution)}}{\mathrm{K}^{+}, dA^{-} + \mathrm{H_2O}} \qquad (2)$$

The first example illustrating this method is the resolution of tetrahydroquinaldine[238] by α-bromocamphor-π-sulfonic acid in the presence of hydrochloric acid. Among other examples, that of the resolution of *dl*-benzoylalanine is particularly striking: two equivalents of this racemic acid treated by one equivalent of strychnine and one equivalent of potassium hydroxide in water yield practically the theoretical quantity of the strychnine salt of (+)-benzoylalanine.

A variant suggested by Pope and Peachey[238] consists of the direct combination of a salt of the racemic acid (a sodium or ammonium salt, for example) with half an equivalent of the hydrochloride (or other salt) of the optically active base or, conversely, the hydrochloride of the racemic base with a salt of the optically active acid. Thus, by combining racemic tetrahydroquinaldine hydrochloride with a half-quantity of ammonium (+)-α-bromocamphor-π-sulfonate in water, they directly obtained the pure (−)-amine α-bromocamphor-π-sulfonate. This example is summarized in the following equation:

$$dBH^{+}lBH^{+}, 2\mathrm{Cl}^{-} + \mathrm{NH_4^{+}}, dA^{-} \rightarrow lBdA\downarrow + \underset{\text{(In solution)}}{dBH^{+}, \mathrm{Cl}^{-} + \mathrm{NH_4Cl}} \qquad (3)$$

By comparison with the first described process, the variant introduced one supplementary achiral salt (here NH_4Cl).

In Table 10, we have compiled a number of cases of resolutions in which the method of Pope and Peachey was used in one or another of its variant forms. It is not always easy to compare the results obtained by the Pope and Peachey procedure with those obtained with classical conditions. In particular, we call attention to the fact that authors who have used the Pope and Peachey method have rarely specified the *quantities of solvent* employed while these clearly must have a decisive influence on the yields and the purities of the isolated salts. In fact, the validity of

Table 10 Resolutions carried out by the method of Pope and Peachey[a]

Variant[b]			Achiral		Refs.
			Acid	Base	
	Racemic Base	Optically Active Acid			
A	Tetrahydroquinaldine (2)	α-Bromocamphor-π-sulfonic acid (1)	HCl (1)		238
B	Tetrahydroquinaldine (2)	α-Bromocamphor-π-sulfonic acid (1)	HCl (2)	NH_3 (1)	238
B	Tetrahydrotoluquinaldine (2)	α-Bromocamphor-π-sulfonic acid (1)	HCl (2)	NH_3 (1)	240
B	Pavine (2)	α-Bromocamphor-π-sulfonic acid (1)	HCl (2)	NH_3 (1)	241
B[c]	Pavine (2)	α-Bromocamphor-π-sulfonic acid (1)	d-Camphor-π-sulfonic acid (2)	NH_3 (1)	241
B	1-Hydroxy-2-hydrindamine (2)	α-Bromocamphor-π-sulfonic acid (1)	HCl (2)	NH_3 (1)	242
A	α-Methylbenzylamine (2)	Tartaric acid (1)	Acetic acid (1)		243
A	α-Methylbenzylamine (2)	Malic acid (1)	Acetic acid (1)		243
A	α-Methylbenzylamine (2)	Malic acid (1)	Nitric acid (1)		243
A	Deoxyephedrine (2)	Tartaric acid (1)	HCl (1)		244
A	*p*-Bromodeoxyephedrine (2)	Tartaric acid (1)	HCl (1)		245
A	*p*-Aminodeoxyephedrine (2)	Tartaric acid (1)	HCl (1)		245
A	1-(*m*-Nitrophenyl)-2-(methylamino)ethanol (2)	Tartaric acid (1?)	? (1?)		246
	Racemic Acid	Optically Active Base			
A	*N*-Benzoylalanine (2)	Strychnine (1)		KOH (1)	239
A	*N*-Benzoylalanine (2)	Brucine (1)		KOH (1)	239
A	Camphorsulfonic acid (2)	Brucine (1)		NH_3 (1)	227
B	α,γ-Dihydroxy-β,β-dimethylbutyric acid (2)	Quinine (1)	HCl (1)	NaOH (2)	247
B	1-Methylcyclohexylidene-4-acetic acid (2)	Brucine (1)	HCl (1)	NaOH (2)	248
A	α-(2,4-Dichlorophenoxy)propionic acid (2)	*threo*-1-(*p*-Nitrophenyl)-2-amino-1,3-propanediol (1)		NaOH (1)	249
A	2-(6-Methoxy-2-naphthyl)propionic acid (2)	Cinchonidine (1)		KOH (1)	250
A	Butan-2-ol hydrogen phthalate (2)	Brucine (1)		$(C_2H_5)_3N$ (1)	251
A	Octan-2-ol hydrogen phthalate	Brucine (1)		$(C_2H_5)_3N$ (1)	251
A	1-Phenylethanol hydrogen phthalate (2)	Brucine (1)		$(C_2H_5)_3N$ (1)	251

[a] The stoichiometry is given by the numbers in parentheses. M. Leclercq and J. Jacques, *Nouv. J. Chim.*, 1979, **3**, 629. Reproduced by permission of Gauthiers-Villars, Paris.
[b] Variant A: racemic base treated with one half-equivalent of optically active acid and one half-equivalent of achiral acid or conversely, racemic acid combined with one half-equivalent of optically active base and one half-equivalent of achiral base. Variant B: hydrochloride of racemic base treated with one half-equivalent of a salt of an optically active acid, or the inverse.
[c] In this example, the supplementary acid is not achiral.

these resolution procedures can be justified theoretically, through analysis of the solubility/dissociation equilibria involved in such diastereomeric salt systems.[252] We return to this point in the following section.

5.1.19 Dissociation and Solubility of Salts. Interpretation of "Nonstoichiometric" Resolutions

The dissociation of diastereomeric salts, which is a complex process, plays a major role in the phenomena examined in the preceding sections. Dissociation intervenes in large measure not only in the behavior of dissolved species, for example, optical rotation, but also affects the relative solubilites of diastereomeric salts. In order to carry out resolutions as economically as possible, it is obviously important to know by what mechanism the dissociation of dissolved salts can influence the quantity and the purity of the crystals that precipitate. This question has been studied recently in a quantitative manner by Leclercq and Jacques.[252] The following discussion summarizes their essential conclusions.

When a solvent dissolves a substance without "reacting" with it, the solubility of the solute is effectively determined by the melting point and the enthalpy of fusion of the *crystals*. This is no longer the case when the solute gives rise to several species, either through dissociation or association.

Suppose than an acid AH and a base B are both present in a given solvent and that part of the salt AHB precipitates. The following species may then be present in the supernatant solution: AHB, which represents the undissociated salt; AH, the free acid; B, the free base; A^-, the ionized acid; and BH^+, the protonated base (the protons being derived from the acid or from the solvent).

The principal equilibria which one may reasonably envisage are represented by expressions (1) to (4), where concentrations are expressed in moles or number of ions per liter:

$$\mathrm{AH} \rightleftharpoons \mathrm{A^-} + \mathrm{H^+} \quad \text{with } K_a^A = \frac{[\mathrm{A^-}][\mathrm{H^+}]}{[\mathrm{AH}]} \quad \text{and } \mathrm{p}K_a^A = -\log K_a^A \tag{1}$$

$$\mathrm{BH^+} \rightleftharpoons \mathrm{B} + \mathrm{H^+} \quad \text{with } K_a^B = \frac{[\mathrm{B}][\mathrm{H^+}]}{[\mathrm{BH^+}]} \quad \text{and } \mathrm{p}K_a^B = -\log K_a^B \tag{2}$$

$$(\mathrm{AHB})_{\mathrm{solid}} \rightleftharpoons \mathrm{AHB}_{\mathrm{solution}} \quad \text{with } K_s = [\mathrm{AHB}] \tag{3}$$

$$\mathrm{AHB} \rightleftharpoons \mathrm{AH} + \mathrm{B} \quad \text{with } K_d = \frac{[\mathrm{AH}][\mathrm{B}]}{[\mathrm{AHB}]} \tag{4}$$

Expressions (1) and (2) describe the dissociations of the acids and bases under the conditions (solvent, temperature) employed. The equilibrium constants K_a^A and K_a^B as well as the corresponding pK_a values may be determined by potentiometry. We have listed some data of this type in Table 11 for the common acidic and basic resolving agents. The values do not, of course, depend upon the sense of chirality (+ or −) of the compounds.

Equilibria (3) and (4) define the solubility constant K_s and the dissociation

Table 11 pK_a Values of some resolving agents in 95% ethanol (10°C)[a]

Name	$(pK_a)_1$	$(pK_a)_2$
Acids		
Camphorsulfonic acid	2.0	
Sulfodehydroabietic acid	2.30	9.55
Binaphthylphosphoric acid	2.50	
Dibenzoyltartaric acid	5.85	7.50
Mandelic acid	6.80	
N-Acetylleucine	7.60	
Hydratropic acid	8.45	
α-Hydroxy-2,4-dichlorobenzylphosphonic acid, monomethyl ester	4.35	
Bases		
Ephedrine	9.70	
α-Methylbenzylamine	9.50	
Dehydroabietylamine	9.50	
Quinidine	8.10	2.75
Brucine	7.30	1.30

[a] M. Leclercq and J. Jacques, *Nouv. J. Chim.*, 1979, 3, 629. Reproduced by permission of Gauthiers-Villars.

constant K_d of the dissolved salt.* If in the system being examined $[AH]_0$ and $[B]_0$ represent the total concentrations of acid and base and W the quantity of precipitated salt (in mol·L^{-1}), one may make the following accounting:

$$[AH]_0 = [AH] + [A^-] + [AHB] + W \tag{5}$$

$$[B]_0 = [B] + [BH^+] + [AHB] + W \tag{6}$$

(a) Influence of pH on the solubility of a dissociable organic salt

In the case of a pure salt, K_a^A, K_a^B, $[AH]_0$ and $[BH]_0$ being given, it suffices to measure the pH of the solution and the quantity of solid W precipitated after attainment of equilibrium to arrive at the constants K_s and K_d. From eqs. (5), (1), and (3), one may write

$$[AH]_0 - K_s - W = [AH]\left(1 + \frac{K_a^A}{[H^+]}\right)$$

and, in the same way from eqs. (6), (3) and (2):

$$[B]_0 - K_s - W = [B]\left(1 + \frac{[H^+]}{K_a^B}\right)$$

*Even if it would seem more logical to represent the dissociation of the salt by equilibrium AHB $\rightleftharpoons$ A^- + BH^+, it amounts to the same thing, and it turns out to be more convenient for the purpose of manipulation, to use equilibrium (4).

And finally, by observing that $[\mathrm{AH}][\mathrm{B}] = K_s K_d$, one obtains eq. (7):

$$([\mathrm{AH}]_0 - K_s - W)([\mathrm{B}]_0 - K_s - W) = K_s K_d \left(1 + \frac{[\mathrm{H}^+]}{K_a^B} + \frac{K_a^A}{[\mathrm{H}^+]} + \frac{K_a^A}{K_a^B}\right) \quad (7)$$

If one carries out two series of measurements under conditions in which at least one of the concentrations $[\mathrm{AH}]_0'$, $[\mathrm{B}_0]'$ or $[\mathrm{H}^+]'$ is made to vary, the latter if necessary through addition of base or acid, one may further write

$$\frac{([\mathrm{AH}]_0 - K_s - W)([\mathrm{B}]_0 - K_s - W)}{([\mathrm{AH}]_0' - K_s - W')([\mathrm{B}]_0' - K_s - W')} = \frac{1 + \dfrac{[\mathrm{H}^+]}{K_a^B} + \dfrac{K_a^A}{[\mathrm{H}^+]} + \dfrac{K_a^A}{K_a^B}}{1 + \dfrac{[\mathrm{H}^+]'}{K_a^B} + \dfrac{K_a^A}{[\mathrm{H}^+]'} + \dfrac{K_a^A}{K_a^B}} \quad (8)$$

This equation may be rewritten in the form

$$xK_s^2 + yK_s + z = 0 \quad (9)$$

whereupon its solution gives access to K_s and consequently to K_d.

Measurements made by pairing an optically active base with one and the other enantiomer of a chiral acid allow the comparison of the solubility constant K_s and

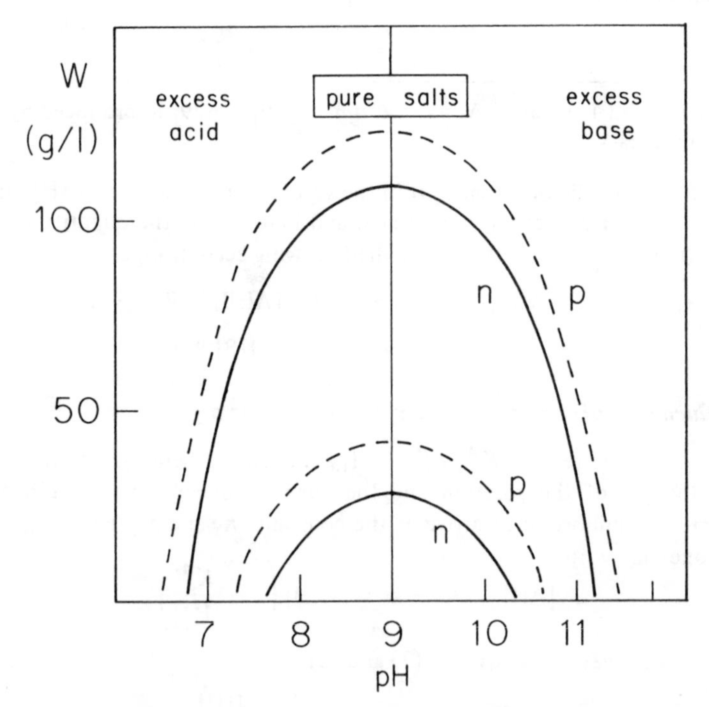

Figure 23 Weight (W) of α-methylbenzylamine hydratropates p and n precipitated as a function of pH of the solution. W represents the quantity of salt (in grams) obtained from 1 L solution containing initially 0.6 mol (top curves) and 0.3 mol (bottom curves) of salts (in 95% ethanol at 10°C). M. Leclercq and J. Jacques, *Nouv. J. Chim.*, 1979, 3, 629. Reproduced by permission of Gauthiers-Villars, Paris.

the dissociation constant K_d of the diastereomeric salts p and n. By way of example, the values obtained for the α-methylbenzylamine hydratropates in 95% ethanol at 10°C and in mol·L^{-1} are as follows: for salt p, $K_s = 0.74 \times 10^{-3}$ and $K_d = 3.7 \times 10^{-3}$; for salt n, $K_s = 1.06 \times 10^{-3}$ and $K_d = 3.9 \times 10^{-3}$. Note that the dissociation constants of the two diastereomers are very close to one another.

For a system whose total acid and base concentrations are known, the knowledge of K_s and K_d allows one to draw the curve that gives the quantity of precipitated salt (W) as a function of pH, eq. (7). Figure 23 represents such experimental curves for the α-methylbenzylamine hydratropates p and n. Two series of curves are given corresponding to an initial concentration of 0.6 mol·L^{-1} (top) and 0.3 mol·L^{-1} (bottom) of pure salt. One immediately sees that the solubility of each salt p or n can increase dramatically when one moves away, in either direction, from the pH corresponding to pure salt by introducing either hydrochloric acid or potassium hydroxide or even an excess of one or the other constituent inorganic ions.

(b) Influence of pH on the solubility of a mixture of diastereomeric salts p *and* n

In the preceding paragraphs we have examined the solubility of pure salts. Let us now examine the influence of pH on the relative solubilities of the two diastereomic salts obtained in the resolution of racemic acid AH_R (consisting of AH_L and AH_D; the initial concentration is $[AH_R]_0$) with optically active base B.

The following species are present in a solution containing an equimolar quantity of p and n salts in the presence of an excess of mineral or organic base or acid:

$$[AH_D], [A_D^-], [AH_L], [A_L^-], [B], [BH^+], [AH_DB], \text{and } [AH_LB]$$

The equilibria that describe the interaction of these species in solution are much more complex than for a single pure salt. Neverthelesss, once can as a first approximation, write equations analogous to eq. (7) in which the subscripts p and n characterize the various constants as well as the quantities of the corresponding precipitates W_p and W_n.

For the species derived from acid D, we have

$$([AH_D]_0 - K_{s_p} - W_p)([B]_0 - K_{s_p} - K_{s_n} - W_p - W_n)$$
$$= K_{s_p}K_{d_p}\left(1 + \frac{[H^+]}{K_a^B} + \frac{K_a^A}{[H^+]} + \frac{K_a^A}{K_a^B}\right) \quad (10)$$

and similarly, for the species deriving from acid L we have

$$([AH_L]_0 - K_{s_n} - W_n)([B]_0 - K_{s_p} - K_{s_n} - W_p - W_n)$$
$$= K_{s_n}K_{d_n}\left(1 + \frac{[H^+]}{K_a^B} + \frac{K_a^A}{[H^+]} + \frac{K_a^A}{K_a^B}\right) \quad (11)$$

Taking into account the fact that $[AH_L]_0 = [AH_D]_0 = 0.5[AH_R]_0$, these two expressions give eq. (12) in which we have arbitrarily assumed that salt p is less soluble than salt n ($W_p > W_n$). Note that, while W_p and W_n are both pH dependent, their difference is not:

$$W_p - W_n = \left(\frac{K_{s_p}K_{d_p}}{K_{s_n}K_{d_n}} - 1\right) W_n - 0.5[\mathrm{AH}_R]_0\left(\frac{K_{sp}K_{dp}}{K_{sn}K_{dn}} - 1\right) + K_{s_p}\left(\frac{K_{d_p}}{K_{d_n}} - 1\right) \tag{12}$$

In order to maximize the yield of a resolution, the quantity of the less soluble salt which precipitates should be maximal, that is, this quantity should tend to one half of the racemic substrate taken. At the same time, the quantity of the more soluble precipitating salt should be minimal.

The pH which conforms to the second condition just given can be calculated if enough data are available. However, this problem can be dealt with more easily in an empirical manner. For example, from solutions of racemic hydratropic acid and optically active α-methylbenzylamine of known concentrations and whose pH values are varied, one obtains variable quantities of precipitates whose compositions must be determined. The pH may be modified either by playing on the stoichiometry of the organic acid or base or by respective addition of potassium hydroxide or triethylamine or of hydrochloric acid.

Figure 24 shows the experimental points obtained which fall quite satisfactorily on the straight line $W_p - W_n = f(W_n)$, which corresponds to eq. (12).

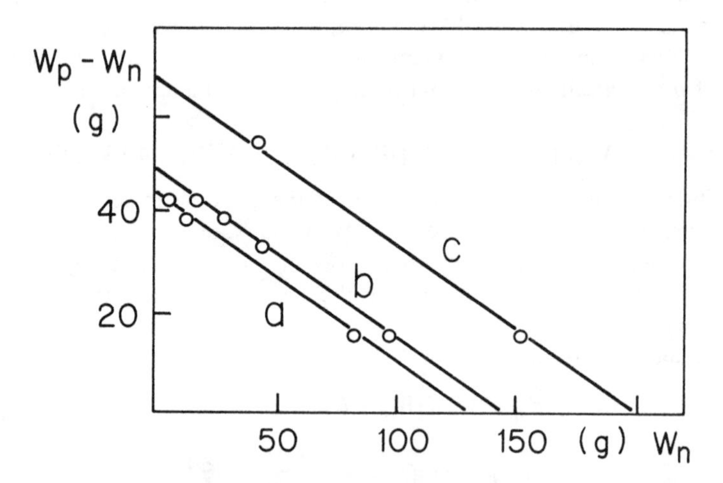

Figure 24 Experimental verification of the relationship $W_p - W_n = f(W_n)$. The straight lines have been calculated by means of eq. (12) for concentrations of racemic hydratropic acid of 1 mol · L^{-1} (a), 1.1 mol · L^{-1} (b), and 1.5 mol · L^{-1} (c). Experimental points for α-methylbenzylamine hydratropates are shown by open circles (95% ethanol at 10°C). M. Leclercq and J. Jacques, *Nouv. J. Chim.*, 1979, **3**, 629. Reproduced by permission of Gauthiers-Villars, Paris.

In a real resolution, by making successive corrections it is always possible to arrive at a pH such that the more soluble salt does not crystallize. Equation (12) may then be simplified as follows:

$$W_p = 0.5[\mathrm{AH}_R]_0 \left(1 - \frac{K_{s_p} K_{d_p}}{K_{s_n} K_{d_n}}\right) + K_{s_p}\left(\frac{K_{d_p}}{K_{d_n}} - 1\right), \tag{13}$$

or

$$\frac{W_p}{0.5[\mathrm{AH}_R]_0} = \alpha + \frac{\beta}{[\mathrm{AH}_R]_0} \tag{14}$$

The resolution yield may be written in the form of eq. (14), which implies that this relationship depends directly upon the initial concentration of racemic acid (which is not completely unexpected) and, in particular, that an increase in this concentration *may be favorable* or *unfavorable* according to the relative value of the dissociation constants of the two diastereomeric salts employed (β positive or negative).

This analysis clarifies a number of observations made in connection with resolutions successfully carried out with nonstoichiometric amounts of reactants. The following conclusions may be drawn from these results:

1 The method of half-quantities and that of Pope and Peachey rest upon the same physicochemical principles.

2 These principles are related to the importance the pH has on the solubility of the salts which *always increases* upon addition of base or acid.

3 Since the dissociation constants of the diastereomeric salts generally do not differ from one another very much, the *difference* in their solubility is little influenced by changes in pH.

4 Just as changes in concentration may allow the crystallization of but one of the two diastereomeric salts, changes in pH may have the same effect.

5 And finally, the advantages of these nonstoichiometric methods rest on the economy in resolving agent and, mainly, in solvent whose quantity may on occasion be reduced *as much as tenfold*. An illustration is given in Table 12.

Table 12 Example of a resolution with a nonstoichiometric amount of resolving agent[a]

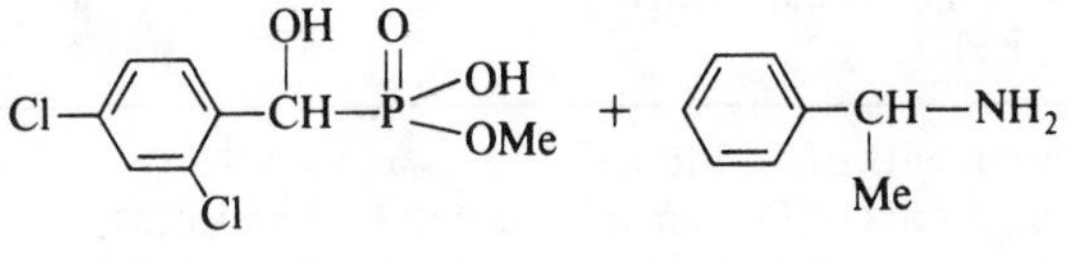

OH O
Cl— —CH—P(—OH)(—OMe) + —CH—NH$_2$
Cl Me

(±) Acid (mmol)	(−) Base (mmol)	Ethanol (mL)	Yield of (+) Acid (%)
10	10	275	77
10	5	34	83

[a] From ref. 252.

5.1.20 Optical Rotations of Diastereomeric Salts and their Constituents

A simple way of following the progress of a resolution would consist of the observation of changes in the optical rotations of the mixtures of the diastereomeric salts *p* and n obtained after each recrystallization. These measurements could even directly give the optical purity of the substance being resolved if one knew (or if one could calculate *a priori*) the rotations of the pure salts and if the specific rotations of their mixtures varied linearly with composition. To what extent are these two conditions actually realized?

In 1873, Landolt[253] showed that in a series of tartrates bearing a variety of achiral counterions, the latter had but little influence on the values of the molecular rotation. Shortly thereafter, Oudemans[254–257] and Tykociner[258] came to similar conclusions with regard to other optically active acids as well as to alkaloid salts with various achiral counterions. Some of these results are shown in Table 13.[259] All of these measurements were carried out in relatively *dilute aqueous* solutions. From these data, it is possible to deduce a value of the molecular rotation $[M]$ which is characteristic of the chiral ion under these conditions. Thus, in water and in the concentrations given in Table 13, tartrate dianion has $[M]_D = +61.5°$, quinate anion* has $[M]_D = -93°$, and strychninium cation has $[M]_D = -114°$. Note that the spread between the highest and lowest values of Table 13 attains a magnitude of 7 to 10% and is therefore not negligible.

Table 13 Molecular rotations of chiral ions in the presence of achiral counterions in water[a]

Tartaric Acid 0.5*M*[a]		Quinic Acid 0.14*M*[b]		Strychnine 0.025*M*[c]	
Salt	$[M]_D$ (deg)	Salt	$[M]_D$ (deg)	Salt	$[M]_D$ (deg)
$Li_2C_4H_4O_6$	+58.1	$KC_7H_{11}O_6$	−93.7	$C_{21}H_{23}N_2O_2Cl$	−114
$(NH_4)_2C_4H_4O_6$	+63.0	$NaC_7H_{11}O_6$	−93.9	$C_{21}H_{23}N_2O_2(NO_3)$	−114
$Na_2C_4H_4O_6$	+59.9	$(NH_4)C_7H_{11}O_6$	−92	$C_{21}H_{23}N_2O_2(HSO_4)$	−118
$K_2C_4H_4O_6$	+64.4	$Sr(C_7H_{11}O_6)_2$	−93.5	$C_{21}H_{23}N_2O_2(H_2PO_4)$	−115
$Na(NH_4)C_4H_4O_6$	+61.7	$Ca(C_7H_{12}O_6)_2$	−93.5	$C_{21}H_{23}N_2O_2$ formate	−113.7
$K(NH_4)C_4H_4O_6$	+63.8	$Mg(C_7H_{11}O_6)_2$	−91.8	$C_{21}H_{23}N_2O_2$ acetate	−113.7
$KNaC_4H_4O_6$	+62.3	$Zn(C_7H_{11}O_6)_2$	−97.9	$C_{21}H_{23}N_2O_2$ oxalate	−110.7
$KAsOC_4H_4O_6$	+58.8	$Ba(C_7H_{11}O_6)_2$	−89.5		
$MgC_4H_4O_6$	+61.7				

[a] From the measurements of Landolt.[259]
[b] From the measurements of Oudemans, quoted by Landolt.[259]
[c] From the measurements of Tykociner, quoted by Landolt.[259]

The values of $[M]_D$ for a given salt also vary substantially with concentration. This phenomenon is clearly related to the extent of dissociation measured, for example, by conductimetry, as is evident from the data of Walden on α-bromocamphor-π-sulfonate anion[259, 260] (Table 14).

In addition to these diverse observations, Walden demonstrated that the molecular rotations of *diastereomeric salts* in dilute aqueous solution could be

*This refers to the conjugate base of quinic acid, the naturally occurring 1,3,4,5-tetrahydroxycyclohexanecarboxylic acid, **37**.

Table 14 Molecular rotation of α-bromocamphor-π-sulfonate anion in water as a function of concentration[a]

	Concentration ($mol \cdot L^{-1}$)	$[M]_D$ (deg)	Dissociation (%)
Free acid	0.481	+287	68.5
	0.0333	+273	92.7
	0.0167	+269	94.4
	0.0083	+270	95.5
Potassium salt	0.0333	+273	83.6
	0.0167	+269	87.2
	0.0083	+269	90.3
Thallium salt	0.0333	+273	83.9
	0.0167	+272	87.3
	0.0083	+271	90.5
Zinc salt	0.0333	+272	71.5
	0.0167	+269	77.2
	0.0083	+270	81.8
Barium salt	0.0333	+272	69.8
	0.0167	+271	74.8
	0.0083	+269	79.4

[a] The extent of dissociation was measured by conductimetry.

equated to the sum of the molecular rotations of the constituent ions.[260] The values of $[M_A]$ and $[M_B]$ of the ions of an acid and of a base and those corresponding to a pair of diastereomers *p* and *n*, respectively, $[M_1]$ and $[M_2]$ formed, for example, by combination of a racemic acid and an optically active base are related by the following expressions:

$$[M_1] = [M_B] + [M_A] \quad (1a)$$

$$[M_2] = [M_B] - [M_A] \quad (1b)$$

$$[M_A] = \frac{[M_1] - [M_2]}{2} \quad (2a)$$

$$[M_B] = \frac{[M_1] + [M_2]}{2} \quad (2b)$$

The validity of these additivity relationships has been tested in a number of cases; several examples follow. Morphine α-bromocamphor-π-sulfonate has a molecular rotation $[M]_D = -100°$ (water)[260] in keeping with that calculated from the molecular rotations of morphinium ion, $[M_B]_D = -370°$ and of α-bromocamphor-π-sulfonate ion, $[M_A]_D = +270°$, eq. (1a). According to Pope and Peachey,[238] the salts *p* and *n* of (+)-10-camphorsulfonic acid with *d*- and *l*-tetrahydroquinaldines have, respectively, $[M_1]_D = +173.3°$ and $[M_2]_D = -69.5°$ in 2% aqueous solution. The calculated molecular rotation of (+)-10-camphorsulfonate ion, $[M_A]_D = 1/2([M_1]_D + [M_2]_D) = +51.9°$, is found to be quite close to that of ammonium (+)-10-camphorsulfonate, $[M_A]_D = +51.7°$ (H_2O, $c = 2\%$). Similarly, the value calculated for tetrahydroquinaldine ion, $[M_B]_D = 1/2([M_1]_D - [M_2]_D) = 121.4°$, is not very different from that found for the hydrochloride measured under the same conditions: 121.7°. Other examples of such calculations may be found in the publications of Pope and his co-workers.[229, 261, 262] Actually, the agreement between calculated and observed values is not always as good as has just been implied, as is evident from the data in Table 15.[242]

Table 15 Additivity of molecular rotations of diastereomeric ions

Salts		$[M]_D$ (deg) Observed	$[M]_D$ (deg) Calculated
d- and *l*-Hydroxy-2-hydrindamine *d*-α-bromocamphor-π-sulfonate	*p*	+307	+340
	n	+221	+218
d- and *l*-1-Hydroxy-2-hydrindamine *d*-10-camphorsulfonate	*p*	+121	+111
	n	−9.7	−10.1
d- and *l*-Tetrahydroquinaldine *d*-α-bromocamphor-π-sulfonate	*p*	+400	+413
	n	+147	+145

All of these tests of the validity of the additivity relationships (1) and (2) were carried out on salts of strong sulfonic acids. It is clear that in dilute aqueous solutions of sulfonic acid salts we can expect nearly total dissociation into ions A^- and BH^+ (and the consequent additivity of the ionic molecular rotations). However, as we have seen in the preceding section, other species such as AH and B or the non-dissociated salts $[A^-BH^+]$ may predominate in the case of salts of weaker acids and bases even in water and, *a fortiori*, in other solvents such as alcohols or chloroform.[263]

From data such as those given in Table 15, it follows that changes in the rotations of diastereomer salt mixtures may *qualitatively* provide a measure of the progress of a resolution. However, it is only in special cases and under carefully specified conditions that one may hope to obtain more quantitative information on the composition of such mixtures.

REFERENCES 5.1

1 L. Pasteur, *C. R. Acad. Sci.*, 1853, **37**, 162.
2 L. Pasteur, *Ann. Chim.* (Paris), 1853, 3, **38**, 437.
3 L. Pasteur, *Leçons de Chimie professées en 1860*, Soc. Chim. de Paris, Paris, 1861.
4 S. H. Wilen, *Topics Stereochem.*, 1971, **6**, 107.
5 L. Pasteur, *C. R. Acad. Sci.*, 1853, **37**, 110.
6 R. B. Woodward and W. E. Doering, *J. Am. Chem. Soc.*, 1945, **67**, 860.
7 R. B. Turner and R. B. Woodward, "The Chemistry of the Cinchona Alkaloids," in *The*

Alkaloids, R. H. F. Manske and H. L. Holmes, Eds., Vol. 3, Academic Press, New York, 1953, Chapter 16.

8 J. Gutzwiller and M. R. Uskoković, *Helv. Chim. Acta*, 1973, **56**, 1494.

9 E. Fourneau and G. Sandulesco, *Bull. Soc. Chim. Fr.*, 1922, **31**, 988.

10 See, for example, A. McKenzie and H. B. P. Humphries, *J. Chem. Soc.*, 1910, **97**, 121.

11 K. Mislow, *J. Am. Chem. Soc.*, 1951, **73**, 3954.

12 K. Mislow and M. Heffler, *J. Am. Chem. Soc.*, 1952, **74**, 3668.

13 J. Jacobus, M. Raban, and K. Mislow, *J. Org. Chem.*, 1968, **33**, 1142.

14 C. A. McKenzie, *Experimental Organic Chemistry*, 3rd ed., Prentice Hall, New York, 1967, p. 257.

15 E. T. Kaiser and F. W. Carson, *J. Am. Chem. Soc.*, 1964, **86**, 2922.

16 W. A. Bonner, J. A. Zderic, and G. A. Casaletto, *J. Am. Chem. Soc.*, 1952, **74**, 5086.

17 B. Sjöberg and S. Sjöberg, *Ark. Kemi*, 1964, **23**, 447.

18 W. J. Gottstein and L. C. Cheney, *J. Org. Chem.*, 1965, **30**, 2072.

19 J. W. Westley, R. H. Evans, and J. F. Blount, *J. Am. Chem. Soc.*, 1977, **99**, 6057.

20 W. Markwald and R. Meth, *Ber.*, 1905, **38**, 801.

21 J. Jacques, C. Fouquey, and R. Viterbo, *Tetrahedron Lett.*, 1971, 4617.

22 D. A. Jaeger, M. D. Broadhurst, and D. J. Cram, *J. Am. Chem. Soc.*, 1979, **101**, 717.

23 E. Fisher, *Ber.*, 1899, **32**, 2451.

24 J. P. Greenstein and M. Winitz, *Chemistry of the Amino Acids*, Vol. 1, Wiley, New York, 1961, p. 715.

25 S. H. Wilen, in *Tables of Resolving Agents and Optical Resolutions*, E. L. Eliel, Ed., University of Notre Dame Press, Notre Dame, Indiana, 1972, p. 268.

26 G. Gal, J. M. Chemerda, D. F. Reinhold, and R. M. Purick, *J. Org. Chem.*, 1977, **42**, 142.

27 (a) S. Tatsuoka, M. Honjo, and T. Kinoshita, *J. Pharm. Soc. Jpn.*, 1951, **71**, 718. (b) S. Tatsuoka, M. Honjo, and H. Miyazaki, *J. Pharm. Soc. Jpn.*, 1951, **71**, 1277.

28 (a) C. P. Berg, *J. Biol. Chem.*, 1936, **115**, 9. (b) F. J. Kearley and A. W. Ingersoll, *J. Am. Chem. Soc.*, 1951, **73**, 4604.

29 F. L. Pyman, *J. Chem. Soc.*, 1911, **99**, 1386.

30 (a) F. H. Radke, R. B. Fearing, and S. W. Fox, *J. Am. Chem. Soc.*, 1954, **76**, 2801. (b) Spanish Patent, 357033 (1970). *Chem. Abstr.*, 1971, **74**, 87369b. (c) J. Weijlard, K. Pfister III, E. F. Swanezy, C. A. Robinson, and M. Tishler, *J. Am. Chem. Soc.*, 1951, **73**, 1216. (d) P. Pratesi, A. La Manna, and L. Fontanella, *Farmaco, Ed. Sci.*, 1955, **10**, 673. *Chem. Abstr.*, 1956, **50**, 10057. (e) D. Shapiro, H. Segal, and H. M. Flowers, *J. Am. Chem. Soc.*, 1958, **80**, 1194. (f) C. A. Grob and E. F. Jenny, *Helv. Chim. Acta*, 1952, **35**, 2106. (g) H. E. Carter and D. Shapiro, *J. Am. Chem. Soc.*, 1953, **75**, 5132.

31 K. Harada, *Bull. Chem. Soc. Jpn.*, 1964, **37**, 1383.

32 (a) G. P. Wheeler and A. W. Ingersoll, *J. Am. Chem. Soc.*, 1951, **73**, 4604. (b) M. Betti and M. Mayer, *Ber.*, 1908, **41**, 2071. (c) S. Gronowitz, I. Sjögren, L. Wernstedt, and B. Sjöberg, *Ark. Kemi*, 1965, **23**, 129.

33 G. Triem, *Ber.*, 1938, **71**, 1522.

34 S. G. Traynor, B. J. Kane, M. F. Betkonski, and L. M. Hirschy, *J. Org. Chem.*, 1979, **44**, 1557.

35 A. Garnier, A. Collet, L. Faury, J. P. Albertini, J. M. Pastor, and L. Tosi, to be published.

36 (a) H. C. Beyerman, *Rec. Trav. Chim.*, 1959, **78**, 134. (b) A. Shafi'ee and G. Hite, *J. Med. Chem.*, 1969, **12**, 266.

37 R. L. Peck and A. R. Day, *J. Heterocycl. Chem.*, 1969, **6**, 181.

38 E. Erlenmeyer, Jr., and A. Arnold, *Liebigs Ann. Chem.*, 1904, **337**, 307.

39 A. W. Ingersoll, in *Organic Reactions*, Vol. 2, R. Adams, Ed., Wiley, New York, 1944, Chapter 9.

40 B. A. Klyashchitskii and V. I. Shvets, *Uspekhi Khim.*, 1972, **41**, 1315. *Russ. Chem. Rev.*, 1972, **41**, 592.

41 R. H. Pickard and W. O. Littlebury, *J. Chem. Soc.*, 1907, **91**, 1973.

42 R. K. Hill and J. W. Morgan, *J. Org. Chem.*, 1968, **33**, 927.

43 G. A. C. Gough, H. Hunter, and J. Kenyon, *J. Chem. Soc.*, 1926, 2052.

44 R. Bäckstrom and B. Sjöberg, *Ark. Kemi*, 1967, **26**, 549.

45 F. A. Abd Elhafez and D. J. Cram, *J. Am. Chem. Soc.*, 1952, **74**, 5846.

46 A. Viola, G. F. Dudding and R. J. Proverb, *J. Am. Chem. Soc.*, 1977, **99**, 7390.

47 W. Reeve, F. J. Bianchi, and J. R. McKee, *J. Org. Chem.*, 1975, **40**, 339.

48 J. G. Molotkovsky and L. D. Bergelson, *Tetrahedron Lett.*, 1971, 4791.

49 F. Bergel, Swiss Patent 124126 (1944). *Chem. Zentralbl.*, 1950, **1**, 2386.

50 J. Kenyon and K. Thaker, *J. Chem. Soc.*, 1957, 2531.

51 W. von E. Doering and H. H. Zeiss, *J. Am. Chem. Soc.*, 1948, **70**, 3966. *Ibid.*, 1950, **72**, 147.

52 H. H. Zeiss, *J. Am. Chem. Soc.*, 1951, **73**, 2391.

53 A. G. Davies, J. Kenyon, and L. W. F. Salame, *J. Chem. Soc.*, 1957, 3148.

54 B. Bielawski and A. Chrzaszczewska, *Lodz Tow. Nauk Wydz. III Acta Chim.*, 1966, **11**, 105. *Chem. Abstr.*, 1967, **67**, 2837.

55 K. A. Thaker and N. S. Dave, *J. Sci. Ind. Res.*, 1962, **21B**, 374.

56 H. L. Holmes and D. J. Currie, *Can. J. Chem.*, 1969, **47**, 4076.

57 K. G. Rutherford, J. M. Prokipcak, and D. P. C. Fung, *J. Org. Chem.*, 1963, **28**, 582.

58 D. Scheffel, P. J. Abbott, G. J. Fitzpatrick, and M. D. Schiavelli, *J. Am. Chem. Soc.*, 1977, **99**, 3769.

59 G. Hofle and W. Steglich, *Synthesis*, 1972, 619.

60 A. L. Henne and R. L. Pelley, *J. Am. Chem. Soc.*, 1952, **74**, 1426.

61 H. S. Mosher, J. E. Stevenot, and D. O. Kimble, *J. Am. Chem. Soc.*, 1956, **78**, 4374.

62 W. H. Pirkle, S. D. Beare, and T. G. Burlingame, *J. Org. Chem.*, 1969, **34**, 470.

63 C. Van Der Stelt, W. J. Heus, and W. T. Nauta, *Arzneim.-Forsch.*, 1969, **19**, 2010. *Chem. Abstr.*, 1970, **72**, 54599.

64 B. L. Murr, C. Santiago, and S. Wang, *J. Amer. Chem. Soc.*, 1969, **91**, 3827.

65 K. G. Rutherford, J. L. H. Batiste, and J. M. Propipcak, *Can. J. Chem.*, 1969, **47**, 4074.

66 E. S. Wallis and F. H. Adams, *J. Am. Chem. Soc.*, 1933, **55**, 3838.

67 L. W. Feller, Ph.D. Thesis, Johns Hopkins University, 1969. *Diss. Abstr. Int. B.*, 1969, **30**, 557.

68 J. W. C. Crawford, *J. Chem. Soc.*, 1965, 4280.

69 E. W. Cantrall, C. Krieger and R. B. Brownfield (to American Cyanamid Co.), Ger. Offen., 1,942,453 (1970). *Chem. Abstr.*, 1970, **72**, 111702m.

70 W. C. Agosta, *J. Am. Chem. Soc.*, 1967, **89**, 3926.

71 A. J. Fry and W. E. Britton, *J. Org. Chem.*, 1973, **38**, 4016.

72 L. H. Sarett, G. E. Arth, R. M. Lukes, R. E. Beyler, G. I. Poos, W. F. Johns, and J. M. Constantin, *J. Am. Chem. Soc.*, 1952, **74**, 4974.

73 C. Neuberg and M. Federer, *Ber.*, 1905, **38**, 868.

74 J. K. Shillington, G. S. Denning, Jr., W. B. Greenough III, T. Hill, Jr., and O. B. Ramsay, *J. Am. Chem. Soc.*, 1958, **80**, 6551.

75 H. Kaehler, F. Nerdel, G. Engemann, and K. Schwerin, *Liebigs Ann. Chem.*, 1972, **757**, 15.

76 E. Touboul, M. J. Brienne, and J. Jacques, *J. Chem. Res.* (S), 1977, 106; (M), 1977, 1182.

77 R. Adams and J. D. Garber, *J. Am. Chem. Soc.*, 1949, **71**, 522.

78 G. Adolphen, E. J. Eisenbraun, G. W. Keen, and P. W. K. Flanagan, *Org. Prep. Proc.*, 1970, **2**, 93.

79 W. R. Adams, O. L. Chapman, J. B. Siega, and W. J. Welstead, Jr., *J. Am. Chem. Soc.*, 1966, **88**, 162.

80 N. J. Leonard and J. V. Paukstelis, *J. Org. Chem.*, 1963, **28**, 3021.

81 V. L. King, Dissertation, University of Zürich, 1912.

82 A. Werner, *Ber.*, 1911, **44**, 1887.

83 See G. B. Kauffman, *Classics in Coordination Chemistry*, Part 1, *The Selected Papers of Alfred Werner*, Dover, New York, 1968.

84 M. Delépine, *C. R. Acad. Sci.*, 1914, **159**, 239.

85 D. H. Busch and J. C. Bailar, Jr., *J. Am. Chem. Soc.*, 1953, **75**, 4574.

86 R. D. Gillard, P. R. Mitchell, and C. F. Weick, *J. Chem. Soc. Dalton*, 1974, 1035.

87 G. B. Kauffman and E. V. Lindley, Jr., *J. Chem. Educ.*, 1974, **51**, 424.

88 J. C. Bailar, Jr., *Inorg. Synth.*, 1946, **2**, 222.

89 A. Werner, *Ber.*, 1911, **44**, 3272.

90 A. Werner, *Ber.*, 1911, **44**, 2445.

91 A. Werner, *Ber.*, 1912, **45**, 1228.

92 A. Werner and A. P. Smirnoff, *Helv. Chim. Acta*, 1920, **3**, 472.

93 C. J. Dippel and F. M. Jaeger, *Rec. Trav. Chim.*, 1931, **50**, 547.

94 A. Werner, *Ber.*, 1912, **45**, 121.

95 J. A. Broomhead, F. P. Dwyer, and J. W. Hogarth, *Inorg. Synth.*, 1966, **6**, 183.

96 Y. Kushi, M. Kuramoto, and H. Yoneda, *Chem. Lett.* 1976, 135.

97 F. Galsbøl, *Inorg. Synth.*, 1970, **12**, 269.

98 G. Schlessinger, *Inorg. Synth.*, 1970, **12**, 267.

99 D. A. House, *J. Chem. Educ.*, 1976, **53**, 124.

100 G. B. Kauffman and L. T. Takahashi, *J. Chem. Educ.*, 1962, **39**, 481.

101 C. F. Liu and J. Doyle, *Chem. Comm.*, 1967, 412

102 S. F. Mason and J. W. Wood, *Chem. Comm.*, 1968, 1512.

103 F. P. Dwyer and F. L. Garvan, *Inorg. Synth.*, 1966, **6**, 195.

104 J. H. Worrell, *Inorg. Synth.*, 1972, **13**, 195.

105 F. P. Dwyer and F. L. Garvan, *Inorg. Synth.*, 1966, **6**, 192.

106 A. Werner and M. Basyrin, *Ber.*, 1913, **46**, 3229.

107 French Patent 1,360,884 (1964). *Chem. Abstr.*, 1964, **61**, 16154f.

108 F. P. Dwyer and F. L. Garvan, *J. Am. Chem. Soc.*, 1959, **81**, 2955.

109 (a) A. Werner, *Helv. Chim. Acta*, 1918, **1**, 5; (b) Y. Saito, *Topics Stereochem.*, 1978, **10**, 95.

110 F. M. Jaeger and H. B. Blumendal, *Z. Anorg. Chem.*, 1928, **175**, 161, 198, 200, 230.

111 A. P. Smirnoff, *Helv. Chim. Acta*, 1920, **3**, 177.

112 S. Kirschner, Y.-K. Wei, and J. C. Bailar, Jr., *J. Am. Chem. Soc.*, 1957, **79**, 5877.

113 F. P. Dwyer, F. L. Garvan, and A. Shulman, *J. Am. Chem. Soc.*, 1959, **81**, 290.

114 F. M. Jaeger, *Optical Activity and High Temperature Measurements*, McGraw–Hill, New York, 1930, p. 158.

115 E. J. Corey and J. C. Bailar, Jr., *J. Am. Chem. Soc.*, 1959, **81**, 2620.

116 P. H. Boyle, *Q. Rev. Chem. Soc.*, 1971, **25**, 323.

117 M. S. Newman, W. B. Lutz, and D. Lednicer, *J. Am. Chem. Soc.*, 1955, 77, 3420.

118 M. S. Newman and W. B. Lutz, *J. Am. Chem. Soc.*, 1956, **78**, 2469.

119 P. Block, Jr., and M. S. Newman, *Org. Synth.*, 1968, **48**, 120.

120 M. Green and R. F. Hudson, *J. Chem. Soc.*, 1958, 3129.

121 D. T. Longone and M. T. Reetz, *Chem. Comm.*, 1967, 46.

122 G. Goedicke and H. Stegemeyer, *Tetrahedron Lett.*, 1970, 937.

123 D. A. Lightner, D. T. Hefelfinger, T. W. Powers, G. W. Frank, and K. N. Trueblood, *J. Am. Chem. Soc.*, 1972, **94**, 3492.

124 M. S. Newman, R. G. Mentzer, and G. Slomp, *J. Am. Chem. Soc.*, 1963, **85**, 4018.

125 M. S. Newman, R. W. Wotring, Jr., A. Pandit, and P. M. Chakrabarti, *J. Org. Chem.*, 1966, **31**, 4293.

126 T. Sato, S. Akabori, M. Kainosho, and K. Hata, *Bull. Chem. Soc. Jpn.*, 1968, **41**, 218.

127 H. Rau and O. Schuster, *Angew. Chem.*, 1976, 88, 90. *Angew. Chem. Int. Ed.*, 1976, **15**, 114.

128 F. I. Carrol, B. Berrang, and C. P. Linn, *Chem. Ind.* (London), 1975, 477. *J. Med. Chem.*, 1978, **21**, 326.

129 H. Wynberg and K. Lammertsma, *J. Am. Chem. Soc.*, 1973, **95**, 7913.

130 L. H. Klemm and D. Reed, *J. Chromatogr.*, 1960, **3**, 364.

131 L. H. Klemm, K. B. Desai, and J. R. Spooner, Jr., *J. Chromatogr.*, 1964, **14**, 300.

132 G. Wittig and K. D. Ruempler, *Liebigs Ann. Chem.*, 1971, **751**, 1.

133 M. S. Newman and H. Junjappa, *J. Org. Chem.*, 1971, **36**, 2606.

134 (a) M. Hagan, *Clathrate Inclusion Compounds*, Reinhold, London, 1962. (b) S. G. Franck, *J. Pharm. Sci.*, 1975, **64**, 10.

135 D. D. MacNicol, J. J. McKendrick, and D. R. Wilson, *Chem. Soc. Rev.*, 1978, 7, 65.

136 C. Asselineau and J. Asselineau, *Ann. Chim.* (Paris), 1964, **9**, 461.

137 W. Schlenk, Jr., *Experientia*, 1952, 8, 337.

138 W. Schlenk, Jr., in *Methoden der Organischen Chemie* (Houben-Weyl), 4th ed., E. Müller, Ed., Vol. I/1, Thieme, Stuttgart 1958, p. 410.

139 W. Schlenk, Jr., *Angew. Chem.*, 1960, **72**, 845.

140 W. Schlenk, Jr., *Chem. Ber.*, 1968, **101**, 2445.

141 W. Schlenk, Jr., *Liebigs Ann. Chem.*, 1973, 1145.

142 W. Schlenk, Jr., *Liebigs Ann. Chem.*, 1973, 1157.

143 W. Schlenk, Jr., *Liebigs Ann. Chem.*, 1973, 1179.

144 W. Schlenk, Jr., *Liebigs Ann. Chem.*, 1973, 1195.

145 W. Schlenk, Jr., *Liebigs Ann. Chem.*, 1949, **565**, 204.

146 For example, W. Schlenk (to BASF A.-G.), German Patent 1,080,557 (1958). *Chem. Abstr.*, 1962, **56**, 2334g. German Patent, 1,074,583 (1960). *Chem. Abstr.*, 1961, **55**, 13318c. German Patent 1,076,686 (1960). *Chem. Abstr.*, 1961, **55**, 13321c.

147 E. Angelescu and G. Nicolau, *Anal. Univ. Bucuresti, Ser. Stiint. Nat.*, 1964, **13**, 91.

148 A. E. Smith, *Acta Crystallogr.*, 1952, **5**, 224.

149 H. U. Von Lenne, *Acta Crystallogr.*, 1954, 7, 1.

150 W. Kutzelnigg and R. Mecke, *Z. Elekrochem.*, 1961, **65**, 109.

151 R. K. McMullan, W. Saenger, J. Fayos, and D. Mootz, *Carbohydr. Res.*, 1973, **31**, 37.

152 W. Saenger, in *Environmental Effects on Molecular Structure and Properties*, B. Pullman, Ed., D. Reidel Publishing Company, Dordrecht, Netherlands, 1976, p. 265–305. (b) *idem., Angew. Chem.*, 1980, **92**, 343; *Angew. Chem. Int. Ed.*, 1980, **19**, 344.

153 D. French, *Adv. Carbohydr. Chem.*, 1957, **12**, 189.

154 J. Hamilton, L. Steinrauf, and R. L. Vanetten, *Acta Crystallogr.*, 1968, **B24**, 1560.

155 F. Cramer and W. Dietsche, *Chem. Ber.*, 1959, **92**, 378.

156 F. Cramer and H. Hettler, *Naturwissenschaften*, 1967, **54**, 625.

157 H. P. Benschop and G. R. Van den Berg, *Chem. Comm.*, 1970, 1431.

158 M. Mikołajczyk, J. Drabowicz, and F. Cramer, *Chem. Comm.*, 1971, 317.

159 M. Mikołajczyk and J. Drabowicz, *J. Am. Chem. Soc.*, 1978, **100**, 2510.

160 M. Mikołajczyk and J. Drabowicz, *Tetrahedron Lett.*, 1972, 2379.

161 J. Drabowicz, Doctoral Thesis, Lodz, Poland, 1974. M. Mikołajczyk and J. Drabowicz, *J. Chem. Soc. Chem. Comm.*, 1976, 220.

162 J. Knabe and N. S. Agarwal, *Deut, Apoth.-Zeit.*, 1973, **113**, 1449.

163 W. Baker, B. Gilbert, and W. D. Ollis, *J. Chem. Soc.*, 1952, 1443.

164 A. P. Downings, W. D. Ollis, and I. O. Sutherland, *J. Chem. Soc.* (B), 1970, 24.

165 A. P. Downings, W. D. Ollis, and I. O. Sutherland, *Chem. Comm.*, 1968, 329.

166 A. C. D. Newman and H. M. Powell, *J. Chem. Soc.*, 1952, 3747.

167 D. Laxton and H. M. Powell, *J. Chem. Soc.*, 1958, 2339.

168 H. M. Powell, *Nature*, 1952, **170**, 155.

169 M. K. Hargreaves and B. Modarai, *Chem. Comm.*, 1969, 16.

170 M. K. Hargreaves and B. Modarai, *J. Chem. Soc.* (C), 1971, 1013.

171 (a) R. Arad-Yellin, B. S. Green, and M. Knossow, *J. Am. Chem. Soc.*, 1980, **102**, 1157. (b) R. Arad-Yellin, B. S. Green, M. Knossow, and G. Tsoucaris, *Tetrahedron Lett.*, 1980, **21**, 387.

172 W. C. Herndon, *J. Chem. Educ.*, 1967, **44**, 724.

173 H. Sobotka and A. Goldberg, *Biochem. J.* 1932, **26**, 905.

174 D. J. Collins and J. J. Hobbs, *Austr. J. Chem.*, 1970, **23**, 119.

175 R. C. Kelly, I. Schletter, S. J. Stein, and W. Wierenga, *J. Am. Chem. Soc.*, 1979, **101**, 1054.

176 W. Baker, A. J. Floyd, J. F. W. McOmie, G. Pope, A. S. Weaving, and J. H. Wild, *J. Chem. Soc.*, 1956, 2010.

177 R. M. Barrer and V. H. Shanson, *J. Chem. Soc. Chem. Comm.*, 1976, 333.

178 J. L. Flippen, J. Karle, and I. L. Karle, *J. Am. Chem. Soc.*, 1970, **92**, 3749.

179 A. D. U. Hardy, J. J. McKendrick, and D. D. MacNicol, *J. Chem. Soc. Chem. Comm.*, 1974, 972.

180 M. J. Brienne and J. Jacques, *Tetrahedron Lett.*, 1975, 2349.

181 A. D. U. Hardy, J. J. McKendrick, and D. D. MacNicol, *J. Chem. Soc. Chem. Comm.*, 1976, 355.

182 A. Collett and J. Jacques, *J. Chem. Soc. Chem. Comm.*, 1976, 708.

183 A. Collet and J. Jacques, *Israel J. Chem.*, 1976/1977, **15**, 82.

184 D. D. MacNicol, private communication; see ref. 135.

185 D. D. MacNicol and F. B. Wilson, *J. Chem. Soc. D*, 1971, 786.

186 M. Delépine, *Bull. Soc. Chim. Fr.*, 1921, **29**, 656.

187 V. C. Arsenijevic, *C. R. Acad. Sci.*, 1957, **245**, 317.

188 L. S. Arsenijevic and A. F. Damansky, *C. R. Acad. Sci.*, 1959, **248**, 3723.

189 M. C. Brianso, *Acta Crystallogr.*, 1976, **B32**, 3040.

190 M. C. Brianso, M. Leclercq, and J. Jacques, *Acta Crystallogr.*, 1979, **B35**, 2751.

191 M. C. Brianso, *Acta Crystallogr.*, 1978, **B34**, 679.

192 S. W. Bergman, *Ark. Kemi, Min. Geol.*, 1927, **9**, No. 42.

193 E. Graf and H. Boedekker, *Liebigs Ann. Chem.*, 1958, **613**, 111.

194 M. Leclercq and J. Jacques, *Bull. Soc. Chim. Fr.*, 1975, 2052.
195 A. H. Beckett and N. H. Choulis, *J. Pharm. Sci.*, 1966, **55**, 1155.
196 C. Jarowski and W. H. Hartung, *J. Org. Chem.*, 1943, **8**, 564.
197 S. Bergman, *Ark. Kemi, Min. Geol.*, 1926, **9**, No. 34.
198 A. W. Ingersoll, S. H. Babcock, and F. B. Burns, *J. Am. Chem. Soc.*, 1933, **55**, 411.
199 A. W. Ingersoll and J. R. Little, *J. Am. Chem. Soc.*, 1934, **56**, 2123.
200 Y. Shimura and K. Tsutsui, *Bull. Chem. Soc. Jpn.*, 1977, **50**, 145.
201 L. Westman, *Ark. Kemi,* 1958, **12**, 167.
202 M. Matell, *Ark. Kemi,* 1951, **3**, 129.
203 D. Varech and J. Jacques, *Tetrahedron*, 1972, **28**, 5671.
204 M. J. Brienne, C. Ouannès, and J. Jacques, *Bull. Soc. Chim. Fr.*, 1967, 613.
205 W. H. Mills and K. A. C. Elliott, *J. Chem. Soc.*, 1928, 1291.
206 R. R. Sealock, M. E. Speeter, and R. S. Schweet, *J. Am. Chem. Soc.*, 1951, **73**, 5386.
207 J. A. Berson and D. A. Ben-Efraim, *J. Am. Chem. Soc.*, 1959, **81**, 4083.
208 B. Sjöberg, *Ark. Kemi*, 1956, **9**, 295.
209 A. Ladenburg, *Liebigs Ann. Chem.*, 1908, **364**, 227.
210 A. Ladenburg, *Ber.*, 1894, **27**, 75.
211 A. Ladenburg and O. Bobertag, *Ber.*, 1903, **36**, 1649.
212 G. Bettoni, R. Perrone, and V. Tortorella, *Gazzetta Chim. Ital.*, 1972, **102**, 196.
213 H. Rapoport, A. R. Williams, and M. E. Cisney, *J. Am. Chem. Soc.*, 1951, **73**, 1414.
214 A. Fredga and G. Ekstedt, *Ark. Kemi*, 1965, **23**, 123.
215 A. Findlay, *The Phase Rule*, 9th ed., revised by A. N. Campbell and N. O. Smith, Dover, New York, 1951, p. 353.
216 M. Matell, *Ark. Kemi*, 1955, **8**, 79.
217 W. J. Pope and J. Read, *J. Chem. Soc.*, 1910, **97**, 987.
218 M. Delépine and F. Lareze, *Bull. Soc. Chim. Fr.*, 1955, 104.
219 A. W. Ingersoll, *J. Am. Chem. Soc.*, 1925, **47**, 1168.
220 A. W. Ingersoll, *J. Am. Chem. Soc.*, 1928, **50**, 2264.
221 A. W. Ingersoll and E. G. White, *J. Am. Chem. Soc.*, 1932, **54**, 274.
222 (a) A. W. Ingersoll and F. B. Burns, *J. Am. Chem. Soc.*, 1932, **54**, 4712. (b) R. Kuhn and K. Vogler, *Z. Naturforsch.*, 1951, **6b**, 232. (c) A. J. Zambito, W. L. Peretz, and E. E. Howe, *J. Am. Chem. Soc.*, 1949, **71**, 2541.
223 (a) W. Marckwald, *Ber.*, 1896, **29**, 42. (b) W. Marckwald, *Ber.*, 1896, **29**, 43.
224 E. L. Eliel, *Stereochemistry of Carbon Compounds*, McGraw–Hill, New York, 1962, p. 50.
225 M. A. Jermyn, *Aust. J. Chem.*, 1967, **20**, 2283.
226 G. J. W. Bremer, *Ber.*, 1880, **13**, 351.
227 J. Read and W. G. Reid, *J. Soc. Chem. Ind.*, 1928, **47**, 8T.
228 G. Wittig, H. J. Cristau, and H. Braun, *Angew. Chem.*, 1967, **79**, 721. *Angew. Chem. Int. Ed.*, 1967, **6**, 700.
229 W. J. Pope and J. Read, *J. Chem. Soc.*, 1910, **97**, 988.
230 M. Delépine and F. Lareze, *Bull. Soc. Chim. Fr.*, 1955, 104.
231 J. P. Vigneron and V. Bloy, *Bull. Soc. Chim. Fr.*, 1976, 649.
232 D. R. Palmer (to Beecham Group Ltd.), German Offen., 2,147,620 (1972). *Chem. Abstr.*, 1972, **77**, 34938s.
233 C. T. Holdrege (to Bristol-Myers Co.), U.S. Patent, 3,796,748 (1974). *Chem. Abstr.*, 1974, **80**, 121327x.

234 R. R. Lorenz (to Sterling Drug Inc.), German Offen., 2,345,302 (1974). *Chem. Abstr,*, 1974, **80**, 133823. U.S. Patent, 3,832,388 (1974).
235 W. J. Pope and J. B. Whitworth, *Proc. Roy. Soc.* (London), 1931, **A134**, 357.
236 D. E. Pearson and A. A. Rosenberg, *J. Med. Chem.*, 1975, **18**, 523.
237 M. D. Armstrong, *J. Am. Chem. Soc.*, 1951, **73**, 4456.
238 W. J. Pope and S. J. Peachey, *J. Chem. Soc.*, 1899, **75**, 1066.
239 W. J. Pope and C. S. Gibson, *J. Chem. Soc.*, 1912, **101**, 939.
240 W. J. Pope and E. M. Rich, *J. Chem. Soc.*, 1899, **75**, 1093.
241 W. J. Pope and C. S. Gibson, *J. Chem. Soc.*, 1899, **97**, 2207.
242 W. J. Pope and J. Read, *J. Chem. Soc.*, 1912, **101**, 758.
243 E. Ott, *Liebigs Ann. Chem.*, 1931, **488**, 198.
244 I. Rusznak, R. Soos, E. Fogassy, M. Acs, and Z. Ecsery (to Chinoin Gyogyszer es Vegyeszeti Termelek Gyara Rt.) Hungarian Teljes, 12,210 (1976). *Chem. Abstr.*, 1976, **85**, 192335n.
245 I. Rusznak, R. Soos, E. Fogassy, M. Acs, and Z. Ecsery (to Chinoin Gyogyszer es Vegyeszeti Termelek Gyara Rt.) Hungarian Teljes, 12,209 (1976). *Chem. Abstr.*, 1976, **85**, 192336p.
246 V. A. Mikhalev, M. I. Dorokhova, N. E. Smolina, T. S. Gorokhova, and O. Ya. Tikhonova, USSR Patent, 250,158 (1969). *Chem. Abstr.*, 1970, **72**, 78645v.
247 E. T. Stiller, S. A. Harris, J. Finkelstein, J. C. Keresztesy, and K. Folkers, *J. Am. Chem. Soc.*, 1942, **62**, 1785.
248 W. H. Perkin, W. J. Pope, and O. Wallach, *J. Chem. Soc.*, 1909, **95**, 1789.
249 N. N. Dykhanov, and N. I. Al'nikova, *Mater. Vses. Nauchn. Simp. Sovrem. Probl. Samoochishcheniya Regul: Kach. Vody*, 5th 1975, **1**, 65. *Chem. Abstr.*, 1976, **85**, 138465s.
250 P. Gallegra (to Syntex Corp.), U.S. Patent, 3,988,365 (1976). *Chem. Abstr.*, 1977, **86**, 55204a.
251 S. H. Wilen, R. Davidson, R. Spector, and H. Stefanou, *Chem. Comm.*, 1969, 603.
252 M. Leclercq and J. Jacques, *Nouv. J. Chim.*, 1979, **3**, 629.
253 H. Landolt, *Ber.*, 1873, **6**, 1077.
254 A. C. Oudemans, Jr., *Liebigs Ann. Chem.*, 1876, **182**, 33.
255 A. C. Oudemans, Jr., *Rec. Trav. Chim.*, 1885, **4**, 166.
256 A. C. Oudemans, Jr., *Rec. Trav. Chim.*, 1882, **1**, 18.
257 A. C. Oudemans, Jr., *Liebigs Ann. Chem.*, 1879, **197**, 48.
258 H. Tykociner, *Rec. Trav. Chim.*, 1882, **1**, 144.
259 H. Landolt, *Das optische Drehungsvermögen*, 2nd ed., F. Vieweg and Son, Braunschweig, 1898, p. 191.
260 P. Walden. *Z. Phys. Chem.*, 1894, **15**, 196.
261 W. J. Pope and J. Read, *J. Chem. Soc.*, 1910, **97**, 2199.
262 W. J. Pope and C. S. Gibson, *J. Chem. Soc.*, 1910, **97**, 2211.
263 M. Guetté and J. P. Guetté, *Bull. Soc. Chim. Fr.*, 1977, 769.
264 Y. Kushi, M. Kuramoto, and H. Yoneda, *Chem. Lett.*, 1976, 135.
265 Y. Kushi, M. Kuramoto, and H. Yoneda, *Chem. Lett.*, 1976, 339.
266 Y. Kushi, M. Kuramoto, and H. Yoneda, *Chem. Lett.*, 1976, 663.
267 M. Kuramoto, Y. Kushi, and H. Yoneda, *Chem. Lett.*, 1976, 1133.
268 M. Kuramoto, Y. Kushi, and H. Yoneda, *Bull. Chem. Soc. Jpn.*, 1978, **51**, 3251.
269 T. Tada, Y. Kushi, and H. Yoneda, *Chem. Lett.*, 1977, 379.
270 M. Kuramoto, Y. Kushi, and H. Yoneda, *Bull. Chem. Soc. Jpn.*, 1978, **51**, 3196.

271 F. Mikes, G. Boshart, and E. Gil-Av, *J. Chem. Soc. Chem. Comm.*, 1976, 99.
272 F. Mikes, G. Boshart, and E. Gil-Av, *J. Chromatogr.*, 1976, **122**, 205.
273 I. S. Krull, *Adv. Chromatogr.*, 1978, **16**, 196.
274 G. Blaschke, *Angew. Chem.*, 1980, **92**, 14. *Angew. Chem. Int. Ed.*, 1980, **19**, 13.
275 F. Mikes and G. Boshart, *J. Chem. Soc. Chem. Comm.*, 1978, 173.
276 Unpublished results of M. J. Brienne, A. Collet, and M. Leclercq.

5.2 COVALENT COMPOUNDS

While earlier proposals and even experimental attempts at carrying out resolutions involving covalent diastereomers are to be found in the literature,[1–4] the first successful resolution of an amine through formation and separation of covalent derivatives was carried out by Erlenmeyer, Jr.,[5] in 1903. He showed that *threo*-1,2-diphenyl-2-aminoethanol, **1**, forms two diastereomeric Schiff bases with helicin (salicylaldehyde β-D-glucoside), **2**, and that these are separable by crystallization. Unfortunately, decomposition of these diastereomers (with dilute HCl) to recover the resolved amine is accompanied by destruction of helicin.

C_6H_5–CH(OH)–CH(NH_2)–C_6H_5

1

2-(CH=O)C_6H_4–O–(β-D-glucose)

2

Again in 1903, Neuberg[6] suggested the use of diastereomeric hydrazones in the resolution of aldehydes and ketones. From racemic arabinose and *l*-menthylhydrazine, he isolated a menthylhydrazone from which he regenerated a nearly optically pure arabinose enantiomer by action of formaldehyde.

The successful formation of two covalent diastereomers from a racemate requires that a number of conditions be met if the process is to be useful in a resolution. It is preferable that the preparation of these compounds be easy, namely, that if at all possible they be formed in one step. It is essential that the reaction that is chosen does not provoke appreciable and uncontrollable racemization of the resolving agent (e.g., in the preparation of an ester from an optically active acid chloride which may easily be racemized through conversion into a ketene, as in the case of hydratropic or α-phenylbutyric acids). In cases in which separation is dependent upon recrystallization, the diastereomers formed must have different solubilities, and at least one of the two must be crystalline. This restriction clearly is irrelevant when separation is carried out chromatographically, a process usually effective with virtually all types of chromatography.

Finally, the problems associated with the cleavage of the separated diastereomers are more serious for covalent diastereomers than for the dissociable compounds previously examined if the resolving agent is to be recovered easily and without decomposition or loss of chirality.

We first describe the covalent derivatives of the various functional groups which figure in resolutions together with the reagents employed in their formation, the separation methods, and the regeneration of the resolved substrates and the resolving agents. We call attention to the fact that chromatographic separation of covalent diastereomers by gas, thin-layer, dry-column, flash, and medium- and high-pressure liquid chromatography for preparative purposes is now carried out at least as often as is separation by crystallization. In the following section, no special distinction has been made between chromatographic and nonchromatographic methods in our discussion. Whether a separation is carried out by crystallization or by chromatography will be dictated in part by the physical state of the product (crystalline or not) as well as by the scale of the resolution, and so on. A list of representative examples of covalent diastereomers resolved by chromatographic methods is given in Table 3 (Section 5.2.7).

5.2.1 Covalent Derivatives of Acids

Among the useful covalent diastereomeric derivatives for the resolution of acids, we first consider *esters*, which may be obtained through reaction of an optically active alcohol either directly with a racemic acid (with or without the presence of a catalyst), with its acid chloride derivative, or through transesterification of a racemic ester in the presence of an optically active alcohol. This approach has not given rise to a large number of applications. The optically active alcohols used are essentially limited to menthol[7–10] and 2-octanol.[11] α-Methylfenchol has also been used recently as a resolving agent.[12] Menthol is also effective in the resolution of organophosphorus acids.[13]

A second possibility consists in the conversion of a racemic acid into an *amide*. This technique has been particularly developed in recent years by Helmchen et al.[14, 15] for the resolution of acids of the type $R_1R_2CHCOOH$ through amides formed with primary amines $R_3R_4CHNH_2$. The amines used for the purpose are, in particular, α-(2-naphthyl)ethylamine, α-methylbenzylamine, and methyl phenylglycinate. The amides are separated by chromatography on silica gel on an analytical (TLC, HPLC) as well as preparative scale.[16] The resolution of organophosphorus acids via amides formed with L-proline methyl ester has been reported.[17]

Haas and Prelog have described the resolution of an acid by formation of its amide with dehydroabietylamine.[18] Since the hydrolysis of amides is usually quite difficult, regeneration of the acid may be effected by rearrangement of the *N*-nitroso derivative,[18] a process which unfortunately destroys the amine resolving agent. S. Rendic et al. have applied a similar strategy to the resolution of the antiarthritic agent ketoprofen [2-(3-benzoylphenyl)propionic acid] as its amide with α-methylbenzylamine. The amides were separated by chromatography on silica.[19] The cited difficulty in cleaving the resolved amides has now been overcome by Helmchen and co-workers in the case of the resolving agents α-methylbenzylamine and phenylglycinol. Relatively mild acidic conditions regenerate resolved acids and even lactones from the corresponding separated amides.[20]

5.2.2 Covalent Derivatives of Amines

The transformation of a racemic amine into a mixture of diastereomeric *amides* by reaction with an optically active acid (or one of its derivatives) is easily achievable. The incorporation of such a reaction into a workable resolution is another matter. Yet we have already cited the historic[3] resolution of α-methylbenzylamine by crystallization of its amide with quinic acid. The principal limitation of such a method is the difficulty in hydrolyzing the amide so as to regenerate the resolved amine (cleavage via the *N*-nitroso derivative obviously being precluded here). To the extent that this difficulty is overcome, the process is actually an interesting one for amines (or for acids for that matter) given the fact, demonstrated by Herlinger, Kleinman, and Ugi,[21] that diastereomeric amides virtually always crystallize and exhibit solubilities which are sufficiently different so as to permit their separation (see Section 5.2.6). The ease with which amides derived from acids or lactones are separated by chromatography[16] has led to the development of a straightforward resolution procedure for amines via amides prepared from resolved 3-phenylbutyrolactone.[22] The separated amides are cleaved by simple acid hydrolysis.

Amines may also be resolved through conversion into urethane or urea derivatives. A mixture of diastereomeric carbamates (urethanes) may be prepared by reaction of the chloroformate ester of an optically active alcohol, for example, menthyl chloroformate, **3**, with a racemic amine. After separation of the diastereomers by crystallization or by chromatography, the resolved amine is easily regenerated by acidic hydrolysis under mild conditions[23] or by cleavage with lithium aluminum hydride to the amine *N*-methyl derivative.[24]

(±) [1,3-diphenylisoindoline]NH + Cl—C(=O)—O—Menthyl ⟶ [1,3-diphenylisoindoline]N—C(=O)—O—Menthyl *(p,n)*

3

Reaction of optically active α-methylbenzylisocyanate, **4**, with racemic amines gives mixture of diastereomeric ureas which are also useful in resolutions. Recovery of the resolved amine may be effected by pyrolysis.[25] Other known optically active isocyanates such as those derived from menthylamine,[26] dehydroabietylamine,[27] and α-(1-naphthyl)ethylamine[28] should be useful in the same way.

$$(\pm)\ R{-}NH_2 + Ph{-}\overset{*}{C}H(CH_3){-}N{=}C{=}O \longrightarrow (p,n)\ R{-}NH{-}C(=O){-}NH{-}\overset{*}{C}H(CH_3){-}Ph$$

4

It would seem that covalent Schiff bases might be good candidates for resolutions of amines, particularly those formed from carbohydrates aldehydes. However, we are unaware of useful resolutions based on this suggestion. Along these lines,

Helferich and Portz have reported the formation of a crystalline adduct between (+)-2,3,4,6-tetra-*O*-acetyl-D-glucose, **5**, as a mixture of α and β anomers and (+)-α-methylbenzylamine.[29] According to the authors, this adduct is different from the Schiff base **6** or from the tautomeric *N-d*-glucopyranoside **7**. The latter may be obtained from either the dextrorotatory or the levorotatory amine by simple heating with **5**. On the other hand, the "complex" alluded to above is formed with the dextrorotatory amine only. The latter may be separated virtually quantitatively by simply mixing the racemic amine and **5** in ether.[30] The isolated crystals may be instantaneously decomposed by hydrochloric acid at 0°C.

While Mislow[31] has suggested that the adduct is a lattice host clathrate, in all likelihood it is nothing more than the carbinolamine **8**. Such compounds, which are postulated to be intermediates during imine synthesis along with other nitrogen-containing carbonyl derivatives, are only rarely isolated[32a] because of their relative instability.

Unfortunately, tetraacetylglucose, **5**, which form similar adducts with several primary and secondary amines,[33, 34] is relatively inaccessible, and this may be responsible for the lack of use of this otherwise promising resolution method.

CH=O
H—OAc
AcO—H
H—OAc
H—OH
CH_2OAc

5

CH_3
CH=N—CH—Ph
*
H—OAc
AcO—H
H—OAc
H—OH
CH_2OAc

6

⇌

CH_3
H, NH—CH—Ph
*
C
H—OAc
AcO—H
H—OAc
H—O—
CH_2OAc

7

CH_3
HO, NH—CH—Ph
*
CH
H—OAc
AcO—H
H—OAc
H—OH
CH_2OAc

8

5.2.3 Covalent Derivatives of Alcohols, Thiols, and Phenols

Probably more so than for any other functionality, covalent derivatives of alcohols are convenient mediators of resolutions of the latter. In particular, in the past two decades, diastereomeric *esters* have figured more and more prominently in the resolution of organic compounds containing hydroxyl or sulfhydryl groups.

ω-Camphanic acid, **9**, which has been recommended by Gerlach as a resolving agent,[35] has often been successfully used for the resolution of alcohols and phenols. Its levorotatory chloride is commercially available or may be relatively easily prepared from camphoric acid.[32b] Since both enantiomers of the latter are commercially available, either form of **9** may readily be obtained.

(+) Camphor ⟶ (+) [COOH, COOH] ⟶ (−) [COOH, O, O]

9

Camphanates are formed from the acid chloride and racemic alcohols or phenols; the diastereomeric esters may be separated by crystallization[36–39] or by chromatography.[35, 40–42] Regeneration of the resolved alcohols is effected by saponification or by reduction with lithium aluminum hydride; in either process, the camphanic acid is destroyed. This fact constitutes a significant limitation to the use of this resolving agent on any but a small scale.

An indirect resolution of glycols with ω-camphanic acid has been reported. The resolution of *threo*-hydrobenzoin involves the stereospecific syn ring opening of the appropriate *trans*-epoxide by camphanic (or mandelic) acid to yield a mixture of *threo p* and *n* monocamphanate (or monomandelate) esters. After separation and cleavage of the monoesters, one obtains the resolved diols.[43]

(±) [Ar, R, O, H, Ar′ epoxide] + R*—COOH ⟶ *(p,n) threo* Ar—CH(OCOR*)—CR(OH)—Ar′

Menthoxyacetic acid, **10**,[26,44] introduced as a resolving agent by Read and Grubb in 1931,[45] forms esters of alcohols and phenols which occasionally may be separated by crystallization and more often are separated by chromatography.[46–51]

Steroidal alcohols form crystalline esters with (+)-*O,O′*-diacetyltartaric acid monomethyl ester, **11**, which may serve to resolve the former.[52] The use of esters of the camphorsulfonic acids **12** and **13** in resolutions has also been described.[53, 54]

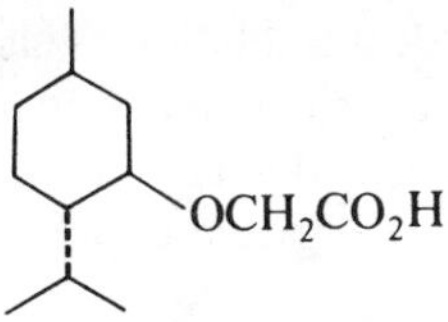

OCH$_2$CO$_2$H

(–)-**10**

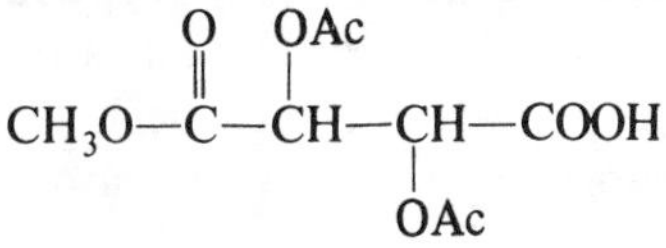

$$CH_3O-\overset{O}{\overset{\|}{C}}-\overset{OAc}{\overset{|}{CH}}-\underset{OAc}{\underset{|}{CH}}-COOH$$

11

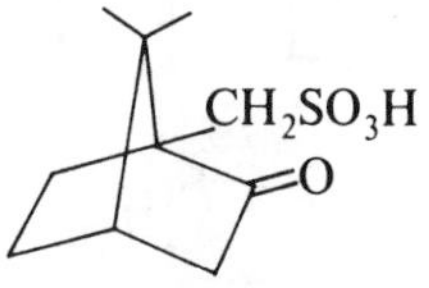

CH$_2$SO$_3$H, O

12

HO$_3$S, O, Br

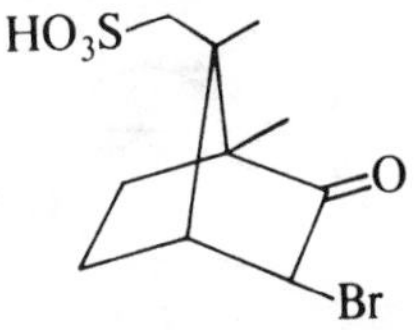

13

3β-Acetoxy-Δ^5-etienic acid, **14**, which may be prepared from the more readily available pregnenolone acetate by the hypobromite oxidation of Serullas–Lieben,[55] reacts with alcohols via the acid chloride to form diastereomeric esters which may be separated by crystallization[56–59] or, if need be, by chromatography on silica gel.[60–63]

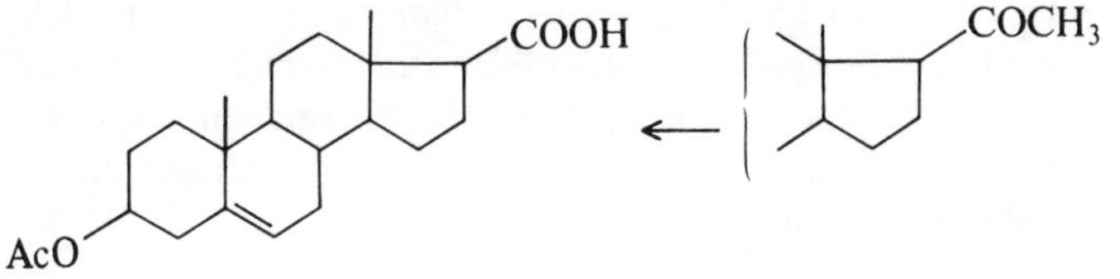

COOH, AcO ← COCH$_3$

14

Regeneration of alcohols resolved through their etienate esters may be attended by complications. Since saponification of these esters leads to sodium 3β-hydroxy-Δ^5-etienate which is insoluble in water, it is preferable to cleave the esters with lithium aluminum hydride. This, however, implies a supplementary separation of the resolved alcohol from the product of the reduction of the etienic acid. Also, when the hydride reduction of the ester is precluded by the structure of the alcohol being resolved,[60] the crude saponification product is acidified and the etienic acid is separated from the resolved alcohol by chromatography on alumina.

Other esters occasionally used are lactates, separable by gas chromatography in the case of 3,3-dimethyl-2-butanol,[64] and esters of *N*-tosylamino acids.[65,66] The latter may be prepared by azeotropic distillation of mixtures of racemic alcohol, amino acid (e.g., valine, alanine), and *p*-toluenesulfonic acid and separated by crystallization.

$$(\pm)\ R-OH + (L)\ CH_3-\underset{NH_2}{\underset{|}{CH}}-COOH + TsOH \longrightarrow (p,n)\ CH_3-\underset{NH-Ts}{\underset{|}{CH}}-COOR$$

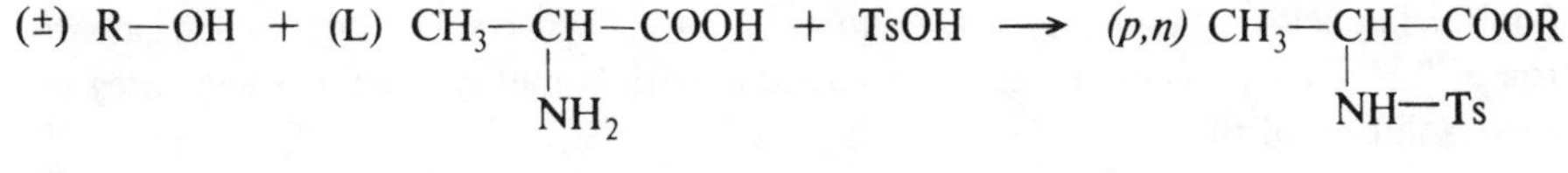

Brooks and colleagues have reported the analytical resolution of alcohols (and amines) through formation of esters (amides) with the terpenoids *trans*-chrysanthemic acid, **15**, and drimanoic acid, **16**. These esters and amides are separated by gas chromatography.[67] Acid **15** has also served as resolving agent in preparative resolutions.[68]

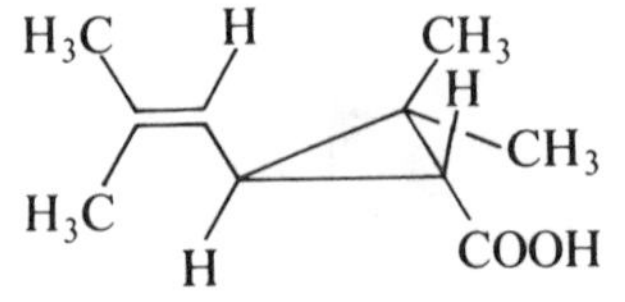

15 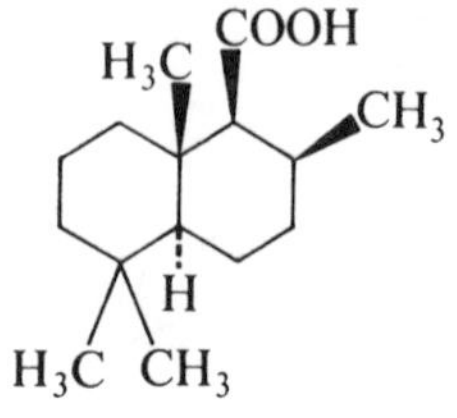 **16**

A major advance in alcohol resolutions is the use of carbamate (urethane) esters suggested by one of us in 1971.[64] Such esters, derived from *l*-menthyl isocyanate, had actually been employed in resolutions as far back as 1906 by Pickard and Littlebury.[26, 69, 70] In 1974, Pirkle and his co-workers introduced the use of α-(1-naphthyl)ethyl isocyanate **17** derived from α-(1-naphthyl)ethylamine for the resolution of alcohols through their carbamates.[28] In a variant route, the alcohol to be resolved is converted with phosgene to its chloroformate, and the latter is treated with active α-(1-naphthyl)ethylamine to form the mixture of carbamates.[71] Other chiral derivatizing agents, for example, **18**, have been employed as well.[72] The diastereomeric carbamates are separated by chromatography on alumina in a procedure applicable on a gram scale.[73]

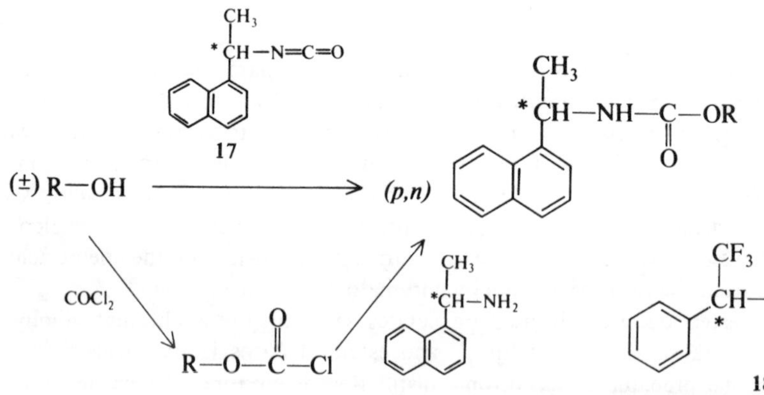

17

(±) R—OH → (*p,n*)

$COCl_2$

18

The resolved alcohols are obtained by cleavage with sodium ethoxide. Silanolysis (cleavage with trichlorosilane) is a milder alternative cleavage procedure.[74]

A variety of alcohols, some in connection with natural product total syntheses (gibberellic acid,[75] lasalocid,[76] disparlure[77]), have been resolved via the carbamate route.[78–83] This procedure has been applied (with α-methylbenzyl isocyanate) to the resolution of thiols.[84]

Carbonates derived from menthyl chloroformate and racemic alcohols have been used for analytical resolutions of alcohols (and amines) by gas chromatography[85] as well as in preparative resolutions by crystallization.[86]

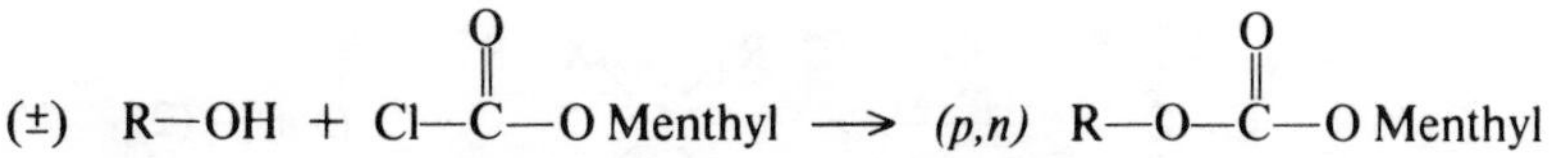

$$(\pm)\ R{-}OH + Cl{-}\overset{O}{\overset{\|}{C}}{-}O\ \text{Menthyl} \longrightarrow (p,n)\ R{-}O{-}\overset{O}{\overset{\|}{C}}{-}O\ \text{Menthyl}$$

α-Methoxy-α-trifluoromethylphenylacetic acid (MTPA), **19**, is a useful chiral derivatizing reagent for the determination of enantiomeric purity of alcohols and amines by nmr spectroscopy.[87] The esters are separable by gas chromatography; a number of preparative resolutions based on this reagent have been reported.[61, 88, 89]

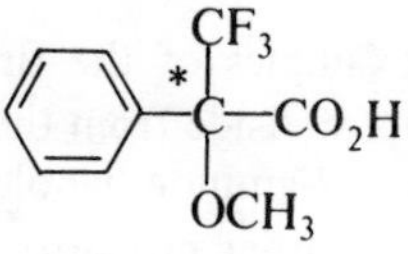

$$C_6H_5{-}\overset{*}{C}(CF_3)(OCH_3){-}CO_2H$$

19

Other methods used on occasion include the formation and separation of glucosides as in the resolution of *trans*-1,2-cyclopentanediol,[90] phenols,[91] and terpenols[92, 93] and the resolution of diols such as propylene glycol by fractional distillation of their dioxolane (ketal) derivatives formed with (−)-menthone or with (+)-camphor.[94] Resolution of alcohols via diastereomeric *lactonic ethers* obtained from the chiral hemiacetal **19a** (the cyclic form of (1R)-*cis*-caronaldehyde) has been reported.[152] This reagent is available as a byproduct of the industrial synthesis of chrysanthemic acid and appears to be especially useful for the resolution of base-sensitive alcohols, e.g., cyanohydrins.

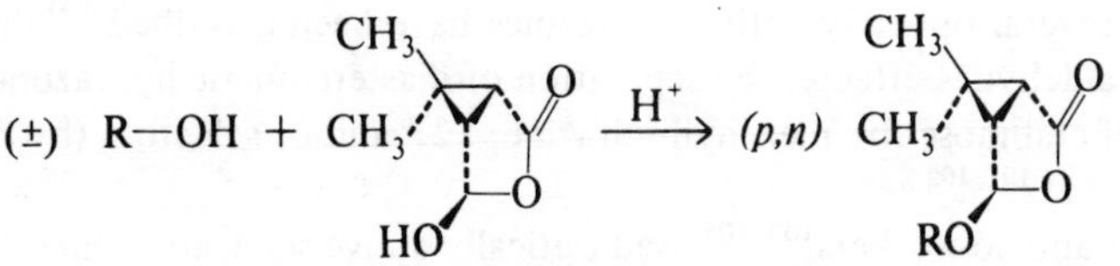

$$(\pm)\ R{-}OH + \mathbf{19a}\ (\text{HO}) \xrightarrow{H^+} (p,n)\ (\text{RO})$$

19a

5.2.4 Covalent Derivatives of Aldehydes, Ketones, and Sulfoxides

For the purposes of resolution, aldehydes and ketones may be transformed into three principal types of covalent derivatives: (1) those in which the carbonyl functional group is transformed into a C=N group; (2) those in which it is transformed into a ketal or a thioketal; and (3) those in which it is transformed into heterocyclic derivatives:

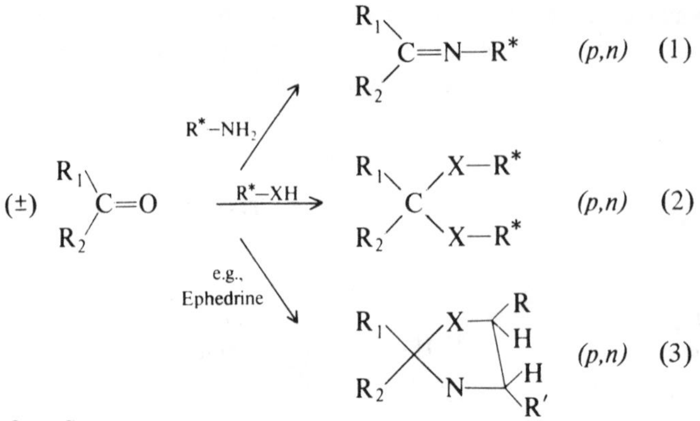

$$(\pm)\ \begin{matrix}R_1\\ \ \ C{=}O\\ R_2\end{matrix} \xrightarrow{R^*-NH_2} \begin{matrix}R_1\\ \ \ C{=}N{-}R^*\\ R_2\end{matrix} \quad (p,n) \quad (1)$$

$$\xrightarrow{R^*-XH} \begin{matrix}R_1 & & X{-}R^*\\ & C & \\ R_2 & & X{-}R^*\end{matrix} \quad (p,n) \quad (2)$$

$$\xrightarrow{\text{e.g., Ephedrine}} \quad (p,n) \quad (3)$$

where X = O or S.

Schiff bases are obvious examples of the first category, yet they have been relatively little used in resolutions. Aside from the historical example of the helicin resolution[5] of *threo*-1,2-diphenyl-2-aminoethanol, **1** (see p. 328), imine derivatives of 2-amino-1-butanol, **20**, and of other primary amines (dehydroabietylamine and α-methylbenzylamine) or their hydrochloride salts have been used in the resolution of steroidal ketones by crystallization.[64, 95] Among other optically active primary amines forming Schiff bases, Betti's base, **21**, which is easily accessible[96] and readily resolved with tartaric acid,[97] has been used in the resolution of hydratropic[98] and *p*-methoxyhydratropic[99] aldehydes.

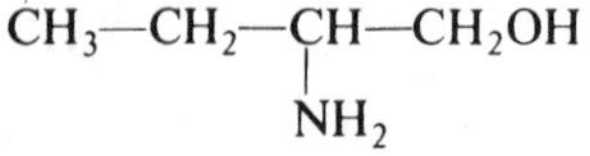

$$CH_3{-}CH_2{-}CH(NH_2){-}CH_2OH$$

20

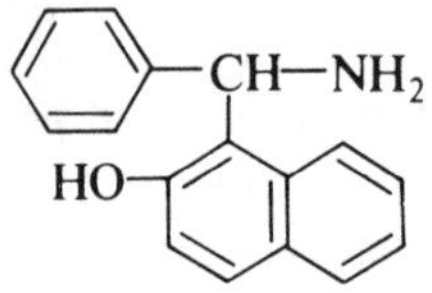

21

While several optically active hydrazines have been described,[100] the only resolutions of aldehydes effected by separation of diastereomeric hydrazones appear to be those of arabinose (by menthylhydrazine, **22**)[6] and of galactose (by amylphenylhydrazine, **23**).[101, 102]

Wilson and co-workers[103–105] used optically active semicarbazones derived from α-methylbenzyl-, **24**, α-ethylbenzyl-, **25**, and menthylsemicarbazides, **26**, in the resolution of ketones such as benzoin. These reagents do not seem to have been used subsequently.

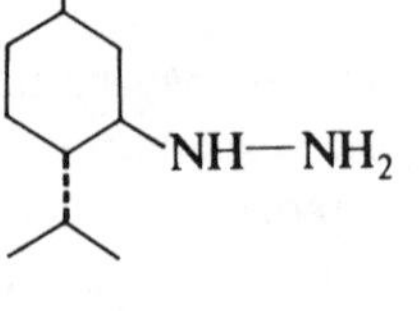

22

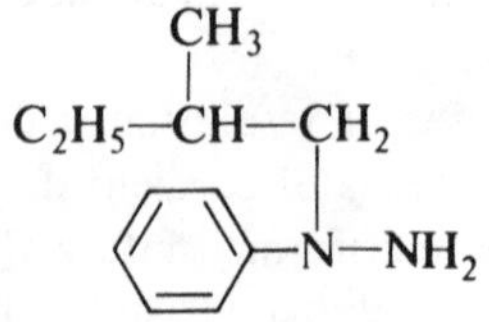

23

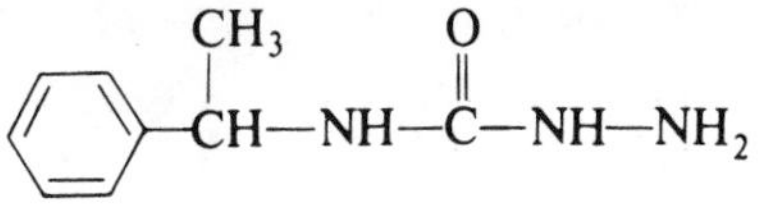

$$C_6H_5-CH(CH_3)-NH-C(=O)-NH-NH_2$$

24

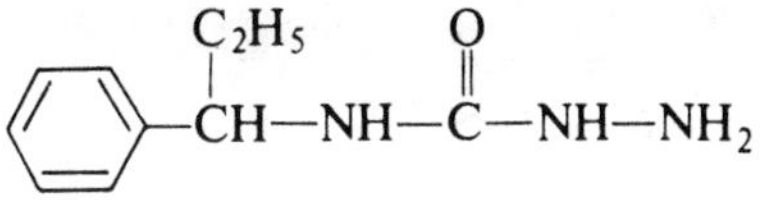

$$C_6H_5-CH(C_2H_5)-NH-C(=O)-NH-NH_2$$

25

$$\text{(menthyl)}-NH-C(=O)-NH-NH_2$$

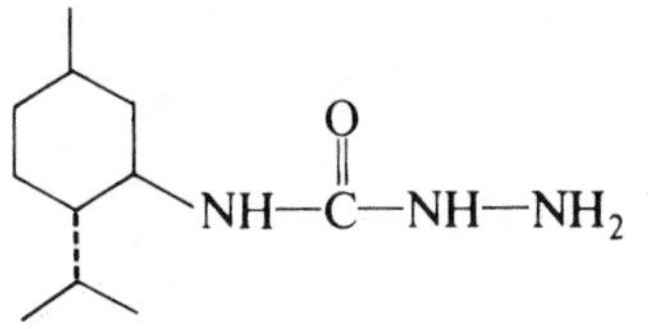

26

A particularly interesting reagent developed by Woodward, Kohman, and Harris for the resolution of ketones[100] is menthyl *N*-aminocarbamate, **27**, known simply as "menthydrazide."

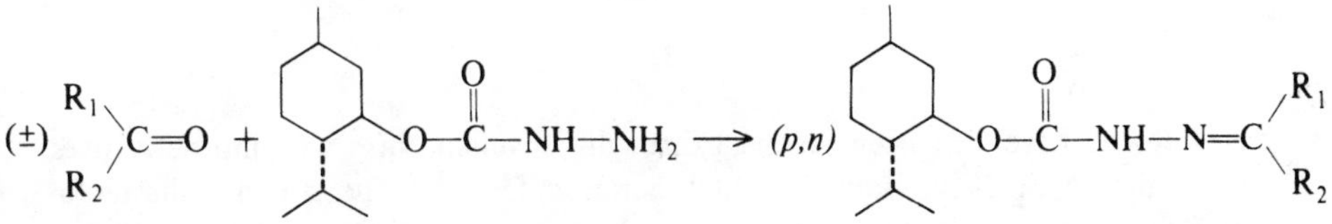

$$(\pm)\ R_1R_2C{=}O + \text{(menthyl)}-O-C(=O)-NH-NH_2 \longrightarrow (p,n)\ \text{(menthyl)}-O-C(=O)-NH-N{=}CR_1R_2$$

27

Resolving agent **27** is easily prepared via menthyl carbonate. The "menthydrazone" derivatives of aldehydes and ketones are obtained in buffered acetic acid/sodium acetate medium. This procedure has been applied to the following resolutions, among others: camphor,[64, 100] α-ionone,[98] 1,1,1-bromochlorofluoroacetone,[106] and various metallocenic aldehydes and ketones.[107, 108]

While menthydrazone cleavage may be carried out by hydrolysis in the presence of dilute sulfuric acid,[100] the possibility that this method may be attended by partial racemization of the carbonyl derivative has given rise to alternative cleavage procedures that avoid this disadvantage. Touboul and Dana[109] hydrolyze menthydrazones in acetone in the presence of phthalic acid; acetone menthydrazone is formed while the resolved carbonyl compound is freed without racemization. Schlögl carries out the same cleavage with formaldehyde in methanolic phosphoric acid.[110]

A number of hydrazides have also been used in resolutions. For example, tartramazide, **28**, has figured in the resolution of 2-phenylcyclopentanone,[111, 112] and mandelic acid hydrazide, **29**, has been used in that of β-phenylindanone.[113] Oxalic acid hydrazides such as 5-(α-methylbenzyl)semioxamazide, **30**,[114] and 5-methyl-5-(α-methyl-β-phenylethyl)semioxamazide, **31** (prepared from deoxyephedrine), have also been used in resolutions, the latter for 2-flavanone.[115] The report of the successful resolution of chiral (arene)tricarbonylchromium aldehydes with **30** and of unsuccessful resolution attempts with other resolving agents is highly instructive.[116]

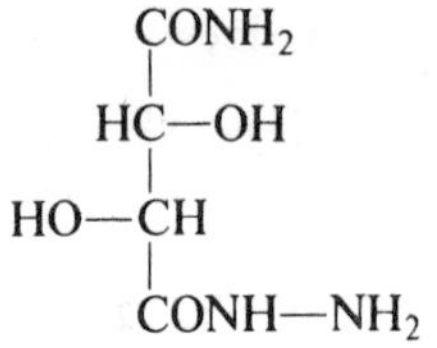

$$\begin{array}{c} CONH_2 \\ | \\ HC{-}OH \\ | \\ HO{-}CH \\ | \\ CONH{-}NH_2 \end{array}$$

28

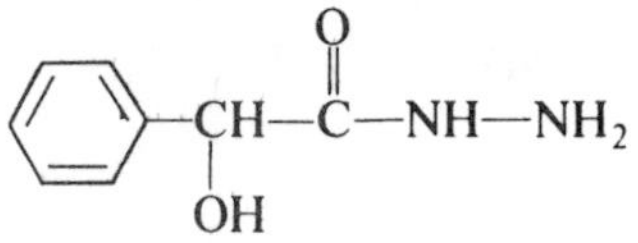

$$C_6H_5{-}\underset{OH}{CH}{-}\overset{O}{\overset{\|}{C}}{-}NH{-}NH_2$$

29

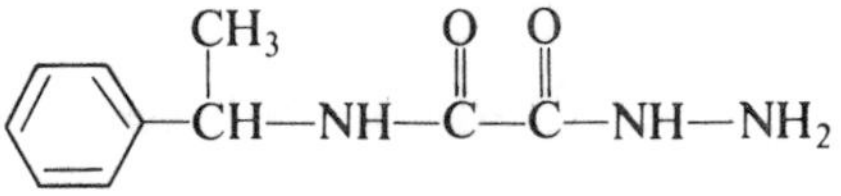

$$C_6H_5{-}\overset{CH_3}{CH}{-}NH{-}\overset{O}{\overset{\|}{C}}{-}\overset{O}{\overset{\|}{C}}{-}NH{-}NH_2$$

30

$$C_6H_5{-}CH_2{-}\overset{CH_3}{CH}{-}\underset{CH_3}{N}{-}\overset{O}{\overset{\|}{C}}{-}\overset{O}{\overset{\|}{C}}{-}NH{-}NH_2$$

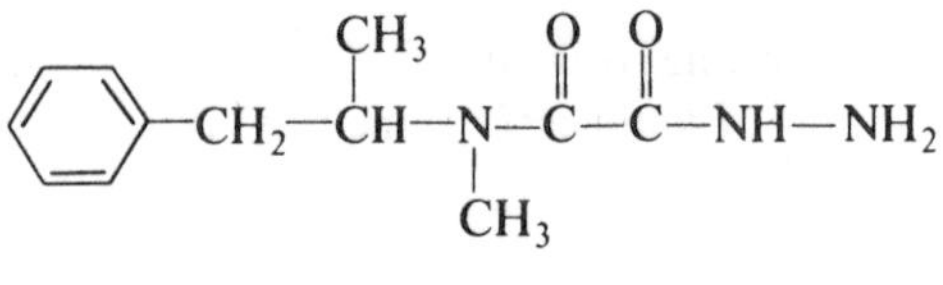

31

Ketones have also been resolved through conversion into oximino derivatives by reaction with 2-aminooxy-4-methylvaleric acid, **32**.[117] The *p* and *n* diastereomeric *O*-substituted oximes were separated by column chromatography and the isolated diastereomers cleaved to the ketone enantiomers by reaction with titanium trichloride.

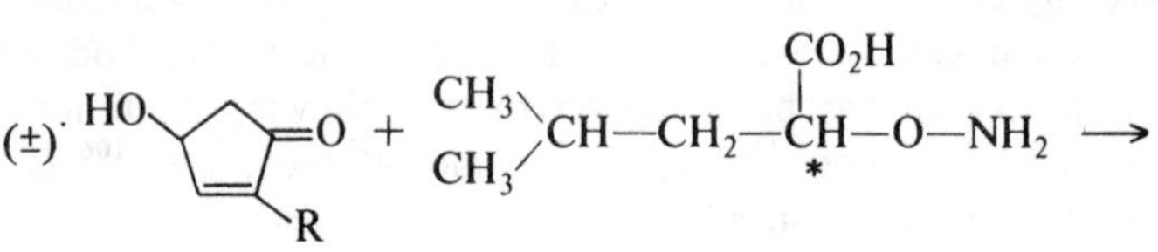

$$(\pm)\ \text{HO-cyclopentenone(R)}{=}O + (CH_3)_2CH{-}CH_2{-}\overset{CO_2H}{\underset{*}{CH}}{-}O{-}NH_2 \longrightarrow$$

32

$$(p,n)\ \text{HO-cyclopentene(R)}{=}N{-}O{-}\underset{CO_2H}{CH}{-}CH_2{-}CH(CH_3)_2$$

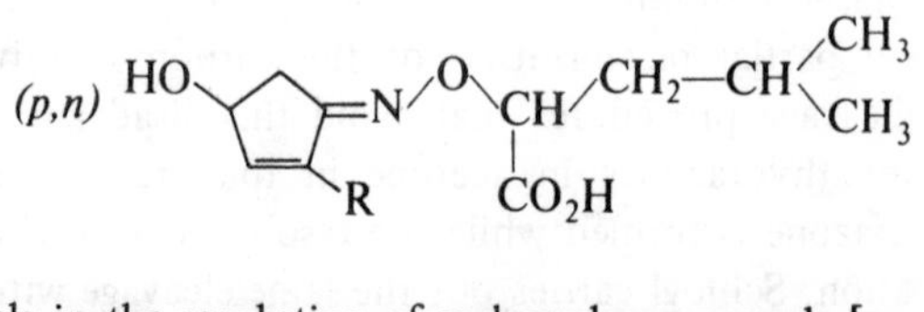

The use of ketals and thioketals in the resolution of carbonyl compounds [eq. (2)] was introduced by Casanova and Corey. The classic example is the resolution of racemic camphor via the diastereomeric dioxolanes formed with (—)-2,3-butanediol.[118] The ketals are separable by gas chromatography. Thioketals derived from (+)-2,3-butanedithiol (prepared from the diol) also figure in resolutions of ketones[119, 120] and of diketones.[121] In the latter case, the dithioketals are separated by column chromatography. Cleavage of the derivatives with mercuric chloride affords the resolved ketone, while Raney nickel gives the corresponding optically active hydrocarbons.[119] The separation of ketals of optically active propylene glycol by distillation has also been reported.[94]

A novel resolution procedure applicable to aldehydes and ketones involves the stereospecific synthesis of oxazolidines [eq. (3)] by reaction with *l*-ephedrine.[122] Fractional crystallization suffices to separate the diastereomers[123–125] which are cleaved, for example, by acid hydrolysis. Thiazolidines, which figure in the reciprocal resolution of *vic*-aminothiols with D-glucose[126–128] should be similarly useful in the resolution of carbonyl compounds.

Chiral sulfoxides bearing α-hydrogens may be resolved by conversion to β-hydroxyalkyl sulfoxides formed by reaction of the sulfoxide lithio derivative with camphor. The diastereomers are separated by crystallization or by chromatography and easily cleaved to the resolved sulfoxide by treatment with base.[129]

(±) $\xrightarrow[\text{2. (+)-camphor}]{\text{1. } n\text{-BuLi}}$ (p,n)

5.2.5 Resolution of Olefins, Sulfoxides, and Phosphines via Diastereomeric Zerovalent Complexes

The resolution of substances able to enter into coordination may be carried out by engaging them as ligands in organometallic zerovalent complexes* provided that the latter are diastereomeric. This is exemplified in particular by olefins such as *trans*-cyclooctene, **33**, which was successfully resolved by Cope and his co-workers[130,131] as follows:

33 **34** $\longrightarrow$ **35**

$[Cl_3Pt \leftarrow (CH_2{=}CH_2)]^- K^+ \longrightarrow [Ph{-}\overset{*}{C}H(CH_3){-}NH_2 \rightarrow PtCl_2 \leftarrow (CH_2{=}CH_2)]$

33 + 35 $\longrightarrow$ $[Ph{-}\overset{*}{C}H(CH_3){-}NH_2 \rightarrow PtCl_2 \leftarrow \text{(trans-cyclooctene)}]$ $\xrightarrow{\text{KCN}}$ (+) or (−) **33**

36

*Resolutions of complex *ions* or involving the formation of complex ions are discussed in Section 5.1.6.

Zeiss' salt, **34**, which is obtained by reaction of ethylene with potassium tetrachloroplatinate(II), K_2PtCl_4, is converted into the optically active complex (+)-*trans*-dichloro(ethylen)(α-methylbenzylamine)platinum(II), (+)-**35**, by treatment of **34** with (+)-α-methylbenzylamine.

The ethylene ligand in **35** is easily displaced by racemic *trans*-cyclooctene forming a mixture of diastereomers **36** the less soluble of which may be obtained in pure form by low-temperature crystallization. Its decomposition with potassium cyanide gives (−)-*trans*-cyclooctene, $[\alpha]_D^{25} = -458°$ (neat). The same series of reactions with (+)-α-methylbenzylamine replaced by the levorotatory isomer leads to the (+) olefin. *cis,trans*-1,5-Cyclooctadiene, **37**,[132] and *endo*-dicyclopentadiene, **38**,[133] have been similarly resolved.

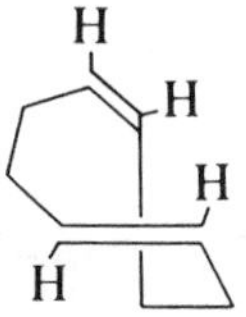

37 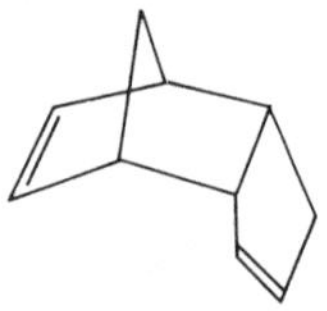 **38**

The optical stability of the trans cycloolefins **33** and **37** is due to ring strain which prevents rotation of the trans double bond. Enlargement of the ring allows this rotation to take place and hence leads to compounds that are easily racemizable and, at the limit, unresolvable. Nevertheless, such optically labile olefins form diastereomeric complexes such as **36**, as in the cases of **33** and **37**, which are sufficiently stable to be isolated by crystallization. In the case of *trans*-cyclononadiene,[134] rapid decomposition of the less soluble diastereomer, **36**, followed by cooling to −80°C allows one to isolate an optically active olefin with an estimated half-life of 4 minutes at 0°C. In the case of *trans*-cyclodecene[134] and of olefins **39**[135] and **40**,[136] separation of the diastereomers **36** does not allow the regeneration of resolved compounds as a consequence of their too rapid racemization.

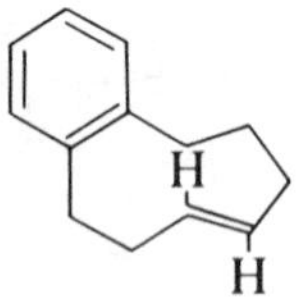

39 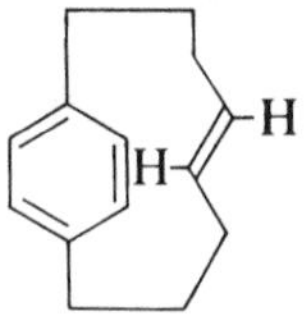 **40** 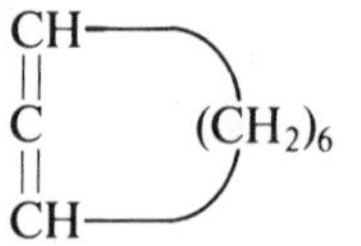 **41**

Resolution of the optically stable cyclic allene **41** leads to optically impure product evidently because the diastereomers cocrystallize.[137]

Acyclic olefins bearing chiral centers in α- or β-positions may also be resolved through use of optically active platinum complexes.[138] Thus, 3,4-dimethyl-1-pentene, **42**, is obtained with an optical purity of 93% after one crystallization of the cis complex **43**.

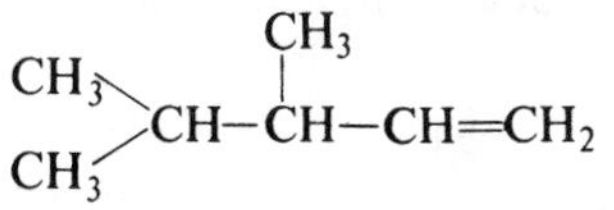

42 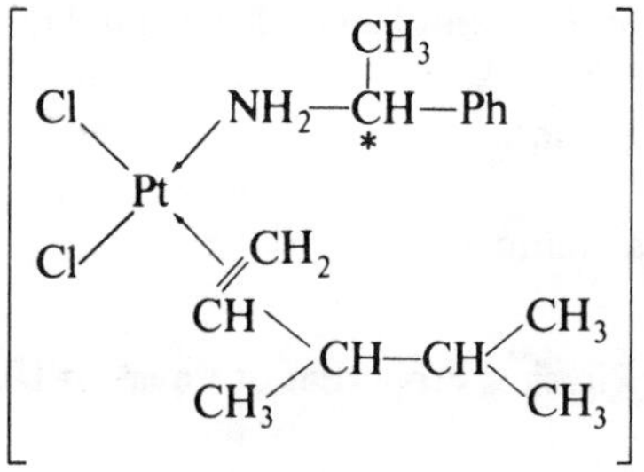 **43**

Resolution of sulfoxides through formation of diastereomeric platinum complexes may be carried out by an analogous route. Displacement of ethylene from complex **35** by ethyl *p*-tolyl sulfoxide gives a crystalline mixture of diastereomers **44**. After their separation and decomposition with NaCN, the sulfoxide enantiomers are obtained.[139]

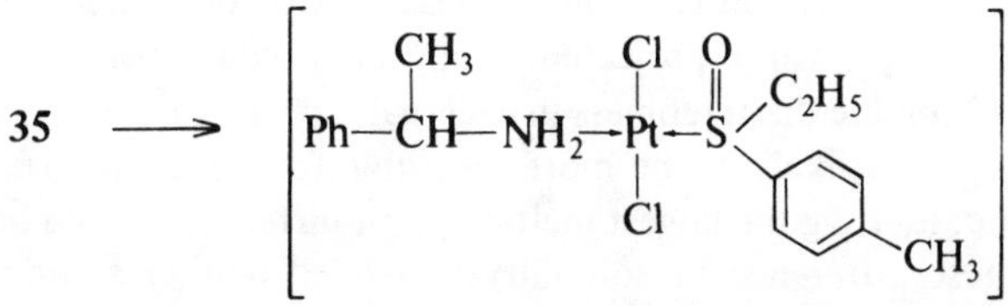

35 ⟶

44

The resolution of methylphenyl-*tert*-butylphosphine, **45**, follows the same principle, but the series of reactions resulting in the diastereomer mixture **46** is different. The racemic phosphine is complexed to platinum prior to introduction of the optically active amine (deoxyephedrine).[140]

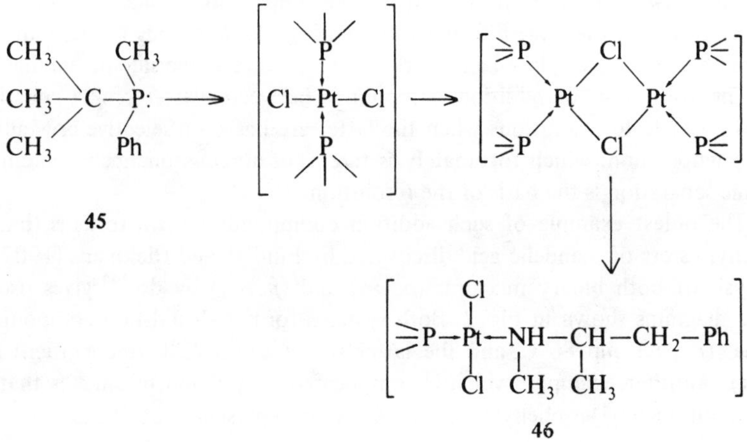

45

46

The use of palladium complexes in the resolution of tertiary phosphines must also be envisaged.[141]

Incorporation of one of the chiral ligands of the diastereomeric complex into a polymer gives rise to the possibility of carrying out a chromatographic resolution in which the (insoluble) polymer acts as stationary phase. This process is called *ligand* or *ligand exchange chromatography*. It has been adapted particularly to the resolution of α-amino acids.[155,168–170]

5.2.6 Physical Properties of Covalent Diastereomers and Their Mixtures

It is possible to get a generally correct idea of the relative solubilities of crystalline diastereomer pairs by examining the difference between the melting points of the p and n derivatives. By way of example, we have assembled some physical constants of covalent diastereomer pairs in Tables 1 and 2. The diastereomers p and n cited in Table 1 were for the most part obtained in connection with resolutions. While solubility data were not determined for the diastereomer pairs, the observed difference in their melting points is consistent with the fact that their separation was feasible.

The diastereomeric amides of Table 2, prepared by Ugi et al.,[21] originate in the direct combination of previously resolved constitutents. Comparison of the melting points of such p,n pairs in a projected resolution would allow one to choose (1) which of the less soluble diastereomers in each pair contains the *desired* enantiomer, and (2) which pair is likely to be more favorable for a separation based on solubility: the pair exhibiting the largest melting point difference would be expected to show the greatest difference in solubility (provided neither addition compound (p,n) nor solid solution intervenes).

Precise relative solubility data, thermodynamic properties such as enthalpies of fusion, and specific heats as well as crystallographic data generally are lacking for covalent diastereomer pairs. There is also little information available on the behavior of mixtures of such diastereomer pairs and, in particular, regarding the occurrence of addition compounds and mixed crystals. The most common behavior of such mixtures may be illustrated by the simple binary phase diagram given in Fig. 1. Note that the large difference in melting points of this diastereomer pair is responsible for the displacement of the eutectic toward one side of the diagram.[171]

The formation of addition compounds between diastereomers constitutes a major obstacle to resolutions when the latter are based on selective crystallization. This phenomenon, which fortunately is rare, is of no consequence when chromatographic separation is the basis of the resolution.

The oldest example of such addition compounds known to us is that of the menthyl esters of mandelic acid discovered by Findlay and Hickmans (1907).[142,143] Analysis of both binary mixtures (p_-,n_-) and (p_-,n_+) by dsc[144] gives rise to the phase diagrams shown in Fig. 2. Both systems form 1–1 addition compounds, the former (p_-,n_-), mp 86°C, and the latter (p_+,n_-), mp 72°C (incongruent melting point). Another example, which is complicated by polymorphism,* is that of the *N*-(α-methylbenzyl)-α-phenylbutyramide system represented by Fig. 3.

*The frequency of polymorphism, particularly in the case of diastereomeric amides, merits attention.

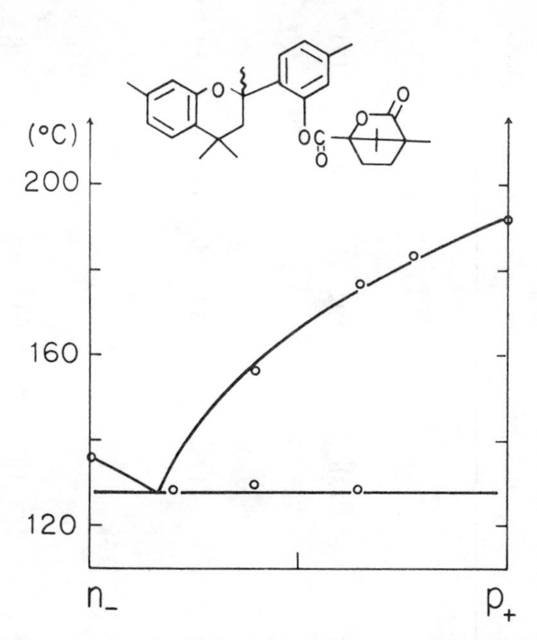

Figure 1 Binary phase diagram of p and n diastereomers exhibiting a eutectic. Experimental points (dsc) are given as open circles. The liquidus curves (solid lines) are calculated ($T_p^f = 192$°C, $\Delta H_p^f = 11.2$ kcal · mol^{-1}; $T_n^f = 136$°C, $\Delta H_n^f = 6.55$ kcal · mol^{-1}).[171]

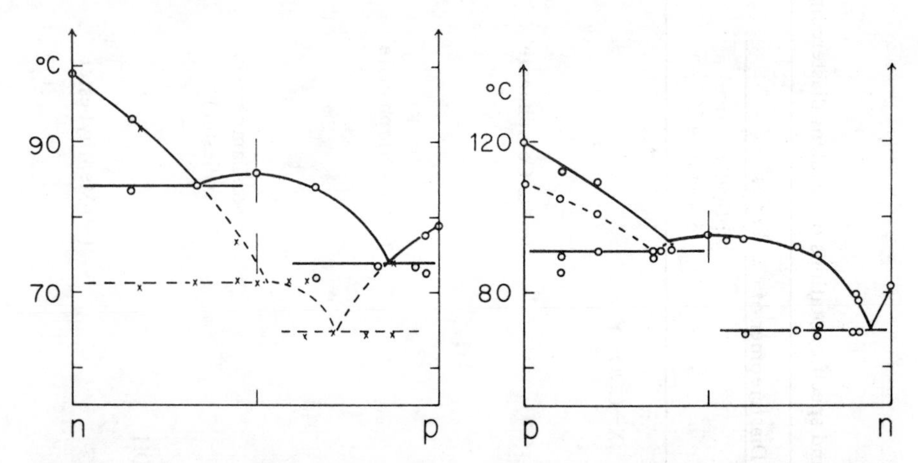

Figure 2 Phase diagrams of the menthyl mandelates (p, n) mixtures. Solid line (—o—) corresponds to diastereomers p_- [(−) acid combined with (−)-menthol, $[\alpha]_{578} = -143°$ (ethanol)], and n_- [(+) acid/(−)-menthol, $[\alpha]_{578} = -12°$ (ethanol)]. Dotted line (---x---) corresponds to p_- and n_+ [(−) acid/(+)-menthol, $[\alpha]_{578} = +12°$].[144]

Figure 3 Phase diagram of the *N*-(α-methylbenzyl)-α-phenylbutyramides p [(+) acid combined with (+) amine] and n [(+) acid combined with (−) amine]. Experimental points are indicated as open circle. Diastereomer p is polymorphic.[144]

Table 1 Melting points and specific rotations of covalent diastereomer pairs[a]

Diastereomer of		mp (°C)		[α] (deg)		
Racemate X	Optically active Substance Y	*p*	*n*	*p*	*n*	Refs.
X—COO—Y						
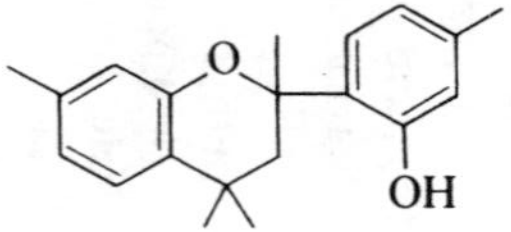	(+)-ω-Camphanate	195	136	+197	−194	171
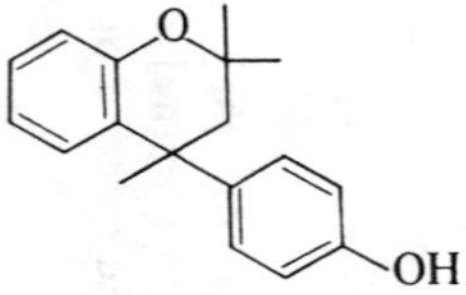	(−)-ω-Camphanate	177	147	−57	+40.5	39
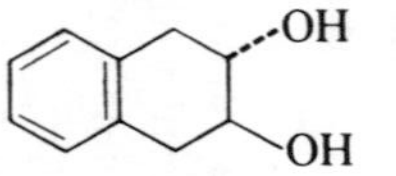	(−)-ω-Camphanate (diester)	178	ca. 109 (solvated)	−64.5	+65	36
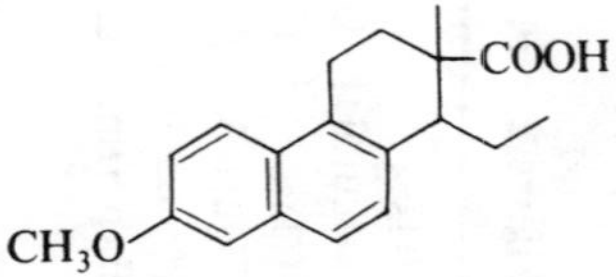	(−)-Menthyl ester	113	164			147

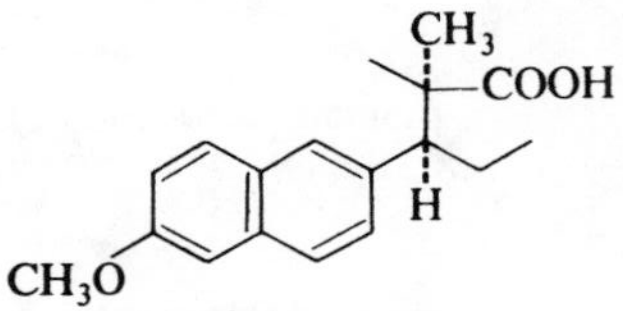 CH_3, COOH, H, CH_3O	(−)-Menthyl ester	134–136	72–76	−74	−6.8	148
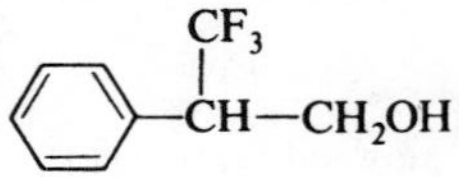 CF_3, CH—CH_2OH	(−)-ω-Camphanate	110–111	99–100	+36	−41	37
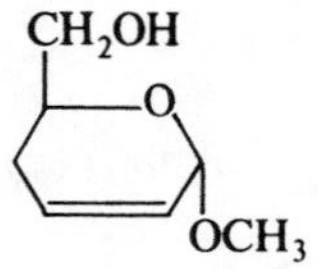 CH_2OH, O, OCH_3	(−)-ω-Camphanate	102–103.5	72–75	−30	+26	38
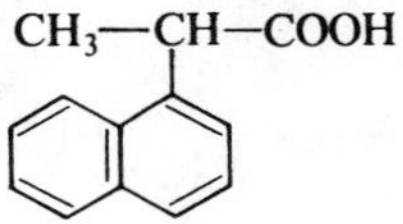 CH_3—CH—COOH	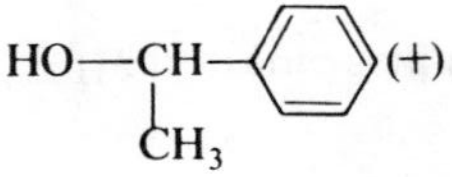 HO—CH(CH_3)— (+)	60	37			171
X—C=N—Y						
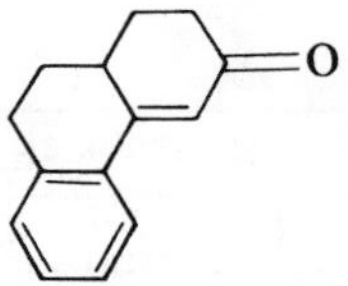 O	(−)-Menthydrazone	184–186	206–209	−126	+186	109
Camphor	(−)-Menthydrazone	193–194	177–178	−29	−68	110

[a] Rotations were measured under variable conditions which are identical for each (p, n) pair.

Table 2 Melting points and specific rotations of pairs of diastereomeric amides of hydratropic acid, $Ph{-}CH(CH_3)CO_2H$[a]

Diastereomers of X–CONH–Y		mp (°C)		$[\alpha]^{20}_{546}$ (MeOH) (deg)	
Racemate X	Y (R or S)	*p*	*n*	*p*	*n*
(±) $C_6H_5{-}CH(CH_3){-}NH_2$	Hydratropic acid (R)	118	134	+55	−134
(±) $C_6H_5{-}CH(C_6H_{11}){-}NH_2$	Hydratropic acid (R)	97	141	−20	−25.5
(±) Hydratropic acid	$H_2N{-}CH(C_2H_5){-}C_6H_5$ (S)	83[b]	121[b]	−67	−147
(±) Hydratropic acid	$H_2N{-}CH(CH_3){-}CH_2{-}C_6H_5$ (S)	117	87	+8	−30
(±) Hydratropic acid	$H_2N{-}CH(CH_3){-}C_{10}H_7$ (2-naphthyl) (R)	107	134	+117	+163
(±) Hydratropic acid	$H_2N{-}CH(CH_3){-}CO_2C_2H_5$ (S)	37	88	−9.3	−108

(±) Hydratropic acid	$H_2N-CH(CO_2C_2H_5)-CH_2-CH(CH_3)_2$ (S)	99	38	−20	−84
(±) Hydratropic acid	$H_2N-CH(CO_2C_2H_5)-CH(CH_3)_2$ (S)	61	82	−17	−69
(±) Hydratropic acid	$H_2N-CH(CO_2C_2H_5)-C_6H_5$ (R)	120	112	−103	−117
(±) Hydratropic acid	$H_2N-CH(CO_2C_2H_5)-CH_2-C_6H_5$ (S)	87	90	+4	−23

[a] The p, n nomenclature is used here in the sense (R, R) or (S, S) for p and (R, S) or (S, R) for n, according to Ugi.[145] The melting points and specific rotations are rounded off. Data from H. Herlinger, H. Kleinmann, and I. Ugi, *Liebigs Ann. Chem.*, 1967, **706**, 37. Reprinted by permission of Verlag Chemie GMBH.

[b] This compound is actually polymorphic (mp 72 and 81°C for p), as is its diastereomer n, mp 110, 120.5, and 122°C.[146]

Partial or total miscibility of covalent diastereomers, which constitutes another potential obstacle to resolution by crystallization, has not to our knowledge been explicitly described.

5.2.7 Chromatographic Behavior of Covalent Diastereomers

Although the first cases of chromatographic resolution mediated by covalent diastereomers were reported over 40 years ago,[149] it is particularly during the past 10 years, beginning around 1970, that a large number of diastereomer pairs have been subjected to chromatography in cases where at least one of the diastereomers fails to crystallize or with a view to bypassing crystallization altogether.[150, 151] A particular advantage of chromatographic resolutions is that it provides *both* diastereomers generally in a state of high purity.

Although there are disadvantages attending this sort of resolution, there are none beyond those discussed at the beginning of Section 5.2. We believe that entering into an argument about the relative merits of chromatographic versus nonchromatographic methods is not fruitful; suffice it to say that both have their place in part as a function of scale or of desired objective. From a reading of Sections 5.2.1 to 5.2.5, it is evident that virtually all types of covalent diastereomer mixtures can and have been subjected to chromatography. Moreover, every type of chromatography (classical gravity column, thin-layer, dry-column, and high- and medium-pressure liquid chromatography) has been applied to such separations. Recent improvements, notably in medium-pressure liquid chromatography,[73, 153, 154] have measurably facilitated and improved the efficiency of preparative chromatographic resolutions, and this fact alone will surely increase the frequency of such separations.

The choice of resolving agent or chiral derivatizing agent (CDA)[71] has been strongly influenced by resolutions successfully effected by crystallization. Also, other CDAs have been developed which more closely match the requirements of models proposed to rationalize the diastereomer separation and which have the potential for predicting the ease of separation. Table 3 provides a representative list of examples of chromatographic resolutions mediated by covalent diastereomers.

Two limiting mechanisms are considered as the major ones controlling separation of diastereomers in adsorption chromatography. The first involves differential solvation of solutes by the mobile phase (solvent). The second emphasizes differential adsorption of solutes independently of the solvent.[71] While both processes must operate during chromatographic separation of covalent diastereomers, the second concept by itself has been surprisingly useful in rationalizing some resolutions by liquid chromatography.

Helmchen and his associates have carried out a number of studies which point the way to the "directed resolution" of enantiomers, that is, chromatographic separation of diastereomers with preestimation of separation effects and concomitant determination of enantiomeric purity and of absolute configuration.[14, 15] The four postulates proposed by them for assessing structural features which enhance separation of *amides* on silica gel are the following[16]:

1 Conformations are the same in solution and when adsorbed.
2 Diastereomers bind to surfaces (silica gel) preferentially via hydrogen bonding.
3 Significant diastereoselectivity is expected only for bound states involving two-point interaction (**1**, Scheme 1). This can be perturbed in favor of less strongly bound states by substituent groups with weaker affinity for the surface.
4 Diastereomers with "chromatographically large" substituents R and A staggered relative to the common amide plane (**2**) are better able to shield the amide group from the surface and are consequently eluted ahead of those in which these substituents are on the same face of the common plane (configuration rule).

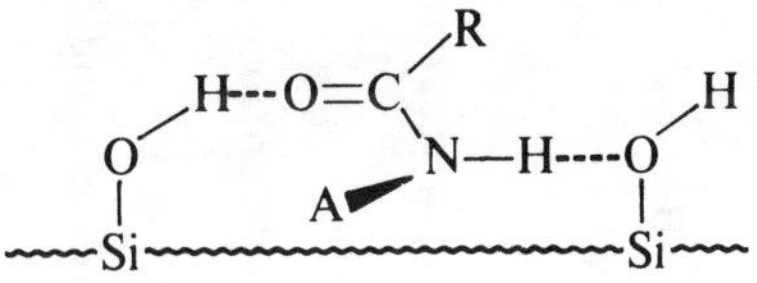

1

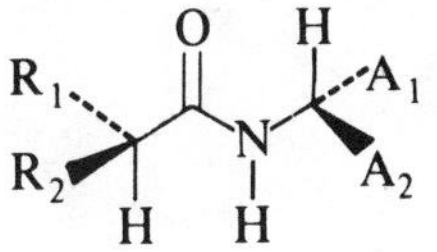

2 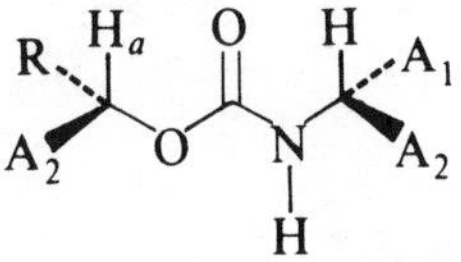 **3**

Chromatographic separation of diastereomeric amides implicated in the resolution of acids,[20] lactones,[20] and amines[22] takes advantage of the presence of a minimum of two functional groups (C=O and NH) capable of hydrogen bonding to Si–OH groups. Separation factors of corresponding diastereomeric esters are generally smaller than those of amides.[72] It would seem that, in some cases, separation of such amides can also be effected by crystallization with monitoring of the separation carried out by TLC.[20]

Wide application of this method, particularly on a preparative scale, would be limited in part by the difficulty in executing the cleavage of amides (however, see Section 7.4.2.). This obstacle has now been overcome by suitable choice of the CDA, namely, incorporation of a hydroxyl group γ or δ to the amide carbonyl group (either in the acid or the amine moiety), such that amide hydrolysis is assisted by neighboring group participation which permits the cleavage to take place under mild conditions (0.5 to 1.0*N* mineral acid at 50 to 90°C):[20]

Pirkle, Hoekstra, and Hauske have developed an efficient analogous process for the resolution of alcohols mediated by diastereomeric carbamates (**3**) on alumina[28,71] which has also been adapted to epoxides.[77] In spite of some doubts about the equivalent partition of conformations in solution and while being adsorbed, they have inferred that carbamates of type **3** do, in fact, follow Helm-

Table 3 Selected chromatographic resolutions mediated by covalent diastereomers

Substrate Functional Group	Example	Derivative	Resolving Agent (CDA)	Chromatographic Mode	Refs.
Acid	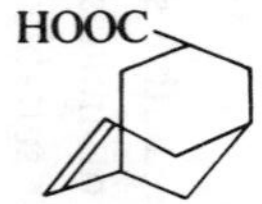 HOOC	Amide	$C_6H_5-CH(NH_2)-CH_3$	HPLC (anal.)	156
Acid	$CH_3CH_2CH(C_6H_5)COOH$	Amide	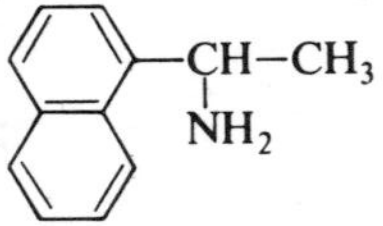 $-CH(NH_2)-CH_3$	HPLC	157
Acid	$Ph-CH(Ph)-CH(CH_3)-COOH$	Amide	$C_6H_5-CH(NH_2)-CH_2-OH$	MPLC (prep.)	20
Amino acid	$CH_3-CH(NH_2)-COOH$	Carbamate	Menthyl chloroformate	GC (anal.)	85
Amino acid	$(CH_3)_2CH-CH(NH_2)-COOCH_3$	Amide	Chrysanthemic and drimanoic acids	GC (anal.)	67
Amine	$C_6H_5-CH(NH_2)CH_2CH_3$	Amide	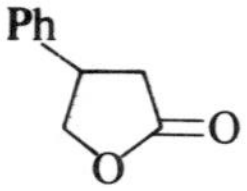 Ph, O, =O	MPLC (prep.)	22
Amine	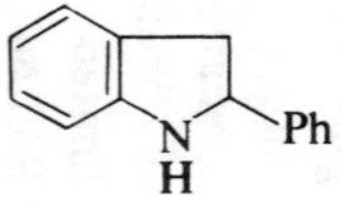 N, H, Ph	Carbamate	Menthyl chloroformate	LC	24

Amine		Amide	$CH_3CH_2CH(Ph)COOH$	HPLC	157
Amine	$Ph-CH_2-CH(NH_2)CH_3$	Amide	Drimanoic acid	GC (anal.)	67
Alcohol	$CH_3C{\equiv}C\,C(CH_3)(OH)-CH(OH)CH_2CH_3$	Ester	$C_6H_5CH(OCH_3)-C(=O)Cl$	MPLC	158
Alcohol	OH, R; (a) = OH, (b) = O=	Ester	Camphanic acid	LC	(a) 35 (b) 41
Alcohol	Menthol and 2-octanol	Ester	Chrysanthemic and drimanoic acids	GC (anal.)	67
Alcohol (thiol)	$CF_3-CH(C_{10}H_7)-OH$	Carbamate	$CH_3-CH(C_{10}H_7)-NCO$	MPLC	28
Alcohol	$C_4H_3O-CH(CH_3)-CH(OCH_3)-CH(CH_3)-CH_2OH$	Carbamate	$C_6H_5-CH(CH_3)-NCO$	MPLC	78
Alcohol	$n-C_5H_{11}-CH(OH)-CH{=}CH_2$	Carbamate	$C_6H_5-CH(CH_3)-NCO$	TLC	159

Table 3 Continued

Substrate Functional Group	Example	Derivative	Resolving Agent (CDA)	Chromatographic Mode	Refs.
Alcohol	$CH_3CH(CH_3)-CH(CH_3)-OH$	Ester	$CH_3CH(OH)COOH$	LC	160
Alcohol	2-Octanol	Ester	$CH_3CH(OAc)COOH$	GC (anal.)	161
Alcohol	$C_6H_5CH(CH_3)-OH$	Ester	3β-Acetoxy-Δ^5-etienic acid	GC	162
Alcohol	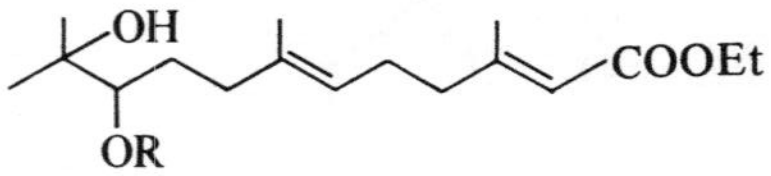 OH, OR, COOEt (structure)	Ester	3β-Acetoxy-Δ^5-etienic acid	TLC (prep.)	61
Alcohol	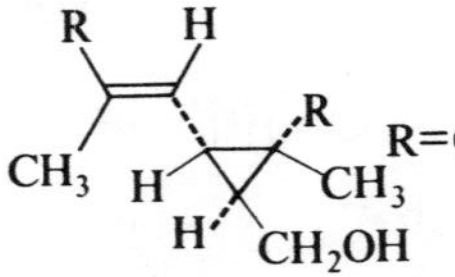 R, H, CH_3, H, R, CH_3, H, CH_2OH (structure); $R=C_{11}H_{19}$	Ester	3β-Acetoxy-Δ^5-etienic acid	HPLC	63
Alcohol	$(CH_3)_3C-CH(OH)CH_3$	Ester	$CF_3CONH-CH(CH_3)COOH$	GC (prep.)	163

Alcohol	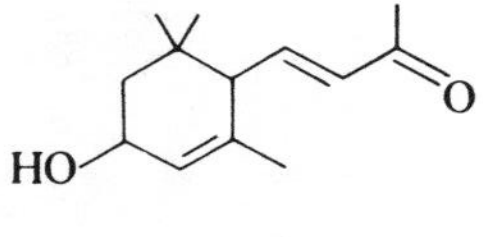	Ester	$C_6H_5-C(CF_3)(OCH_3)-COOH$ (MTPA)	HPLC	88
Alcohol	Borneol	Glycoside	Acetobromoglucose	GC	92, 93
Alcohol	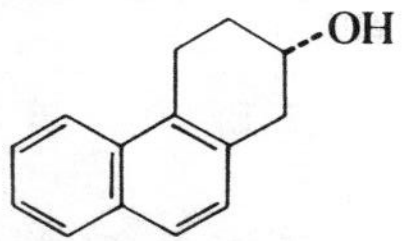	Ester	Menthoxyacetyl chloride	LC	51
Alcohol	$CH_3CH(OH)-COOCH_3$	Ester	Menthyl chloroformate	GC (anal.)	85, 162
Ester (lactone)	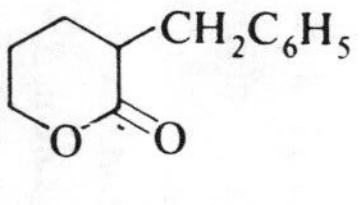 $CH_2C_6H_5$	Amide	$C_6H_5-CH(NH_2)-CH_3$	MPLC	20
Ketone	Camphor	Ketal	$CH_3-CH(OH)-CH(OH)-CH_3$	GC (prep.)	118
Ketone	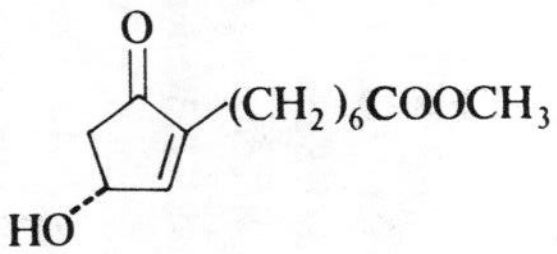 $(CH_2)_6COOCH_3$	Oximino derivative	$CH_3CH(CH_3)CH_2CH(ONH_2)COOH$	LC	117
Ketone	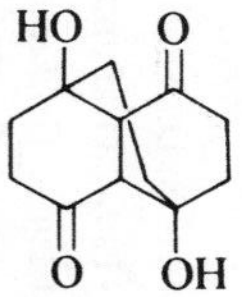	Dithioketal	$CH_3-CH(SH)-CH(SH)-CH_3$	LC	121

chen's first postulate. Pirkle and Hauske have also emphasized the possible effects upon separation of stereochemically dependent intramolecular interactions including carbinyl hydrogen (see H_a in **3**) bonding to the carbonyl group which is particularly significant where $R = CF_3$ or C_3F_7.[71,165]

Gas-chromatographic resolution evidently partakes more of the second of the two limiting mechanisms, but this must be recast in the form of differential solvation of the solutes in the stationary phase. To this must be added a variable and unknown contribution to selective retention from adsorption on the polar solid support. The latter tends to increase when the liquid phase loading is decreased.[164]

While the rational development of diastereomer separation by gas chromatography began somewhat earlier than the corresponding resolution by liquid chromatography under pressure, it would seem that separation is not as efficient as with LC in spite of the application of special reagents. Temperature effects may be partially responsible for this. Many similarities remain between the two separation modes, for example, greater ease of separation of amides over esters[167] and the importance of conformations of the diastereomers separated. The enhanced separation afforded by conformationally rigid derivatives attests to the latter point.[71,164,166]

Since gas-chromatographic resolution is of greater interest to analytical than to preparative resolutions, specific examples are discussed in Section 7.5.

REFERENCES 5.2

1 A. Hantzsch, *Grundriss der Stereochemie*, 2nd ed., Barth, Leipzig, 1904.
2 W. Markwald and A. McKenzie, *Ber.*, 1899, **32**, 2130.
3 W. Markwald and R. Meth, *Ber.*, 1905, **38**, 801.
4 P. Frankland and T. S. Price, *J. Chem. Soc.*, 1897, **71**, 253, 696.
5 E. Erlenmeyer, Jr., *Ber.*, 1903, **36**, 976.
6 C. Neuberg, *Ber.*, 1903, **36**, 1192.
7 E. J. Eisenbraun, G. H. Adolphen, K. S. Schorno, and R. N. Morris, *J. Org. Chem.*, 1971, **36**, 414.
8 British Patent 622,892 (1949) (to Ciba Ltd.). *Chem. Abstr.*, 1950, **44**, 665.
9 V. Libert, Doctoral Dissertation, Université Libre de Bruxelles, 1977.
10 S. D. Allen and O. Schnepp, *J. Chem. Phys.*, 1973, **59**, 4547.
11 E. Graf and H. Boeddeker, *Liebigs Ann. Chem.*, 1958, **613**, 111.
12 A. Gossaner and J.-P. Weller, *J. Am. Chem. Soc.*, 1978, **100**, 5928.
13 For an example, see B. D. Vineygard, W. S. Knowles, M. J. Sabacky, G. L. Bachman, and D. J. Weinkauff, *J. Am. Chem. Soc.*, 1977, **99**, 5946 and references therein.
14 G. Helmchen, R. Ott, and K. Sauber, *Tetrahedron Lett.*, 1972, 3873.
15 G. Helmchen, H. Völter, and W. Schüle, *Tetrahedron Lett.*, 1977, 1417.
16 G. Helmchen, G. Nill, D. Flockerzi, W. Schüle, and M. S. K. Youssef, *Angew. Chem.*, 1979, **91**, 64. *Angew. Chem., Int. Ed.*, 1979, **18**, 62.
17 T. Koizumi, Y. Kobayashi, H. Amitani, and E. Yoshii, *J. Org. Chem.*, 1977, **42**, 3459.
18 G. Haas and V. Prelog, *Helv. Chim. Acta*, 1969, **52**, 1202.
19 S. Rendic, V. Sunjic, F. Kajfez, N. Blazevic, and T. Alebic-Kolbah, *Chimia*, 1975, **29**, 170.

20 G. Helmchen, G. Nill, D. Flockerzi, and M. S. K. Youssef, *Angew, Chem.*, 1979, **91**, 65. *Angew. Chem. Int. Ed.*, 1979, **18**, 63.

21 H. Herlinger, K. Kleinmann, and I. Ugi, *Liebigs Ann. Chem.*, 1967, **706**, 37.

22 G. Helmchen and G. Nill, *Angew. Chem.*, 1979, **91**, 66. *Angew. Chem. Int. Ed.*, 1979, **18**, 65.

23 L. A. Carpino, *Chem. Comm.*, 1966, 858.

24 J. F. Nicoud and H. B. Kagan, *Israel J. Chem.*, 1976/77, **15**, 78.

25 R. B. Woodward, *Pure Appl. Chem.*, 1968, **17**, 519.

26 A. W. Ingersoll, *Organic Reactions*, Vol. 2, R. Adams, Ed., Wiley, New York, 1944, Chapter 9.

27 H. H. Zeiss and W. B. Martin, Jr., *J. Am. Chem. Soc.*, 1953, **75**, 5935.

28 W. H. Pirkle and M. S. Hoekstra, *J. Org. Chem.*, 1974, **39**, 3904.

29 B. Helferich and W. Portz, *Chem. Ber.*, 1953, **86**, 1034.

30 P. Axerio and M. Bemporad, *Afinidad*, 1975, **32**, 469.

31 K. Mislow, "Stereoisomerism," in *Comprehensive Biochemistry*, Vol. 1, M. Florkin and E. H. Stotz, Eds., Elsevier, Amsterdam, 1962, p. 222.

32 (a) See, for example, P. K. Chang and T. L. V. Ulbricht, *J. Am. Chem. Soc.*, 1958, **80**, 976 and references therein. (b) S. H. Wilen, A. Collet, and J. Jacques, *Tetrahedron*, 1977, **33**, 2725.

33 J. E. Hodge and C. E. Rist, *J. Am. Chem. Soc.*, 1952, **74**, 1494.

34 J. F. Tocanne and C. Asselineau, *Bull. Soc. Chim. Fr.*, 1965, 3348.

35 H. Gerlach, *Helv. Chim. Acta*, 1968, **51**, 1587.

36 A. Collet, M. J. Brienne, and J. Jacques, *Bull. Soc. Chim. Fr.*, 1972, 336.

37 J. Jurczak, A. Konował, and Z. Krawczyk, *Synthesis*, 1977, 258.

38 A. Konował, J. Jurczak, and A. Zamojski, *Tetrahedron*, 1976, **32**, 2957.

39 M. J. Brienne and J. Jacques, *Tetrahedron Lett.*, 1975, 2349.

40 R. Haller and U. Werner, *Arch. Pharm.* (Weinheim, Ger.), 1977, **310**, 349.

41 H. Kuritani, F. Iwata, M. Sumiyoshi, and K. Shingu, *J. Chem. Soc. Chem. Comm.*, 1977, 542.

42 A. Kotynski and W. J. Stec, *J. Chem. Res.*, 1978, (S), 41.

43 M. J. Brienne and A. Collet, *J. Chem. Res.*, 1978, (S), 60; (M), 772.

44 T. G. Cochran and A. C. Huitric, *J. Org. Chem.*, 1971, **36**, 3046.

45 J. Read and W. J. Grubb, *J. Chem. Soc.*, 1931, 188.

46 D. R. Galpin and A. C. Huitric, *J. Org. Chem.*, 1968, **33**, 921.

47 D. R. Galpin and A. C. Huitric, *J. Pharm. Sci.*, 1968, **57**, 447.

48 D. Taub, N. N. Girotra, R. D. Hoffsommer, C. H. Kuo, H. L. Slates, S. Weber, and N. L. Wendler, *Tetrahedron*, 1968, **24**, 2443.

49 M. N. Akhtar and D. R. Boyd, *J. Chem. Soc. Chem. Comm.*, 1975, 916.

50 D. R. Boyd, J. D. Neill, and M. E. Stubbs, *J. Chem. Soc. Chem. Comm.*, 1977, 873.

51 M. Koreeda, M. N. Akhtar, D. R. Boyd, J. D. Neill, D. T. Gibson, and D. M. Jerina, *J. Org. Chem.*, 1978, **43**, 1023.

52 M. Hübner, K. Ponsold, H. J. Siemann, and S. Schwartz, *Z. Chem.*, 1968, 8, 380.

53 W. T. Borden and E. J. Corey, *Tetrahedron Lett.* 1969, 313.

54 H. H. Inhoffen, D. Kopp, S. Marić, J. Bekurdts, and R. Selimoglu, *Tetrahedron Lett.*, 1970, 999.

55 J. Staunton and E. J. Eisenbraun, *Organic Syntheses* Coll. Vol. V, H. E. Baumgarten, Ed., Wiley, New York, 1973, p. 8.

56 R. B. Woodward and T. J. Katz, *Tetrahedron*, 1959, **5**, 70.

57 C. Djerassi, E. J. Warawa, R. E. Wolff, and E. J. Eisenbraun, *J. Org. Chem.*, 1960, **25**, 917.

58 C. Djerassi and J. Staunton, *J. Am. Chem. Soc.*, 1961, **83**, 736.

59 D. M. Feigl and H. S. Mosher, *J. Org. Chem.*, 1968, **33**, 4242.

60 J. B. Heather, R. S. D. Mittal, and C. J. Sih, *J. Am. Chem. Soc.*, 1976, **98**, 3661.

61 K. Imai, S. Marumo, and T. Ohtaki, *Tetrahedron Lett.*, 1976, 1211.

62 J. A. Nelson, M. R. Czarny, T. A. Spencer, J. S. Limanek, K. R. McCrae, and T. Y. Chang, *J. Am. Chem. Soc.*, 1978, **100**, 4900.

63 L. J. Altman, R. C. Kowerski, and D. R. Laungani, *J. Am. Chem. Soc.*, 1978, **100**, 6174.

64 S. H. Wilen, *Topics Stereochem.*, 1971, **6**, 107.

65 M. A. Jermyn, *Aust. J. Chem.*, 1967, **20**, 2283.

66 B. Halpern and J. W. Westley, *Aust. J. Chem.*, 1966, **19**, 1533.

68 F. B. Laforge, *J. Org. Chem.*, 1952, **17**, 1635.

69 R. H. Pickard, W. O. Littlebury, and A. Neville, *J. Chem. Soc.*, 1906, **89**, 93.

70 R. H. Pickard, and W. O. Littlebury, *J. Chem. Soc.*, 1906, **89**, 467, 1254.

71 W. H. Pirkle and J. R. Hauske, *J. Org. Chem.*, 1977, **42**, 1839.

72 W. H. Pirkle and J. R. Hauske, *J. Org. Chem.*, 1977, **42**, 2436.

73 W. H. Pirkle and R. W. Anderson, *J. Org. Chem.*, 1974, **39**, 3901.

74 W. H. Pirkle and J. R. Hauske, *J. Org. Chem.*, 1977, **42**, 2781.

75 E. J. Corey, R. L. Danheiser, S. Chandrasekaran, G. E. Keck, B. Gopalan, S. D. Larsen, P. Siret, and J. L. Gras, *J. Am. Chem. Soc.*, 1978, **100**, 8034.

76 T. Nakata, G. Schmid, B. Vranesic, M. Okigawa, T. Smith-Palmer, and Y. Kishi, *J. Am. Chem. Soc.*, 1978, **100**, 2933.

77 W. H. Pirkle and P. L. Rinaldi, *J. Org. Chem.*, 1979, **44**, 1025.

78 G. Schmid, T. Fukuyama, K. Akasaka, and Y. Kishi, *J. Am. Chem. Soc.*, 1979, **101**, 259.

79 W. H. Pirkle, J. R. Hauske, C. A. Eckert, and B. A. Scott, *J. Org. Chem.*, 1979, **42**, 3101.

80 W. H. Pirkle and C. W. Boeder, *J. Org. Chem.*, 1978, **43**, 1950.

81 W. H. Pirkle and C. W. Boeder, *J. Org. Chem.*, 1978, **43**, 2091.

82 W. H. Pirkle and P. L. Rinaldi, *J. Org. Chem.*, 1978, **43**, 3803.

83 W. H. Pirkle and P. E. Adams, *J. Org. Chem.*, 1979, **44**, 2169.

84 D. M. Mulvey and H. Jones, *J. Heterocycl. Chem.*, 1978, **15**, 233.

85 J. W. Westley and B. Halpern, *J. Org. Chem.*, 1968, **33**, 3978.

86 J. F. Nicoud, C. Eskenazi, and H. B. Kagan, *J. Org. Chem.*, 1977, **42**, 427.

87 J. A. Dale, D. L. Dull, and H. S. Mosher, *J. Org. Chem.*, 1969, **24**, 2543.

88 M. Koreeda, G. Weiss, and K. Nakanishi, *J. Am. Chem. Soc.*, 1973, **95**, 239.

89 K. J. Judy, D. A. Schooley, L. L. Dunham, M. S. Hall, B. J. Bergot, and J. R. Siddall, *Proc. Natl. Acad. Sci., U.S.A.*, 1973, **70**, 1509.

90 B. Helferich and R. Hiltmann, *Ber.*, 1937, **70**, 308.

91 G. Wagner and S. Böhme, *Arch. Pharm.* (Weinheim, Ger.), 1964, **297**, 257.

92 I. Sakata and K. O. Shimizu, *Agric. Biol. Chem.* (Japan), 1979, **43**, 411.

93 I. Sakata and H. Iwamura, *Agric. Biol. Chem.* (Japan), 1979, **43**, 307.

94 W. L. Howard and J. D. Burger (to Dow Chemical Co.), U.S. Patent, 3,491,152 (1970). *Chem. Abstr.*, 1970, **72**, 110779e.

95 E. W. Cantrall, C. Krieger, and R. B. Brownfield, (to American Cyanamid Co.) German Offen., 1,942,453 (1970). *Chem. Abstr.*, 1970, **72**, 111702m.

96 M. Betti, *Organic Syntheses*, Coll. Vol. I, 2nd ed., H. Gilman and A. H. Blatt, Eds., Wiley, New York, 1941, p. 381.

97 M. Betti, *Gazzetta Chim. Ital.*, 1906, **36**(II), 392.

98 H. Sobotka, E. Bloch, H. Cahnmann, E. Feldbau, and E. Rosen, *J. Am. Chem. Soc.*, 1943, **65**, 2061.

99 M. Betti, *Ber.*, 1930, **63**, 874.

100 R. B. Woodward, T. P. Kohman, and G. C. Harris, *J. Am. Chem. Soc.*, 1941, **63**, 120.

101 C. Neuberg and M. Federer, *Ber.*, 1905, **38**, 868.

102 C. Neuberg and M. Federer, *Ber.*, 1905, **38**, 866.

103 I. V. Hopper and F. J. Wilson, *J. Chem. Soc.*, 1928, 2483.

104 A. B. Crawford and F. J. Wilson, *J. Chem. Soc.*, 1934, 1122.

105 A. J. Little, J. M'Lean, and F. J. Wilson, *J. Chem. Soc.*, 1940, 336.

106 M. K. Hargreaves and B. Modarai, *J. Chem. Soc. D*, 1969, 16.

107 J. B. Thomson, *Tetrahedron Lett.*, 1959, 26.

108 O. Hofer and K. Schlögl, *Tetrahedron Lett.*, 1967, 3485.

109 E. Touboul and G. Dana, *Bull. Soc. Chim. Fr.*, 1974, 2269.

110 K. Schlögl and M. Fried, *Monatsh. Chem.*, 1964, **95**, 558.

111 F. Nerdel and E. Henkel, *Chem. Ber.*, 1952, **85**, 1138.

112 K. Mislow and C. L. Hamermesh, *J. Am. Chem. Soc.*, 1955, 77, 1590.

113 K. Bott, *Tetrahedron Lett.*, 1965, 4569.

114 N. J. Leonard and J. H. Boyer, *J. Org. Chem.*, 1950, **15**, 42.

115 M. Kotake and G. Nakaminani, *Proc. Jpn. Acad.*, 1953, **29**, 56.

116 A. Solladié-Cavallo, G. Solladié, and E. Tsamo, *J. Org. Chem.*, 1979, **44**, 4189.

117 R. Pappo, P. Collins, and C. Jung, *Tetrahedron Lett.*, 1973, 943.

118 J. Casanova, Jr., and E. J. Corey, *Chem. Ind.* (London), 1961, 1664.

119 E. J. Corey and R. B. Mitra, *J. Am. Chem. Soc.*, 1962, **84**, 2938.

120 W. Ten Hoeve and H. Wynberg, *J. Org. Chem.*, 1979, **44**, 1508.

121 H. Buding and H. Musso, *Angew. Chem.*, 1978, **90**, 899. *Angew. Chem., Int. Ed.*, 1978, **17**, 851.

122 L. Neelakantan, *J. Org. Chem.*, 1971, **36**, 2257.

123 R. Kelly and V. Van Rheenen, *Tetrahedron Lett.*, 1973, 1709.

124 D. Olliero, J.-M. Ruxer, A. Solladié-Cavallo, and G. Solladié, *J. Chem. Soc. Chem. Comm.*, 1976, 276.

125 P. E. Eaton and B. Leipzig, *J. Org. Chem.*, 1978, **43**, 2783.

126 T. Taguchi, T. Takatori, and M. Kojima, *Chem. Pharm. Bull.*, 1962, **10**, 245. *Chem. Abstr.*, 1963, **58**, 4536.

127 J. R. Piper and T. P. Johnston, *J. Org. Chem.*, 1964, **29** 1657.

128 T. Takatori, M. Kojima, and T. Taguchi, *Yakugaku Zasshi*, 1968, 88, 366. *Chem. Abstr.*, 1968, **69**, 76717.

129 R. F. Bryan, F. A. Carey, O. D. Dailey, Jr., R. J. Maher, and R. W. Miller, *J. Org. Chem.*, 1978, **43**, 90.

130 A. C. Cope, C. N. Ganellin, and H. W. Johnson, Jr., *J. Am. Chem. Soc.*, 1962, **84**, 3191.

131 A. C. Cope, C. R. Ganellin, H. W. Johnson, Jr., T. V. Van Auken, and H. J. S. Winkler, *J. Am. Chem. Soc.*, 1963, **85**, 3276.

132 A. C. Cope, J. K. Hecht, H. W. Johnson, Jr., H. Keller, and H. J. S. Winkler, *J. Am. Chem. Soc.*, 1966, **88**, 761.

133 A. Panunzi, A. Derenzi, and G. Paiaro, *Inorg. Chim. Acta*, 1967, **1**, 475.

134 A. C. Cope, K. Banholzer, H. Keller, B. A. Pawson, J. J. Wang, and H. J. S. Winkler, *J. Am. Chem. Soc.*, 1965, **87**, 3644.

135 A. C. Cope and M. W. Fordice, *J. Am. Chem. Soc.*, 1967, **89**, 6187.

136 A. C. Cope and B. A. Pawson, *J. Am. Chem. Soc.*, 1968, **90**, 636.

137 A. C. Cope, W. R. Moore, R. D. Bach and H. J. S. Winkler, *J. Am. Chem. Soc.*, 1970, **92**, 1243.

138 R. Lazzaroni, P. Salvadori, and P. Pino, *Tetrahedron Lett.*, 1968, 2507.

139 A. C. Cope and E. A. Caress, *J. Am. Chem. Soc.*, 1966, **88**, 1711.

140 T. H. Chan, *Chem. Comm.*, 1968, 895.

141 S. Otsuka, A. Nakamura, T. Kano, and K. Tani, *J. Am. Chem. Soc.*, 1971, **93**, 4301.

142 A. Findlay and E. M. Hickmans, *J. Chem. Soc.*, 1907, **91**, 905.

143 A. Findlay, *The Phase Rule*, 9th ed., revised by A. N. Campbell and N. O. Smith, Dover, New York, 1951, p. 191.

144 M. J. Brienne, unpublished results.

145 I. Ugi, *Z. Naturforsch.*, 1965, **20b**, 405.

146 L. Tanguy, unpublished results.

147 R. Rometsch and K. Miesher, *Helv. Chim. Acta*, 1946, **29**, 1231.

148 R. Gay and A. Horeau, *Tetrahedron*, 1959, 7, 90.

149 A. Stoll and A. Hofmann, *Z. Physiol. Chem.*, 1938, **251**, 155.

150 I. S. Krull, *Adv. Chromatogr.*, 1978, **16**, 175.

151 T. Tamegai, M. Ohmae, K. Kawabe, and M. Tomoeda, *J. Liq. Chromatogr.*, 1979, **2**, 1229.

152 J. J. Martel, J. P. Demoute, A. P. Tèche, and J. R. Tessier, *Pestic. Sci.*, 1980, **11**, 188.

153 W. C. Still, M. Kahn, and A. Mitra, *J. Org. Chem.*, 1978, **43**, 2923.

154 Reference 16, footnote 2.

155 B. Lefebvre, R. Audebert, and C. Quivoron, *Israel J. Chem.*, 1976/77, **15**, 69.

156 H. Numan and H. Wynberg, *J. Org. Chem.*, 1978, **43**, 2232.

157 R. Eberhardt, C. Glotzmann, H. Lehner, and K. Schlögl, *Tetrahedron Lett.*, 1974, 4365.

158 E. J. Corey, P. B. Hopkins, S. Kim. S.-e. Yoo, K. P. Nambiar, and J. R. Falck, *J. Am. Chem. Soc.*, 1979, **101**, 7131.

159 W. Freytag and K. H. Ney, *J. Chromatogr.*, 1969, **41**, 473.

160 R. E. Leitch, H. E. Rothbart, and W. Reiman, *Talanta*, 1968, **15**, 213.

161 H. C. Rose, R. L. Stern, and B. L. Karger, *Anal. Chem.*, 1966, **38**, 469.

162 M. W. Anders and M. J. Cooper, *Anal Chem.*, 1971, **43**, 1093.

163 G. S. Ayers, J. H. Mossholder, and R. E. Monroe, *J. Chromatogr.*, 1970, **51**, 407.

164 B. L. Karger, R. L. Stern, W. Keane, B. Halpern, and J. W. Westley, *Anal. Chem.*, 1967, **39**, 228.

165 W. H. Pirkle and J. R. Hauske, *J. Org. Chem.*, 1976, **41**, 801.

166 C. J. W. Brooks, M. T. Gilbert, and J. D. Gilbert, *Anal. Chem.*, 1973, **45**, 896.

167 B. L. Karger, S. Herliczek, and R. L. Stern, *J. Chem. Soc. D*, 1969, 625.

168 S. V. Rogozhin and V. A. Davankov, *Usp. Khim.*, 1968, **37**, 1327. *Russ. Chem. Rev.*, 1968, **37**, 565.

169 I. S. Krull, *Adv. Chromatogr.*, 1978, **16**, 196.

170 G. Blaschke, *Angew. Chem.*, 1980, **92**, 14. *Angew. Chem., Int. Ed.*, 1980, **19**, 13.

171 Unpublished results of M. J. Brienne, A. Collet, M. Leclercq, and S. H. Wilen.

5.3 STRUCTURE–PROPERTY CORRELATIONS OF DIASTEREOMERS

Chemists have for a long time been interested in the possibility of predicting some of the properties of p,n diastereomer pairs so as to facilitate their separation. There has also been much interest in the possibility of deducing information on the configurations (especially absolute configurations) of the compounds bound to the resolving agent, that is, the substrates, from diastereomer properties.

In keeping with our objectives, we are particularly concerned with correlations that may exist between relative configurations and solubilities of diastereomers (principally those of salts). Such a correlation was attempted at the turn of the century and summarized in the "rules" of Winther and Werner. More recently, such considerations have given rise to the notion of "quasi-enantiomeric" resolving agents. Correlations between relative configurations and the order of chromatographic elution of some covalent diastereomers (Section 5.2.7) and between configurations and nmr chemical shifts are also known. They are not further treated here.

5.3.1 Rules of Winther and Werner

In 1895, the possibility of carrying out correlations of configurations between analogous compounds belonging to a given series on the basis of the stereospecifity of enzymatic reactions was no more than a hypothesis. At that time, Winther suggested[1] that selective affinity could also apply to processes that occur during resolutions based on solubility differences between diastereomer salts.

After examining the behavior of a limited number of organic acids toward a variety of alkaloids in different solvents, Winther concluded that the alkaloids may be grouped in two series: (1) quinine, quinotoxine, quinidine, strychnine, and brucine, and (2) cinchonine, cinchonidine, and morphine. This classification rests essentially on the observation that the first series of bases forms the *less soluble* diastereomeric salts with (+)-tartaric and (+)-2,3-dibromopropionic acids, while the second series of bases forms the less soluble salts with the corresponding (−)-acids.

From these few data, Winther sought to establish a correlation between configurations and solubilities as follows: For a series of structurally related compounds combined with a given resolving agent, the less soluble salts (or the more soluble ones) always correspond to enantiomers of like configuration. As a corollary, Winther was led to propound the following rule: two acids that are precipitated by the same base must have like configurations.

This rule led its author to propose correct relative configurations for (+)-tartaric, (−)-lactic, (−)-malic, (+)-methoxysuccinic, (+)-ethoxysuccinic, (−)-mandelic, and (−)-isopropylphenylglycolic acids. For other acids subjected to a similar analysis, Winther felt unable to comment with any certainty on their configuration based upon the results of resolutions.

At a later time (1928), when the assignment of absolute configurations was still very uncertain, Holmberg[2] examined the enantiomeric acids giving the less soluble salts with (+)-α-methylbenzylamine. He noted apparent solubility regularities among numerous related compounds and suggested that the correlations of configuration which they made possible appeared to confirm those made by other methods. However, "having lived with too many deceptions in the domain of dynamic stereochemistry," Holmberg declined "to augment confusion in this domain by taking these conclusions as definitive."

The prudence of Holmberg was well taken inasmuch as the results do not permit making any valid deduction and, in fact, lead one to erroneous assignments of configuration.

By according to the rule of Winther a rigor it never pretended, it is easy to expose all of its faults and finally to discredit it altogether. Both Matell[3] and Petterson[4] had no difficulty in turning up exceptions to the rule, especially among the aryloxypropionic acids. By way of example, Table 1 gives the results of resolutions of a series of acids carried out with alkaloids and with α-methylbenzylamine.

Is the "rule of Winther" still useful in any way today? To compare the relative solubility of diastereomeric salts, it is essential to do so under identical conditions. Numerous cases are known in which the order of solubilities is *inverted* with a change in solvent. One example of this is the case of α-phenyladipic acid whose less soluble strychnine, quinine, and brucine salts afford *either* enantiomer according to whether one operates in aqueous ethanol or anhydrous methanol.[19b] As we have already seen (Section 5.1.12), solvation of salts plays a decisive role in such cases.

When comparable experimental conditions are used, the regularities foreseen by Winther are in fact observed. For example, in the series of ten substituted phenylhydracrylic acids, **1**, the less soluble brucine salts have the same absolute configuration in nine cases.[5] A similar frequency is observed (eight out of ten cases) for the ephedrine salts of a series of substituted mandelic acids, **2**.[6] Pickard and Kenyon[7–9] have also found regularities of the same type in connection with resolutions of numerous hydrogen phthalate and succinate esters of secondary alcohols R–CH(OH)–R′.

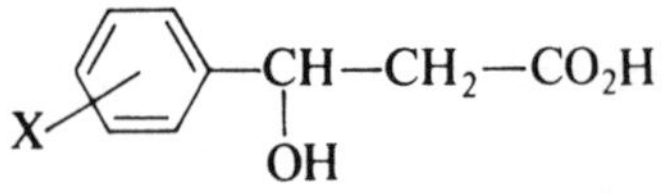

$X-C_6H_4-CH(OH)-CH_2-CO_2H$ **1**

$X-C_6H_4-CH(OH)-CO_2H$ **2**

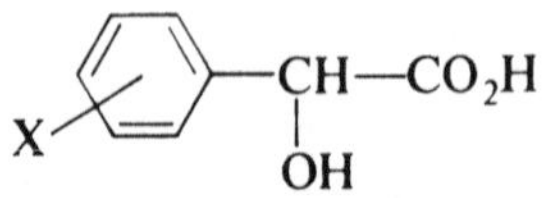

X = H, or F, Cl, Br *(o,m,p)*

The "rule of Winther," including even its exceptions, is in fact quite comprehensible in modern terms. It is not unexpected that in a series of conformationally and structurally related compounds (salts included) one observes analogous crystal packing. This *isomorphism*, in turn, has consequences in the realm of solubility relationships. Conversely, when even subtle changes in structure modify confor-

Table 1 Resolution of acids of type Ar–CH(R)COOH and Ar–X–CH(R)COOH[a]

Acid	Brucine	Cinchonine	Cinchonidine	Quinine	Quinidine	Morphine	Strychnine	α-Methylbenzylamine
α-(2,4-Dichlorophenoxy)propionic	D	Oil	D	$\underline{\underline{D}}$	Oil	Oil	O	Oil
α-(3,4-Dichlorophenoxy)propionic	$\underline{L}$	$\underline{L}$	O	L	Oil	Oil	$\underline{\underline{D}}$	D
α-(2-Methyl 4-chlorophenoxy)propionic	Oil	Oil	Oil	$\underline{D}$	Oil	Oil	L	$\underline{D}$
α-(1-Chloro 2-naphthoxy)propionic	Oil	L	$\underline{\underline{D}}$	$\underline{\underline{L}}$	L	$\underline{\underline{L}}$	Oil	$\underline{D}$
α-(1-Naphthylmethyl)propionic	Oil	$\underline{L}$	Oil	L	Oil	Oil	Oil	Oil
α-(2-Naphthylmethyl)propionic	L	Oil	$\underline{D}$	L	Oil	Oil	Oil	$\underline{L}$
α-(1-Naphthylthio)propionic	Oil	Oil	Oil	(–)?	Oil	Oil	Oil	Oil
α-(2-Naphthylthio)propionic	Oil	L	Oil	$\underline{D}$	Oil	Oil	Oil	$\underline{\underline{D}}$
α-Phenoxybutyric	Oil	L	Oil	L	Oil	$\underline{D}$	$\underline{\underline{L}}$	Oil
α-(1-Naphthoxy)butyric	Oil	$\underline{\underline{L}}$	$\underline{\underline{L}}$	Oil	Oil	Oil	$\underline{L}$	Oil
α-(2-Naphthoxy)butyric	Oil	Oil	D	$\underline{\underline{D}}$	Oil	Oil	Oil	$\underline{L}$
α-(2,4-Dichlorophenoxy)butyric	Oil	Oil	$\underline{D}$	$\underline{L}$	Oil	Oil	$\underline{\underline{D}}$	$\underline{\underline{D}}$
α-Phenoxy-*n*-caproic	Oil	Oil	D	L	Oil	Oil	Oil	$\underline{D}$
α-(2-Naphthoxy)-*n*-caproic	Oil	Oil	$\underline{D}$	L	Oil	O	Oil	D

[a] The resolutions were carried out in aqueous alcohol. The results identify the composition of the first crystalline fraction obtained according to the following conventions: O, no resolution; D or L, feeble excess of D or L; $\underline{D}$ or $\underline{L}$, moderate excess, and $\underline{\underline{D}}$ or $\underline{\underline{L}}$, large excess (D or L are configurational descriptors according to the convention of Fisher). M. Matell, *Kgl. Lantbrukshögskolans Ann.*, 1953, **20**, 205. Reproduced by permission of the editor.

mations (as, for example, in the case of homologues such as hydratropic and α-phenylbutyric acids), all predictions concerning the structure of crystals and consequently their solubilities become hazardous. The application of the "rule of Winther" is thus but an illustration of the complexities in the relationship between molecular structure and crystal packing which remains an unsolved problem.

Even if the rule is not systematically useful to establish the stereochemical correlations sought by Winther, examination of Table 1 inescapably shows a correlation between the nature of the salts and the success of the resolutions. For the acids examined, quinine and α-methylbenzylamine are effective in promoting useful resolutions in virtually all cases, while quinidine and brucine are only exceptionally so. For other acids of the same *type*, we have here a possible way *a priori* of guiding the choice of resolving agent.

In 1912, Werner[10] proposed a rule which does for coordination compounds, or complexes, what the rule of Winther does for organic compounds. This rule, which is sometimes called the rule of "corresponding solubilities,"[11] or rule of "the less soluble diastereomers,"[12] may be stated as follows: If one combines chiral complex ions having the same type of structure but possessing varying central atoms (Cr, Co, Rh, . . . etc.) with the same optically active substance, for example (+)-tartaric acid, the ions having the same configuration will give diastereomeric salts having comparable solubilities, that is, uniformly higher or lower than the salts obtained with (−)-tartaric acid. It has been observed that in the oversimplified form just given, this rule, like that of Winther, leads to false conclusions.[12, 13]

As we have just seen, in order for a relationship to exist, that is, for the stereospecific interactions in the crystal to be really comparable, whether for salts composed of organic ions or for complex ions, the organic salts or complex ions must be isomorphous. This excludes the comparison of complexes whose ions are not isoelectronic or whose geometry is very different, for example, $[Co(en)_2(NO_2)_2]^+$ and $[Co(en)_2(SCN)_2]^+$ or, further, those whose degree of solvation differs.

Garbett and Gillard have recently recast the rule of Werner in a form earlier suggested by Jaeger[11, 12]: "if two ions form isomorphous less-soluble diastereomers with the same resolving agent, then they have related configurations." Such a formulation certainly makes the best of the original rules of Winther and Werner by specifying in a precise manner the limits of their validity.

Finally, the data of Table 2 show quite satisfactorily that, with the stated reservations, the rule of the "less soluble diastereomers" may sometimes provide valid correlations which have the merit of simplicity.

5.3.2 Quasi-Enantiomeric Resolving Agents

During the course of their total synthesis of strychnine,[14] Woodward and his associates called attention to the existence of "quasi-enantiomeric" relationships that exist between the several cinchona bark alkaloids by stating that " . . . quinine and quinidine, in the vicinity of the salt-forming basic nitrogen atom, are enantiomeric . . ." This is true also for the pair cinchonidine/cinchonine. These enantiomeric relationships, which are illustrated in Fig. 1, hold only to an approximate extent

Table 2 Resolutions of the complex ions $[M(en)_2XY]^{n+}$. Configuration of the enantiomer forming the less soluble diastereomeric salt with different optically active anions[a]

Complex ion	(+)-Bromocamphorsulfonate	(+)-Camphorsulfonate	(+)-Antimonyl tartrate
$[Co(en)_3]^{3+}$	u	u	D
$[Co(en)_2Cl_2]^+$	D	L	D
$[Co(en)_2BrCl]^+$	D	L	D
$[Co(en)_2Br_2]^+$	D	O	D
$[Co(en)_2(N_3)_2]^+$	u	u	u
$[Co(en)_2(CN)_2]^+$	u	u	u
$[Co(en)_2(NO_2)_2]^+$	D	L	L
$[Co(en)_2(NCS)_2]^+$	L	u	L
$[Co(en)_2(NH_3)_2]^{3+}$	O	O	D
$[Co(en)_2(NH_3)Cl]^{2+}$	D	L	L
$[Co(en)_2(NH_3)Br]^{2+}$	D	O	L
$[Co(en)_2(NH_3)(NCS)]^{2+}$	D	u	u
$[Co(en)_2(NH_3)(NO_2]^{2+}$	L	O	L
$[Co(en)_2(NO_2)(NCS)]^+$	L	u	L
$[Co(en)_2(NO_2)(ONO)]^+$	D	u	u
$[Co(en)_2(NO_2)Cl]^+$	D	L	D
$[Co(en)_2(NO_2)Br]^+$	D	L	D
$[Co(en)_2(NO_2)F]^+$	u	u	u
$[Co(en)_2(NCS)Cl]^+$	D	L	L
$[Co(en)_2(H_2O)Cl]^{2+}$	D	L	L
$[Co(en)_2(H_2O)Br]^{2+}$	u	u	u
$[Co(en)_2(H_2O)_2]^{3+}$	D	u	u
$[Co(en)_2(sal)]^+$	u	u	u
$[Cr(en)_2Cl_2]^+$	D	u	D
$[Rh(en)_2Cl_2]^+$	D	u	u
$[Ir(en)_2(NO_2)_2]^+$	O	L	u

[a] O, No resolution; u, Unknown configuration. K. Garbett and R. D. Gillard, *J. Chem. Soc. A*, 1966, 802. Reproduced by permission of the Royal Society of Chemistry.

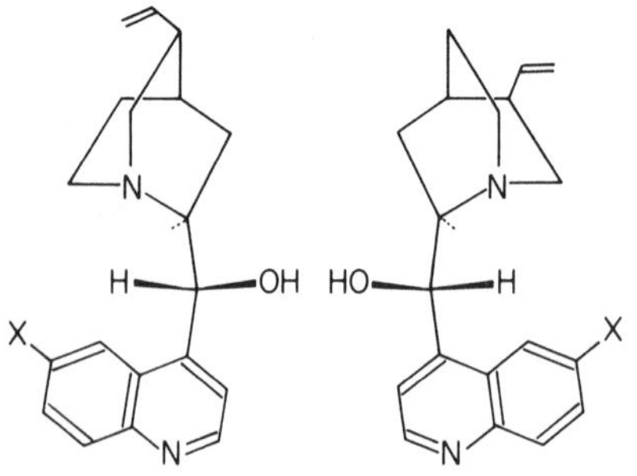

Quinine	$X = OCH_3$	Quinidine
Cinchonidine	X = H	Cinchonine

Figure 1 Stereochemical relationships in the cinchona bark alkaloids.

but suffice to circumvent the limitation imposed by the absence of enantiomers among naturally occurring resolving agents, at least for these two sets of optically active bases. This requires the combined use of the "quasi-enantiomeric" pairs.

Table 3 lists a number of cases where it is possible to use the pairs quinine/

Table 3 Examples of resolutions in which pairs of "quasi-enantiomeric" alkaloid resolving agents give both enantiomers of an acid

Resolved Acid	Alkaloids		Refs.
	Quinine	Quinidine	
(2-benzimidazolyl)–S–$CH(CH_3)$–CO_2H	–	+	15
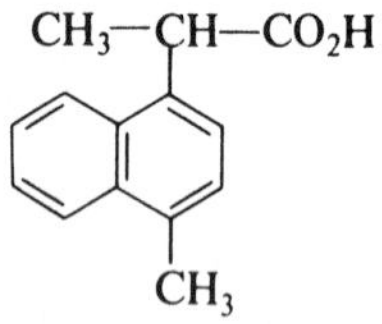	Cinchonidine	Quinidine	
CH_3–CH(CO_2H)–(4-methyl-1-naphthyl)	–	+	16
(2,3-dimethylphenyl)–O–$CH(CH_3)$–CO_2H	+	–	17
O_2N, COOH, S, S, HOOC, NO_2 (bithienyl dicarboxylic acid)	–	+	18

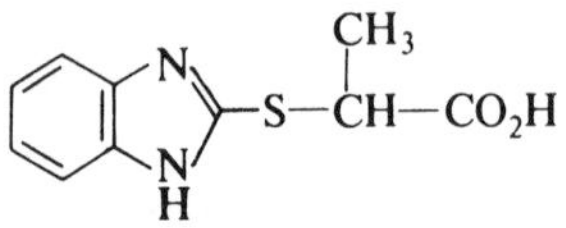

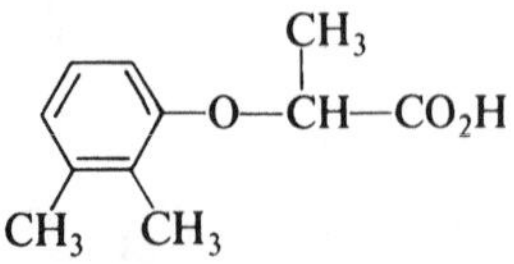

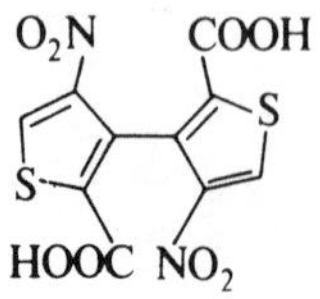

Table 3 Continued

Resolved Acid	Alkaloids		Refs.
	Quinine	Cinchonine	
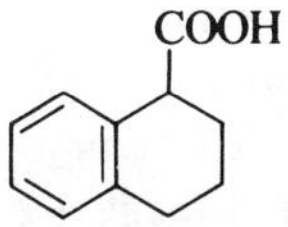 COOH	–	+	19a
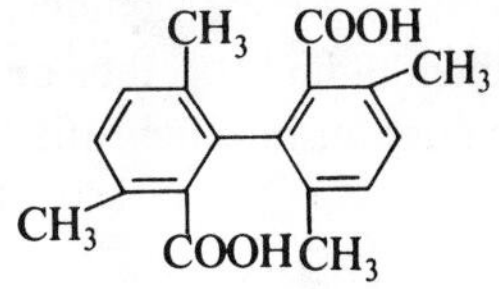 CH_3 COOH CH_3 CH_3 COOH CH_3	–	+	20
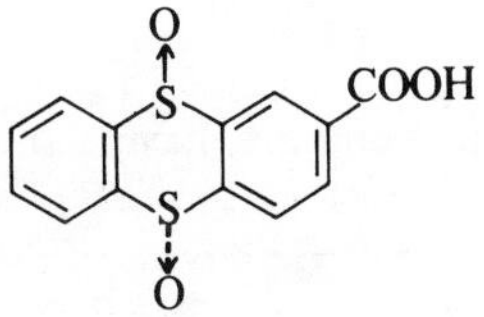 O S COOH S O	–	+	21, 22
	Cinchonidine	Cinchonine	
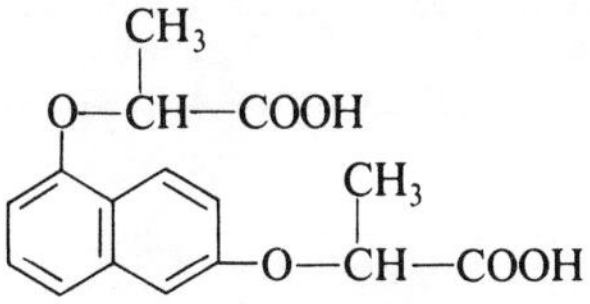 CH_3 O–CH–COOH CH_3 O–CH–COOH	–	+	23
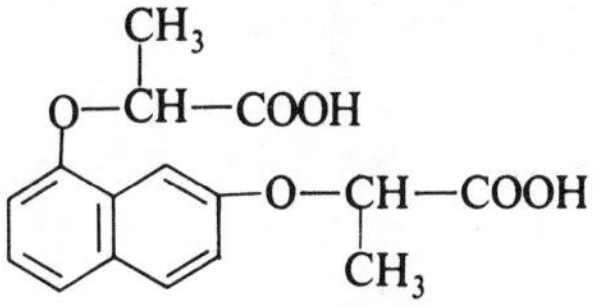 CH_3 O–CH–COOH O–CH–COOH CH_3	+	–	24
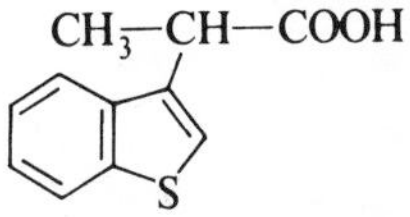 CH_3–CH–COOH S	–	+	25
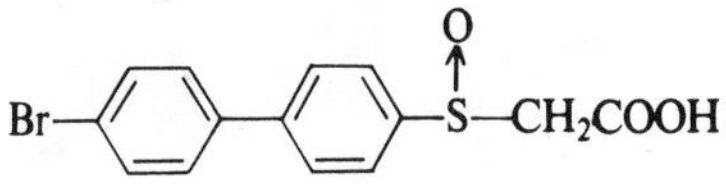 Br– O –S–CH_2COOH	+	–	26
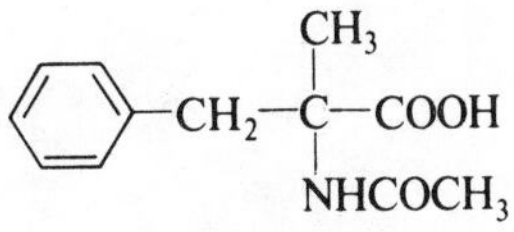 CH_3 –CH_2–C–COOH $NHCOCH_3$	–	+	27

quinidine, cinchonidine/quinidine, quinine/cinchonine, and cinchonidine/cinchonine to obtain *both* enantiomers of the resolved acid.

Surprisingly, use of the stereochemically *related* alkaloids, quinine and cinchonidine, provides access to acids of opposite configurations at least as often as do the cited "quasi-enantiomeric" pairs (Table 4)! This occurrence is sufficiently common so as to recommend the use of these two bases together whenever one of them gives a crystalline salt. Moreover, it is interesting to note that Winther grouped quinine and quinidine together on one hand and cinchonine and cinchonidine together on the other, which is contradictory to the quasi-enantiomeric relationships identified above.

Table 4 Examples of resolutions in which quinine and cinchonidine, having the same configuration, give both enantiomers of an acid

Resolved Acid	Quinine	Cinchonidine	Refs.
$CH_3-CH_2-CH_2-CH(CH_3)-CH_2-COOH$	+	−	28, 29
$C_6H_5-CH(CH(CH_3)_2)-CH_2-COOH$	+	−	30
H CH_3 COOH	−	+	31
HO H COOH H	+	−	32
Cl Cl $O-CH(CH_3)-COOH$	+	−	33
$CH_3O-C(=O)-CH(CH_3)-CH_2-COOH$	−	+	34
$CH_3O-C(=O)-CH_2-CH(CH_3)-CH_2-COOH$	−	+	35
$CH_3CH_2O-C(=O)-C(CH_3)(CH_2CH_3)-COOH$	+	−	36

Table 4 Continued

Resolved Acid	Quinine	Cinchonidine	Refs.
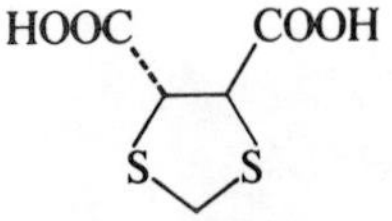 HOOC COOH S S	–	+	37
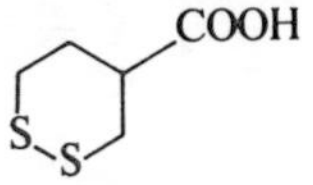 COOH S S	–	+	38
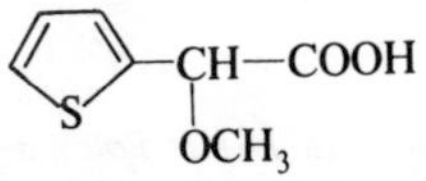 CH—COOH S OCH_3	+	–	39
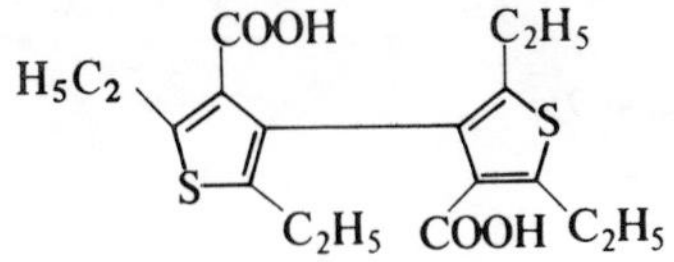 COOH C_2H_5 H_5C_2 S S C_2H_5 COOH C_2H_5	–	+	20
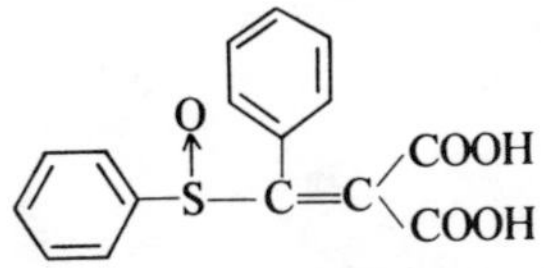 O ↑ S—C=C COOH COOH	+	–	40

REFERENCES 5.3

1 C. Winther, *Ber.*, 1895, **28**, 3000.

2 R. Holmberg, *Z. Phys. Chem.*, 1928, **137A**, 18

3 M. Matell, *Stereochemical Studies on Plant Growth Substances*, Lantbrukshögskolans Bibliothek, Uppsala, 1953. *Kgl. Lantbrukshögskolans Ann.*, 1953, **20**, 205.

4 K. Petterson, *Arkiv. Kemi*, 1956, **10**, 297.

5 A. Collet and J. Jacques, *Bull. Soc. Chim. Fr.*, 1972, 3857.

6 A. Collet and J. Jacques, *Bull. Soc. Chim. Fr.*, 1973, 3330.

7 R. H. Pickard and J. Kenyon, *J. Chem. Soc.*, 1911, **99**, 45.

8 R. H. Pickard and J. Kenyon, *J. Chem. Soc.*, 1912, **101**, 620.

9 R. H. Pickard and J. Kenyon, *J. Chem. Soc.*, 1913, **103**, 1923.

10 A. Werner, *Bull. Soc. Chim. Fr.*, 1912, **11**, 1.

11 F. M. Jaeger, *Bull. Soc. Chim. Fr.*, 1937, **4**, 1201.

12 K. Garbett and R. D. Gillard, *J. Chem. Soc. A*, 1966, 802.

13 K. Garbett and R. D. Gillard, *J. Chem. Soc.*, 1965, 6084.

14 R. B. Woodward, M. P. Cava, W. D. Ollis, A. Hunger, H. U. Daeniker, and K. Schenker, *Tetrahedron*, 1963, **19**, 247.

15 A. Fredga and K. Stadell, unpublished work; see S. H. Wilen, in *Tables of Resolving Agents and Optical Resolutions*, E. L. Eliel, Ed., University of Notre Dame Press, Notre Dame, Indiana, 1972, p. 114.

16 B. Sjöberg, *Ark. Kemi*, 1958, **12**, 573.

17 A. Fredga and K. Olsson, *Ark. Kemi*, 1969, **30**, 409.

18 S. Gronowitz, *Ark. Kemi*, 1957, **11**, 361.

19 (a) L. Westman, *Ark. Kemi*, 1958, **12**, 161. (b) *Ibid.*, 1958, **12**, 167.

20 S. Gronowitz and J. E. Skramstad, *Ark. Kemi*, 1968, **28**, 115.

21 J. Chickos and K. Mislow, *J. Am. Chem. Soc.*, 1967, **89**, 4815.

22 M. Janczewski, M. Dec, and W. Charmas, *Roczniki Chem.*, 1966, **40**, 1021.

23 J. Suszko and M. Kielczewski, *Roczniki Chem.*, 1967, **41**, 1291.

24 J. Suszko and M. Kielczewski, *Roczniki Chem.*, 1967, **41**, 1565.

25 B. Sjöberg, *Ark. Kemi*, 1957, **11**, 439.

26 M. Janczewski and W. Charmas, *Ann. Univ. Mariae Curie-Sklodowska Lublin-Polonia*, 1963, Sect. AA, **18**, 81. *Chem. Abstr.*, 1967, **66**, 37570.

27 F. W. Bollinger, *J. Med. Chem.*, 1971, **14**, 373.

28 J. Cason and R. A. Coad, *J. Am. Chem. Soc.*, 1950, **72**, 4695.

29 F. I. Carroll and R. Meck, *J. Org. Chem.*, 1969, **34**, 2676.

30 G. Sörlin and G. Bergson, *Ark. Kemi*, 1968, **29**, 593.

31 D. J. Cram and L. Gosser, *J. Am. Chem. Soc.*, 1964, **86**, 5445.

32 D. S. Noyce and D. B. Denney, *J. Am. Chem. Soc.*, 1952, **74**, 5912.

33 A. Fredga, *Croat. Chem. Acta*, 1957, **29**, 313.

34 G. Ställberg, *Ark. Kemi*, 1958, **12**, 79.

35 R. P. Linstead, J. C. Lunt, and B. C. L. Weedon, *J. Chem. Soc.*, 1950, 3333.

36 J. Kenyon and W. A. Ross, *J. Chem. Soc.*, 1951, 3407.

37 M.-O. Hedblom, *Ark. Kemi*, 1969, **31**, 489.

38 G. Claeson, *Ark. Kemi*, 1969, **30**, 511.

39 T. Raźnikiewicz, *Ark. Kemi*, 1962, **18**, 467.

40 O. Bohman and S. Allenmark, *Ark. Kemi*, 1969, **31**, 299.

CHAPTER 6

Crystallization-Induced Asymmetric Transformations

By definition, the theoretical maximum yield in a resolution cannot exceed 50% for each of the two pure enantiomers. However, if the enantiomers are easily racemizable, this limitation need no longer apply. That is, if racemization in solution or in the molten state is facile, preferential crystallization of one of the two and displacement of the resulting equilibrium could, under some conditions, lead to a total transformation of the initial racemate into a *single* enantiomer. Similarly, and more generally, formation of interconvertible diastereomers may lead to the preferential and possibly even to sole crystallization of one of the two.

The necessary instability of the compounds involved in this special type of crystallization process may, however, also make the isolation of pure enantiomers or diastereomers difficult. In spite of this potential drawback, the use of crystallization-induced asymmetric transformations is very appealing, particularly when separations on an industrial scale are contemplated.

Since asymmetric transformations[1,2] have often been defined in confusing ways, we believe that it is worthwhile to restate the meaning of appropriate terms in a precise way here. The general case, defined by Scheme 1, is commonly observed in diastereomers. The less common, even rare case, that of asymmetric transformation of enantiomers, is dealt with in Section 6.1.

When a racemate, *dlA*, is combined with an equimolecular amount of optically active reagent, *lB*, one obtains initially either upon simple dissolution (salts) or upon reaction (covalent compounds) an equimolar solution of the two diastereomers *dAlB* and *lAlB*, *provided* that sufficient time is allowed (particularly in the second case).*

If the two diastereomers are stable to interconversion under the reaction conditions, then separation of the 1–1 mixture – by whatever method – constitutes a resolution in the usual sense of the word.

* If insufficient time is allowed, a *kinetic resolution* may ensue and the amounts of the two products need no longer be equal.

SCHEME 1

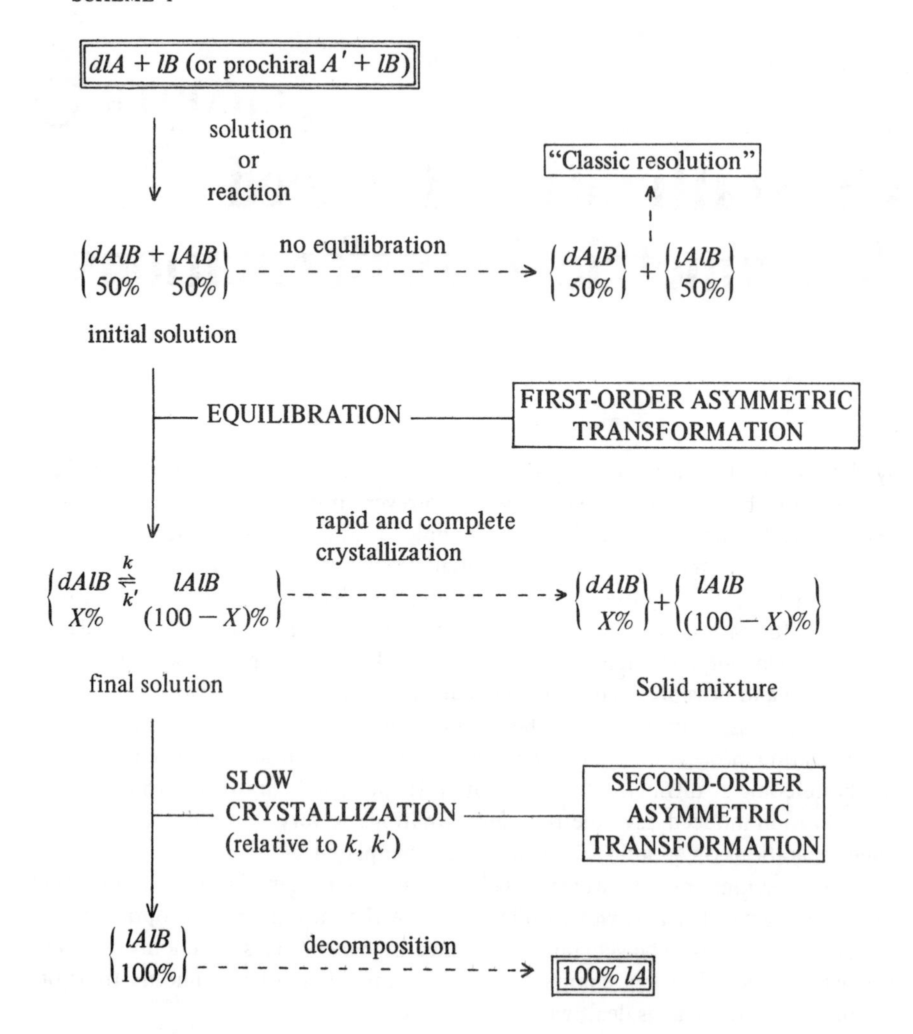

However, if substrate A is relatively easily racemized in the presence of lB, then after a given length of time a solution will be obtained which contains $X\%$ of $dAlB$ and $(100 - X)\%$ of $lAlB$. This equilibrium corresponds to the definition of "first-order asymmetric transformation." If the solution is chilled so as to effect a rapid and total precipitation of the solutes, a solid mixture of the two diastereomers may be obtained which reflects the proportion present at equilibrium in solution. Only exceptionally does this *asymmetric activation* lead to just one of the two possible species.

In a "second-order asymmetric transformation," the diastereomer equilibrium is continually displaced by crystallization of one of the two species until the totality of the diastereomer mass is present as one pure diastereomer. Note that the compound which crystallizes is not necessarily that which predominates at equilibrium (in solution or melt).

While the first possibility corresponds to an *asymmetric equilibration in solution*, the second may be described as a *crystallization-induced asymmetric disequilibration*. The latter supposes that the rate of crystallization of the less soluble diastereomer be slower than the rate of equilibration of the two species in solution, two phenomena which, *a priori*, are unrelated to one another. Incidentally, in the classical definitions, the word "order" is improperly used; it stems from an incorrect translation of the German word "Art" whose correct meaning in context is "kind," as in "asymmetric transformation of the first kind."[2–4]

The asymmetric transformation in solution (of "first order"), which reflects essentially the difference in stability between two isomeric entities, intervenes, for instance, in the mutarotation of sugars as well as in the mechanism of the Pfeiffer effect.[5,6] The latter is the name ascribed to the variation in the magnitude of the optical activity of a chiral ion in the presence of another racemic or achiral ion that is also labile.

We shall confine further analysis to the several known cases of crystallization-induced asymmetric transformation (of "second order") which are associated with resolutions of some special significance and which permit the direct or indirect isolation of one enantiomer from a racemate.

REFERENCES 6

1 E. E. Turner and M. M. Harris, *Q. Rev. Chem. Soc.,* 1947, 299.
2 M. M. Harris, *Progr. Stereochem.*, 1958, **2**, 157.
3 R. Kuhn, *Ber.*, 1932, **65**, 49.
4 M. M. Jamison and E. E. Turner, *J. Chem. Soc.*, 1942, 437.
5 P. Pfeiffer and K. Quehl, *Ber.*, 1931, **64**, 2267.
6 P. E. Schipper, *Inorg. Chim. Acta*, 1975, **12**, 199.

6.1 ASYMMETRIC DISEQUILIBRATION OF A RACEMATE. TOTAL "SPONTANEOUS RESOLUTION"

An asymmetric transformation may occasionally be observed when the interconvertible species are not diastereomers but *enantiomers*. It is clear that in this situation the equilibrium in solution or in the molten state (which corresponds to a "first-order" asymmetric transformation) furnishes an *equimolecular* mixture of right- and left-handed molecules. Rapid precipitation (relative to the rate of racemization) can yield but one product which is itself racemic.

However, if the substance in question does not form a racemic compound, slow crystallization of one of the enantiomers, whether spontaneous or provoked, may lead to the exclusive isolation of the latter. This special type of "second-order" asymmetric transformation is rather rare since it requires that the crystals formed from the racemic solution be those of the enantiomers (conglomerate). A good illustration of this phenomenon is the spontaneous resolution of *N,N,N*-methylethylallylanilinium iodide, **1**. Upon slow evaporation over the course of several months, a chloroform solution of this quaternary ammonium salt deposits crystals exhibiting large rotations while the mother liquor remains virtually racemic.[1]

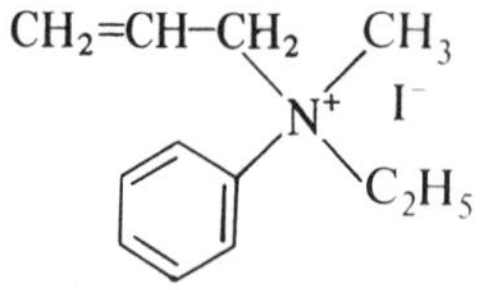

1 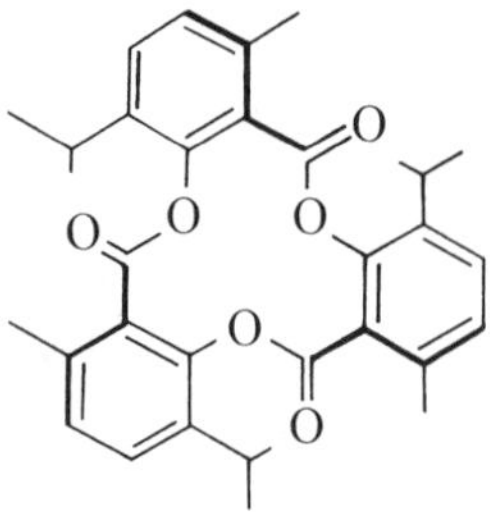 **2** 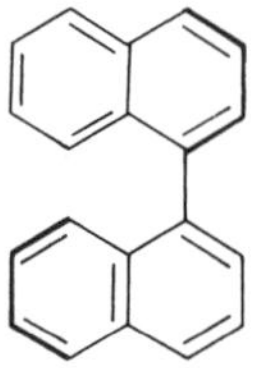 **3**

Tri-*o*-thymotide, **2**, whose racemization is relatively rapid in solution ($\Delta G^{\ddagger}$ = 20.9 kcal · mol^{-1} at 0°C), forms optically active crystalline inclusion compounds by the same process[2–4] (see Section 5.1.8).

Binaphthyl, **3**, has been studied by Pincock et al. in considerable detail. Conditions were found which permit virtually total transformation of the racemate into one enantiomer.[5–9,15]

One of the very rare examples of a fully inorganic (carbon-free) optically active compound, $(NH_4)_2Pt(S_5)_3$, was recently obtained through such an asymmetric transformation.[10]

Application of this process to the resolution α-amino-ϵ-caprolactam (ACL), an important precursor of L-lysine, demonstrates the usefulness of this optical activation route in a case where the substrate is not optically labile under ordinary conditions.[11,12] Racemic ACL reacts with nickel(II) chloride to form a mixture of two enantiomeric complexes formulated as $(\text{L-ACL})_3NiCl_2$ and $(\text{D-ACL})_3NiCl_2$. These compounds are easily interconvertible in ethanol solution in the presence of ethoxide ion which functions as a catalyst. When a supersaturated solution $[(\text{D-ACL})_3NiCl_2 \rightleftharpoons (\text{L-ACL})_3NiCl_2]$ is seeded with crystals of $(\text{L-ACL})_3NiCl_2$, a nearly complete asymmetric transformation occurs which leads to the isolation of the $(\text{L-ACL})_3NiCl_2$ crystalline complex. The latter may be decomposed instantly into optically pure L-ACL by reaction with hydrogen chloride in methanol solution.

In addition to the generation of enantiomorphous crystals from chiral-racemic molecules, it will be recalled (Section 1.2.3) that formally achiral molecules also are known to generate crystals of one handedness by processes entirely analogous to those described above. Recent studies by B. S. Green, M. Lahav and their colleagues have shown that reactions proceeding in, or influenced by, such chiral crystals may lead to formation of new chiral substances with moderate to high enantiomeric excess.[13,14] Such asymmetric syntheses, called "topochemically controlled reactions" can have quite useful consequences as has been alluded to in Section 4.4.

REFERENCES 6.1

1 E. Havinga, *Biochim. Biophys. Acta*, 1954, **13**, 171.

2 W. Baker, B. Gilbert, and W. D. Ollis, *J. Chem. Soc.*, 1952, 1443.

3 A. P. Downings, W. D. Ollis, and I. O. Sutherland, *J. Chem. Soc.*, (B), 1970, 24; *Chem. Comm.*, 1968, 329.

4 A. C. D. Newman and H. M. Powell, *J. Chem. Soc.*, 1952, 3747.

5 K. R. Wilson and R. E. Pincock, *J. Am. Chem. Soc.*, 1975, 97, 1474.

6 R. E. Pincock and K. R. Wilson, *J. Am. Chem. Soc.*, 1971, 93, 1291.

7 R. E. Pincock, R. R. Perkins, A. S. Ma, and K. R. Wilson, *Science*, 1971, **174**, 1018.

8 R. E. Pincock and K. R. Wilson, *J. Chem. Educ.*, 1973, **50**, 455.

9 K. R. Wilson and R. E. Pincock, *Can. J. Chem.*, 1977, **55**, 889.

10 R. D. Gillard and F. L. Wimmer, *J. Chem. Soc. Chem. Comm.*, 1978, 936.

11 S. Sifniades, W. J. Boyle, Jr., and J. F. Van Peppen, *J. Am. Chem. Soc.*, 1976, **98**, 3738.

12 W. J. Boyle, Jr., S. Sifniades, and J. F. Van Peppen, *J. Org. Chem.*, 1979, **44**, 4841.

13 B. S. Green, M. Lahav, and D. Rabinovich, *Acc. Chem. Res.*, 1979, **12**, 191.

14 L. Addadi and M. Lahav, *Pure Appl. Chem.*, 1979, **51**, 1269.

15 See also (a) R. Kuroda and S. F. Mason, *J. Chem. Soc., Perkin II*, 1981, 167. (b) R. B. Kress, E. N. Duesler, M. C. Etter, I. C. Paul, and D. Y. Curtin, *J. Am. Chem. Soc.*, 1980, **102**, 7709.

6.2 ASYMMETRIC TRANSFORMATION OF DIASTEREOMERIC SALTS

Leuchs was the first to describe transformation of diastereomeric salts. During the resolution of (±)-2-(*o*-carboxybenzyl)-1-indanone, **1**, reported in 1913, the brucine salt which is isolated by crystallization from acetone in 94% yield contains *exclusively* the dextrorotatory acid.[1]

CO_2H O NO_2 OMe CO_2H OMe

1 **2**

This type of "total" resolution is applicable in a general way to all compounds racemizable in solution under conditions in which crystallization of a salt formed by reaction with a resolving agent is possible. A few examples follow:

1 Atropisomers (biphenyls, binaphthyls and related compounds); the resolution of **2** by brucine is an example.[2]

2 Substances whose optical activity is due to slow inversion about a trivalent heteroatom (N, Sb, . . .) such as acid **3**, resolved with cinchonidine,[3] or acid **4**, resolved with strychnine.[4]

3 Compounds such as the indanone **1**, which are readily equilibrated by enolization, and acids **4** and **6** whose α-protons are highly mobile.[5,6] We will later see other examples in this category which have given rise to industrial scale applications.

4 Compounds such as acid **7** whose chirality is due to syn-anti isomerism.[7]

5 Werner complexes such as cation $[Co(en)_3]^{3+}$ which is resolved by tartaric acid.[8]

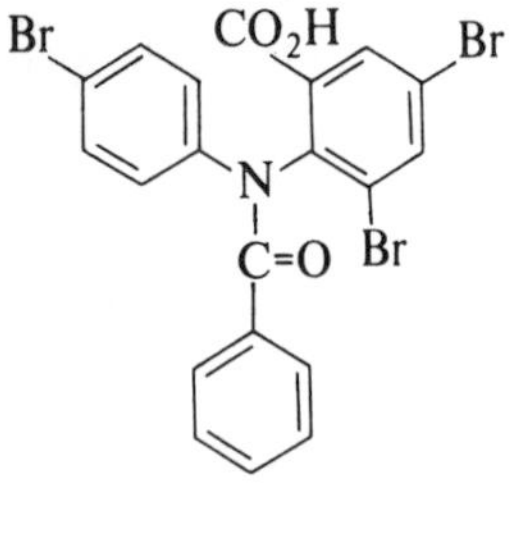

3 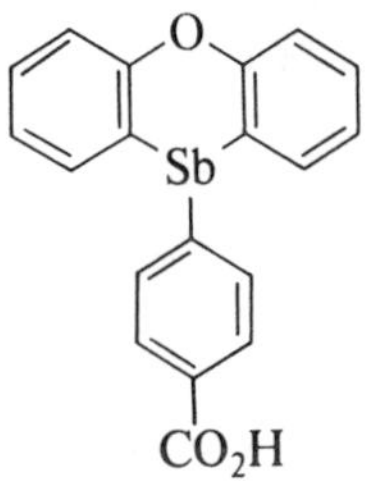 **4** 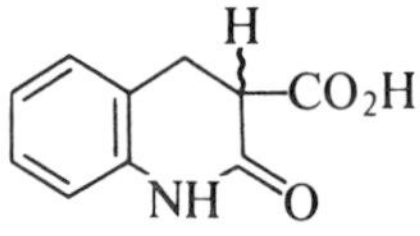 **5** 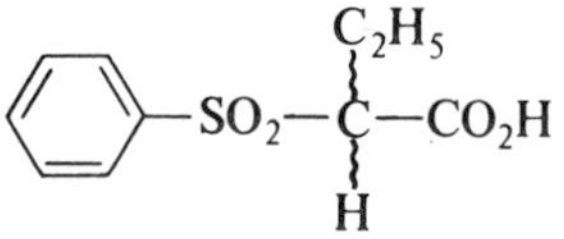 **6**

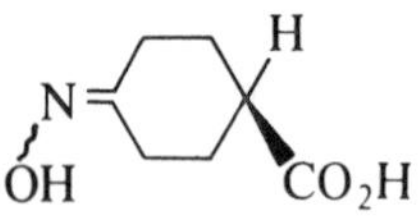

7 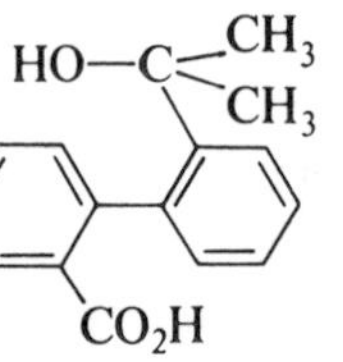 **8**

The resolution of acid **8**, described by Corbellini and Angeletti,[9] illustrates the fact that there is not necessarily any relation between the asymmetric equilibrium in solution and the crystallization-induced asymmetric disequilibration. While the brucine salt which predominates in chloroform solution at equilibrium is that of the (+) acid (58%), that which crystallizes in nearly quantitative yield is that of the (−) acid.

Several applications of crystallization-induced transformations to the synthesis of amino acids are indicative of the potential utility of this process on an industrial scale. D- and L-Phenylglycines and some of their derivatives (D-*p*-hydroxyphenylglycine in particular) may be obtained in this way. In the procedure described by Clark, Phillips, and Steer,[10] an ester of the racemic amino acid is treated with one equivalent of (+)-tartaric acid in an alcohol in the presence of an aldehyde or ketone. The latter catalyze the racemization of the amino acid ester through formation of a Schiff base intermediate. After a crystallization period of the order of one to seven days at 20 to 25°C has elapsed, one of the diastereomers of the ester is isolated nearly pure and in yields which may exceed 95%. The most significant results are summarized in Table 1.

Table 1 Asymmetric transformation of phenylglycine derivatives[a]

Amino Ester	Solvent	Carbonyl Compound (mol. equiv.)	Crystn. Time (hours)	Yield (%)	Optical Purity (%)
$Ph-CH(NH_2)-CO_2Me$	EtOH	PhCHO (1)	24	85 (D)	99
$Ph-CH(NH_2)-CO_2Me$	EtOH	Me_2CO (8)	20	93 (D)	99
$Ph-CH(NH_2)-CO_2iPr$	EtOH	PhCHO (0.55)	144	62 (L)	98
p-$HO-Ph-CH(NH_2)-CO_2Me$	$MeOH-C_6H_6$	PhCHO (1)	48	80 (D)	98

[a] J. C. Clark, G. H. Phillipps, and M. R. Steer, *J. Chem. Soc., Perkin I*, 1976, 475. Reproduced by permission of the Royal Society of Chemistry and the authors.

Lonyai[11] has described an alternative process in which the substrate is 2-phenylaminoacetonitrile (prepared from benzaldehyde). When this precursor of the amino acid is treated with (+)-tartaric acid and acetone in toluene, the salt corresponding to the amino acid D isomer is obtained in 12 hours. The isolated salt is finally hydrolyzed to D-phenylglycine with an overall yield in excess of 60% relative to benzaldehyde. Here, too, acetone intervenes so as to facilitate racemization of the substrate during the course of the asymmetric transformation.

REFERENCES 6.2

1 H. Leuchs and J. Wutke, *Ber.*, 1913, **46**, 2420.

2 H. C. Yuan and R. Adams, *J. Am. Chem. Soc.*, 1932, **54**, 8966.

3 M. M. Jamison and E. E. Turner, *J. Chem. Soc.*, 1938, 1646.

4 I. G. M. Campbell, *J. Chem. Soc.*, 1947, 4.

5 H. Leuchs, *Ber.*, 1921, **54**, 830.

6 W. C. Ashley and R. L. Shriner, *J. Am. Chem. Soc.*, 1932, **54**, 4410.

7 W. H. Mills and A. M. Bain, *J. Chem. Soc.*, 1914, **105**, 64.

8 J. A. Broomhead, F. P. Dwyer, and J. W. Hogarth, *Inorg. Synth.*, 1960, **6**, 186.

9 A. Corbellini and A. Angeletti, *Atti R. Accad. Lincei*, 1932, **15**, 968.

10 J. C. Clark, G. H. Phillipps, and M. R. Steer, *J. Chem. Soc., Perkin I*, 1976, 475.

11 P. Lonyai, G. Toth, F. Garamszegi, G. Lehoczky, A. Hunyadi, and G. Csermely (to Chinoin Gyogyszer es Vegyeszeti Termedek Gyara Rt.), Hungarian Teljes 13579 (1977). *Chem. Abstr.*, 1978, **89**, 6557w.

6.3 ASYMMETRIC TRANSFORMATION OF COVALENT DIASTEREOMERS

Some *covalent diastereomers* may be equilibrated in solution to give mixtures whose compositions may vary widely. Ratios exceeding 70–30 seem to be infrequently encountered, however, as seen in the data of McKenzie and Smith on the menthyl and bornyl phenylhalogenoacetates[1,2] as well as in those of Weiges et al., which are summarized in Table 1. Moreover, it is observed in these amino-nitriles (**1**) (prepared by Strecker synthesis with the optically active auxiliary reagent α-methylbenzylamine) that the diastereomer isolated during the crystallization-equilibration process is the one that is less stable in solution, that is, the less abundant one at equilibrium.[3]

Table 1 Equilibration of diastereomeric α-methyl-α-amino-nitriles **1** in $CDCl_3$[a]

R–C6H3(R′)–$(CH_2)_n$–C(CH_3)(C≡N)–NH–C*H(CH_3)–C6H5

1

			Diastereomeric Ratio at Equilibrium (%)	
n	R	R′	*p*	*n*
1	H	H	45	55
1	OCH_3	H	68	32
1	OCH_3	OCH_3	75	25
2	H	H	45	55
2	OCH_3	H	60	40
2	OCH_3	OCH_3	55	45

[a] From nmr measurements of K. Weiges, K. Gries, B. Stemmle, and W. Schrank, *Chem. Ber.*, 1977, **110**, 2098. Reproduced by permission of Verlag Chemie GMBH.

A similar process was used by Bucourt et al. in a total synthesis of steroids (Scheme 1). The prochiral diketone **2** forms two diastereomeric tartramazones on reaction with tartramic acid hydrazide (the monoamide monohydrazide of natural tartaric acid). By crystallization under equilibrating conditions (methanol/acetic acid) only isomer **3** is isolated (75% yield); subsequent cyclization leads to steroids having the natural configuration.[4]

SCHEME 1

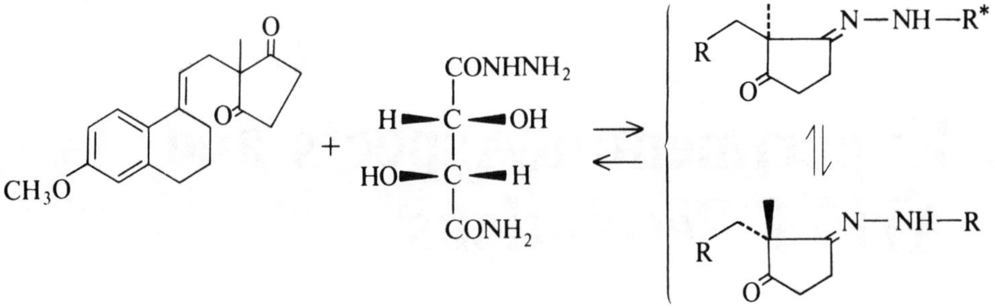

2

3

Insoluble

REFERENCES 6.3

1. M. M. Harris, *Progr. Stereochem.*, 1958, **2**, 157.
2. A. McKenzie and I. A. Smith, *J. Chem. Soc.*, 1924, 1582. *Ber.*, 1925, **58**, 894.
3. K. Weiges, K. Gries, B. Stemmle, and W. Schrank, *Chem. Ber.*, 1977, **110**, 2098.
4. R. Bucourt, L. Nedelec, J. -C. Gasc, and J. Weill-Raynal, *Bull. Soc. Chim. Fr.*, 1967, 561.

CHAPTER 7

Experimental Aspects and Art of Resolutions

The information and concepts found in the preceding chapters on the nature and properties of enantiomers, diastereomers, and their mixtures should permit a better understanding of, and ability to carry out, resolutions. The practical details about how one actually begins the resolution and how one controls and monitors the progress and carries the final purification to a successful conclusion are given in this chapter. This should normally suffice to achieve success, as would be true in any other unit process associated with the practice of organic chemistry.

Yet many practitioners believe that there is a mystique and an aura of "art" to resolutions. While we do subscribe to the proposition that such techniques as recrystallization require some practice to be carried out well, we believe, and think organic chemists would generally agree, that virtually anyone can learn to carry out a purification implicit in a recrystallization. So, too, with resolutions.

We hope that this chapter will complete the demystification. In our experience, there need be few – if any – failures in intelligently and systematically executed resolutions.

7.1 CHOICE OF RESOLUTION METHOD

Even if it is difficult to predict with certainty how the resolution of a racemate should be undertaken in order to achieve success efficiently, a number of elements must be taken into consideration quite early in the design: (a) the quantity and (b) the structure of the compound to be resolved.

7.1.1 Choice of Method as a Function of Scale

It must be evident that on a scale smaller than 100 mg, a new resolution carried to completion using crystallization techniques may be achieved, but rarely without difficulty. Chromatographic resolution techniques are more suitable here. The choice to be made is that of a covalent diastereomeric derivative from which regeneration of the enantiomers is easy and attended by little risk of racemization.

With the chromatographic techniques presently available, few mixtures will withstand separation trials carried out with perserverance.

For amounts ranging from one to hundreds of grams, chromatography is less useful in resolutions, although at the lower end of the scale, techniques such as flash chromatography[1] are applicable in many cases. We remain doubtful of the claim made by Pirkle to the effect that "most 'first time' resolutions of enantiomers will soon be effected almost solely by liquid chromatographic techniques."[2] Many laboratories are neither equipped with the necessary facilities for carrying out preparative liquid chromatography nor do they have the experience required to quickly achieve such separations successfully. The investment of time required for a satisfactory chromatographic resolution on a modest scale would then be such as to lead many researchers to undertake traditional resolutions involving crystallization first.

On the intermediate scale indicated, separation of dissociable diastereomeric compounds by crystallization would seem, for the time being, to offer the best chance of success and convenience. On this scale, the choice of resolving agent is still either influenced by their cost or by the need to eventually recover them.

For quantities of substance which range from a kilogram to tons, use of diastereomeric salts in resolutions becomes in turn less and less attractive as a consequence of the relatively high cost of resolving agents (bases in particular). Only resolution through direct crystallization of enantiomers is likely to answer to industrial requirements.* It is necessary, however, that the racemate employed (or one of its derivatives) be a conglomerate.

7.1.2 Choice of Method According to the Structure of the Substrate

It is clear that the structure of the substrate has at least as much importance in the choice of the resolution method as does the scale.

Use of covalent derivatives in the resolution of chiral acids and bases is relatively rare. Salts are easy to prepare, and their components are easy to regenerate on virtually any scale. In contrast, resolution with covalent diastereomers is accompanied by additional risks (racemization, low yields) in connection with their formation and decomposition. Such derivatives are preferred in the resolution of alcohols, phenols, ketones, and so on, where salts cannot be directly formed. Moreover, an advantage of covalent diastereomers is that they may frequently be separated either by chromatography or by crystallization.

Other structural peculiarities also figure in the choice of resolution method. In the case of enolizable ketones a detour is sometimes advantageous: they may be optically activated by resolution of the corresponding alcohol.

* In some well-defined cases, such as the amino acids, fermentation processes and even asymmetric synthesis are applicable on an industrial scale.

REFERENCES 7.1

1 W. C. Still, M. Kahn, and A. Mitra, *J. Org. Chem.*, 1978, **43**, 2923.
2 W. H. Pirkle and J. R. Hauske, *J. Org. Chem.*, 1977, **42**, 1839.

7.2 OBTAINING CRYSTALLINE DIASTEREOMERS

Once it has been determined that a resolution requires the formation of diastereomers, there still remains the matter of choosing a resolving agent from among the many available. For covalent diastereomers, the choices are quite varied, and it is difficult to give more specific advice than that which may be gleaned from Section 5.2.

When a suitable derivative has been prepared, the separation (by chromatography or by crystallization) with which one is confronted is fundamentally no different from that of most mixtures ordinarily encountered in the practice of organic chemistry. It is otherwise if the projected resolution requires the crystallization of diastereomeric salts.

Finding the "best" diastereomeric derivative constitutes the first step and is undoubtedly the most difficult to rationalize in a resolution.

7.2.1 Systematic Trials. Choice of Resolving Agent

The search for a suitable resolving agent in separations mediated by diastereomeric salts may be carried out in a systematic manner,[1] as a result of the ease of formation and decomposition of this type of diastereomer. Moreover, the crystallization and separation of diastereomeric salts of a wide range of acidic and basic resolving agents are operationally quite similar. We do not consider it to be a good strategy to begin a resolution by combining a relatively large quantity of racemate, even as little as 1 g, with a single resolving agent chosen more or less at random or on the basis of its immediate availability and to follow this with the persistent and repeated recrystallization of the salt which has hopefully formed. While this way of proceeding may occasionally lead to the expected result, it is statistically much more likely to result in a product that is optically impure or even completely racemic.

Taking advantage of the experience of several groups of specialists in this type of separation, notably that of Fredga and his collaborators (these being perhaps responsible for the largest number of published resolutions of acids), it is possible to systematize the process that leads most rapidly to the selection of usable diastereomer salts.

It is essential that laboratories where resolutions are carried out have available at least a half-dozen optically active bases and as many acids with which systematic trials with small quantities of substrate may be undertaken. The process comprises three steps:

Table 1 Measurement of rotations of small samples[a]

Case a

Molecular weight 150: $[\alpha]_D = +20^\circ$

Substrate taken: 150 mg (0.001 mol)

Substrate recovered from the diastereomer mixture: 30 mg (40% of 0.0005 mol maximum)

Rotation observed: $\alpha_D = +0.300^\circ$ ($c = 0.030$ g/2.0 mL = 1.5 g/100 mL) if optically pure

Case b

Molecular weight 450; $[\alpha]_D = +20^\circ$

Substrate taken: 45 mg (0.0001 mol)

Substrate recovered: 9 mg

Rotation observed: $\alpha_D = +0.090^\circ$ ($c = 0.009$ g/2.0 mL) if optically pure

[a] S. H. Wilen, A. Collet, and J. Jacques, *Tetrahedron*, 1977, **33**, 2725. Reproduced by permission of Pergamon Press Ltd.

1 Combining 10^{-4} to 10^{-3} mole of the racemic substrate with an equivalent of optically active base or acid in the presence of a quantity of solvent of the order of 1 mL (see Section 7.2.2). The same operation is simultaneously carried out with several other resolving agents.
2 When a crystalline salt is obtained (see Section 7.2.3), it is isolated, dried, and weighed. In order for the test to be significant, the quantity of crystals obtained must obviously not exceed approximately one half of the total weight of the two diastereomers that may be formed; otherwise, one must use more solvent.
3 After having decomposed the salt, the liberated substrate is isolated to permit the measurement of its specific rotation.

The possibility of working with such a small amount of substance (of the order of 100 mg substrate) is tied to the sensitivity of modern photoelectric polarimeters (minimum $\pm 0.005^\circ$) which is quite adequate for the rotations typically observed during the course of resolution trials (Table 1). Since the specific rotation by itself does not generally reveal the optical purity of the product, other criteria of purity of the supposedly resolved substance should be applied: melting points (or more usefully the fusion interval, measured e.g., through differential scanning calorimetry) in particular.

The results are compiled into a crystallization–rotation matrix (for example, Table 2), which provides a semiquantitative summary of the resolution trials. These results may be classified into three situations; (a) no crystalline diastereomers obtained; (b) crystalline salt obtained giving evidence of weak or no resolution; and (c) crystalline salt obtained corresponding to a net resolution which may be fair or good.

While the first of these situations precludes a resolution in the short term, it does not necessarily mean that the mixture would not form a crystalline product

Table 2 Resolution trials for the phenylglyceric acids[a]

X–C₆H₄–CH(OH)–CH(OH)–COOH

	X	1	2	3	4	5	6	7	8	9	10	11	12	13
Threo	H	$\underline{\underline{l}}$	0		Oil		0			0	Oil			0
	o–Cl	0	0			0		*l*			$\underline{l}$	$\underline{\underline{d}}$		
	m–Cl											$\underline{\underline{d}}$		
	p–Cl	$\underline{\underline{l}}$												
Erythro	H					0							$\underline{\underline{d}}$	
	o–Cl	0	$\underline{\underline{l}}$			Oil							*l*	0
	m–Cl		0	$\underline{\underline{l}}$		Oil			0		Oil		0	
	p–Cl		*d*		Oil	$\underline{\underline{l}}$	0		$\underline{d}$				$\underline{\underline{d}}$	

[a] Samples of acid (100 mg, ca. 5.0 mmol) are treated with 1 equivalent base in 1 mL ethanol. The crystals obtained are directly decomposed and the rotation of the acid measured. Key: 0 = no resolution; *d* or *l*, $\underline{d}$ or $\underline{l}$, $\underline{\underline{d}}$ or $\underline{\underline{l}}$ = weak, fair, or good resolution. From ref. 13. 1 = (+) α-methylbenzylamine; 2 = (+)-*threo*-1-*p*-nitrophenyl-2-amino-1,3-propanediol; 3 = (+)-*threo*-1-*p*-nitrophenyl-2-dimethyl-amino-1,3-propanediol; 4 = (−)-ephedrine; 5 = (−)-deoxyephedrine; 6 = (−)-amphetamine; 7 = dehydroabietylamine; 8 = funtumine (3α-amino-5α-pregnan-20-one); 9 = quinine; 10 = cinchonine; 11 = cinchonidine; 12 = strychnine; 13 = brucine.

were one to continue waiting for a longer period of time or, for example, if the trial were carried out in another laboratory.

The second situation may correspond to the formation of a 1–1 addition compound between salts *p* and *n* (whence $[\alpha]_{\text{substrate}} = 0°$), to their cocrystallization (formation of a solid solution), or to an insufficient difference in solubility. All three of these possibilities render the resolution impractical.

The last of these situations constitutes the favorable case, since it indicates that the diastereomer mixture has a eutectic sufficiently removed from an equimolar composition so as to make separation of the diastereomers possible.

Table 2 gives some examples of the use of such resolution trials. Moreover, the problem is somewhat different if one is dealing with an isolated case or with a series of related substances. In the latter case, it is sometimes possible to limit the number of systematic trials to one or two members of the series and subsequently to orient the choice of resolving agent used with the other members of the series on the basis of the first results (see, for example, the *threo*-phenylglyceric acids in Table 2). However, as we have seen in connection with the results of Winther and Werner (Section 5.3.1), this "limited" approach may not always be successful.

Table 2 shows that of the 33 trials, nine give usable results. By the same token, the 112 trials cited by Matell and collected in Table 1 of Section 5.3.1 furnish 26 favorable cases. It is not possible to generalize the proportion of success (which approximates one fourth in the two series of trials), particularly because one rarely finds descriptions of resolution failures in the literature.

Although resolution trials such as those described in Table 2 suggest that even for closely related compounds it is difficult to find any pattern in the choice of usable resolving agents, nonetheless analysis of reported resolutions as in *Tables of Resolving Agents and Optical Resolutions*[3] improves the probability of success substantially over the random use of resolving agents in the trial matrix illustrated. For example, brucine is the preeminent resolving agent for the successful resolution of numerous alcohols via their hydrogen phthalate esters.

While it remains true that this first step in resolutions is still virtually totally unpredictable, experience shows that if one conducts the search patiently and systematically, it is quite rare not to finally achieve success.

Finally, we mention a variant in the way in which resolving agents are chosen in some specific cases. It consists in the use in the first trials not of the racemic substrate but of one of its enantiomers. This is possible, for instance, when a total synthesis of a natural product leads to a racemate which is less accessible than one of the enantiomers.[2] However, note that while the formation of a relatively insoluble crystalline salt with this pure enantiomer and some resolving agent is a necessary condition for the sucess of the resolution, it is not a sufficient one. It is necessary that the racemate and the same resolving agent lead neither to an addition compound between the salts nor to their cocrystallization.

7.2.2 Choice of Crystallization Solvent

We have already seen (Section 5.1) that the nature of the solvent has only a small effect on the *ratio* of solubilities of the two diastereomers, except in the case of differential solvation of the *p* and *n* diastereomers. One also knows that good laboratory practice calls for the use of crystallization solvents in which the solute is neither too soluble nor too insoluble. Given the ionic nature of the salts used in resolutions, one can predict that the best solvents will be polar ones. While there are exceptions to this generalization for diastereomeric salts, and it is probably less true for covalent diastereomers to be separated by crystallization, the solubility of salts in nonpolar solvents such as hexane or benzene is generally too low to permit their use. Moreover, such solvents sometimes even prevent the formation of salts; certain alkaloids crystallize from them in the pure state even in the presence of acids.

Table 3 gives a statistical analysis of solvents used in over 800 resolutions[3] involving the formation of salts. It is immediately evident that alcohols and acetone (anhydrous or aqueous), water, and mixtures of solvents containing an alcohol figure in about 90% of the cases. The distinction between anhydrous and aqueous solvents is important to the extent that one knows that the tendency of forming solvates is greater with water (hydrates) than with alcohols and other solvents.[4] Surprisingly, some resolutions take a different course according to whether they are carried out in absolute ethanol or in 96% ethanol, for example. Moreover, hydrate formation sometimes facilitates the crystallization of salts.

Table 3 Statistical summary of solvents used in 819 resolutions involving formation and separation of diastereomeric salts[a]

Solvent	Type	Cases	Subtotal	Total	%
Pure or Aqueous Solvents					
Ethanol	Anhydrous	139	323	660 cases	(81%)
	Aqueous[b]	184			
Methanol	Anhydrous	80	103		
	Aqueous	23			
Other alcohols[c]			7		
Acetone	Anhydrous	81	133		
	Aqueous	52			
Water			94		
Ethyl acetate			37	78 cases	
Benzene			9		
Dioxane			10		
Diethyl ether			5		
Chloroform			6		
Miscellaneous[d]			11		
Mixture of Solvents Containing at Least One Alcohol					
Alcohol[e] +	Acetone		19	61 cases	(7%)
	Ethyl acetate		13		
	Diethyl ether		14		
	Benzene		5		
	Miscellaneous		10		

Mixtures of Solvents Containing Neither an Alcohol nor Water: 20

[a] Found in ref. 3.
[b] 96% Ethanol or ethanol diluted with variable amounts of water.
[c] 1-Propanol, 2-propanol, 1-butanol.
[d] Acetonitrile, DMF, pyridine, hexane, THF, 2-butanone, diisobutylketone.
[e] Ethanol or methanol in almost all cases.

Table 4, derived from the same data, attempts to identify the most appropriate solvents for each of the common resolving agents. This list does not imply that other solvents than those given are useless. What it does do is to list, in decreasing order of importance, those solvents in which the largest number of resolutions have been carried out. One can see that ethanol (absolute or aqueous) ranks high in virtually all cases. This is why we recommend that resolution trials involving salts as described in the preceding paragraphs always be begun using 96% ethanol as solvent.

Table 4 Choice of solvents favoring the crystallization of salts according to the resolving agent used

Resolving Agent	Solvent
Bases	
Brucine	Acetone; aqueous acetone; ethanol; aqueous ethanol; water; methanol; ethyl acetate
Strychnine	Aqueous ethanol; aqueous acetone; water; ethanol
Quinine	Aqueous ethanol; ethanol; acetone; methanol; aqueous acetone
Quinidine	Ethanol; aqueous ethanol; acetone
Cinchonine	Aqueous ethanol; acetone; dioxane
Cinchonidine	Aqueous ethanol; ethanol; aqueous acetone; acetone; ethyl acetate
α-Methylbenzylamine	Aqueous ethanol; ethanol; ethyl acetate; water
Ephedrine	Ethanol; benzene; aqueous ethanol; chloroform
Amphetamine	Aqueous ethanol
Dehydroabietylamine	Aqueous ethanol; aqueous methanol; ethanol
Acids	
Tartaric	Ethanol; methanol; water; acetone
Dibenzoyltartaric	Ethanol; methanol
Camphorsulfonic	Water; ethanol; aqueous ethanol; ethyl acetate
Bromocamphorsulfonic	Water; ethanol; methanol; aqueous acetone
Mandelic	Water; alcohol/acetone or alcohol/ethyl acetate mixtures

Mixtures of solvents are sometimes used (Table 3) in order to adjust the solubility of salts to a convenient range. For example, ether is added to reduce the solubility of a salt in ethanol. We wish to emphasize that the common practice of mixing solvents *ad libitum* and of not recording their proportions is to be avoided. The ability to repeat a resolution closely depends upon a precise knowledge of the relative quantities of solvent constituents used. It is indeed preferable to work with pure solvents, if possible. In addition to methanol and ethanol, for example, isopropyl alcohol, butanol and methoxyethanol (methylcellosolve) provide a wide range of dissolving power that satisfies common solubility requirements without having to have recourse to mixtures.

It is also important to remember that one must sometimes change solvents in the successive recrystallizations of a resolution since the purified salt becomes less and less soluble. By the same token, one can occasionally obtain one diastereomer with a first solvent and the second by recrystallizing the product isolated from the mother liquor in another solvent.

There remains the question of how best to carry out the resolution itself. This requires the combination of the racemic substrate and the resolving agent in the chosen solvent. A common technique consists in the separate dissolution of the acid and the base, followed by the combination of the two solutions. Most resolving agents are sufficiently soluble in methanol or ethanol; but some, such as binaphthylphosphoric acid or cinchonine, are only very slightly so.*

Also, the substrate itself is sometimes little or not at all soluble in the resolving solvent. When one or the other of the two reactants is little soluble, application of the aforementioned technique leads to solutions that may be too dilute. One must then either concentrate the solution or preferably use a second technique consisting in the dissolution or suspension of one of the reactants in a smaller quantity of solvent and the addition to this of the other reactant. Brief heating generally suffices to effect total dissolution if it has not already occurred in the cold. Application of this mutual solubilizing effect is generally preferable to the first-mentioned approach, particularly for resolution trials carried out on small quantities.

7.2.3 Isolation of the First Crystallization Seeds

To obtain the first crystals of any new compound, even when it is relatively pure, is a well-known problem among chemists. From experience, we know that the presence of even a very small impurity in some types of compounds can hinder the formation as well as the growth of the first nuclei. The formation of crystals from equimolar mixtures of diastereomers is understandably even more difficult.

No infallible recipe exists for overcoming the resistance of a diastereomer to crystallize for the first time. We can only provide some general guidance which we believe is valid in a particular case of interest to us.

The effect of scratching with a glass stirring rod or with a metal spatula is well known. We can also suggest that at this stage in the manipulations there is no point in being meticulous. It is almost desirable to leave reaction vessels unstoppered and accessible to laboratory dust. It is certain that in the absence of seeds of the compound itself, microscopic crystals of isomorphic compounds can induce the desired crystallization (see Section 4.4). It has even been observed that granite dust, in which several crystalline systems are represented, may be beneficial in seeding crystallizations.

The probability of nucleation also depends upon the amount of substance brought into play; in a given time and at a given temperature, there is 1000 times less chance to see crystals appear in 100 mg than in 100 g of substance.

From the work of Tammann one may conclude that the most favorable temperature for the appearance of crystals in a pure substance in the supercooled liquid state lies in or near a range of 60 to 130° below the melting point. These results, which remain valid also for supersaturated solutions, show that it is not always advantageous to cool samples too much in the hope of crystallizing them. Also, the large increase in viscosity associated with the cooling of solutions, particularly concentrated salt solutions, unfavorably affects the growth of the first micro-

* Mixing two solvents sometimes increases the solubilizing power of each one separately. For example, cinchonine is nearly eight times more soluble in an ethanol–chloroform mixture than in absolute ethanol and more than 20 times more soluble in this mixture than in chloroform alone.[5]

crystals. The rate of crystal growth is maximal in the vicinity of 25° below the melting point.

Finally, it is good practice to submit samples to alternate cycles of freezing at ca. −20°C and of warming to room temperature, so as to favor nucleation as well as the growth of crystals.

It is desirable to try numerous solvents, including aqueous solvents, to induce crystallization. Recall that occasionally the formation of hydrates is itself the condition of crystallization. In practice, the following technique may be adopted advantageously: If the trial solution prepared in ca. 1 mL 96% ethanol does not spontaneously crystallize or does not crystallize upon scratching, several drops are removed to a watch glass where the solvent evaporates rather quickly. It can then be replaced by another while one attempts to induce crystallization through scratching. In this manner one can quickly try out a large number of solvents.

7.2.4 Neutral Salts and Acid (or Basic) Salts

In the resolution of bases by diacids (of which tartaric acid is the prime example) or, conversely, in the resolution of diacids by an optically active base, one is always faced by a choice of stoichiometry: should the reactants be combined in equimolar proportions (1–1) or should the neutralization of the diacid by base be assured through 1–2 molar proportions? It seems that resolutions effected with tartaric acid and its diacid derivatives are most often carried out with a stoichiometry that leads to formation of *acid salts* (1–1).

When reactants are combined in both proportions, one sometimes observes the crystallization of a neutral salt *or* of an acid salt, according to the conditions. In such cases, it is often found that the neutral salt contains one enantiomer of the substrate while the acid salt contains the other; this circumstance obviously facilitates their total separation. Some quite old observations illustrate this possibility. For example, in connection with the resolution of *dl-N*-benzoylaspartic acid with brucine, Fischer found that the less soluble acid salt contains the (−) acid, while of the two neutral salts it is the salt containing the (+) acid that precipitates.[6] Read and Reid,[7] while completing the work begun by Marckwald[8] on the resolution of racemic (tartaric) acid, showed that the less soluble acid tartrate of cinchonine contains the (−) acid while the less soluble neutral tartrate contains the (+) acid. More recently, it has been found[9,10] that the diacids *dl*-**1** and **2**, when combined

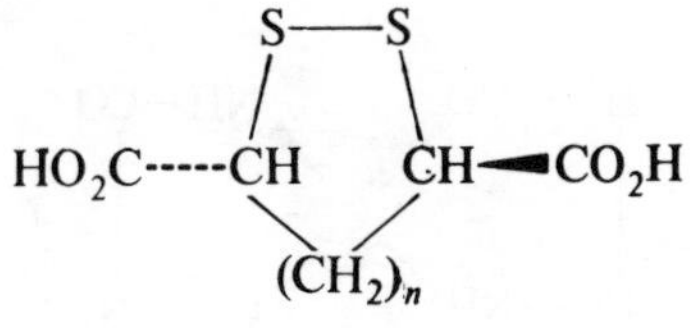

S—S

HO_2C-----CH   CH◄CO_2H

$(CH_2)_n$

1 $n = 1$

2 $n = 3$

with one equivalent of brucine (acid salts), furnish the (+) enantiomers, while with two equivalents of the same base (neutral salts) they give the (−) enantiomers.*

Finally, Leigh[12] has described several resolutions of amines by *O,O'*-dibenzoyl and *O,O'*-di-*p*-toluoyl-*d*-tartaric acids in which the formation of neutral or acid salts is taken advantage of to obtain expeditious separations. One of the enantiomers is first precipitated through use of 0.5 (or even 0.25) mole acid per mole racemic base (neutral salt). The same quantity of acid added to the mother liquors provokes the crystallization of the acid salt of the other enantiomer.

References 7.2

1 S. H. Wilen, A. Collet, and J. Jacques, *Tetrahedron*, 1977, **33**, 2725.

2 R. B. Woodward, M. P. Cava, W. D. Ollis, A. Hunger, H. U. Daeniker, and K. Schenker, *Tetrahedron*, 1963, **19**, 247.

3 S. H. Wilen, in *Tables of Resolving Agents and Optical Resolutions*, E. L. Eliel, Ed., University of Notre Dame Press, Notre Dame, Indiana, 1972.

4 J. H. Hildebrand and R. L. Scott, *The Solubility of Nonelectrolytes*, 3rd ed., Reinhold, New York, 1950; Dover, New York, 1964, p. 183.

5 A. C. Oudemans, Jr., *Liebigs Ann. Chem.*, 1873, **166**, 65.

6 E. Fischer, *Ber.*, 1899, **32**, 2451.

7 J. Read and W. G. Reid, *J. Soc. Chem. Ind.*, 1928, **47**, 8T.

8 W. Markwald, *Ber.*, 1896, **29**, 42.

9 L. Schotte, *Ark. Kemi*, 1956, **9**, 413.

10 L. Schotte, *Ark. Kemi*, 1956, **9**, 429.

11 W. J. Pope and J. B. Whitworth, *Proc. Roy. Soc.* (London) 1931, **A134**, 357.

12 T. Leigh, *Chem. Ind.* (London), 1977, 36.

13 A. Collet, *Bull. Soc. Chim. Fr.*, 1975, 215, and unpublished results.

* These data may be related to those of Pope and Whitworth,[11] already cited in Section 5.1.17, who found that the spirohydantoin **3** first precipitates an "acid salt" (1–1) of the (−)-hydantoin in the presence of two equivalents of brucine and that subsequently the "neutral salt" (2–1) containing the (+)-hydantoin crystallizes.

HN—CO NH—CO

C

CO—NH CO—NH

3

7.3 PURIFICATION OF DIASTEREOMERS

In our description of the resolution process beginning with a racemic substrate, we have now reached the stage in which a mixture of two separable diastereomeric derivatives, more or less enriched in one, is at hand. If purification by crystallization is required, then one must establish optimal conditions for this task and find a way to control its progress.

7.3.1 Optimal Conditions for Crystallization. Quantity of Solvent

As we have already seen, when dealing with a system containing a simple eutectic, knowledge of the solubility diagram of the two diastereomers allows one to determine the conditions under which the less soluble one may be isolated pure and even to stipulate the conditions permitting maximal recovery of the pure enantiomer (Section 5.1.12). In fact, it is not even necessary to know the entire diagram to obtain the desired information; it suffices to determine the composition and the solubility of the *eutectic*. These data may, in principle, be deduced from an examination of the mother liquors obtained during the course of the first crystallization of the diastereomers derived from the racemate (Fig. 1). If the compound precipitated is not yet pure (e.g., A_s), the solution from which it originates must have composition E, or at least one whose composition approaches E, since one does not usually wait long enough for solubility equilibrium to be attained.

To place point E on the diagram, one must determine the *concentration* (line CC'; by evaporation of the solvent, for example) and the *enantiomeric composition* e of the solution. If the resolution is the first one undertaken on the substrate, the latter information may be obtained either through analysis of the diastereomer mixture or of the enantiomer mixture derived from it (Section 7.5), or indirectly by pushing the resolution to completion. This may be achieved by carrying out one or more recrystallizations of the mixture first isolated, A_s, with enough solvent so as to obtain a low yield.

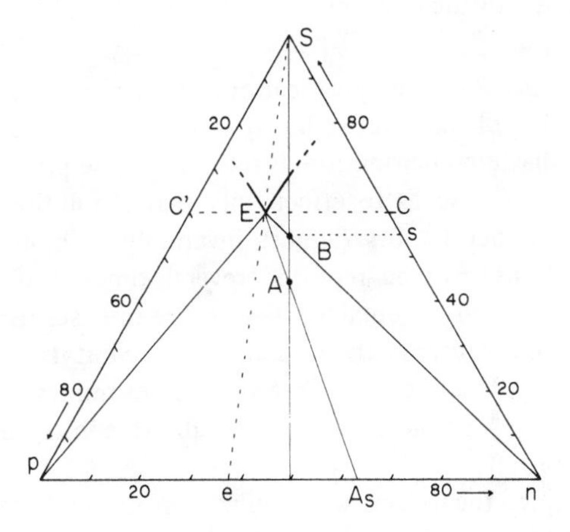

Figure 1 Separation of crystalline diastereomers. Ideal case.

Once point E has been located on the diagram, the indicated construction in Fig. 1 gives the optimal amount of solvent (s) which permits one to obtain the maximum quantity of the less soluble diastereomer in one crystallization directly *from the racemic substrate*. The diagram even allows one to estimate the yield of this diastereomer since EB/En represents the proportion of pure solid n present at equilibrium in a mixture of original composition B. For example, in Fig. 1, B consists of 22.5% p, 22.5% n, and 55% solvent. At equilibrium, the crystals of n amount to $EB/En \simeq 9\%$ of the whole system, that is, $9/22.5 = 40\%$ of the diastereomer n in the original mixture.

In practice, the purity of the diastereomer obtained in such a process depends upon the possible existence of solid solutions at the extremities of the diagram and upon their relative importance (Fig. 2). Unless solid solutions dominate the diagram, one or two recrystallizations of the diastereomer will normally suffice to obtain it in a pure state or at least in a state of purity which allows the final purification to be carried out on the recovered *enantiomer* itself (Section 7.6.2).

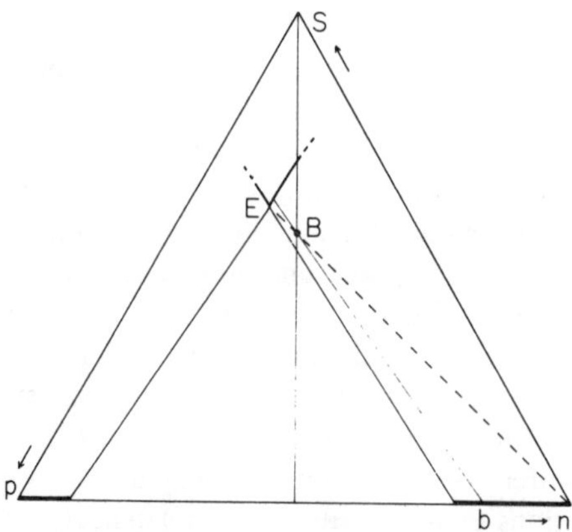

Figure 2 Separation of crystalline diastereomers. Real case: Solid solutions at the extremities p and n of the diagram.

The literature also describes purifications of diastereomers based on selective dissolution or leaching of mixtures rather than on recrystallization. In one variant of this process, the diastereomer mixture is refluxed in the presence of a sufficiently small quantity of solvent so as to effect only partial solution of the solid. After separation of the mother liquors (almost invariably without cooling), the same sequence of operations may be repeated several times until the purity of the remaining solid is deemed acceptable. This procedure seems to have been first applied by Kantor and Hauser to the resolution of *sec*-butyl alcohol via the brucine salt of its hydrogen phthalate ester.[1] Several other examples of this type of purification may be cited.[2–4] In another variant, a diastereomer mixture is repeatedly leached with cold solvent.[5]

In this technique, the selective solubilization of the eutectic mediating the enrichment of the less soluble diastereomer may be difficult to carry out reproducibly since it depends not only on the solubilities (in warm solvent) of the two isomers *p* and *n* but also on their rates of solution which, in turn, are related to the mechanical state of the solid. It is thus evident that the experimental conditions, including the state of subdivision (particle size) of the solid, interpenetration of the two crystalline species *p* and *n* (mutual inclusion), quantity of solvent, and length of reflux, all may affect the successful outcome of the purification.

Even with these reservations, the technique may still be quite useful, particularly in the purification of relatively insoluble diastereomers, since it may serve to reduce the quantity of solvent required. A similar result may also be obtained, but in a more controllable manner, by use of nonstoichiometric quantities of the resolving agent (Sections 5.1.17 to 5.1.19).

7.3.2 Crystallization with Seeding

When the diastereomers *p* and *n* form a eutectic, seeding with crystals of one or the other species may facilitate their separation. In the same manner as in the course of resolution by entrainment of a pair of enantiomers (Section 4.2), this crystallization technique is possible only if the solution is supersaturated with respect to *both* diastereomers; the quantity of solvent required is thus smaller than that defined in the preceding section.

There are even cases in which seeding represents the sole means of ensuring a correct separation. For example, the *p* and *n* salts might have very similar solubilities, such as for the β-ethoxy-β-phenylethylamine acid tartrate salts whose recrystallization leads to only mediocre enrichment. Brode and Wernert observed during this work that the two diastereomers have sufficiently different crystal aspects to permit their separation by triage. Each being obtained pure in small quantity, it was possible to carry out a successful separation through seeding.[6]

In other cases, seeding can help reduce the problems resulting from the existence of polymorphism. Colles and Gibson have observed that the resolution of *p*-nitrobenzoylalanine by strychnine is very difficult due to the formation of a double salt alongside the *p* and *n* salts. The difficulty may be completely obviated by seeding an equimolar solution of the two salts with crystals of the less soluble salt. After separation of the crystals, the mother liquors are seeded with the other salt.[7] A case related to the foregoing has been described by Brode and Hill:[8] without special precautions, *dl*-methadone forms an acid tartrate which is a double salt, mp 135–138°C (rosettes). Nevertheless, slow crystallization allows the isolation of crystals of the *n* salt, mp 149.5–151°C (needles). Subsequently, seeding with the *n* salt allows one to collect this diastereomer pure. The mother liquors are then seeded with the double salt, and the residual solution is enriched in nearly pure *p* salt.

Even in the absence of the problems just cited, one should view seeding as a procedure free of disadvantages. Seeding facilitates the obtention of the more soluble salt since it allows the mother liquors to exceed the composition of the eutectic.[9,10]

We observe that a systematic process for carrying out the separation by entrainment of diastereomer mixtures forming eutectics is possible in principle and would be especially useful with derivatives of nearly equal solubilities which are difficult to separate under quasi-equilibrium conditions.

7.3.3 Influence of Temperature

Solubility differences between diastereomeric salts sometimes vary with temperature (variation in the position of the eutectic), and advantage can be taken of this in optimizing a resolution. For example, in the case of the cinchonine mandelate salts, the difference in the solubilities of the diastereomers diminishes as the temperature is raised. Here, it is desirable to carry out the resolution at the lowest possible temperature.[11]

This influence of temperature, which is generally small and unpredictable, can have even more important consequences when the existence of solvated forms *inverts* the relative solubilities of the two diastereomers or is responsible for the apparition of *double salts*. The bases of these phenomena have been examined in Section 5.1 along with the remedies appropriate to each specific case. For example, in the resolution of β-pipecoline with *d*-tartaric acid, the double salt, being less solvated than the simple salts, tends to form at elevated temperature. As a result, resolution is possible only if crystallization of the salts is carried out at room temperature.[12,13] In other cases, raising the temperature has an opposite effect.

In a general way, it is useful to know the degree of solvation of salts (e.g., as determined by elemental analysis) to understand their behavior as a function of crystallization temperature.

7.3.4 Fractional Crystallization

The formation of solid solutions of diastereomers can be a real stumbling block in a resolution. The first step to take in the face of such a problem is to seek an alternative resolution route. It is only in the fortunately rare cases in which it is truly not possible to do otherwise that one must confront the problem of separating two diastereomers which cocrystallize. Fractional crystallization techniques are then indispensable if one is to have a chance of attaining success. In any event, the yield in such cases is generally poor.

The application of fractional crystallization methods to the separation of solid solutions has been described numerous times. In particular, we call attention to the excellent review of Tipson, which we commend to the reader.[14] An example is given by the resolution of the hydrogen phthalate ester of isopropyl-*n*-amylcarbinol with strychnine, which involves an extremely laborious separation. From 570 g of a mixture of the two diastereomeric salts, a series of 120 fractional crystallizations, part of which is shown in schematic form in Fig. 3, leads to 40.9 g of one of the salts which was considered pure.[15]

Freudenberg and Lwowski have less completely described the case of the cinchonidine α-chlorobutyrate salts.[16] From 800 g racemic acid, and after 159 systematic crystallizations, they obtained the following: 10.9 g acid with

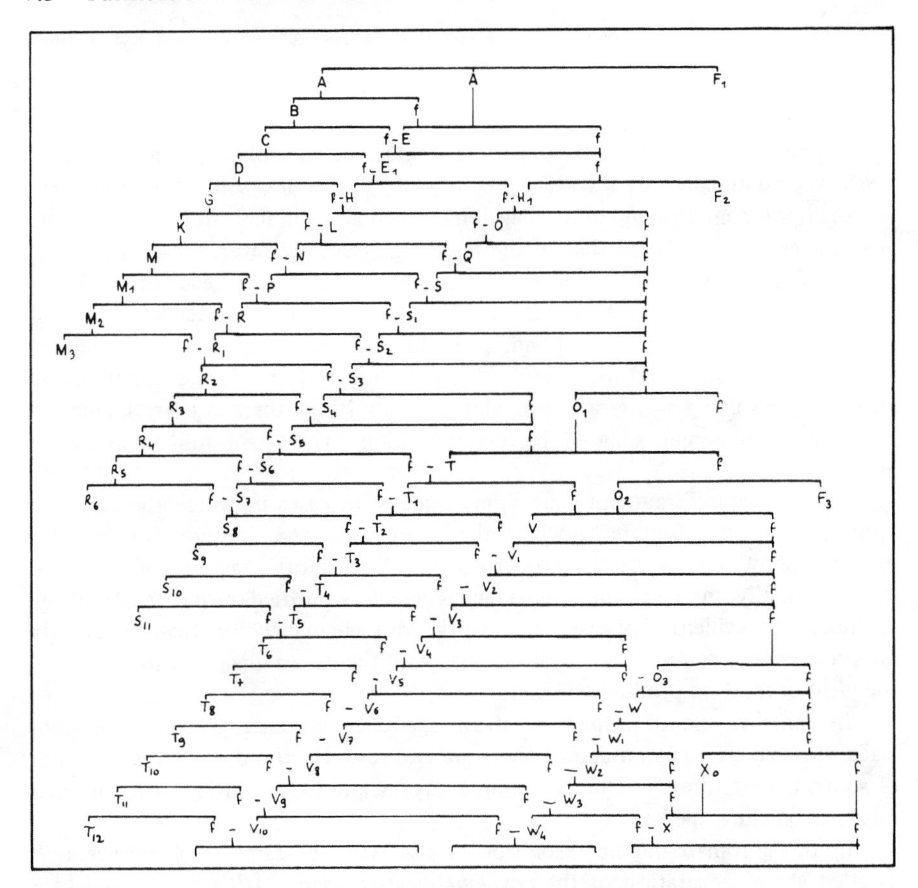

Figure 3 Part of the fractionation of the strychnine salts of the hydrogen phthalates of isopropyl-*n*-amylcarbinol. R. H. Pickard and J. Kenyon, *J. Chem. Soc.*, 1912, **101**, 620. Reproduced by permission of the Royal Society of Chemistry.

$[\alpha]_D^{25} = +16.5$ to 17.2°, 30 g acid with $[\alpha]_D^{25} = +14.2^\circ$, and 18 g acid with $[\alpha]_D^{25} = -14.4^\circ$, without having attained optical purity.

Examples in which 10 or more recrystallizations of salts are required are not rare.[17,18] However, we suggest that, when a given enantiomeric purity has been attained after a reasonable number of recrystallizations (three to four), a preferred strategy is to pursue the separation not on the diastereomers but rather on the enantiomers themselves if these are crystalline (which is not the case for the phthalates of Pickard and Kenyon or the acid of Freudenberg and Lwowski). We return to this possibility in Section 7.6.2.

7.3.5 Use of Optically Impure Resolving Agents

There is a widespread presumption among organic chemists who carry out resolutions that the enantiomeric purity of a substance resolved by the crystallization of diastereomers may not exceed that of the resolving agent employed.[19] In reality, this presumption is incorrect.

Suppose that in the resolution of compound *dlA*, resolving agent *lB* is used which is contaminated by a certain quantity of *dB*. The less soluble diastereomer, say *dAlB*, will contain some of its enantiomer *lAdB* and will have the same aggregate enantiomeric purity P_0 as that of the resolving agent. However, the enantiomeric purity of the crystals isolated from the original mixture will be higher or lower than P_0 according to whether this value is greater or smaller, respectively, than the enantiomeric purity corresponding to the eutectic of the solubility diagram for the enantiomer mixture (*dAlB*, *lAdB*).[20] Since, after all, the enantiomeric purities of resolving agents are generally close to 100%, there is every likelihood that the diastereomer isolated by crystallization may attain full enantiomeric purity.

The preceding argument may apply equally to cases in which the resolving agent is chemically impure or where the racemate is itself impure. One can, for example, show, by use of the properties of the solubility diagrams of mixtures, that an initially impure racemic substance is generally purified during its resolution. It should be evident, however, that work with chemically or enantiomerically impure compounds may lead to difficulties with their crystallization. In any event, the yield of resolved product will be lower.

In contrast, use of impure resolving agents in chromatographic resolutions leads to a less clear-cut outcome. There are two cases to consider: (1) resolution of diastereomer mixtures on achiral columns or layers, and (2) resolution of enantiomer mixtures on chiral media.

In the chromatographic resolution of *dlA* with the same resolving agent *lB* as cited above, separation of the *covalent* diastereomers *dAlB* (say n_+) and *lAlB* (say p_-) may be achieved on an achiral stationary phase so as to afford a given separation. The presence of some *dB* in the dominant *lB* leads to the formation of n_- and p_+ diastereomers that have identical retention characteristics to their enantiomers n_+ and p_- (above). Hence here, enantiomerically impure resolving agents will lead to enantiomerically impure separated diastereomers. If the latter are oils or liquids, further purification will require cleavage to enantiomers whose crystallization (as solid derivatives if need be) is subject to the phase properties alluded to earlier. If the diastereomers are crystalline, then crystallization may be invoked to purify the mixture as described above.

On the other hand, the chromatographic resolution of enantiomers, for example, *dlA*, on a chiral medium (column or layer) containing a single optically active substance *lB* may be achieved through apparent formation of *dissociable* diastereomers, that is, *p* and *n* complexes which have different stability constants[21] and consequently have differing retention characteristics. The chromatographic treatment of *A* on columns containing *B* is described by Fig. 4.

The lack of separation of *dA* and *lA* on columns containing racemic stationary phase *dlB* stems from the ability of the transient diastereomers to dissociate. Enantiomers *dA* and *lA* experience the same average environment (diastereomeric interaction), and identical retention behavior is the result.[22] Enantiomerically

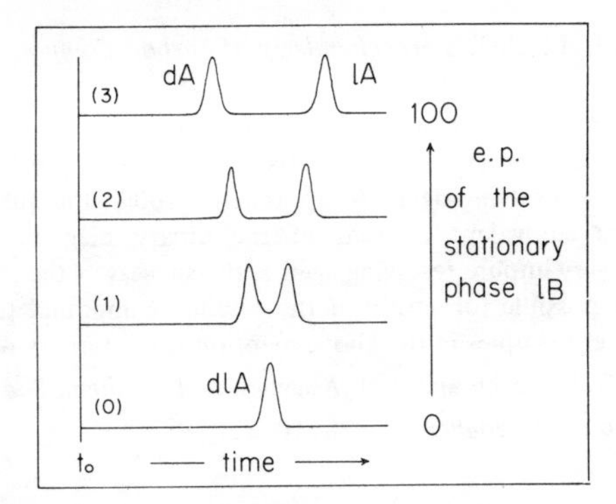

Figure 4 Chromatographic separation of enantiomers on a chiral medium.

impure *lB* affords intermediate behavior since the environments experienced by *dA* and *lA* are not completely averaged out.

Therefore, provided that separation in optimal case 3 of Fig. 4 is adequate, and that the enantiomeric purity of the chiral stationary component is not too greatly reduced then, as suggested by case 2, separation of *dA* and *lA* without contamination may still be achieved on enantiomerically impure columns.*

REFERENCES AND NOTES 7.3

1 S. W. Kantor and C. R. Hauser, *J. Am. Chem. Soc.*, 1953, **75**, 1744.

2 S. H. Wilen, R. Davidson, R. Spector, and H. Stefanou, *J. Chem. Soc. D*, 1969, 603.

3 M. R. Uskoković, D. L. Pruess, C. W. Despreaux, S. Shiuey, G. Pizzolato, and J. Gutzwiller, *Helv. Chim. Acta*, 1973, **56**, 2834.

4 W. A. McBlain and F. H. Wolfe, *Tetrahedron Lett.*, 1975, 6351.

5 W. J. Gensler and H. N. Schlein, *J. Am. Chem. Soc.*, 1956, 78, 169.

6 W. R. Brode and I. J. Wernert, *J. Am. Chem. Soc.*, 1933, **55**, 1685.

7 W. M. Colles and C. S. Gibson, *J. Chem. Soc.*, 1928, 100.

8 W. R. Brode and M. W. Hill, *J. Org. Chem.*, 1948, **13**, 191.

9 A. Fredga, *Ark. Kemi*, 1947, **24B**, 1.

10 D. R. Palmer, German Offen., 2,147,620 (1972) (to Beecham Group Ltd.). *Chem. Abstr.*, 1972, **77**, 34938s.

11 E. Rimbach, *Ber.*, 1899, **32**, 2385.

12 A. Ladenburg, *Liebigs Ann. Chem.*, 1908, **364**, 227.

13 G. Bettoni, R. Perrone, and V. Tortorella, *Gazz. Chim. Ital.*, 1972, **102**, 196.

14 R. S. Tipson, *Anal. Chem.*, 1950, **22**, 628.

15 R. H. Pickard and J. Kenyon, *J. Chem. Soc.*, 1912, **101**, 620.

16 K. Freudenberg and W. Lwowski, *Liebigs Ann. Chem.*, 1955, **597**, 141.

17 P. W. B. Harrison, J. Kenyon, and J. R. Shepherd, *J. Chem. Soc.*, 1926, 658.

18 D. J. Cram, *J. Am. Chem. Soc.*, 1952, **74**, 2152.

* This analysis is formally similar to that which obtains for the determination of enantiomeric purity by nmr spectroscopy in chiral solvents. See Section 7.5.

19 See, for example, E. L. Eliel, *Stereochemistry of Carbon Compounds*, McGraw–Hill, New York, 1962, p. 49.

20 This is a simplified way of reasoning. Strictly speaking, one would have to consider a quaternary phase diagram of reciprocal salts (*dAlB*, *dAdB*, *lAdB*, *lAlB*). As the enantiomeric purity of the resolving agent *lB* approaches 100%, the quaternary diagram representation becomes equivalent to that of the binary diagram of pure diastereomers (*dAlB*, *lAlB*). For very impure resolving agents, the success of the resolution is not certain. It depends on the possible formation of the racemic compound [*dAdBlAlB*] and on the area which the latter occupies in the above-mentioned quaternary diagram.

21 See, for example, B. L. Beach and R. J. Angelici, *J. Am. Chem. Soc.*, 1969, **91**, 6296.

22 G. R. Sullivan, *Topics Stereochem.*, 1978, **10**, 320.

7.4 RECOVERY OF ENANTIOMERS FROM DIASTEREOMERS

Except in the case of resolution by direct crystallization (Chapter 4), the enantiomers are finally obtained only following the cleavage of the diastereomers whose separation was the basis of the resolution. In some cases, the return to enantiomers during a resolution may be the most direct way of increasing the enantiomeric enrichment through further crystallizations. It may also be the simplest way of monitoring the progress of a resolution.

Diastereomer cleavage can take place under a wide variety of conditions. Nevertheless, these reactions must aim to satisfy the following requirements: (a) they must be simple, selective, and occur in high yield; (b) they must be nonracemizing; and (c) they should allow the resolving agent to be recovered, especially when the latter is rare and costly.

7.4.1 Decomposition of Diastereomeric Salts

The decomposition of diastereomeric salts is based on an obvious principle: reaction with an achiral acid or base and separation of the new components of the resulting mixture (Scheme 1).

BH^+A^- (Diastereomeric salt)

(a) $\xrightarrow{HCl}$ HA (*Free acid*) + BH^+Cl^- (*Base hydrochloride*)

(b) $\xrightarrow{NaOH}$ Na^+A^- (*Acid salt*) + B (*Free base*) + H_2O

In these routine operations so familiar to organic chemists, the choice of experimental conditions is mostly a matter of common sense. The controlling variables are solubilities of the various components present in water and in typical extraction solvents.

(a) Decomposition by mineral acids

The decomposition of a salt by mineral acid (dilute HCl or H_2SO_4) is indicated whenever the hydrochloride or the sulfate of the base is soluble in water in which the acid HA is insoluble. Acid HA, which may be either the desired product or the resolving agent, is *extracted*[1] with an appropriate solvent (ether, ethyl acetate, chloroform,* etc.), or it is *filtered* if crystalline.[2] Note that, in order to avoid precipitation of the amine mineral salt BH^+, X^- (hydrochloride, sulfate, and so on) as a consequence of the common ion effect, it is never desirable to decompose a salt with acid which is too concentrated or to use too large an excess. Use of 1.5 molar equivalent of N or $2N$ acid generally suffices.

Occasionally, nonmineral acids (and bases) find use in the cleavage of diastereomeric salts. For example, the α-(1-naphthyl)ethylamine salt of *cis*-2-methylglycidic acid is conveniently cleaved and isolated quantitatively in one operation by treatment with methanesulfonic acid in ether.[3]

(b) Decomposition by inorganic bases

The decomposition of diastereomeric salts may be carried out equally well with base (NaOH, KOH, Na_2CO_3, NH_3, etc.). This technique is recommended, in particular, whenever the hydrochloride or sulfate salt of the amine is relatively insoluble in water, a situation found with several of the common resolving agents such as dehydroabietylamine, α-(1-naphthyl)ethylamine, quinidine, and strychnine. Table 1 lists some solubility data which are useful in this regard. The free amine is either extracted[4] with chloroform or ether, or filtered[5] if it is crystalline. This process assumes that the salt of the acid HA used (NaA, KA, NH_4A) is water soluble.

Table 1 Solubilities of the most important basic resolving agents and their inorganic salts

	Free Bases		Inorganic Salts in Water, at room temp.	
Name	Chloroform	Diethyl Ether	Hydrochloride	Sulfate
Strychnine	1 g/5 mL		1 g/40 mL	1 g/35 mL
Brucine	1 g/5 mL		Soluble	1 g/75 mL
Quinine	1 g/1.2 mL		1 g/16 mL	1 g/9 mL
Quinidine	1 g/1.6 mL		1 g/60 mL	1 g/90 mL
Cinchonine	1 g/110 mL		1 g/20 mL	1 g/65 mL
Cinchonidine	Soluble		1 g/25 mL	1 g/70 mL
Morphine	Insoluble		1 g/17.5 mL	1 g/15.5 mL
Dehydroabietylamine	Soluble	Soluble	Slightly sol.	
threo-1-*p*-Nitrophenyl-2-amino-1,3-propanediol			Soluble	
Ephedrine	Soluble	Soluble	1 g/3 mL	1 g/1.2 mL
Deoxyephedrine	Soluble	Soluble	Soluble	
Amphetamine	Soluble	Soluble	Soluble	
α-Methylbenzylamine	Soluble	Soluble	Soluble	
α-(1-Naphthyl)ethylamine	Soluble	Soluble	Slightly sol.	

* Certain amine hydrochlorides are soluble in chloroform. Therefore, use of this extraction solvent is to be avoided if possible.

In most cases, at least one of the variants (*a*) or (*b*) is usable; often, both are. It is useful to know the peculiarities of some of the resolving agents in order to avoid difficulties at the point where their salts must be decomposed. We have already mentioned the low solubility in water of the hydrochlorides of several basic resolving agents. Conversely, as a free base, cinchonine, for example, is practically insoluble in chloroform (as well as other extraction solvents). Consequently, when cinchonine salts are decomposed with base, the free base is preferably isolated by filtration, particularly if the scale is relatively large.

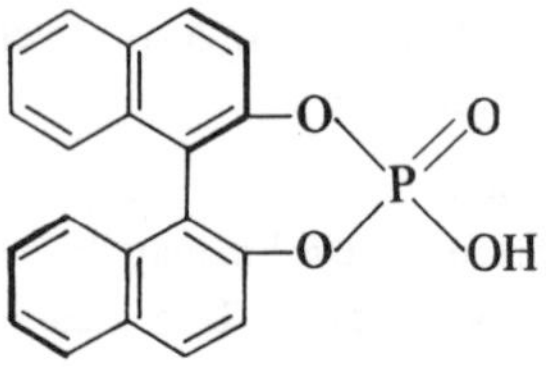

O O P OH O

1

Most acidic resolving agents do not pose special problems. Binaphthylphosphoric acid, **1**, however, merits special mention. This acid is insoluble in water as well as in the usual organic solvents. Its inorganic salts are also rather insoluble in water. The decomposition of its diastereomeric salts is carried out preferably with HCl in water or in an aqueous alcohol. The precipitated acid is filtered and the amine hydrochloride remains in solution.[6] The amine is finally isolated by addition of sodium hydroxide followed by either filtration or extraction depending upon the circumstance. Nevertheless, this procedure may be difficult to use if the amine hydrochloride is insoluble under these conditions.

Another peculiarity of acid **1** is that its sodium salt is soluble in ethanol. This property may be profitably put to use in the resolution of amino acids by this reagent. For example, *o*-tyrosine binaphthylphosphate (*p* salt) is decomposed as follows:[7] An ethanol suspension of the salt is treated with one equivalent of 6*N* sodium hydroxide. The freed amino acid is insoluble in alcohol and crystallizes while the sodium salt of **1** remains in solution.

(c) Decomposition by alumina

An interesting variant of method (*b*) consists of the decomposition of the diastereomeric salt by passage through a column of basic alumina.[8] The acid is retained by the alumina and the freed base is recovered by simple elution. This process is indicated when the desired amine is fragile or is soluble in water.

(d) Decomposition by ion exchange resins

Finally, the problems resulting from the water solubility of some organic compounds which prevents their separation from inorganic salts formed in the course of processes of type (*a*) or (*b*) may be overcome by use of ion exchange resins.[9,10] Here, too, there are a number of variants. The two principal cases are illustrated in Scheme 2.

SCHEME 2

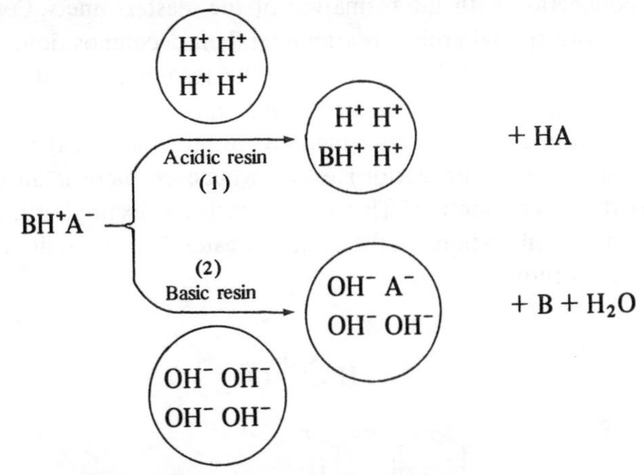

1 Passage of a solution of a diastereomeric salt BH^+A^- in a polar solvent (water, alcohol, acetone, etc.) through a *cation exchanger* in the H^+ form liberates acid HA in the solution, while the base is retained on the resin. The acid is isolated by simple evaporation of the solvent and the base may be subsequently recovered in the form BH^+ by treating the resin with mineral acid.[10,11]

2 Conversely, passage of the diastereomer salt solution through an *anion exchanger* in the OH^- form liberates base B in the solution while the acid is retained on the resin. The latter may be recovered in the form A^- by washing the resin with an inorganic base.

The operations described immediately above may be carried out either by elution of a column of resin or by stirring a solution (or suspension) of the salt to be decomposed in the presence of an excess of resin which is subsequently separated by filtration. Many commercial resins are useful in these separations. However, as a consequence of the relatively large size of the ions involved, it is preferable to use resins having a low degree of crosslinking (e.g., 2%) such as Dowex 50 W-X2 (H^+) or Dowex 1-X2 (OH^-) or their equivalents in other brands.[12]

Ion exchange resins may be used to good advantage even in cases in which methods (*a*) and (*b*) are applicable. Also, they are equally useful in small and large-scale resolutions. In particular, it is often quite convenient to use resins to decompose salts obtained in very small quantity during preliminary resolution trials.

The use of ion exchange resins is particularly indicated in the following cases: (1) recovery of water-soluble resolving agents such as the camphorsulfonic acids,[11,13] or even of (—)-tartaric acid whose relative rarity may justify regeneration after use;[14] (2) resolution of complex ions; (3) resolutions of salts (e.g., quaternary ammonium ions).

7.4.2 Decomposition of Covalent Diastereomers

The methods applicable to the cleavage of covalent diastereomers are more or less specific for each category of resolving agent. They have already been examined in

Section 5.2 in connection with the formation of the diastereomers. Consequently, we deal here only with special problems attending their decomposition.

Diastereomeric *esters*, which are useful in the resolution of alcohols, thiols, phenols, and occasionally acids, most often are simply saponified so as to yield a mixture of basic and neutral products which are, in principle, easily separable. The risk of racemization (or epimerization) exists whenever there is an asymmetric center alpha to the ester function. This is observed, for example, in connection with the difficult saponification of the menthyl ester **2** with sodium hydroxide in methoxyethanol at reflux.[15]

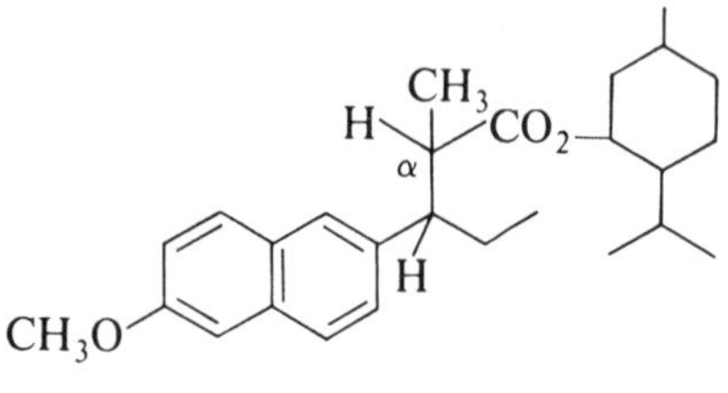

2

There is also a risk of racemizing – or destroying – the neutral part of a diastereomeric allylic or benzylic ester or, in general, of fragile substrates. A somewhat *a priori* unexpected example is the camphanate of phenol **3** whose saponification of flavan **4** is accompanied by substantial racemization.[16]

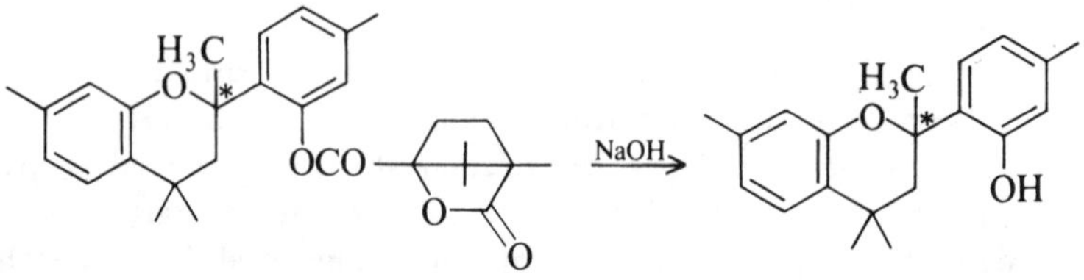

3 **4**

In most cases, racemization may be avoided by carrying out the ester cleavage with lithium aluminum hydride. However, this gives rise to a mixture of two alcohols which must still be separated. With camphanate esters, the reduction product of the acid moiety is water soluble, which may facilitate the isolation of the desired product. However, lithium aluminum hydride reduction – or cleavage of the diastereomeric ester by reaction with a Grignard reagent – is accompanied by a major disadvantage: the resolving agent is destroyed.

While it does not, properly speaking, deal with diastereomers, the regeneration of resolved alcohols from their *hydrogen phthalates* poses the same problems as those just mentioned. Most optically active hydrogen phthalates of primary and secondary alcohols may be saponified without racemization; however, phthalates of secondary allylic or benzylic alcohols which do racemize during alkaline hydrolysis must be cleaved by lithium aluminum hydride to avoid this problem.[17,18]

Saponification of hydrogen phthalates of tertiary alcohols is generally impracticable; regeneration of the resolved alcohol may be carried out by reduction with $LiAlH_4$, by addition of ethylmagnesium bromide, or by action of sodium ethoxide on the acid phthalate or even on the alkaloid salt itself.[19,20]

While the separation of diastereomeric *amides* is often quite easy to carry out, their hydrolysis may constitute a major obstacle in the obtainment of the acids or amines whose resolution they mediate in principle. Certain amides actually resist cleavage under the most drastic conditions. Carpino has cited the case of amide **4**, for example, which resists hydrolysis equally with 48% HBr in acetic acid (24 hour reflux) or with KOH in ethylene glycol.[21]

To overcome this difficulty, one sometimes has recourse to the pyrolytic rearrangement of the *N*-nitroso amide. This reaction, studied by White,[22] leads to an ester which must still be saponified.[23] This long process permits the regeneration only of the acid part of the amide; the amine is transformed into an alcohol:

$$\mathrm{R{-}NH{-}CO{-}R'} \rightarrow \mathrm{R{-}\underset{\displaystyle |}{N}(NO){-}CO{-}R'} \rightarrow \mathrm{R{-}O{-}CO{-}R'} + \mathrm{N_2}$$

In spite of the length of the process, there are several recorded applications.[24,25]

Nevertheless, in certain cases the hydrolysis of amides takes place under relatively mild conditions. In particular, Helmchen and his co-workers have found this to be true for certain hydroxyamides (see Section 5.2.7),[26] and this may well increase the usefulness of amides in resolutions.

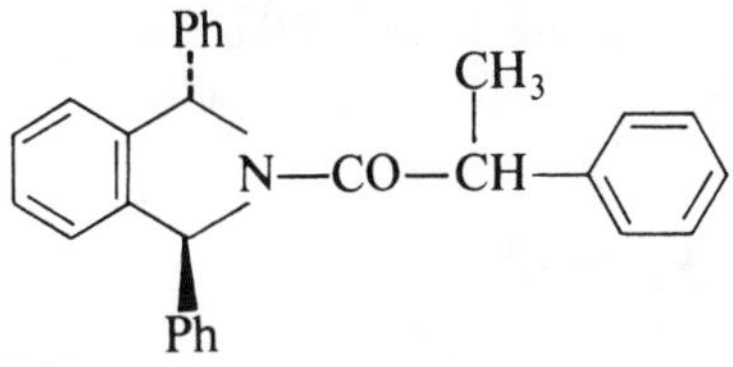

Ph Ph CH$_3$ N—CO—CH— N-CO-O—

4 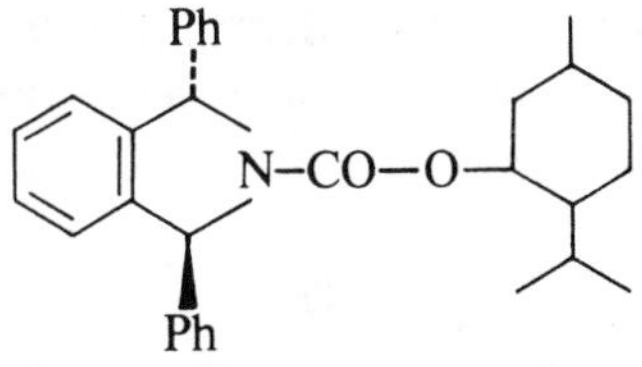 **5**

Diastereomeric carbamates, which also are relatively easily separated, have the advantage that they are more easily cleaved than amides. For example, hydrolysis of **5** to the resolved amine is carried out by passage of hydrogen bromide gas through its solution in boiling acetic acid for 15 minutes.[21]

The regeneration of an alcohol resolved through a carbamate (e.g., **6**) may be effected by transesterification with sodium ethoxide. The by-product is ethyl *N*-1-(1-naphthyl)ethyl carbamate, **7**. The latter is easily reconvertible into optically active isocyanate **8** which may be reused in subsequent resolutions:[27]

Access to optically active ketones, resolved via covalent diastereomers of type **9** and **10**, is attended by the risk of racemization (or epimerization) when these contain asymmetric centers alpha to the carbonyl group. It is, therefore, always

CH_3–*CH–NH–COOR* (1-naphthyl) **6** $\xrightarrow{\text{NaOEt}}$ CH_3–*CH–NH–COOEt (1-naphthyl) **7** + R*OH

7 $\xrightarrow{HSiCl_3}$ CH_3–*CH–N=C=O (1-naphthyl) **8**

9 C*>C=N–R*

10 C*>C<(X_1–R*)(X_2–R*)

→ C*>C=O

desirable to carry out the hydrolysis under the least acidic or basic conditions possible so as to avoid the enolization of the desired ketone. To this end, *exchange* methods employed with menthydrazones, for example, merit wider use than has heretofore been the case.[28,29] This consists in the treatment of the ketone derivative in weakly acidic medium with a large excess of acetone or formaldehyde:

C*>C=N–NH–C(=O)–R* + CH_3–C(=O)–CH_3 $\xrightarrow{\text{Phthalic Acid}}$ C*>C=O + CH_3–C(=N–NHCOR*)–CH_3

When a resolution is incorporated into a multistep synthesis, it may be desirable to eliminate the auxiliary chiral agent at the most opportune stage or, even better, to transform the cleavage reaction into an element of the synthesis. This possibility is illustrated in an elegant manner by the following sequence devised by Woodward and co-workers:[30] The separated menthyl ester **11** is first transformed into **12**. Cleavage of the menthyl group of **12** by ozonolysis leads to an aldehyde intermediate which, by intramolecular Wittig reaction, leads to the cyclized product **13**.

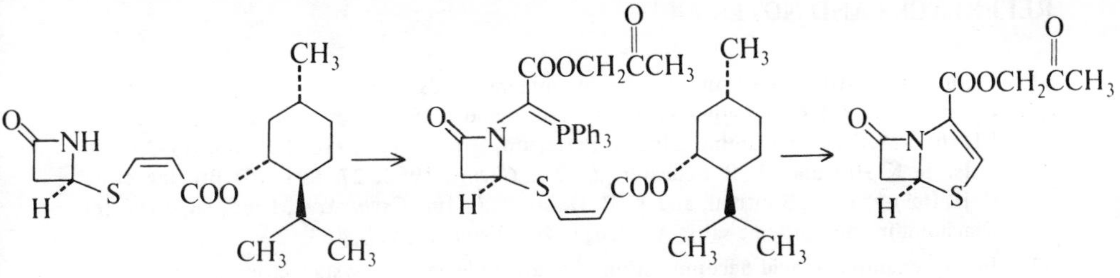

11 12 13

7.4.3 Recovery and Purification of Resolving Agents

The costs of resolving agents (Section 5.1) mandate that virtually all of these be recovered for possible reuse whenever more than just a few grams are used in a resolution. Tartaric acid is one of the few exceptions to this general rule (see, however, the comment on page 399).

The recovery of a resolving agent usually presents no special problem independent of those which attend the associated recovery of enantiomer(s) from the corresponding diastereomers (Section 7.4.2). And resolving agents can be reused essentially indefinitely provided that they do not significantly degrade, that is, become too impure during the course of resolutions and their workups. An apparent exception to such reuse is the alkaloid quinine, which has a tendency to deteriorate on attempted recovery.[31]

Purification of recovered resolving agents prior to reuse may be essential to avoid complications. One or more cycles of dissolution and precipitation with acid and base followed by drying in vacuo as a rule should suffice; see, for example, refs. 32 through 36 on the purification of brucine in this way or by simple recrystallization. In particular, drying of the recovered material should be carried out carefully to avoid problems in subsequent resolutions as a consequence of solvation (see Section 5.1.12).

A number of amine resolving agents are conveniently stored as salts. They must then be converted to free amines prior to use.[37,38] This is particularly indicated for liquid primary amines which rapidly absorb carbon dioxide on exposure to air forming solid carbonates. It may, therefore, be necessary to decompose the carbonate and to distil the sample just before use.[39] In particular, α-(1-naphthyl)ethylamine should be stored as its hydrochloride or, better, as its phenylacetate salt which is a conglomerate. Other useful salts are α-methylbenzylamine hydrogen sulfate and deoxyephedrine hydrochloride (both are conglomerates) and dehydroabietylamine acetate.

Water-soluble resolving agents such as the camphorsulfonic acids require different treatment in their recovery. In particular, it is convenient to handle such acids as their ammonium salts. The acid may then be isolated from said salt by means of an ion exchange resin.[11]

REFERENCES AND NOTES 7.4

1 Examples of decomposition of salts by mineral acids followed by extraction of the organic acid: (a) Cinchonine salts: A. Fredga and T. Svensson, *Ark. Kemi*, 1966, **25**, 81. (b) α-Methylbenzylamine salts: K. Petterson, *Ark. Kemi*, 1956, **10**, 283. (c) Quinine salts: R. K. Hill and W. R. Schearer, *J. Org. Chem.*, 1962, **27**, 921. (d) Brucine salts: C. H. DePuy, F. W. Breitbeil, and K. R. DeBruin, *J. Am. Chem. Soc.*, 1966, 88, 3347. (e) Cinchonidine and brucine salts: A. Fredga, *Ark. Kemi*, 1955, 8, 463.

2 For an example of acid decomposition of a salt, followed by crystallization of the organic acid substrate, see the resolution of binaphthylphosphoric acid: J. Jacques, C. Fouquey, and R. Viterbo, *Tetrahedron Lett.*, 1971, 4617. Also see ref. 6.

3 E. J. Corey, E. T. Trybulski, L. S. Melvin, Jr., K. C. Nicolaou, J. A. Secrist, R. Lett, P. W. Sheldrake, J. R. Falck, D. J. Brunelle, M. F. Haslanger, S. Kim, and S. -e. Yoo, *J. Am. Chem. Soc.*, 1978, **100**, 4618.

4 Decomposition of a salt by base followed by extraction of the amine substrate or resolving agent: (a) M. J. Kalm, *J. Org. Chem.*, 1960, **25**, 1929. (b) H. Biere, C. Rufer, H. Ahrens, O. Loge, and E. Schröder, *J. Med. Chem.*, 1974, **17**, 716. (c) Steam distillation combined with extraction of the amine: A. W. Ingersoll and F. B. Burns, *J. Am. Chem. Soc.*, 1832, **54**, 4712. (d) Resolution with dehydroabietylamine: G. Belluci, G. Berti, A. Borraccini, and F. Macchia, *Tetrahedron*, 1969, **25**, 2979. (e) Resolution with strychnine: W. L. F. Armarego and E. E. Turner, *J. Chem. Soc.*, 1956, 3668.

5 Decomposition of salts by base followed by crystallization of the amine substrate or resolving agent: (a) B. F. Tullar, L. S. Harris, R. L. Perry, A. K. Pierson, A. E. Soria, W. F. Wettereau, and N. F. Albertson, *J. Med. Chem.*, 1967, **10**, 383. (b) R. Kuhn and P. Goldfinger, *Liebigs Ann. Chem.*, 1929, **470**, 183. (c) Resolution with strychnine: M. Matell, *Ark. Kemi*, 1951, **3**, 129.

6 R. Viterbo and J. Jacques, (to Richardson-Merrel S.p.A.) German Offen., 2,212,660 (1972) and British Patent, 1,360,946 (1974); *Chem. Abstr.*, 1973, **78**, 43129b. This procedure can be found in: P. Newman, *Optical Resolution Procedures for Chemical Compounds*, Vol. I, Optical Resolution Information Center, Manhattan College, New York, 1978, p. 556; see also ref. 2.

7 A. Garnier, A. Collet, L. Faury, J. P. Albertini, J. M. Pastor, and L. Tosi, to be published.

8 W. Aschwanden, E. Kyburz, and P. Schönholzer, *Helv. Chim. Acta*, 1976, **59**, 1245.

9 J. F. Tocanne and C. Asselineau, *Bull. Soc. Chim. Fr.*, 1965, 3346.

10 F. W. Bachelor and G. A. Miana, *Can. J. Chem.*, 1967, **45**, 79.

11 S. Gronowitz, I. Sjögren, L. Wernstedt, and B. Sjöberg, *Ark. Kemi*, 1965, **23**, 129.

12 For a review on organic ion exchanges, see P. R. Brown and A. M. Krstulovic, *Separation and Purification*, 3rd ed., Vol. XII of *Techniques of Chemistry*, E. S. Perry and A. Weissberger, Eds., Wiley, New York, 1978, Chapter IV.

13 T. Kunieda, K. Koga, and S. Yamada, *Chem. Pharm. Bull.*, 1967, **15**, 337; see footnote p. 339.

14 G. B. Kauffman and R. D. Myers, *J. Chem. Educ.*, 1975, **52**, 777.

15 R. Gay and A. Horeau, *Tetrahedron*, 1959, 7, 90.

16 A. Collet, unpublished results.

17 M. P.Balfe and J. Kenyon, *Nature*, 1941, **148**, 196.

18 J. English, Jr., and V. Lamberti, *J. Am. Chem. Soc.*, 1952, **74**, 1909.

19 A. G. Davies, J. Kenyon, and L. W. F. Salame, *J. Chem. Soc.*, 1957, 3148.

20 B. Bielawski and A. Chrzaszczewska, *Lodz Tow. Nauk Wydz. III Acta Chim.*, 1966, **11**, 105. *Chem. Abstr.*, 1967, 67, 2837.

21 L. A. Carpino, *Chem. Comm.*, 1966, 858.
22 E. H. White, *J. Am. Chem. Soc.*, 1955, 77, 6008.
23 G. Helmchen and V. Prelog, *Helv. Chim. Acta*, 1972, **55**, 2599.
24 G. Haas and V. Prelog, *Helv. Chim. Acta*, 1969, **52**, 1202.
25 S. Rendic, V. Sunjic, F. Kajfez, N. Blazevic, and T. Alebic-Kolbah, *Chimia*, 1975, **29**, 170.
26 G. Helmchen, G. Nill, D. Flockerzi, and M. S. K. Youssef, *Angew. Chem.*, 1979, **91**, 65. *Angew. Chem. Int. Ed.*, 1979, **18**, 63.
27 W. H. Pirkle and J. R. Hauske, *J. Org. Chem.*, 1977, **42**, 2781.
28 E. Touboul and G. Dana, *Bull. Soc. Chim. Fr.*, 1974, 2269.
29. K. Schlögl and M. Fried, *Monatsh. Chem.*, 1964, **95**, 558.
30 H. R. Pfaender, J. Gosteli, and R. B. Woodward, *J. Am. Chem. Soc.*, 1979, **101**, 6306.
31 E. L. Eliel, *Stereochemistry of Carbon Compounds*, McGraw-Hill, New York, 1962, p. 51.
32 F. Saunders, *J. Am. Chem. Soc.*, 1928, **50**, 1231.
33 A. I. Vogel, *Practical Organic Chemistry*, 3rd ed., Longmans, Green and Co., London, 1957, p. 507.
34 C. H. Depuy, F. W. Breitbeil, and K. R. DeBruin, *J. Am. Chem. Soc.*, 1966, **88**, 3347.
35 B. E. Leach and J. H. Hunter, *Biochem. Prep.*, 1958, **3**, 115. See footnote 8.
36 An alternative method of purification is recrystallization from aqueous acetone.
37 J. Jacobus and T. B. Jones, *J. Am. Chem. Soc.*, 1970, **92**, 4583.
38 K. Oki, K. Suzuki, S. Tuchida, T. Saito, and M. Kotake, *Bull. Chem. Soc. Jpn.*, 1970, **43**, 2554.
39 N. Kornblum, W. D. Gurowitz, J. O. Larson, and D. E. Hardies, *J. Am. Chem. Soc.*, 1960, **82**, 3099.

7.5 MONITORING ENANTIOMERIC PURITY

A variety of methods are available for the determination of enantiomeric purity. The choice of method may be determined by the specific case and/or by the instrumentation available. The use of the most precise analytical methods may be justified, for example, in connection with an asymmetric synthesis in which the most precise value of the enantiomeric or optical purity – in particular, low and very high values – may be needed even if this analysis is long and complicated. On the other hand, the simplest methods generally suffice for monitoring a resolution at its beginning and during its course even if these methods are not the most precise. More sophisticated and precise methods may be required for the final determination of purity of diastereomers and/or enantiomers.

A number of reviews dealing with the determination of enantiomeric purity have been published.[1–4] Consistent with the aims of this book, we shall stress those methods which are better suited to resolutions, without pretending to be exhaustive in our survey.

When the specific rotation of a pure enantiomer, $([\alpha]_\lambda^T)_{max}$, is known, the measurement of the rotation of a partially resolved mixture $([\alpha]_\lambda^T)_x$ carried out *under identical conditions* (solvent, concentration, temperature and wavelength) immediately provides access to its *optical purity*:

$$p_x = \frac{([\alpha]^T_\lambda)_x}{([\alpha]^T_\lambda)_{max}}$$

While this method is quite easy to apply, its precision must nevertheless not be overestimated for several reasons. First of all, it is relatively easy to show that the precision of polarimetric measurements carried out with normal care even with a photoelectric polarimeter rarely exceeds 1 to 2%. Second, the presence of impurities, for example, solvent residues, also frequently constitutes a supplementary source of error in this type of measurement. Moreover, one must be quite certain that the $[\alpha]^T_\lambda$ value used is actually the maximum value of the optical rotation of the compound in question. This is, in fact, far from always the case; and, when determined in solution, as pointed out by Wynberg, "extrapolation of optical rotations, itself dependent on concentration, cannot give entirely exact data."[5] Finally, additional uncertainty can result as a consequence of the nonlinearity of the optical rotation with the composition of the enantiomer mixture $d + l$ (Horeau effect).[6] In this case, the *optical purity* $([\alpha]^T_\lambda)_x/([\alpha]^T_\lambda)_{max}$ is not equal to the *enantiomeric purity* $(d - l)/(d + l)$, which defines the *real*, that is, molar composition of the mixture. Though it would appear that this is a rare phenomenon, one must be aware of the fact that there are very few data bearing on the relationship between optical purity and enantiomeric purity, namely, on either the occurrence or absence of the effect in the general case.

Given these reservations, the comparison of the optical rotation of a partially resolved sample with that of a supposedly pure sample remains one of the simplest ways of providing information of reasonable precision about samples whose optical purities are neither very small nor very large. The validity of the values that this method may provide above the level of 95% optical purity is often more problematic, however, hence the desirability of disposing of methods specifically adapted to the analysis of nearly pure substances.

Comparison of the optical rotation of an enantiomer sample with that of the mother liquor obtained in the recrystallization of the sample also provides information which may be useful in monitoring the progress of the enantiomeric enrichment in the context of the solubility rules taken up in Chapter 3.

With the exception of one case (see p. 418), none of the methods which we now describe requires samples or knowledge of the pure enantiomers or diastereomers.

7.5.1 Analysis of Diastereomer Mixtures

There are two distinct possibilities in this group. In the first, the mixture of diastereomers being separated is itself subjected to analysis. In the second, a mixture of enantiomers of unknown purity is quantitatively transformed by reaction with a reagent specially conceived or adapted for the purpose into a mixture of diastereomers which is suitable for analysis.

(a) Direct analysis of diastereomer mixtures

Quite often it is possible to directly analyze the mixture of diastereomers during the course of a resolution without having to free the enantiomers which they contain. The composition of binary mixtures of covalent diastereomers may be obtained by any one of numerous quantitative analytical methods: gas and high-pressure liquid chromatography, nuclear magnetic resonance spectroscopy, and differential scanning microcalorimetry. Thin-layer chromatography or changes in optical rotation can also provide qualitative measures of composition which often suffice to follow the progress of a resolution.

Covalent diastereomers useful in preparative separations and directly analyzable by nmr include camphorsulfonate,[7] menthoxyacetate,[8] and camphanate[9,10] esters. Without doubt, many other compounds derived from other resolving agents are likely to lend themselves to this type of analysis which can take advantage of the numerous resources of nmr spectroscopy including solvent effects, shift reagents, ^{13}C-nmr, and so on. Also, a variety of covalent diastereomers which are separable by preparative chromatography (gc and hplc) may just as well be *analyzed* by the same techniques.[11] Among these are carbonates and carbamates derived from menthyl chloroformate,[12] carbamates derived from α-methylbenzylisocyanates, α-(1-naphthyl)ethylisocyanates,[13] and amides.[14]

When the mixture of diastereomers is crystalline, the measurement of its *fusion interval*, that is, its melting range, either through a visual method (preferably measured on a hot stage microscope) or better by differential scanning microcalorimetry, provides extremely useful information on the composition of the mixture and on the progress of the separation (see Section 5.2.6). Figure 1 illustrates the most common case corresponding to this possibility. The fusion interval of mixtures of different composition is shown in Fig. 1*a*, while the corresponding dsc traces are shown in Fig. 1*b*.

By the same token, it is frequently possible to follow the purification of *diastereomeric salts* without the need to decompose them. Though the chroma-

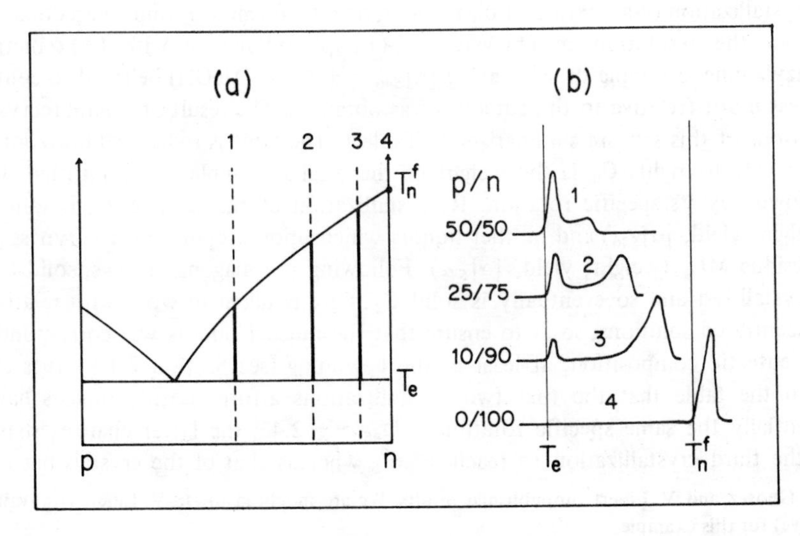

Figure 1 Fusion intervals of diastereomer mixtures (a) and the corresponding dsc scans (b).

tographic analysis of salts by simple methods would seem to be excluded, the measurement of fusion intervals is often feasible, just as for covalent diastereomers, and it provides the same information as for the latter. While it is true that many salts formed with natural alkaloids have high melting points, are solvated, and decompose on heating, the majority of salts formed from other amines (α-methylbenzylamine, ephedrine, etc.) melt normally, and their binary fusion diagrams are indistinguishable from those of covalent diastereomers (see Section 5.1.10).

The direct examination of diastereomeric salts by nmr spectroscopy is often feasible and occasionally such salt mixture may be analyzed in this way.[15-20] The nmr nonequivalence of salts may be observed in ^1H, ^{31}P, and ^{13}C spectra.[16] However, it is necessary to operate in such solvents as $CHCl_3$, CCl_4, C_6H_6, and especially pyridine in which the salts are present in the form of ion pairs or aggregates. This requirement as well as the need to maintain relatively low temperatures leads to reduced solubility which in turn affects the magnitude of nmr signals, a problem which can be overcome by use of Fourier transform spectrometers. An additional limitation is that the salt mixture must exhibit sharp peaks, singlets or doublets which are well separated from other peaks, for the analysis to be feasible. In any event, the process is sufficiently simple for routine trials.

We have already described the information that can be extracted from the measurement and calculation of the optical rotation of diastereomeric salts together with the limitations of this procedure (Section 5.1.20). More simply, recrystallization of diastereomeric salts until constant specific rotation is reached is frequently found in the literature as a way of monitoring the resolution. In most instances this method gives correct results, providing that no solid solution formation complicates the course of the crystallizations; the occurrence of such a situation may be easily detected by comparing the specific rotations of the crystals with those of the mother liquors from which they are isolated during recrystallization. The following example* is typical of the change in specific rotation during the course of the recrystallization of a mixture of diastereomeric salts forming a simple eutectic.

In the resolution of (±)-*N*-acetyl-β-(1-naphthyl)alanine with (−)-α-methylbenzylamine, a sample of salt having $[\alpha]_{546} = +4.65°$ (MeOH) believed to contain excess *n* salt (relative to the eutectic) was obtained. The result of several recrystallizations of this salt are summarized in Table 1. The table is to be read horizontally from left to right; C_0 is the weight of the original sample to be purified. It is followed by its specific rotation. Recrystallization of this sample yields solid C_1 (weight, yield, $[\alpha]_{546}$) and mother liquors which upon evaporation to dryness give a residue ML_1 (weight, yield, $[\alpha]_{546}$). Following the diagonal arrows, solid C_1 is recrystallized and so eventually is solid C_2. Care is taken to work with relatively concentrated solutions, so as to ensure that the mother liquors will correspond to the eutectic composition, at least at the beginning (see Section 7.3.1). It is clear from the table that the first two crystallizations afford mother liquors having essentially the same specific rotation $[\alpha]_{546} \sim -2.4°$; the latter changes sharply at the third crystallization to reach $+5.6°$, whereas that of the crystals has risen

* L. Ghosez and V. Libert, unpublished results. We are indebted to Dr. V. Libert (Louvain-la-Neuve) for this example.

Table 1 Purification of α-methylbenzylamine *N*-acetyl-β-(1-naphthyl)alaninate salts. Monitoring of the diastereomeric excess[a]

Salt Sample	Crystals Obtained	Recrystallization Mother Liquors
$C_0 = 14.3$ g $n > p$ $[\alpha]_{546} = +4.65°$	$C_1 = 12.8$ g (89%) $[\alpha]_{546} = +5.3°$	$ML_1 = 1.2$ g (10%) $[\alpha]_{546} = -2.35°$
$C_1 = 12.6$ g	$C_2 = 11.7$ g (93%) $[\alpha]_{546} = +6.0°$	$ML_2 = 0.6$ g (5%) $[\alpha]_{546} = -2.5°$
$C_2 = 11.7$ g	$C_3 = 11.1$ g (95%) $[\alpha]_{546} = +6.3°$	$ML_3 = 0.4$ g (3%) $[\alpha]_{546} = +5.6°$

[a] All rotations ±0.5° in methanol.

only slightly, from +6.0° to +6.3°. It may be concluded that the first two crystallizations have eliminated the eutectic and that C_2 should be in the vicinity of the pure compound, that is, within the terminal solid solution zone, if not pure. A system forming a solid solution with a large range of composition would not have given rise to such an abrupt change in the specific rotation of the mother liquors; it would only have shown a parallel variation of the rotations of crystals and their mother liquors.

(b) Analysis of enantiomer mixtures following conversion to diastereomers

Equation (1) describes the transformation of an enantiomer mixture $d + l$ of unknown composition into an easily analyzable mixture of diastereomers. Compound A is a chiral derivatizing agent (Section 5.2.7).

$$xd + (1-x)l + A \rightarrow xdA + (1-x)lA \qquad (1)$$

This process, which has been the subject of numerous studies, is necessarily a longer one that that described in (*a*) above and, in our view, would rarely need to be used to monitor the *progress* of a resolution. This in no way diminishes the utility of the process for other purposes.

The chiral derivatizing agents of greatest usage in this context are undoubtedly α-methoxyphenylacetic acid (*O*-methylmandelic acid), **1**,[1,21,22] and α-methoxy-α-trifluoromethylphenylacetic acid, **2** (Mosher's MTPA reagent).[23] A related new reagent is α-methyl-α-methoxy(pentafluorophenyl) acetic acid, **3** (MMPA).[24]

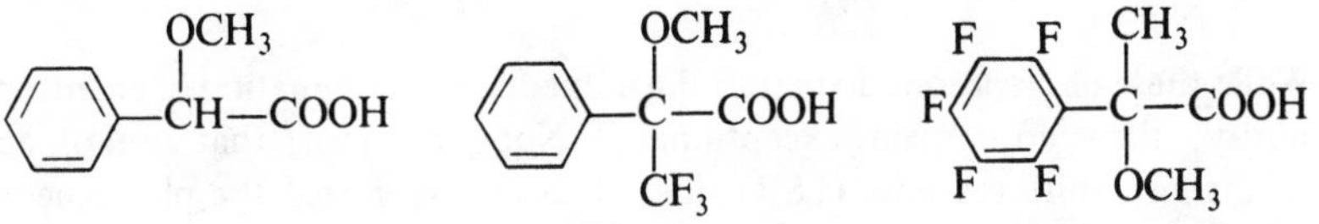

C_6H_5–CH(OCH_3)–COOH **1**

C_6H_5–C(OCH_3)(CF_3)–COOH **2**

C_6F_5–C(CH_3)(OCH_3)–COOH **3**

By means of the esters and amides prepared with these reagents, it is possible to determine the enantiomeric purity of alcohols and amines by nmr spectroscopy as well as by gas and high pressure liquid chromatography. Let us recall that the validity of such measurements implies the fulfillment of a number of conditions:[1] (a) conversion of the $d + l$ mixture into $dA + lA$ [eq. (1)] must be complete; (b) reactant A must itself be optically pure; (c) reaction (1) should provoke racemization neither in the chiral derivatizing agent nor in the substrate; and (4) the diastereomers formed should be stable to equilibration under the conditions of the analysis.

7.5.2 Enantiomeric Composition by Means of Diastereomeric Interactions

When an enantiomer mixture resides in a chiral environment (E^*), *diastereomeric interactions* ($E^* \cdots d$) and ($E^* \cdots l$) are created which may differ sufficiently in energy so as to permit the two species d and l to be separately observed and hence analyzed. Here, too, the analytical techniques of choice are nmr spectroscopy[25] and chromatography.

(a) Methods dependent on nmr spectroscopy

The diastereomeric interactions may be observed by nmr spectroscopy if they are expressed as chemical shift differences for the signals of the two enantiomers.

1 The simplest – but exceptional – case is that in which the chiral environment (E^*) is created by the enantiomers themselves. The homochiral diastereomeric interactions ($d \cdots d$ or $l \cdots l$) may be sufficiently different from the heterochiral ones ($d \cdots l$) so as to give rise to a splitting of nmr signals when both enantiomers are present in unequal amounts. Usković and his associates were the first to observe this phenomenon on dihydroquinine and some of its derivatives.[26] Harger[27] has described similar results for methylphenylphosphonamide, **4**, and its derivatives **5** and **6**:

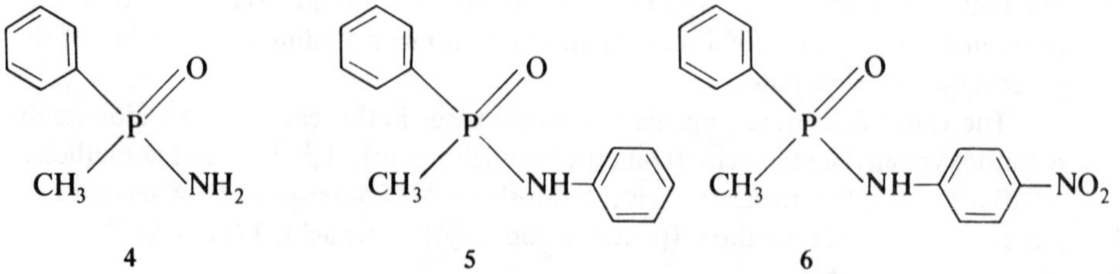

CH_3–P(=O)(Ph)–NH_2 **4** | CH_3–P(=O)(Ph)–NH–C_6H_5 **5** | CH_3–P(=O)(Ph)–NH–C_6H_4–NO_2 **6**

While these observations do permit the immediate determination of enantiomeric purity, they do remain exceptional.[28] Note, however, that use of achiral lanthanum shift reagents (LSR) does occasionally reveal the phenomenon as manifested by spectral nonequivalence.[29]

2 In 1965, Raban and Mislow postulated that an *optically active solvent* could cause the nonequivalence of the nmr spectra of enantiomer pairs.[22a] Pirkle and his collaborators were the first to observe and to use this nonequivalence in the determination of enantiomeric purity. The first observations were made on 2,2,2-trifluoro-1-phenylethanol, 7 (TFPE).[30] The ^{19}F-nmr spectra of racemic or of partially resolved **7** examined in (−)-α-methylbenzylamine solution, **8**, revealed a nonequivalence in the signals of the trifluoromethyl group.

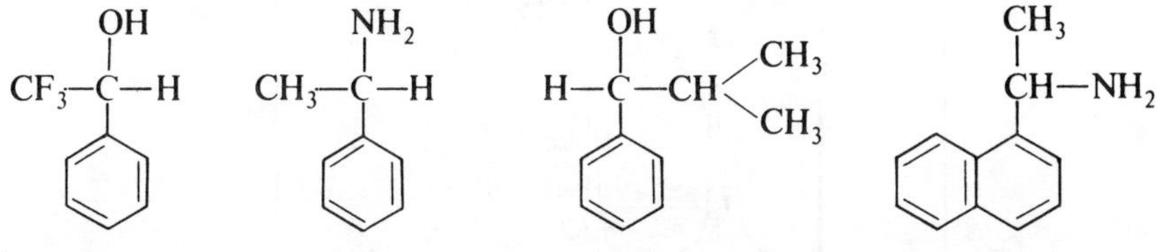

7 8 9 10

Soon thereafter the same type of nonequivalence was observed in ^{1}H-nmr spectra.[31] The signal of the benzylic proton of racemic alcohol **9** which exhibits a doublet in achiral solvents, shows two doublets of equal intensity separated by 6 Hz when the spectrum is measured in (+)-α-(1-naphthyl)ethylamine solution, **10.**

The fact that the two doublets merge into one when racemic amine **10** is the solvent implies that there is a rapid averaging on the nmr time scale of interactions between solute molecules d and l and those of the solvent S_d and S_l. An important consequence of this explanation is that the nonequivalence may be observed even if the solvent is not enantiomerically pure; the magnitude of the nonequivalence ($\Delta\Delta\delta$) is simply proportional to the enantiomeric purity of the solvent.[31]

In practice, the easily accessible resolved TFPE 7[32] has undoubtedly been the most widely used chiral solvent for this type of enantiomeric purity determination, for example, for alcohols,[33] hydroxy acids,[34] amino acids,[35] amines,[36] sulfoxides,[37,38] sulfinamides, sulfinates, sulfites, thiosulfates, phosphine, and amine oxides.[39] The observed nonequivalences $\Delta\Delta\delta$ are generally in the range of 0 to 4 Hz, which practically limits the application of this method to compounds whose nmr spectra contain sharp and well-resolved signals. For an example, see Fig. 2.

The nonequivalence of enantiomers in the presence of TFPE has also been observed in ^{31}P-nmr and ^{13}C-nmr spectra.[40,41] The same phenomenon has been observed in an achiral solvent (CCl_4 or CS_2) with spectra measured in the presence of an optically active cosolute.[42]

3 The principal deficiency of Pirkle's method lies in the degree of separation of the signals arising from the two enantiomers which is very often too small to be of use in quantitative determinations. While the use of achiral lanthanide shift reagents to magnify the separation has been envisaged,[43] the most spec-

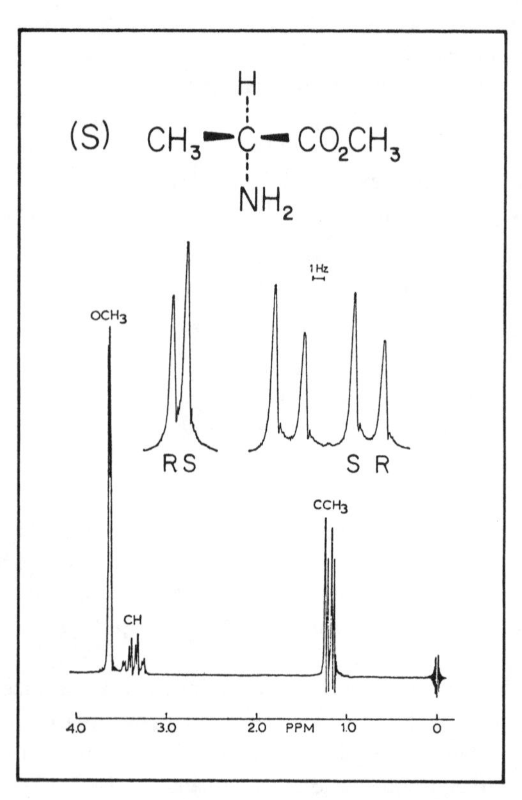

Figure 2 ^{1}H-nmr spectrum of partially resolved (S)-methyl alaninate in (R)-7. The upper traces are expanded *O*-methyl (left) and *C*-methyl (right) resonances. W. H. Pirkle and S. D. Beare, *J. Am. Chem. Soc.*, 1969, **91**, 5150. Reproduced by permission of the publisher and authors. Copyright 1969, American Chemical Society.

tacular progress in enantiomeric purity determinations by nmr techniques is due to the introduction of chiral lanthanide shift reagents (chiral LSRs) which was described nearly simultaneously by Whitesides,[44] Fraser,[45] and Goering[46] and their co-workers. The use of chiral LSRs for the determination of enantiomeric purity has been the subject of an excellent comprehensive review by Sullivan.[3]

Of the three chiral LSRs* originally described, **11**, **12**, and **13**, the last two, which are commercially available, have been the object of the largest number of applications. To these, one must add **14**,* a compound more recently described by Whitesides[47] and whose efficiency with respect to the separation of signals due to the two enantiomers is often considerably greater than that of **12** and **13**.

* **11** = tris(3-*tert*-butylhydroxymethylene-*d*-camphorato)europium(III),
12 = tris(3-heptafluorobutyryl-*d*-camphorato)europium(III),
13 = tris(3-trifluoroacetyl-*d*-camphorato)europium(III), and
14 = tris(*d,d*-dicampholylmethanato)europium(III).

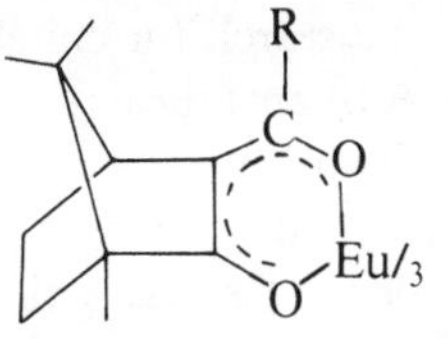

11 R = *t*-butyl

12 R = $CF_2CF_2CF_3$ Eu(hbfc)$_3$

13 R = CF_3 Eu(facam)$_3$

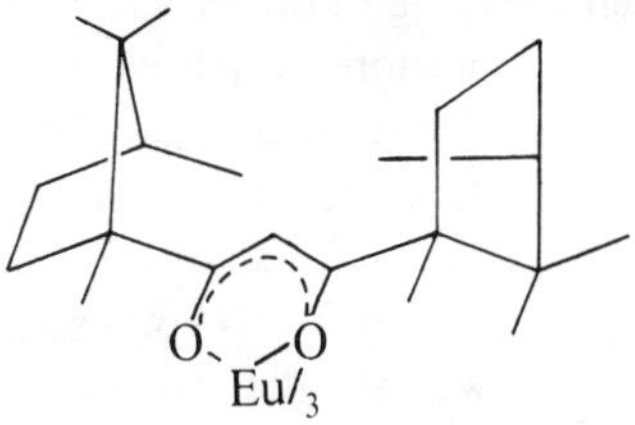

14 Eu(dcm)$_3$

In a general way, it is found that chiral LSRs generate separate signals for the enantiomers of compounds bearing functional groups able to complex with europium, such as alcohols, esters, amines, ketones, ethers, epoxides, amides, and sulfoxides.[3] These are the functional groups which normally exhibit lanthanide-induced shifts (LISs) with achiral LSRs. Indeed, chiral nonequivalence of signals due to enantiomers is superposed upon achiral LISs. The ^{1}H-nmr spectrum is recorded in CCl_4, $CDCl_3$, or CS_2 in the presence of increasing quantities of chiral LSR. The concentration of the substrate is of the order 0.1 to 0.25*M*, while the LSR/substrate ratio necessary to obtain a good separation of signals can vary from case to case between ca. 0.5 and 2. Typically, the difference between the chemical shifts induced by the two enantiomers ($\Delta\Delta\delta$) is of the order 0.1 to 0.5 ppm, that is, 10 to 50 times the displacements observed in chiral solvents (method of Pirkle). Occasionally, even larger differences (up to 4.4 ppm) are observed, especially with Eu(dcm)$_3$, **14**.[47]

The magnitude of $\Delta\Delta\delta$ varies with the protons in a given substrate. Moreover, the enantiomeric purity determination may be carried out even if not *all* of the signals are split. Finally, with functional groups that interact only feebly with the available LSRs, for example, nitro derivatives and nitriles, the peak separation may be enhanced by recording the spectra at low temperatures.[47]

Just as in the case of optically active solvents, the complexation equilibria between chiral LSRs and *d* or *l* substrates give rise to rapid exchanges with respect to the nmr time scale. In consequence, the enantiomeric purity determination of the substrate may be carried out even if the chiral LSR is not enantiomerically pure; only the magnitude of $\Delta\Delta\delta$ would be reduced.[3]

Nonequivalence is also observed in ^{13}C-nmr spectra. Fraser[48] and Williamson[49] have found that chiral praseodymium and ytterbium LSRs are sometimes more effective than those of europium.

(b) Chromatographic methods

Just as is true for analytical methods based on nmr spectroscopy, in order for enantiomer pairs to be differentiated in chromatographic systems diastereomeric interactions must be generated. Hence, in order to observe differences in retention times or in R_f due to enantiomers, either the stationary phase or the mobile phase must be optically active.

Much energy and effort has been expended in the search for chiral stationary phases and conditions which would make possible both analytical and preparative resolutions of enantiomers. The development of such analytical resolutions, which this section addresses, is now in an advanced stage of development. Preparative resolutions of enantiomers on chiral stationary phases is less well developed, although at this writing (1980) some of the studies are beginning to bear fruit. In particular, we cite the studies by Hesse and Hagel[50] and Mannschreck and his associates[51] on the use of microcrystalline cellulose triacetate as a liquid chromatography stationary phase of wide applicability.[52] A notable application is the partial resolution of perchlorotriphenylamine by Mislow and his colleagues.[53]

Pirkle and his associates have been involved in a major effort looking to the rational construction of chiral stationary phases. Some of the most promising results reported involve the resolution of alcohols on a gram scale with liquid chromatography columns containing chiral stationary phases (CSP) prepared by reacting (R)-*N*-(3,5-dinitrobenzoyl)phenylglycine with silica gel functionalized with γ-aminopropyltriethoxysilane, **15**.[54] Another very promising CSP with a wider range of applicability, for example, amines, sulfoxides, and amino acids, developed by the Pirkle group is based on the 2,2,2-trifluoro-1-(9-anthryl)ethanol chiral moiety, **16**.[55] Several reviews have summarized developments in this area.[56–58]

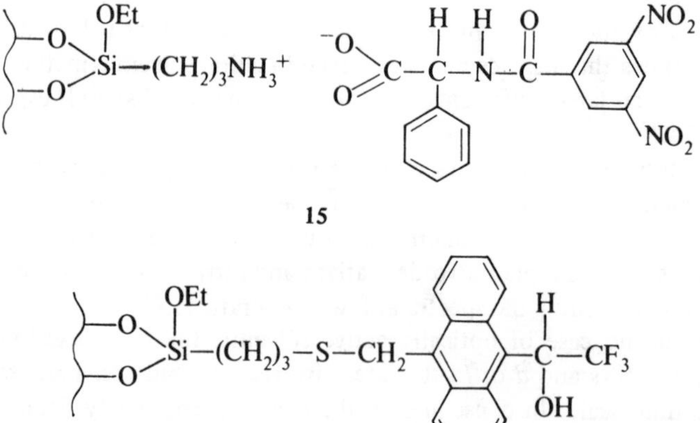

15

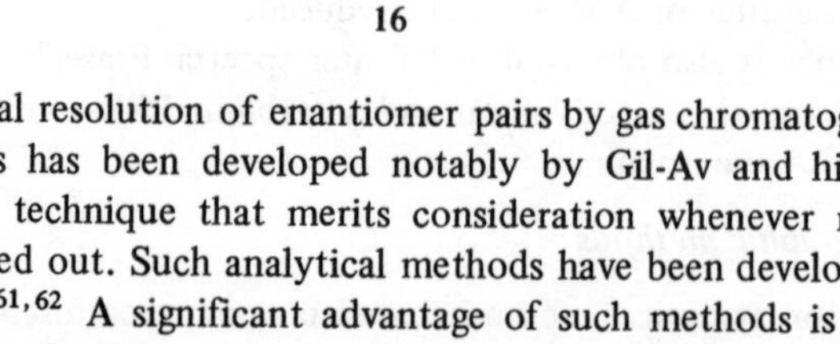

16

The analytical resolution of enantiomer pairs by gas chromatography on chiral stationary phases has been developed notably by Gil-Av and his associates[59,60] as a convenient technique that merits consideration whenever routine analysis needs to be carried out. Such analytical methods have been developed particularly for amino acids.[61,62] A significant advantage of such methods is their ability to effect the analysis of mixtures of several amino acids at one time. The disadvantages include the need to derivatize the amino acids (with achiral reagents) to increase their volatility and the use of relatively harsh conditions implicit in gas chromatography at elevated temperatures.

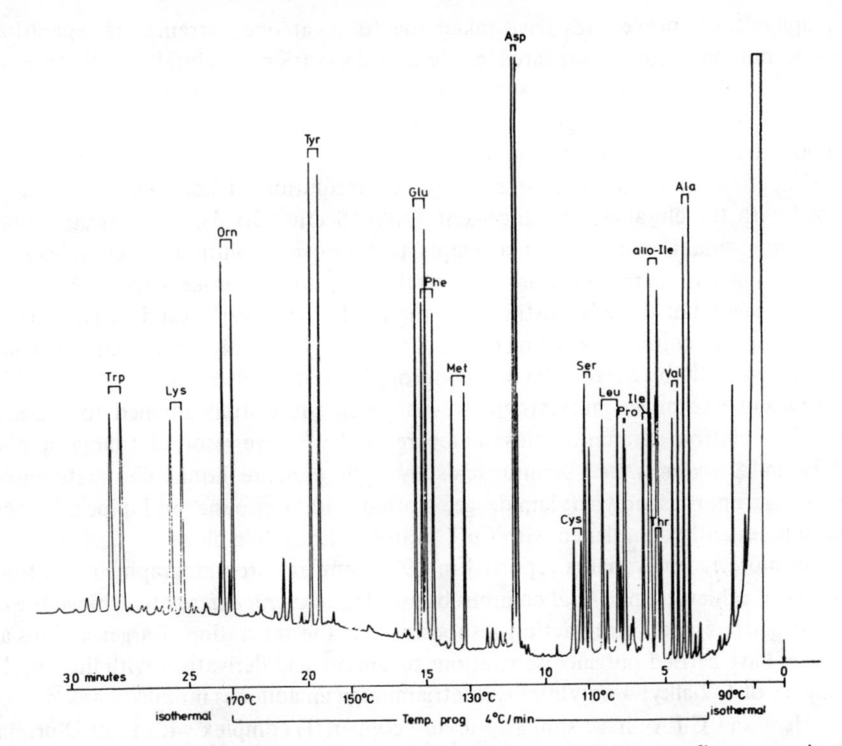

Figure 3 Gas chromatogram of a synthetic mixture of 17-*N*-pentafluoropropionyl-D,L-amino acid isopropyl esters on a 18 m × 0.3 mm open tubular column coated with *N*-propionyl-L-valine-*t*-butylamide polysiloxane. The first peak of each amino acid represents the D enantiomer. H. Frank, G. J. Nicholson, and E. Bayer, *J. Chromatogr. Sci.*, 1977, **15**, 174. Reproduced by permission of Preston Publications, Inc., and the authors.

Figure 3 illustrates the analysis of up to 17 enantiomer pairs of amino acids derivatized as *N*-pentafluoropropionyl amino acid isopropyl esters on a capillary column coated with a liquid phase consisting of *N*-propionyl-L-valine-*tert*-butylamide coupled to a copolymer of dimethylsiloxane and carboxyalkylmethylsiloxane. This stationary phase, developed by Bayer and co-workers,[63] is commercially available (Applied Science Laboratories). It appears to have high thermal stability and to be applicable as well to the analysis of chiral bases, hydroxy acids, and amino alcohols.

An interesting development in the direct determination of the enantiomeric purity of epoxides and related compounds is their gas-chromatographic analysis on stationary phases containing chiral metal coordination compounds of nickel, rhodium, as well as lanthanoid shift reagents. This "complexation" gas chromatography is especially sensitive.[64,65] Several reviews have summarized the substantial literature describing enantiomeric purity determinations by gas chromatography.[66]

A more recent development in chromatographic resolution which bridges the gap between preparative and analytical separations is high-pressure liquid chromatography. Such procedures have taken the form, at one extreme, of separations carried out on columns prepared by *in situ* deposition of chiral agents on polar stationary phases and, at the other, of stationary phases consisting of standard resolving agents or even specially conceived chiral substances grafted onto achiral supports by means of covalent bonds.

The first of these is illustrated by the resolution of helicenes on silica gel doped with the chiral complexing agent TAPA (Section 5.1.7).[67] This is but a high-resolution variant of numerous attempts to improve resolution by complexation by adapting chromatography in its several forms to the process (page 274).[68-70]

The second of these is illustrated by the resolution of carbo- and heterohelicene derivatives on a column consisting of chiral binaphthyl-2,2′-diyl hydrogen phosphate linked covalently by means of a 3-aminopropyl spacer to silica gel.[71a]

Ligand exchange chromatography has been successfully applied to the analytical resolution of amino acids. Lefebvre et al.[71b] have resolved a large number of racemic *underivatized* amino acids by high pressure liquid chromatography on a macroporous polyacrylamide gel grafted with L-proline or L-pipecolic acid and subsequently complexed with Cu^{++}. Water is the mobile phase.

In more recent variants reported in 1979, similar chromatographic resolutions have been achieved on *achiral* columns by turning to *chiral eluents* to achieve the required diastereomeric interaction responsible for the separation. Karger and his associates have carried out such separations on amino acid derivatives with the zinc(II) complex of L-2-alkyl-4-octyldiethylenetriamine in an aqueous mobile phase.[72]

Hare and Gil-Av have similarly used a copper(II) complex with L- or D-proline in an aqueous mobile phase to separate underivatized amino acids on an ion exchange column,[73] and Grushka et al. have described the resolution of underivatized amino acid enantiomers on a reversed-phase (ODS) column with a mobile phase consisting of copper(II) or zinc(II) complexes of the dipeptide L-aspartyl-L-phenylalanine methyl ester (the commercially available artificial sweetener Aspartame) in water containing variable amounts of acetonitrile.[74a] In all of these analytical resolutions, it is the differences in stability and polarity of the diastereomeric complexes formed between the amino acids to be separated and the metal cation that are responsible for the separation.[74b]

It would seem that gas-chromatographic or high-pressure liquid chromatographic analysis as described in this section would rarely be used in routine monitoring of the progress of a typical resolution given the effort and cost required. We observe, however, that such methods may be quite useful and worthwhile to set up in laboratories where resolutions of classes of compounds, for example, amino acids, are carried out frequently or for resolutions carried out on an industrial scale where repetitive monitoring may be in order.

7.5.3 Calorimetric Methods

This technique is directly applicable to mixtures of enantiomers. It has been described in Section 2.7, and we limit ourselves here to the scope of its applications and to some possible drawbacks in its use.

The "direct method" (Section 2.7.1) is usable for mixtures of enantiomeric purities which are neither too high nor too low. It rapidly furnishes information which, without being very precise, is generally sufficient for monitoring the progress of a resolution and to suggest the best strategy for subsequent purification.

The "indirect method" is specifically suited to the determination of high purities, that is, $x > 0.95$ or e.p. > 0.90. It reveals the presence of impurities which lower the melting point by forming a eutectic mixture with the nearly pure substance (even though this eutectic may be visually undetectable). The method thus does not differentiate between chemical and enantiomeric purity. On the other hand, it is inapplicable to enantiomer mixtures which form solid solutions, whether ideal or not.

This method furnishes results which may be quite precise (± ca. 0.1% or less), especially in the range 99.0 to 99.9 mol %. The melting point curve may be obtained and analyzed manually (this normally requires 2 to 3 hours of work per sample), or automatically if the differential scanning microcalorimeter is equipped with a microprocessor. In the latter case, the purity determination can be carried out in a matter of minutes. Most dsc manufacturers can provide programs for the determination of purity for their instruments.

There remain a number of limitations to the use of calorimetric methods:

1 Liquid enantiomers. There are two ways of dealing with them. The first is to carry out the dsc measurements at low temperature, and the second is to convert them to crystalline derivatives.
2 Polymorphism. This is probably the most frequently encountered problem and the one for which no totally effective remedy exists. Prolonged heating of the sample at a temperature below the melting point (annealing) can sometimes eliminate the least stable crystal forms.
3 Thermal instability of the products. Compounds which decompose or which racemize on heating can obviously not be analyzed by microcalorimetry.

7.5.4 Isotope Labeling

Isotopically labeled molecules (with stable or with radioactive isotopes) are often used in the detection and the analysis of compounds of biological origin or in the elucidation of chemical or biochemical reaction mechanisms.

Isotope labeling may be applied to the determination of enantiomeric purity in two quite different ways. In one of these, the labeled *racemate* is added to an enantiomer mixture of unknown purity (isotope dilution method). In the other, one uses a labeled pure *enantiomer*.

(a) The isotope dilution method

The determination of enantiomeric purity by isotope dilution was first applied by Berson and Ben-Efraim in 1959.[75] In its most general form, the method may be described as follows:

To a quantity A (in g) of a partially resolved compound (D > L) of specific rotation $[\alpha]$ and of unknown enantiomeric purity p, one adds a quantity B (in g) of labeled racemate, having an isotope content I_0 (expressed in counts $\cdot$ min$^{-1}\cdot$ g^{-1} for radioactive isotopes or in atom $\cdot$ g^{-1} for stable isotopes such as deuterium). The isotope contents I_D and I_L of the enantiomers in the new mixture $A + B$ are given by

$$I_D = \frac{BI_0}{B + A(1+p)} \quad \text{and} \quad I_L = \frac{BI_0}{B + A(1-p)} \tag{1}$$

These values of I_D and I_L remain constant throughout all subsequent manipulations of the mixture $A + B$. The latter mixture is fractionated by crystallization (or by another means) so as to isolate a sample possessing a new specific rotation $[\alpha']$ and a new enantiomeric purity p'. The isotope content I' of the new mixture is given by

$$I' = I_D \frac{1+p'}{2} + I_L \frac{1-p'}{2} \tag{2}$$

If the values of I_D and I_L from eqs. (1) are substituted into eq. (2) and the latter is rearranged, one obtains

$$I' = I_0 \frac{B^2 + AB - ABpp'}{B^2 + 2AB + A^2(1-p^2)} \tag{3}$$

Since $p = [\alpha]/[\alpha]_{max}$ and $p' = [\alpha']/[\alpha]_{max}$, one arrives at eq. (4):

$$I' = BI_0 \frac{(A+B)[\alpha]^2_{max} - A[\alpha][\alpha']}{(A+B)^2[\alpha]^2_{max} - A^2[\alpha]^2} \tag{4}$$

The maximum value of the specific rotation $[\alpha]_{max}$ of the compound in question is finally given by

$$[\alpha]^2_{max} = \frac{I'A^2[\alpha]^2 - I_0AB[\alpha][\alpha']}{I'(A+B)^2 - I_0B(A+B)} \tag{5}$$

It may be possible to isolate pure racemate (in the form of the racemic compound, for example) upon crystallization of mixture $A + B$. In this case, $[\alpha'] = 0$ and eq. (5) reduces to

$$[\alpha]^2_{max} = \frac{I'A^2[\alpha]^2}{I'(A+B)^2 - I_0B(A+B)} \tag{6}$$

This method for the determination of enantiomeric purity has been relatively little used. For examples, see ref. 1. The precision of the results obtained depends both upon the uncertainty in the specific rotations $[\alpha]$ and $[\alpha']$ and upon the precision of the isotope analysis which varies with the isotope used. Berson and Ben-Efraim[75] have estimated the uncertainty in p to be between 3 and 5% with the radioactive isotope ^{15}N (analyzed by mass spectrometry). Gerlach has assigned a precision of ca. ± 1% in a case involving deuterium labeling with the analysis carried out by mass spectrometry.[76]

(b) The radioactive tracer method

The determination of enantiomeric purity of samples having very high e.p. values (> 99.9 mol %) is possible only by means of very specific methods. The one we describe here is usable if one has available an enantiomer of high purity (>99%) which is labeled with a radioactive tracer. A small quantity of this tracer is added to the other enantiomer, itself nearly pure, and the mixture obtained is subjected to an appropriate purification process (Section 7.6). The residual amount of the minor enantiomer may thus be followed by the decrease in radioactivity of the sample during the purification.

Consider a quantity A (in g) of a mixture D + L of enantiomer purity p (D > L). An amount B (in g) of the minor enantiomer labeled with a radioactive isotope L^*, supposedly enantiomerically pure and having specific activity I_0, is added to the mixture. The specific activity of mixture $A + B$ is $I_1 = BI_0/(A + B)$, while that of enantiomer L becomes, as a result of the dilution of L^*,

$$I_L = \frac{2BI_0}{(1-p)A + B} \tag{7}$$

Subsequent to purification, a sample of enantiomeric purity p' and of specific activity $I' = I_L(1-p')/2$ is obtained. Since I_L is given by eq. (7), one easily can show that

$$p' = 1 - \frac{I'}{I_0}\left[\frac{A(1-p)}{B} + 1\right] \tag{8}$$

It is thus necessary to know at least roughly the initial purity p in order to be able to calculate the order of magnitude of p'. The following specific example is typical: $A = 10\,g$, $p = 0.96$; $B = 0.030\,g$, $I_0 = 10^6$ counts $\cdot$ min$^{-1}\cdot$g^{-1}. The specific activity of mixture $A + B$ is $I_1 \sim 3 \times 10^3$. If, after purification $I' = 10^3$, then one has

$$p' = 1 - \frac{10^3}{10^6}\left[\frac{10 \times 0.04}{0.03} + 1\right] = 0.986$$

With $I' = 10^2$, one would have $p' = 0.998$, and so on. This method has been applied by Tensmeyer, Landis, and Marshall[77] to the analysis of d-α-propoxyphene (Section 7.6).

7.5.5 Quantitative Enzymatic Analysis

The stereospecificity of enzymatic reactions may be put to good advantage in the determination of the enantiomeric purity of biochemically transformable substrates. In practice, this method has been applied solely to the analysis of amino acids for which a number of enzymatic systems are available.[78] For example, L-amino acid oxidase and D-amino acid oxidase catalyze the oxidation of L- and D-amino acids, respectively, to α-keto acids according to the following scheme:

$$\text{L- or D-R-}\underset{\displaystyle NH_2}{\underset{|}{CH}}\text{-COOH} + \tfrac{1}{2}O_2 \xrightarrow{\text{L-or D-oxidase}} \text{R-}\overset{\displaystyle O}{\overset{||}{C}}\text{-COOH} + NH_3 \tag{9}$$

Also, L-amino acid decarboxylase catalyzes the decarboxylation of L-amino acids according to the equation

$$\text{L-R}-\underset{\displaystyle NH_2}{\underset{|}{CH}}-COOH \xrightarrow{\text{L-decarboxylase}} R-CH_2-NH_2 + CO_2 \qquad (10)$$

The analysis is effected by measuring the volume of oxygen absorbed (eq. (9)] or of carbon dioxide evolved [eq. (10)] during the course of the reaction. The enzymatic system is chosen so as to react with the minor enantiomer of the mixture being analyzed. The method is particularly useful with mixtures of high enantiomeric purities. It is possible to detect enantiomer D in L (or L in D) with a precision of the order of 0.1%.

REFERENCES 7.5

1 M. Raban, and K. Mislow, *Topics Stereochem.*, 1967, **2**, 199.

2 J. Campbell, *Aldrichimica Acta*, 1972, **5**, 29.

3 G. E. Sullivan, *Topics Stereochem.*, 1978, **10**, 287.

4 Y. Izumi and A. Tai, *Stereo-differentiating Reactions*, Academic Press, New York, 1977, pp. 218–242.

5 H. Numan and H. Wynberg, *J. Org. Chem.*, 1978, **43**, 2232.

6 A. Horeau, *Tetrahedron Lett.*, 1969, 3121.

7 G. -A. Hoyer, D. Rosenberg, C. Rufer, and A. Seeger, *Tetrahedron Lett.*, 1972, 985.

8 (a) D. R. Galpin and A. C. Huitric, *J. Org. Chem.*, 1968, **33**, 921. (b) T. G. Cochran and A. C. Huitric, *ibid.*, 1971, **36**, 3046.

9 P. Briaucourt, J. -P. Guetté, and A. Horeau, *C. R. Acad. Sci., Ser. C*, 1972, **274**, 1203.

10 H. Gerlach, *J. Chem. Soc. Chem. Comm.*, 1973, 274.

11 E. Gil-Av and D. Nurok, *Adv. Chromatogr.*, 1975, **10**, 99.

12 J. F. Nicoud and H. B. Kagan, *Israel J. Chem.*, 1976/77, **15**, 78.

13 W. H. Pirkle and M. S. Hoekstra, *J. Org. Chem.*, 1974, **39**, 3904.

14 (a) G. Helmchen, G. Nill, D. Flockerzi, W. Schule, and M. S. K. Youssef, *Angew. Chem.*, 1979, **91**, 64. *Angew. Chem., Int. Ed. Engl.*, 1979, **18**, 62. (b) G. Helmchen, G. Nill, D. Flockerzi, and M. S. K. Youssef, *ibid.*, 1979, **91**, 65, and 1979, **18**, 63.

15 (a) J. -P. Guetté, L. Lacombe, and A. Horeau, *C. R. Acad. Sci., Ser. C*, 1968, **276**, 166. (b) A. Horeau and J. -P. Guetté, *ibid.*, 1968, **276**, 257.

16 (a) A. Ejchart and J. Jurczak, *Bull. Acad. Pol. Sci., Ser. Sci. Chim.*, 1970, **18**, 445, (b) A. Ejchart and J. Jurczak, *J. Chem. Soc. D*, 1970, 654. (c) M. Mikołajczyk, A. Ejchart, and J. Jurczak, *Bull. Acad. Pol. Sci., Ser. Sci. Chem.*, 1971, **19**, 721. (d) A. Ejchart and J. Jurczak, *ibid.*, 1971, **19**, 725. (e) A. Ejchart, J. Jurczak, and K. Bankowski, *ibid.*, 1971, **19**, 731. (f) M. Mikołajczyk and J. Omelanczuk, *Tetrahedron Lett.*, 1972, 1539. (g) M. Mikołajczyk, J. Omelanczuk, M. Leitloff, J. Drabowicz, A. Ajchart, and J. Jurczak, *J. Am. Chem. Soc.*, 1978, **100**, 7003. (h) W. Kuchen and J. Kutter, *Z. Naturforsch.*, 1979, **34B**, 1332.

17 L. Mamlok, A. Marquet, and L. Lacombe, *Tetrahedron Lett.*, 1971, 1093.

18 L. Mamlok, A. Marquet, and L. Lacombe, *Bull. Soc. Chim. Fr.*, 1973, 1524.

19 A. Mannschreck and W. Seitz, *Pure Appl. Sci.*, 1971, 310.

20 C. A. R. Baxter and H. C. Richard, *Tetrahedron Lett.*, 1972, 3357.

21 J. A. Dale and H. S. Mosher, *J. Am. Chem. Soc.*, 1968, **90**, 3732.

22 (a) M. Raban and K. Mislow, *Tetrahedron Lett.*, 1965, 4249. (b) J. Jacobus and M. Raban, *J. Chem. Educ.*, 1969, **46**, 351; (c) J. Jacobus, M. Raban, and K. Mislow, *J. Org. Chem.*, 1968, **33**, 1142. (d) H. T. Thomas and K. Mislow, *J. Am. Chem. Soc.*, 1970, **92**, 6292.

23 (a) J. A. Dale and H. S. Mosher, *J. Am. Chem. Soc.*, 1973, **95**, 512 and references cited therein. (b) J. A. Dale, D. L. Dull, and H. S. Mosher, *J. Org. Chem.*, 1969, **34**, 2543. (c) J. A. Dale, D. L. Dull, and H. S. Mosher, *ibid.*, 1970, **35**, 4002. (d) J. A. Dale, D. L. Dull, and H. S. Mosher, *ibid.*, 1971, **36**, 3468. (e) P. Low and W. S. Johnson, *J. Am. Chem. Soc.*, 1971, **93**, 3765. (f) E. Weissberger, *ibid.*, 1974, **96**, 7219.

24 (a) L. R. Pohl and W. F. Trager, *J. Med. Chem.*, 1973, **16**, 475. (b) D. E. Nichols, C. F. Barfknecht, D. B. Rusterholz, R. Bennington, and R. D. Morin, *ibid.*, 1973, **16**, 480. (c) J. Gal, *J. Pharm. Sci.*, 1977, **66**, 169.

25 M. I. Kabachnik, T. A. Mastryukova, E. I. Fedin, M. S. Vaisberg, L. L. Morozov, P. V. Petrovskii, and A. E. Shipov, *Usp. Khim.*, 1978, **47**, 1541. *Russ. Chem. Rev.*, 1978, **47**, 821.

26 T. Williams, R. G. Pitcher, P. Bommer, J. Gutzwiller, and M. Uskoković, *J. Am. Chem. Soc.*, 1969, **91**, 1871.

27 M. J. P. Harger, *J. Chem. Soc., Perkin II*, 1977, 188.

28 A. Horeau and J. -P, Guetté, *Tetrahedron*, 1974, **30**, 1923.

29 (a) K. Ajisaka and M. Kainosho, *J. Am. Chem. Soc.*, 1975, **97**, 1761. (b) W. H. Pirkle and D. L. Sikkenga, *J. Org. Chem.*, 1975, **40**, 3431. (c) J. Rauben, *J. Chem. Soc. Chem. Comm.*, 1979, 68. *J. Am. Chem. Soc.*, 1980, **102**, 2232.

30 W. H. Pirkle, *J. Am. Chem. Soc.*, 1966, **88**, 1837.

31 T. G. Burlingame and W. H. Pirkle, *J. Am. Chem. Soc.*, 1966, 88, 4294.

32 W. H. Pirkle, S. D. Beare, and T. G. Burlingame, *J. Org. Chem.*, 1969, **34**, 470.

33 (a) W. H. Pirkle and T. G. Burlingame, *Tetrahedron Lett.*, 1967, 4039. (b) W. H. Pirkle and S. D. Beare, *J. Am. Chem. Soc.*, 1967, **89**, 5485.

34 W. H. Pirkle and S. D. Beare, *Tetrahedron Lett.*, 1968, 2579.

35 W. H. Pirkle and S. D. Beare, *J. Am. Chem. Soc.*, 1969, **91**, 5150.

36 W. H. Pirkle, T. G. Burlingame, and S. D. Beare, *Tetrahedron Lett.*, 1968, 5849.

37 W. H. Pirkle and S. D. Beare, *J. Am. Chem. Soc.*, 1968, **90**, 6250.

38 W. H. Pirkle and M. S. Pavlin, *J. Chem. Soc. Chem. Comm.*, 1974, 274.

39 W. H. Pirkle, S. D. Beare, and R. L. Muntz, *J. Am. Chem. Soc.*, 1969, **91**, 4575.

40 M. D. Joesten, H. E. Smith, and V. A. Vix, Jr., *J. Chem. Soc. Chem. Comm.*, 1973, 18.

41 W. H. Pirkle and M. S. Hoekstra, *J. Magnetic Resonance*, 1975, **18**, 396.

42 (a) J. C. Jochims, G. Taigel, and A. Seeliger, *Tetrahedron Lett.*, 1967, 1901. (b) F. A. L. Anet, L. M. Sweeting, T. A. Whitney, and D. J. Cram, *ibid.*, 1968, 2617.

43 C. P. R. Jennison and D. McKay, *Can. J. Chem.*, 1973, **51**, 3726.

44 G. M. Whitesides and D. W. Lewis, *J. Am. Chem. Soc.*, 1970, **92**, 6979.

45 R. R. Fraser, M. A. Petit, J. K. Saunders, *J. Chem. Soc. D*, 1971, 1450.

46 H. L. Goering, J. N. Eikenberry, and G. S. Koermer, *J. Am. Chem. Soc.*, 1971, **93**, 5913.

47 M. D. McCreary, D. W. Lewis, D. L. Wernick, and G. M. Whitesides, *J. Am. Chem. Soc.*, 1974, **96**, 1038.

48 R. R. Fraser, J. B. Stothers, and C. T. Tan, *J. Magnetic Resonance,* 1973, **10**, 95.

49 K. L. Williamson, C. P. Beeman, and J. G. Magyar, quoted in ref. 3.

50 (a) G. Hesse and R. Hagel, *Chromatographia*, 1973, **6**, 277. *Ibid.*, 1976, **9**, 62. (b) *Liebigs Ann. Chem.*, 1976, 996.

51 (a) H. Hakli and A. Mannschreck, *Angew. Chem.*, 1977, **89**, 419. *Angew. Chem., Int. Ed.*,

1977, **16**, 405. (b) H. Ahlbrecht, G. Becher, J. Blecher, H.-O. Kalmowski, W. Raab, and A. Mannschreck, *Tetrahedron Lett.*, 1979, 2265. (c) H. Hakli, M. Mintas, and A. Mannschreck, *Chem. Ber.*, 1979, **112**, 2028. (d) M. Mintas, A. Mannschreck, and M. P. Schneider, *J. Chem. Soc. Chem. Comm.*, 1979, 602.

52 (a) See also K. Bertsch and J. C. Jochims, *Tetrahedron Lett.*, 1977, 4379. (b) K. Bertsch, M. S. Rahman, and J. C. Jochims, *Chem. Ber.*, 1979, **112**, 567.

53 K. S. Hayes, M. Nagumo, J. F. Blount, and K. Mislow, *J. Am. Chem. Soc.*, 1980, **102**, 2773.

54 J. M. Finn and W. H. Pirkle, Paper presented at the American Chemical Society 179th National Meeting, Houston, Texas, March 24–28, 1980. Abstract ORGN 14; *Chem. Eng. News*, 1980, **58**(14), 28.

55 W. H. Pirkle and D. W. House, *J. Org. Chem.*, 1979, **44**, 1957.

56 I. S. Krull, *Adv. Chromatogr.*, 1978, **16**, 175.

57 R. Audebert, *J. Liq. Chromatogr.*, 1979, **2**, 1063.

58 G. Blaschke, *Angew. Chem.*, 1980, **92**, 14. *Angew. Chem., Int. Ed.*, 1980, **19**, 13.

59 E. Gil-Av and B. Feibush, *Tetrahedron Lett.*, 1967, 3345.

60 E. Bayer, E. Gil-Av, W. Konig, S. Nakaparksin, J. Oro, and W. Parr, *J. Am. Chem. Soc.*, 1970, **92**, 1738.

61 W. Parr and P. Y. Howard, *J. Chromatogr.*, 1972, **67**, 227.

62 W. Parr and P. Y. Howard, *Angew. Chem.*, 1972, **84**, 586. *Angew. Chem., Int. Ed.*, 1972, **11**, 529.

63 H. Frank, G. J. Nicholson, and E. Bayer, *J. Chromatogr. Sci.*, 1977, **15**, 174.

64 V. Schurig, B. Koppenhoefer, and W. Buerkle, *J. Org. Chem.*, 1980, **45**, 538, and references therein.

65 B. T. Golding, P. J. Sellars, and A. K. Wong, *J. Chem. Soc. D*, 1970, 570.

66 (a) E. Gil-Av and D. Nurok, *Adv. Chromatogr.*, 1974, **10**, 99. (b) D. R. Buss and T. Vermeulen, *Ind. Eng. Chem.*, 1968, **60**, 12. (c) C. H. Lochmuller and R. W. Souter, *J. Chromatogr.*, 1975, **163**, 283. (d) E. Gil-Av and B. Feibush, in *Peptides 1974*, Y. Wolman, Ed., Wiley, New York, 1975.

67 F. Mikes, G. Boschart, and E. Gil-Av, *J. Chem. Soc. Chem. Comm.*, 1976, 99.

68 G. Wittig, K.-D. Rumpler, *Liebigs Ann. Chem.*, 1971, **751**, 1.

69 W. Rebafka and H. A. Staab, *Angew. Chem.*, 1973, **85**, 835; *Angew. Chem., Int. Ed.*, 1973, **12**, 776.

70 A. Krebs, J. Odenthal, and H. Kimling, *Tetrahedron Lett.*, 1975, 4663.

71 (a) F. Mikes and G. Boschart, *J. Chem. Soc. Chem. Comm.*, 1978, 173. (b) B. Lefebvre, R. Audebert, and C. Quivoron, *Israel J. Chem.*, 1976/77, **15**, 69.

72 J. N. LePage, W. Lindner, G. Davies, D. E. Seitz, and B. L. Karger, *Anal. Chem.*, 1979, **51**, 433.

73 P. E. Hare and E. Gil-Av, *Science,* 1979, **204**, 1226.

74 (a) C. Gilon, R. Leshem, Y. Tapuhi, and E. Grushka, *J. Am. Chem. Soc.*, 1979, **101**, 7612. (b) E. Gil-Av, A. Tishbee, and P. E. Hare, *ibid.*, 1980, **102**, 5115.

75 J. A. Berson and D. A. Ben-Efraim, *J. Am. Chem. Soc.*, 1959, **81**, 4083.

76 H. Gerlach, *Helv. Chim. Acta*, 1966, **49**, 2481.

77 L. G. Tensmeyer, P. W. Landis, and F. J. Marshall, *J. Org. Chem.*, 1967, **32**, 2901.

78 J. P. Greenstein and M. Winitz, *The Chemistry of the Amino Acids*, Vol. 2, Wiley, New York, 1961, p. 1738.

7.6 FINAL PURIFICATION. ENRICHMENT OF PARTIALLY RESOLVED ENANTIOMER MIXTURES

As we approach the desired goal of rational attainment of enantiomer purity in separations originating from racemates, we need to address the refinement and completion of resolutions. Now that we are in possession of one or both enantiomers in a state of moderately high purity, how do we attain full enantiomeric purity? To properly deal with this question, we need to define what we mean by purity and to comment upon the degree of purity required by the use to which the resolved substance is destined.

Some of the methods described in this section depend upon the *crystallization* of enantiomer mixtures. They require that the latter have already reached a state of *minimum* purity beyond which the final purification is possible. Other procedures, which we have assembled under the title *chemical purification*, allow, at least in principle, all mixtures containing even a small excess of one enantiomer to be brought to a state of maximum purity. There are very few partially resolved substances, of whatever origin, that are not susceptible to further enantiomeric enrichment by one of these methods. The last mentioned methods may be applied in many instances to the enrichment of compounds arising from *asymmetric syntheses*, whose efficiencies are often only mediocre.

7.6.1 Definitions. Pure and Ultrapure Enantiomers

For enantiomers there are two notions of purity that need be considered – *enantiomeric* purity (absence of the other enantiomer) and *chemical* purity (absence of one or more foreign substances). An ideal definition, in fact, makes no distinction between these two possibilities; an ideally pure compound contains but one molecular species. From an experimental point of view, however, the idea of purity is closely associated with the purification methods and with the sensitivity of the analytical methods applied. This problem is, in fact, more complex than appears at first sight. Several reviews dealing with it have been published, and the reader is referred to them for details.[1–3]

As a rule, organic chemists rarely concern themselves with products whose purity exceeds 99.9 mol% and, *a fortiori*, with "ultrapure" substances, that is, those containing less than 500 ppm impurity. Such purities or ultrapurities may be required for the establishment of physicochemical or pharmaceutical standards as well as for specific uses such as those involving, for example, electronic properties of matter – semiconductors, scintillation counters, and so on.

The preparation of *enantiomerically ultrapure* substances may be of fundamental importance in some areas. When it is important to demonstrate the absence of biological activity of an enantiomer, it is necessary that the sample be free of traces of the other, particularly if the latter is active at very low concentrations (as in the case of estrogens,[4] for example, or insect pheromones). The same may be true if one wishes to reveal a difference in taste and especially in odor; in the latter case, not only enantiomeric but chemical purity as well is essential.[5]

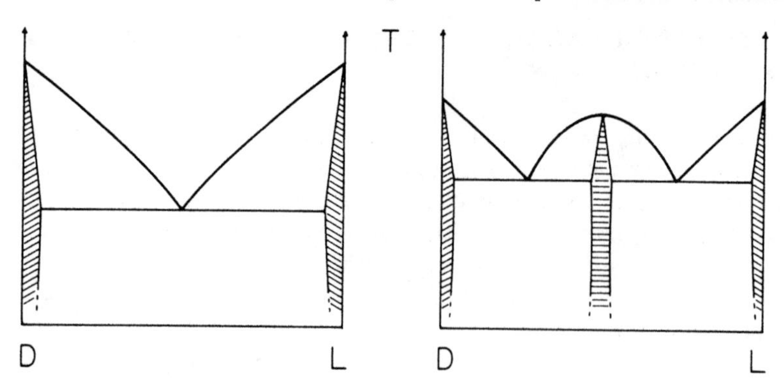

Figure 1 Zones of miscibility (terminal solid solutions) [shading] in real phase diagrams of conglomerates (left) and racemic compounds (right).

Outside of these special cases, the need for enantiomerically ultrapure compounds is rarely justified. In pharmacology, for example, it is certainly more important to eliminate any toxic chemical impurities than to seek to rid the sample of 1 to 2% of the residual "wrong" enantiomer, which is generally inactive anyway.

Taking into account the purification and analytical methods usually employed, the "pure substance" of the organic chemist may be considered to lie between 99.0 and 99.9 mol% and more often closer to the former value. According to these criteria, a "pure enantiomer" would need to have an enantiomeric purity of at least 98%. That is then the final objective of most resolutions.

In the first part of this book, where phase diagrams were described and discussed, little account was taken of the existence of *terminal solid solutions.* In fact, most real phase diagrams of conglomerates and racemic compounds necessarily must look like those shown in Fig. 1, where the zones of miscibility in the solid state are shown in grey. Moreover, other impurities than the residual enantiomer can lead to terminal solid solutions. The experimental delineation of these zones is rarely possible for practical reasons, except in those rare cases when they are abnormally large (Section 2.4.8). It is these solid solutions which impede the attainment of very high purity. In general, and fortunately, they are much less of a hindrance when the goal of the purification process is the ordinary one defined in the preceding paragraph.

7.6.2 Classical Final Purification

When the goal is not ultrapurity, the normal procedure for preparing a pure enantiomer from a partially resolved sample is recrystallization. Success in this operation naturally requires that the initial purity of the mixture be greater than that of the eutectic in the (binary or ternary) phase diagram. When the enantiomers are liquids, it is often possible to transform them into crystalline derivatives by means of reaction with achiral reagents: amine → hydrochloride, sulfate and so on; acid → sodium or achiral amine salt, and so on; alcohol → benzoate, and so on. Alternatively, liquid enantiomers quite often can be crystallized at low temperature.

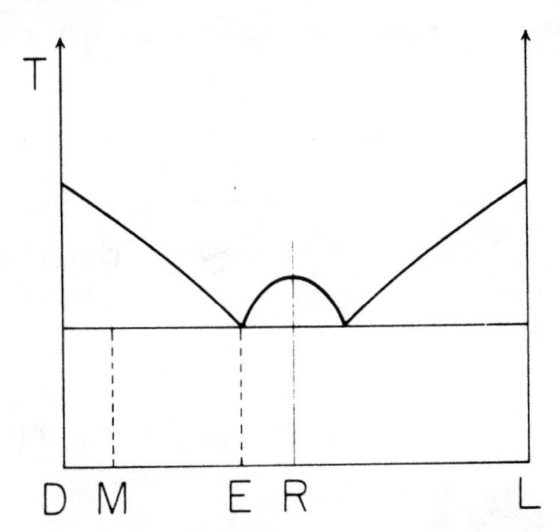

Figure 2 Estimation of the maximum theoretical yield of enantiomer obtainable in a recrystallization: for mixture M, the yield of pure D is ME/DE.

It seems that in the majority of cases the zone of terminal solid solutions rarely exceeds 1% for mixtures of enantiomers. This is indirectly suggested by the observation that it is almost always possible to obtain the pure enantiomer, that is, >98% e.p. through *one* recrystallization of a partially resolved mixture, provided that it is situated on the "good" side of the eutectic. Moreover, the optimal quantity of solvent required may be estimated if one knows the solubility of the eutectic and the initial enantiomeric composition. It is also possible to estimate the maximum theoretical yield (Fig. 2). Given a mixture M, the yield of pure D which may be expected is equal to ME/DE if one neglects the terminal solution near 100% D. This yield is higher the greater the enantiomeric purity of the initial mixture ($M \rightarrow$ D) and the smaller the enantiomeric purity of the eutectic ($E \rightarrow R$). The highest yields are thus obtained with conglomerates ($E = R$). This underscores the value earlier alluded to (Section 2.2.5) of systematic searches for derivatives that form conglomerates or at least form racemic compounds whose eutectics are close to R. Thus, α-methylbenzylamine (a liquid) may be purified by recrystallization of its crystalline sulfate, and α-(1-naphthyl)ethylamine similarly may be purified through its phenylacetate. Both these salts are conglomerates. α-Phenoxypropionic acid, which forms a racemic compound whose eutectic composition is close to that of the pure enantiomer, is quite difficult to purify in a direct manner, even beginning with samples of 90% e.p. On the other hand, the purification of this acid is quite easy via its dicyclohexylamine salt which forms a racemic compound whose eutectic composition is less unfavorable.

A small sampling of other cases of purification by crystallization of partially resolved enantiomers is summarized in Table 1. We note that some authors[6,8] find the ease with which enantiomeric enrichment is effected by simple recrystallization sufficiently remarkable to comment on it.

Rarer are the cases in which recrystallization reduces the enantiomeric enrichment, leading in the extreme to pure racemate. Two examples of such compounds

Table 1 Purification by crystallization of partially resolved enantiomer mixtures. Examples

Compound	Original e.e. (%)	Final e.e. (%)	Recrystallization Solvent	Refs.
$CH_3-C_6H_4-Se-$(3-oxocyclohexyl)	43	~85	Pentane or pentane-ethanol (20:1)	6
$C_6H_5P(O)(CH_3)NH_2$	90	Pure	C_6H_6-$CHCl_3$ (3–1)	7
$CH_3-CH(OTs)-CH_2OTs$	80	Pure	CH_2Cl_2-cyclohexane	8
6-methyl-N-acetyltryptophan (CH_3-indolyl-$CH_2CH(NHAc)COOH$)	82	99.8	Aqueous NH_3	9

are 5-norbornene-2-*endo*-methyl-2-*exo*-carboxylic acid,[10] and (*E*)-dypnone oxide.[11] In both cases, the mother liquors afforded moderately to highly enriched product. When the eutectics are very close to the pure enantiomers (see Section 3.3, Fig. 4), these may be obtained in a nearly pure state from the mother liquors; (+)-α-phenylbutyric acid[12] and (−)-1-phenylethyl hydrogen phthalate[13] have been purified in this way.

The desirability of having the initial enantiomeric purity *M* be as high as possible leads to an important question regarding the resolution strategy: *What is the most opportune moment for the cleavage of the diastereomers into enantiomers*? While it is impossible to give a universally valid answer to this question, at very least several suggestions may be given to guide experimenters in dealing with the problem that the question addresses.

1 If the diastereomers can easily and completely be separated (by chromatography, for example), it is clear that the problem of the best timing for the return to enantiomer does not arise.
2 The typical separation by crystallization of diastereomers which do not form terminal solid solutions is illustrated in Fig. 3. Consider the recrystallization of 1 mole mixture *M*, either via diastereomers (Fig. 3*a*) or via enantiomers (Fig. 3*b*). In this case, it is easy to see that the yield of pure D is equal to ME'/PE' mole in the first case and to ME/DE mole in the second. It is thus

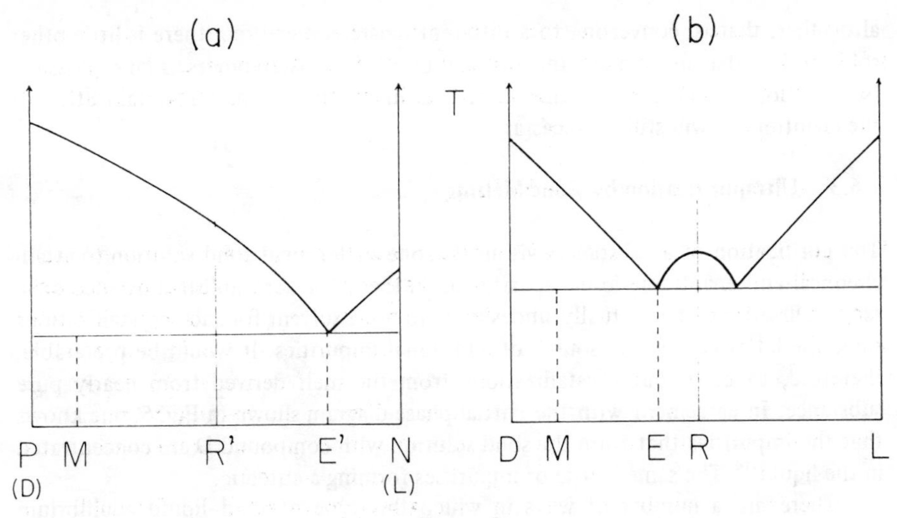

Figure 3.

more profitable to recrystallize the diastereomer mixture until purity is achieved (in principle, one crystallization should suffice), since ME'/PE' is always greater than ME/DE.

3 Contrary to what is generally observed with enantiomer mixtures, the formation of relatively extended zones, say 5 to 10% of solid solutions at the ends of the phase diagram, is common in the case of diastereomer mixtures (Fig. 4*a*). In this case, one observes that after a net initial enrichment, subsequent recrystallizations lead to only mediocre enrichment of the less soluble diastereomer.

It is only if the entire phase diagram is known that it is possible to decide what course is best to follow. If in the phase diagram for the enantiomers the eutectic composition is close to the racemate (Fig. 4*b*) or if the system is that of a conglomerate, then it probably will be preferable to pursue the purification of mixture *M* by *recrystallization of the enantiomers* themselves. In contrast, in the presence of a very stable racemic compound (Fig. 4*c*), other than changing the system

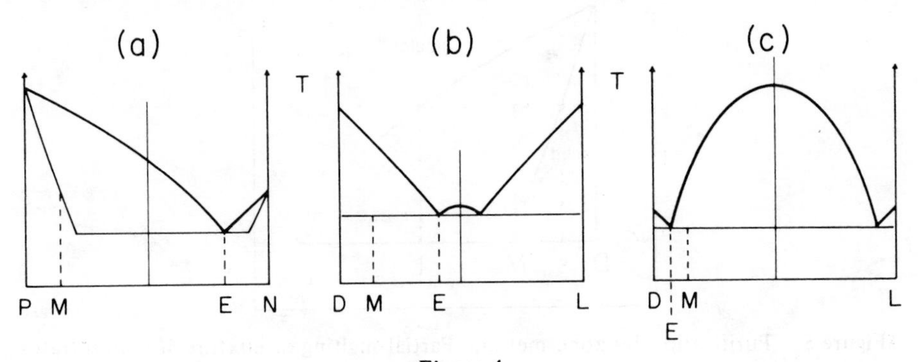

Figure 4.

altogether, that is, converting to a different diastereomer type, there is little other solution but that of pursuing the purification of the diastereomers as far as possible (see Section 7.3.4). Even in this case, it is likely that a final recrystallization of the enantiomers will still be necessary.

7.6.3 Ultrapurification by Zone Melting

The purification of a substance within its zone of terminal solid solution to attain chemical and enantiomeric ultrapurity can in theory be accomplished by successive recrystallizations. It is actually undesirable to use a solvent for such crystallizations since the latter can be the source of additional impurities. It would be preferable, therefore, to carry out crystallizations from the melt derived from nearly pure substance. In agreement with the partial phase diagram shown in Fig. 5, one knows that the impurities that form the solid solution with compound D are concentrated in the liquid.[14] The same is true of impurities forming a eutectic.

There are a number of ways in which this type of solid–liquid equilibrium may be applied. The first consists in growing a monocrystal by slowly raising it from the melt which is maintained at the melting point (Fig. 6). This technique, called "single-pass zone refining," is illustrated by the purification of the enantiomers of **1** and **2** carried out by Tensmeyer, Landis, and Marshall.[15]

$$(CH_3)_2N{-}CH_2{-}CH(CH_3){-}C(C_6H_5)(CH_2C_6H_5){-}OR$$

1 R = H

2 R = $\overset{O}{\overset{\|}{C}}CH_2CH_3$ (α-Propoxyphene)

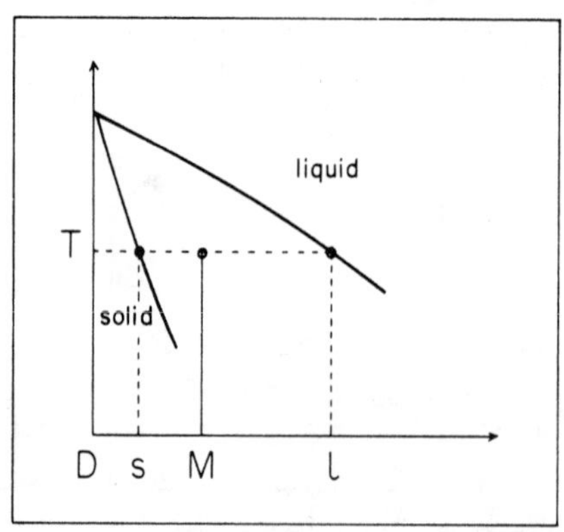

Figure 5 Purification by zone melting. Partial melting of mixture *M* concentrates the impurities in the liquid.

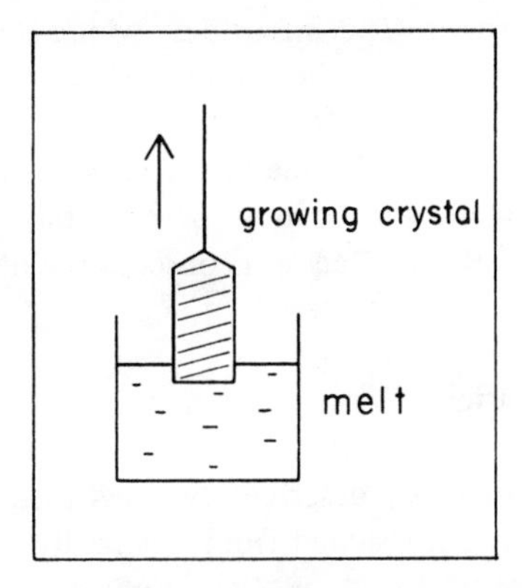

Figure 6 Single-pass zone refining.

A very pure sample of alchohol (—)-**1** was obtained by "pulling" a monocrystal at a rate 0.3 to 2 mm per hour from a melt of 95% purity. A very pure sample of (+)-α-propoxyphene, **2**, was obtained in the same way. These authors also prepared a radioactive sample of (—)-**2*** by esterification of pure alcohol (—)-**1** with propionic-1-^{14}C acid. The use of the radioactive tracer made it possible to demonstrate the efficiency of this purification method [see Section 7.5.4, part (b)]. Thus, starting from a synthetic mixture consisting of 96.9% (+)-**2** and 3.1% (—)-**2***, they obtained a monocrystal of (+)-**2** of 99.8% purity. Subsequently, a sample containing 0.15% of (—)-**2*** was similarly treated; the monocrystal of (+)-**2** isolated had an enantiomeric purity of 99.996%.

The second process based on the solid–liquid equilibrium which is, in principle, applicable to the purification of enantiomers is conventional zone melting. This process is capable of very high efficiency and has been employed in the ultrapurification of numerous compounds.[1] The substance to be purified, contained in a glass tube, is melted at one end and the molten zone is slowly moved along the length of the sample (Fig. 7). The same operation may be repeated numerous times

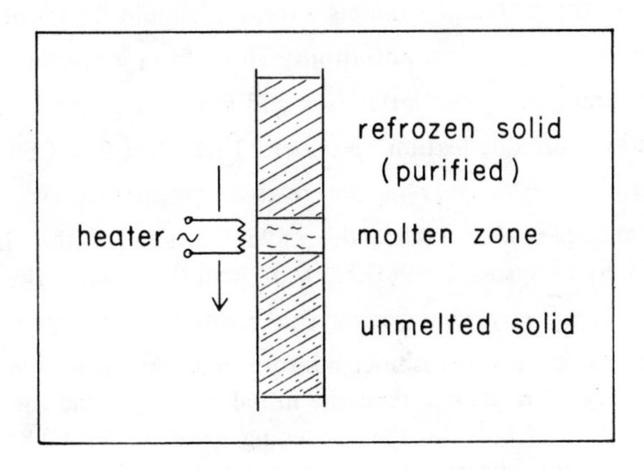

Figure 7 Purification by zone melting.

in a fully automatic manner. The impurities, which are extracted by the liquid during each passage, concentrate in the end of the column.[16]

While many organic compounds have been successfully purified by zone melting, it does not seem as if the method has been applied with success to the specific case of enantiomeric purification. Several trials along these lines cited by Wilcox, Friedenberg, and Back failed as a consequence of the thermal instability of the compounds subjected to the process.[1]

7.6.4 Chemical Purification

Horeau and his collaborators have described two methods that permit the maximum enantiomeric purity to be approached from a partially resolved product.[17,18] The first of these approaches is based on the *duplication* of enantiomer molecules which are present in unequal numbers in a given mixture. The second involves a chiral reagent which, when it reacts with the two enantiomers in a partially resolved mixture, does so with unequal rates.

(a) Purification by duplication

Starting from a mixture of two enantiomers whose functionality is susceptible to "duplication" through an appropriate X function, one obtains a *meso* compound on one hand and a pair of *threo* compounds on the other (Scheme 1).

SCHEME 1

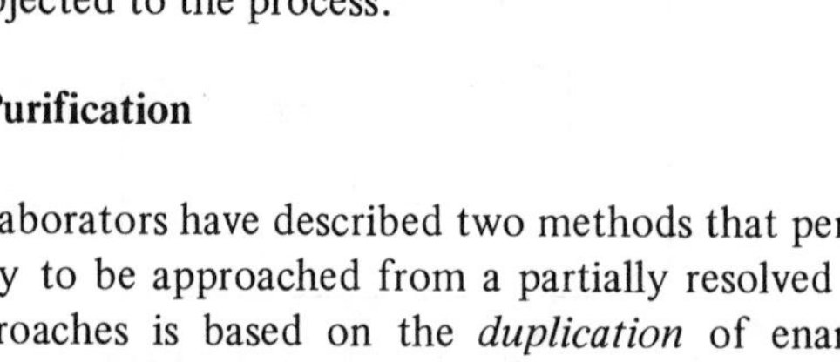

Ⓡ , Ⓢ —duplication, ~X~→ [Ⓡ–X–Ⓢ *meso*; Ⓡ–X–Ⓡ, Ⓢ–X–Ⓢ *threo*]

If a sufficient amount of the X-containing reagent and sufficient time is allowed, the duplication reaction should not be influenced by any stereoselectivity, and the quantities of the different "dimers" formed should be proportional to the product of the "monomer" concentrations. Thus, from a mixture rich in Ⓡ in which the mole fraction of the latter is x and that of Ⓢ is $(1 - x)$, one would obtain a mixture of duplication products Ⓡ–X–Ⓡ, Ⓢ–X–Ⓢ, and Ⓡ–X–Ⓢ (or Ⓢ–X–Ⓡ) in the relative proportions x^2, $(1 - x)^2$, and $2(1 - x)$. For example, if $x = 0.85$, the mole fractions of the three preceding "dimers" would equal, respectively, 0.72, 0.02, and 0.26. After the elimination of compound Ⓡ–X–Ⓢ by whatever physical method is appropriate, followed by cleavage of the remaining threo isomer mixture, one obtains a "monomer" whose enantiomeric purity p' is greater than the initial purity p. One can easily demonstrate that $p' = 2p/(1 + p)^2$. In the preceding example, $x = 0.85$ and $p = 0.7$; consequently, $p' = 0.94$. The process may be repeated by subjecting the "monomer" once again to "duplication"; the resulting $p'' = 2p'/(1 + p'^2)$, or $p'' = 0.998$, and so on.

Horeau and his associates have applied this method to the purification of various optically active alcohols through their carbonates, dimethylalkoxysilanes, phthalates, and malonate esters as well as to the purification of α-phenylbutyric acid via its anhydride (Scheme 2). The threo and meso "duplication" products were separated by gas chromatography (analytical) and by crystallization. Note that crystallization is itself susceptible to a modification of the relative proportions of the two threo enantiomers which can conceivably lead to a greater enrichment than that which is predictable by calculation. The following example is representative of the efficiency of the process: (−)-α-phenylbutyric acid, $p = 0.7$, is converted into its anhydride. The threo isomer isolated after two recrystallizations from pentane is hydrolyzed back to the acid whose enantiomeric purity is found to be $p' = 0.99$.

SCHEME 2

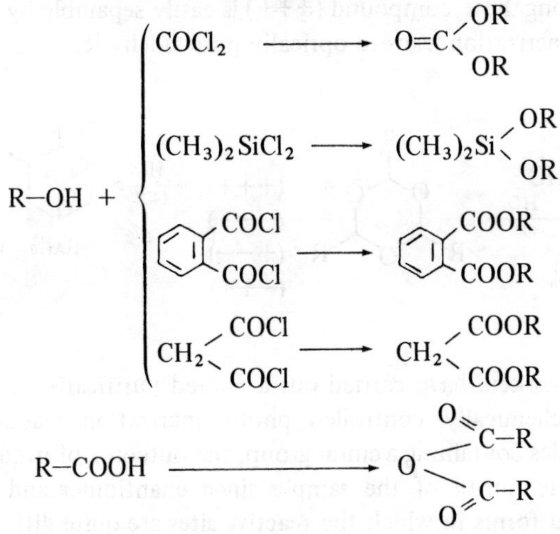

R–OH +
- $COCl_2$ ⟶ $O{=}C(OR)_2$
- $(CH_3)_2SiCl_2$ ⟶ $(CH_3)_2Si(OR)_2$
- $C_6H_4(COCl)_2$ ⟶ $C_6H_4(COOR)_2$
- $CH_2(COCl)_2$ ⟶ $CH_2(COOR)_2$

R–COOH ⟶ $O(C(=O)R)_2$

The purification of enantiomers by "duplication" is hardly limited to alcohols or acids. An interesting example involving a hydrocarbon has been described by Brown and Yoon.[19] When (+)-α-pinene, **3**, of 97.4% optical purity is treated with borane in tetrahydrofuran, a mixture of *threo*- and *meso-sym*-tetra-3-pinanyldiborane (diisopinocampheylborane dimers [IPC_2BH]*) is formed. The threo (++) isomer which precipitates can be easily isolated; after it is decomposed by triethylamine, the liberated (+)-α-pinene is found to have 99.8% o.p. (Scheme 3).[19]
Similarly, optically pure monoisopinocampheylborane* ($IPCBH_2$) can be obtained from α-pinene of 94% o.p. by "duplication" in the presence of *N,N,N,N*-tetramethylethylenediamine (TMED). The (++) complex [TMED, 2 $IPCBH_2$] is also isolated by crystallization.[20]

* The names diisopinocampheylborane (IPC_2BH) and monoisopinocampheylborane ($IPCBH_2$), which ignore the dimeric character of the boranes formed, persist in the literature.[19,20] They are used here for convenience.

SCHEME 3

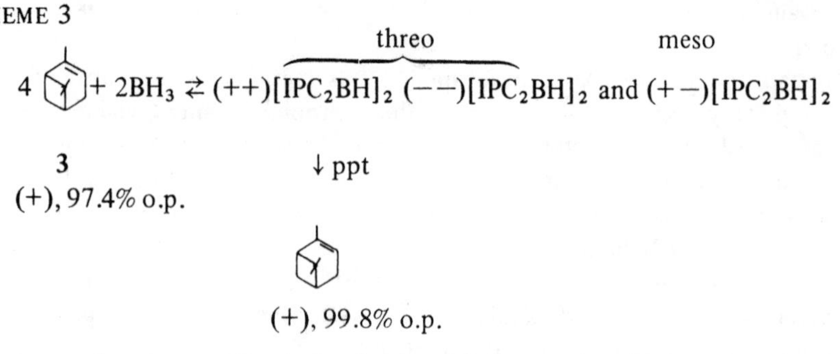

threo meso

$$4\ \mathbf{3} + 2BH_3 \rightleftarrows \overbrace{(++)[IPC_2BH]_2\ (--)[IPC_2BH]_2}^{\text{threo}} \text{ and } \overbrace{(+-)[IPC_2BH]_2}^{\text{meso}}$$

3
(+), 97.4% o.p.

↓ ppt

(+), 99.8% o.p.

An analogous purification of 3-formylpinane, **4**, which is based not on duplication but rather on *triplication*, has been described by Himmele and Siegel[21] (Scheme 4). (+)-Aldehyde **4** of 70% optical purity is trimerized in acid medium forming a mixture of homochiral (+++), (– – –), and heterochiral (++ –), (+ – –) diastereomers. Among these, compound (+++) is easily separable by crystallization. Its thermal depolymerization leads to optically pure aldehyde.

SCHEME 4

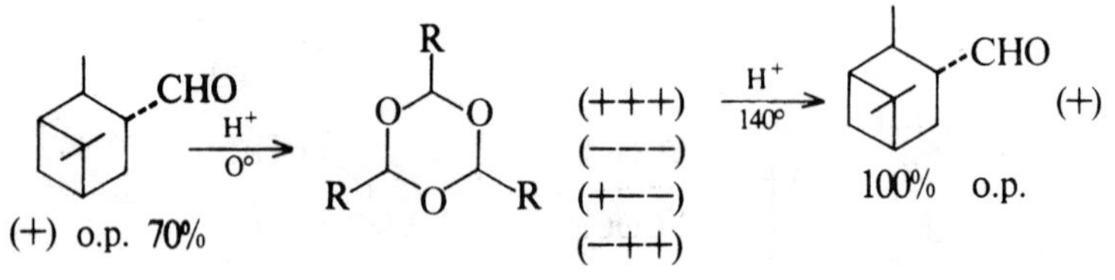

CHO
$\xrightarrow[0^\circ]{H^+}$
(+) o.p. 70%
4

R, O, O, R, O, R
(+++)
(– – –)
(+ – –)
(– ++)
$\xrightarrow[140^\circ]{H^+}$
CHO
(+)
100% o.p.

Lahav and coworkers have carried out a related purification process which is based on a topochemically controlled photodimerization reaction. In photo-dimerizable molecules containing a chiral group, the outcome of irradiation depends on the enantiomeric purity of the sample since enantiomer and racemate may crystallize in crystal forms in which the reactive sites are quite differently oriented and separated. In suitable cases, racemates crystallize in a photoactive form while enantiomers crystallize in a light-stable form.[23]

The 9-anthroate esters of enantiomerically enriched secondary alcohols such as 1-phenylethanol, for example, may be irradiated to form *meso*-dianthracenes (the photodimers). Since only the racemate, present in the photoactive crystal form, dimerizes, residual monomer in the light-stable form is enriched in the predominant enantiomer. For example, irradiation of 0.48 g of 1-phenylethyl 9-anthroate having $[\alpha]_D$ 38.9° and extraction of the unreacted monomer from the sparingly soluble meso-dimer led to recovery of 0.23 g of ester having $[\alpha]_D$ 66.7° corresponding to 95% enantiomeric purity.

(b) Kinetic purification

This last type of enantiomer purification is based not on the separation of diastereomeric compounds, as in the preceding cases, but rather on the *rates of stereoselective* reactions carried out to an *incomplete* extent.[18]

Table 2 Kinetic purification as a function of reagent stereoselectivity and the degree of enrichment desired

Stereoselectivity	Conversion F of substrate (%)[a]	
k_D/k_L	95% → 99% e.p.	99.0% → 99.9% e.p.
2	81	90
3	57	67
4	43	54
5	35	44

[a] Percentage of the enantiomer mixture that must be consumed to achieve the stated enantiomeric enrichment.

Let us consider a chiral reagent C_+ which acts upon an enantiomer mixture D, L of enantiomeric purity p (L > D) with rate constants k_D and k_L, respectively. Horeau has shown that, after fraction F of the mixture has reacted, the new enantiomeric purity p' of the *unreacted* substrate may be deduced by

$$(1-F)^{(k_D/k_L)-1} = \frac{1-p'}{1-p}\left(\frac{1+p}{1+p'}\right)^{(k_D/k_L)-1} \tag{1}$$

In order to increase the enantiomeric purity of the mixture (rich in L), a reagent must be chosen which reacts faster with enantiomer D than with L ($k_D > k_L$). If, on the other hand, $k_D < k_L$, then $p' < p$. Application of this method thus requires a knowledge of the ratio of rate constants which determines the stereoselectivity of the reaction. This ratio may easily be obtained by allowing *racemic* reagent C to react with the racemic substrate:

$$C_\pm + D, L \rightarrow \begin{cases} DC_+, LC_- \ (p_\pm) \\ DC_-, LC_+ \ (n_\pm) \end{cases}$$

The proportions of the two racemic diastereomers formed, $p_\pm$ and $n_\pm$, are equal to the ratio of the rate constants sought. In the example given below, if one finds that more $p_\pm$ is present than $n_\pm$, this signifies that reactant C reacts faster with the substrate of same sign; thus, with C_+, $k_D > k_L$. Conversely, C_-, $k_D < k_L$.

Once k_D/k_L is known, eq. (1) allows the calculation of the conversion F of the substrate required to raise the e.p. from p to p'. The conversions corresponding to increases in purity $p \rightarrow p'$ from 95% to 99% and from 99.0% to 99.9% as a function of stereoselectivities k_D/k_L varying from 2 to 5 calculated by means of eq. (1) are shown in Table 2. We see, for example, that in order to pass from 99.0% to 99.9% e.p., 90% of the substrate must be consumed if $k_D/k_L = 2$ and only 44% if $k_D/k_L = 5$.

This method was used by Horeau in the purification of (−)-phenylmethylcarbinol.[18] The reagent was (−)-α-phenylbutyric anhydride with which $k_D/k_L = 4.7$. Alcohol of 97% enantiomeric purity was esterified with an insufficient quantity of anhydride. When the esterification was 75% complete, the residual alcohol, isolated by chromatography, was found to have enantiomeric purity greater than 99.9%.

In a more recent application of the method, successive esterifications of (−)-mesitylisopropylcarbinol with (+)-α-phenylbutyryl chloride raised the optical purity from 0.1% to 98%.[22]

REFERENCES 7.6

1 W. R. Wilcox, R. Friedenberg, and N. Back, *Chem. Rev*., 1966, **64**, 187.

2 D. P. MacDougall, *Phys. Rev*., 1931, **38**, 2296.

3 H. Eyring, *Anal. Chem*., 1948, **20**, 98.

4 R. Rometsch and K. Miescher, *Helv. Chim. Acta*, 1946, **29**, 1231.

5 G. Ohloff, in *Olfaction and Taste*, Vol. IV, D. Schneider, Ed., Wissenschaftliche Verlagsgesellschaft mbH., Stuttgart, 1972, p. 156.

6 H. Pluim and H. Wynberg, *Tetrahedron Lett*., 1979, 1251.

7 M. J. P. Harger, *J. Chem. Soc. Chem. Comm*., 1976, 520.

8 M. D. Fryzuk and B. Bosnich, *J. Am. Chem. Soc*., 1978, **100**, 5491.

9 U. Hengartner, D. Valentine, Jr., K. K. Johnson, M. E. Larscheid, F. Pigott, F. Scheidl, J. W. Scott, R. C. Sun, J. M. Townsend, and T. H. Williams, *J. Org. Chem*., 1979, **44**, 3741.

10 S. -I. Hashimoto, N. Komeshima, and K. Koga, *J. Chem. Soc. Chem. Comm*., 1979, 437.

11 J. M. Domagala and R. D. Bach, *J. Org. Chem*., 1979, **44**, 3168.

12 A. Horeau in *Chemical Methods*, Vol. 3 of *Stereochemistry, Fundamentals and Methods*, H. B. Kagan, Ed., Georg Thieme Verlag, Stuttgart, 1977, pp. 51 and 76.

13 E. Downer and J. Kenyon, *J. Chem. Soc*., 1939, 1156.

14 A. Findlay, *The Phase Rule*, 9th ed., revised by A. N. Campbell and N. O. Smith, Dover, New York, 1951, p. 162ff.

15 L. G. Tensmeyer, P. W. Landis, and F. J. Marshall, *J. Org. Chem*., 1967, **32**, 2901.

16 (a) E. F. G. Herington, *Zone Melting of Organic Compounds*, Wiley, New York, 1963. (b) H. Schildknecht, *Zone Melting*, Verlag Chemie, Weinheim, and Academic Press, New York and London, 1966.

17 J. P. Vigneron, M. Dhaenens, and A. Horeau, *Tetrahedron*, 1973, **29**, 1055.

18 A. Horeau, *Tetrahedron*, 1975, **31**, 1307.

19 H. C. Brown and N. M. Yoon, *Israel J. Chem*., 1976/1977, **15**, 12.

20 H. C. Brown, J. R. Schwier, and B. Singaram, *J. Org. Chem*., 1978, **43**, 4395.

21 W. Himmele and H. Siegel, *Tetrahedron Lett*., 1976, 911.

22 P. Briaucourt and A. Horeau, *C. R. Acad. Sci., Ser. C*, 1979, **289**, 49.

23 M. Lahav, F. Laub, E. Gati, L. Leiserowitz, and Z. Ludmer, *J. Am. Chem. Soc*., 1976, **98**, 1620.

Index